American Military
Training Aircraft

Also by E.R. Johnson and from McFarland

American Military Transport Aircraft Since 1925 (2013)

United States Naval Aviation, 1919–1941: Aircraft, Airships and Ships Between the Wars (2011)

American Flying Boats and Amphibious Aircraft: An Illustrated History (2010)

American Attack Aircraft Since 1926 (2008, paperback 2012)

American Military Training Aircraft

Fixed and Rotary-Wing Trainers Since 1916

E. R. Johnson

Drawings by Lloyd S. Jones

Foreword by Walter J. Boyne

McFarland & Company, Inc., Publishers
Jefferson, North Carolina

All photographs provided from the personal archives of David W. Ostrowski unless otherwise noted.

LIBRARY OF CONGRESS CATALOGUING-IN-PUBLICATION DATA

Johnson, E. R., 1948–
American military training aircraft :
fixed and rotary-wing trainers since 1916 / E.R. Johnson ;
drawings by Lloyd S. Jones ; foreword by Walter J. Boyne.
p. cm.
Includes bibliographical references and index.

ISBN 978-0-7864-7094-5 (softcover : acid free paper) ♾
ISBN 978-1-4766-1789-3 (ebook)

1. Training planes—History.
2. Training planes—Design and construction.
3. Airplanes, Military—United States—History.
4. Airplanes, Military—United States—Design and construction.
I. Title.

UG1242.T67J65 2015 623.74'620973—dc23 2014046889

BRITISH LIBRARY CATALOGUING DATA ARE AVAILABLE

On the cover: (top) front view of a Curtiss JN-4D "Jenny," the most numerous type of primary trainer to be employed by the U.S. Army Air Service from mid–1917 through the end of World War I; (bottom) side profile of a North American SNJ-5 (AT-6D in USAAF) shown in U.S. Navy markings of Basic Training Unit Two (BTU-2), based at NAAS Corry Field during the early to mid–1950s (illustrations by Lloyd S. Jones)

Printed in the United States of America

McFarland & Company, Inc., Publishers
Box 611, Jefferson, North Carolina 28640
www.mcfarlandpub.com

This book is a tribute to all flight instructors,
civilian and military, past and present, and in particular:

My late grandfather, Capt. Rufus C. Branch,
U.S. Army Air Service, Primary Flight Instructor,
65th Aero Squadron, Park Field, Tennessee, 1918.

The best youth flight instructor I have ever seen,
Maj. Bruce H. Bailey, CFII, AR-107, Arkansas Wing,
Civil Air Patrol, 2001–present.

And my good friend and aviation colleague,
Lt. Col. Gary W. Smith, USMC (ret.), advanced instructor pilot,
VT-25, NAS Chase Field (Beeville), Texas, 1988–1990.

Acknowledgments

Bringing this book from bare outline to a finished manuscript was very much a collaborative effort. I am especially indebted to Lloyd S. Jones and David W. Ostrowski for their unswerving support through the entire course of this project. Lloyd, in order to create a number of the book's three-view drawings, was forced to work from only photographs and reported dimensions; Dave spent an untold number of hours searching for photographic images, which, in turn, he had to identify, reproduce, and format.

I owe special thanks to all of the dedicated researchers who furnish the extensive data available online at the Aerofiles website, the U.S. Air Force Museum website, the Naval Aviation History Office website, and the U.S. Army Aviation website at Fort Rucker. Likewise, I greatly appreciate the unselfish efforts of Joseph F. Baugher for his unstinting work in providing online access to the serial number records of every U.S. military aircraft procured since 1908. And thanks also to Walt Boyne for sharing his personal archive of helicopter photos.

Contents

Foreword by Walter J. Boyne

This addition to E. R. (Buddy) Johnson's excellent series of authoritative books on American military aircraft is, perhaps surprisingly, one of the very most valuable and at the same time, one of the most interesting. This stems from the fact that training aircraft have traditionally received little attention from authors of aviation literature because their non-combat mission is deemed to be less glamorous.

But any pilot knows this is a mistake, for there is nothing so thrilling as that first flight in a training aircraft, from the days of the venerable Curtiss Jenny to today's modern jet and turboprop powered trainers. Fortunately, Johnson has provided an in-depth survey of the American training plane that fills a large gap in current literature on the subject.

Trainers, after all, have been the basis for the excellence of the U.S. military air forces since their inception. America has traditionally provided more and better pilot training than have its adversaries. This paid off in combat in every war, but most particularly during World War II. When Pearl Harbor came, the American air arms had a relatively small number of pilots ready for combat. They had been superbly trained at Pensacola and Randolph Field, but there was now a requirement for the mass production of pilots. The country established a vast training system that ultimately produced more pilots than there were planes for them to fly. The increase in size and numbers in just four years seems almost miraculous today, and in the light of the current problems in budget and procurement, probably could not be duplicated. As a natural part of this incredible build-up, a variety of specialized trainers was produced, and today many of them fly on as venerated "war-birds."

American pilots were soon arriving in combat with many more hours flying time and far better training than their enemy counterparts. This advantage was continually improved upon during and after that great conflict, and has paid off in every subsequent war since. (There have been lapses, as in Vietnam, when overconfidence in air superiority led to a lack of training in air combat maneuverability. It took some time to create TOPGUN and RED FLAG to overcome this.)

What is often overlooked is that this same degree of training resulted in the Cold War American air power that preserved the peace through the most critical period of history, when nuclear weapons by the thousands were prepared and ready for action. The Strategic Air Command is a prime example of a decisive military force that delivered the excellence demanded by its commanders, and was able to do so because of the ultra-high quality training of its crews.

From about 1917 on, the United States invested heavily in a variety of firms to manufacture the trainers. The author carefully follows the development of each one, and in doing so gives a portrait not only of the aircraft, but of the firm, the industry and the air forces at the time. Many of the names will be unfamiliar to most readers, e.g. Huff-Daland, Elias, Mohawk, etc., but each one has its own tale to tell. Other names—Curtiss, Standard, Stearman, Cessna, Beech, etc.—will evoke iconic memories of classic aircraft that won the heart of the American flying public.

Trainers are deceptive in appearance, usually looking far simpler, less complex, and certainly, less fierce than contemporary fighter aircraft. Yet they are demanding to design and build, for they must be structurally strong to stand up to ham-handed students, easy to repair, especially the undercarriage and in tail-dragger days, the wing tips, and still be relatively inexpensive to build. In many cases, trainers turned out to be "overbuilt," i.e., to be able to undergo a few thousand hours training students, then be converted to some arduous task such as crop-dusting, then enjoying a third life as a lovingly restored war-bird.

One fortunate and cost-saving result of being overbuilt is that trainers have shown themselves to be capable of service careers of fifty or more years. Aircraft such as the Beech T-34 and Northrop T-38, both still operational, have served for decades past their expected life. Further, they and others such as the North American T-28 were rugged enough to be converted into effective warplanes.

The most difficult trick is to make them demanding to fly—yet forgiving of errors. Two of the most popular trainers of all time, the British de Havilland Tiger Moth and the magnificent American North American T-6, are splendid examples. Both could be very demanding to fly, but both had the ability to be saved from imminent destruction by rapid and powerful control inputs.

It is a credit to the author that he takes in not only those aircraft originally designed as trainers, but also operational types whose demanding capabilities dictated that a trainer version be made. Thus he covers aircraft like the Lockheed TB-58 Hustler, the Lockheed TP-3 Orion and many more. And he might easily have overlooked the equally important rotary-wing trainer aircraft, where many a front line chopper had to be adapted to the school role.

The reader will admire the way that the author provides all the essential information—specifications, procurement quantities, price, etc.—concisely and in a regular format for quick reference. But there is so much more to the book than hard data, for Johnson has a feel and a love for his subject matter that is borne out in his succinct coverage of such things as company management, mergers and more.

It is very difficult to write a book and it is even more difficult to write a reference book. Johnson has done the extremely difficult: writing a reference book that is fun to read. I believe that any reader will add the rest of Johnson's works to his collection after having read this, and I also believe that anyone who has previously purchased his work will buy this book on the spot.

Walter S. Boyne is a retired U.S. Air Force colonel, a pilot, combat veteran and aviation historian. A former director of the Nation Air and Space Museum, he has written more than 50 books and more than 1,000 magazine articles on aviation topics and was inducted into the Aviation Hall of Fame in 2007. He lives in Silver Spring, Maryland.

Preface

The United States of America did not become the world's foremost military air power by accident. Arriving at this position has been a long and difficult learning process that continues to present. The learning curve—World War I, World War II, the Korean War, the Vietnam War, the Gulf War, and more recently, the War on Terror—has been a steep one. While climbing this curve, the United States has not only produced superior military aircraft in greater numbers than its foes but has—in due course—out-trained them, too. This work is designed to provide the reader with a comprehensive historical survey, including technical specifications, drawings, and photographs, of each type of fixed and rotary-wing aircraft used by U.S. military forces over 98 years to accomplish the first step of the learning process: the training of pilots and aircrews.

For organizational purposes, the book is divided is divided into four chronological series: Series I, covering fixed-wing trainers from 1916 to 1947, subdivided by Part A, USAAS, USAAC, and USAAF aircraft, and Part B, USN, USMC, and USCG aircraft; Series II, examining fixed-wing trainers from 1947 to 1962, subdivided by Part A, USAF and U.S. Army aircraft, and Part B, USN and USMC aircraft; Series III, covering fixed-wing trainers since1962; and Series IV, rotary-wing trainers since 1942. Preceding the series is an historical overview, "Training the American Air Armada," which summarizes how the services have organized and developed their flying training functions from World War I to the present. Integrated within the main series are "Synopses of Trainer Procurement," intended to condense and interrelate the chronology of events. Two appendices cover training gliders, USAAF, USN, and USMC from 1934 to 1945, and USAF since 1957, and fixed and rotary-wing military trainer designation systems. The book ends with a comprehensive glossary and a bibliography of source materials.

* * *

Note: Naval or non-standard designations are indicated in parentheses in the subject headings for certain planes. For example: "Thomas-Morse S-4 (S-5)—1917."

Training the American Air Armada

Author's Note: For reasons of space and brevity, this historical summary will be confined to flight training—pilots and aircrew—as opposed to technical training of ground-based personnel. At the same time, the author recognizes the vital contribution of all technical personnel in military aviation, without whose expertise and support the aircraft described in this book would never leave the ground.

World War I Era—1917–1918

Aviation Section of U.S. Army Signal Corps/Army Air Service. When the United States declared war on Germany on April 4, 1917, the single Army Flying School based at Rockwell Field, California, then a component of the Aviation Section of the U.S. Army Signal Corps, consisted of 35 ground and flight instructors and 51 pilot trainees. By the time World War I ended on November 11, 1918, the Army's—reorganized as the U.S. Army Air Service May 24, 1918—flying training operations had expanded to 32 bases nationwide, headquartered at Kelly Field near San Antonio, Texas. Copying the Canadian model, the Army adopted a "Flying Cadet Program" in which applicants first attended an eight-week ground school at one of eight American universities, followed by a six-to-eight-week primary flying course, usually comprising 40 to 50 hours of flight. Initially, pilot candidates undertook all flight training in Europe, but from late 1917 onwards, all primary training was conducted in the United States, at the end of which, trainees received their wings (i.e., rated as pilots) and were commissioned as 2nd Lieu-

U.S. Army Air Service Training Headquarters, Kelly Field, Texas, at the close of World War I.

tenants in the Army Reserve. Advanced training was a 90-day course—conducted in Europe until the end of 1918—in the type of operational aircraft to be flown in combat. From this humble beginning, over 10,000 men had received their wings as Reserve Military Aviators by the time the war ended.

U.S. Navy, Marine Corps and Coast Guard. At the time America entered World War I in 1917, Naval Aviation, Navy and Marines combined, was little more than a vestigial organization, boasting a strength of only 48 pilots and 54 airplanes. Nevertheless, its flight training operations at Pensacola, Florida, under the supervision of the Bureau of Navigation (BuNav), were well organized, and a cadet program for commissioned aviators in the Naval Reserve Flying Corps had been established in 1916. The training program produced two types of officer-candidate pilot-trainees, Naval Aviator (seaplane) or (dirigible [i.e., blimps]), and in early 1917 was expanded to include enlisted pilots as Naval Aviation Pilots (NAPs). The Navy training syllabus for pilot candidates (seaplane) differed from the Army model: a two-month ground school at the Massachusetts Institute of Technology (MIT), followed by a ten-week primary course involving approximately 40 flight hours in a floatplane and 15 hours in a small flying boat, upon the completion of which the trainee received his wings and a commission (if an officer candidate). Advanced training was carried out in large or small flying boats of the type to be flown operationally and featured advanced topics such as navigation, gunnery, and bombing. Through the balance of 1917, BuNav enlarged the facilities in Pensacola and established four more training sites, plus an airship school at the Goodyear plant in Akron, Ohio. By the time the war ended in November 1918, these efforts had created over 2,500 new pilots, Navy, Marine, and Coast Guard.

Seaplane ramp and canvas hangars at NAS Pensacola, Florida, during World War I, showing Curtiss F Boat in foreground.

Inter-War Era—1919–1941

U.S. Army Air Service/Air Corps. In the wake of the demobilization that followed the Armistice, personnel and aircraft levels within the Army Air Service fell by a factor of 85 percent, averaging about 9,500 to 10,000 in manpower and 750 to 800 aircraft on hand throughout the 1920s. Only two training bases were retained: Brooks Field, Texas (originally Kelly Field No. 5)

North American BT-9s seen over-flying the headquarters building at Randolph Field, Texas, in the late 1930s.

became the Primary Flying School, comprising the 11th School Group with three squadrons; and Kelly Field, the Advanced Flying School as the 10th School Group with six squadrons. Most training was carried out in World War I-era aircraft (e.g., Curtiss JN-4D/H, Vought VE-7, and Airco DH-4) until the mid–1920s when Consolidated PT-1s began taking the place of JN-4s in primary training. On July 2, 1926, the Army Air Service became the Army Air Corps, which fundamentally restructured the Army's air arm within the War Department but had little impact on the overall size of the active force. A five-year expansion plan approved at the same time, however, included a continuation of the Flying Cadet Program in order to buildup a pool of reserve officer pilots, many of whom subsequently served with newly established National Guard units, and led to the reopening of March Field, California, as a second primary training base to cover the overflow. Equally significant, Congress authorized construction of Randolph Field (adjacent to Kelly and Brooks in Texas) with the plan of making it the new Army Air Corps training center.

Soon after Randolph became operational in mid–1930, all primary and basic training was centered there, with Brooks and March Fields being deactivated, while Kelly was retained for advanced training, and by 1932, Randolph accounted for four training squadrons equipped with 250 Consolidated PT-3s (improved PT-1s) and about 90 Douglas BT-1s and -2s (trainer variants of the O-2 and O-32). Between 1931 and 1935, Randolph produced approximately 1,000 new pilots who continued to advanced training at Kelly. Advanced trainers consisted mainly of older operational aircraft—attack, fighter, bomber, and observation types—until the late 1930s. The mid–1930s saw the introduction of new types of training aircraft: in 1936, Boeing-Stearman PT-13s began displacing PT-3s in primary and the first all-metal monoplane trainers, Seversky

Lineup of student pilots at Tuskeegee Institute, Alabama, in late 1941.

BT-8s, were introduced to the basic course; then in 1937, another new monoplane, the North American BT-9 replaced the short-lived BT-8 as the standard basic trainer. Even so, new pilot production at Randolph lagged behind the planned growth of the active force (i.e., from 13,305 personnel in 1930 to 22,387 in 1939), but as a consequence of the Air Corps Expansion Act, in mid–1939—2,200 new pilots per year—the USAAC contracted with nine civilian flying schools to provide primary training in government-owned aircraft (e.g., PT-13s), while Randolph, now having five outlying satellite fields, retained the basic phase. The first purpose-built advanced trainers, North American BC-1s, began entering service at Kelly in late 1938.

In 1940, with war clouds looming, the USAAC boosted its original new pilot goal to 7,000 per year. Training aircraft ordered in unprecedented numbers set into motion mass-production of new trainer types: in primary, the Boeing-Stearman PT-17 (re-engined PT-13), Fairchild PT-19, and Ryan PT-22; in basic, the Vultee BT-13 and North American BT-14 (improved BT-9); and in advanced, the North American BC-1A (re-designated AT-6). The same year, two new Air Corps Training Centers (ACTC) were established: Randolph was retained as the Gulf Coast ACTC, while Maxwell Field. Alabama became the Southeast ACTC and Santa Ana Field, California, the West Coast ACTC. And this was just the beginning.

U.S. Navy, Marine Corps and Coast Guard. In the aftermath of the post–World War I demobilization, though significantly downsized from 1918 levels, Naval Aviation managed to emerge from the process as an important force in being, which, by 1920, still maintained a strength of about 850 aircraft and 30 airships in Navy units and another 88 aircraft allocated to the Marines. Training operations were scaled back to one base at NAS Pensacola, with approximately 290 training aircraft on hand, plus a small number of training airships. Establishment of the U.S. Naval Bureau of Aeronautics (BuAer) in 1921 combined with the commissioning of the USS *Langley* (CV-1), the first aircraft carrier, in 1922, served to have a profound influence on the direction of pilot and aircrew training and the procurement of new aircraft to train them. While flying boats remained a large part of the overall force, new emphasis was placed on training pilots to operate from ships, in both floatplanes and wheeled aircraft equipped with carrier arresting gear. Between 1920 and 1929, over 700 new Navy, Marine, and Coast Guard pilots were trained at Pensacola, and in addition to maintaining an active force of about 500 pilots, Navy and Marine Reserve Squadrons were formed in ten locations.

The decade of the 1930s was a period of growth for Naval Aviation in general and its training operations in particular. Importantly, passage of the Vinson-Trammell Expansion Act of 1934 authorized the Navy to build three new carriers and procure all the aircraft needed to equip them, which in turn, called for training pilots to fill new Carrier Air Group slots. To facilitate the process, Congress passed the Naval Aviation Cadet Act in 1935 (known as the V-5 Program), whereby civilian and enlisted candidates would be trained as pilots with a three-year active service obligation. Applicants, between ages 19 and 25, having two years of college, received a commission upon being designated as Naval Aviators. By 1937 operations at Pensacola had expanded to two primary squadrons operating 160 aircraft and three advanced squadrons with another 148, and augmenting the active force, pilot training was also carried out at 13 different reserve bases having 10 to 14 training aircraft apiece. At the time 1939 drew to a close, a total of 2,842 new aviators had been added to the Navy, Marine, and Coast Guard rosters.

The most far-reaching stimulus, however, came in 1940 with passage of the Two-Ocean Naval Expansion act of 1940, which, among other things, called for 18 new aircraft carriers and thousands of new aircraft to man them. To train the pilots needed, the Navy opened two new training centers, one at NAS Jacksonville, Florida, and a second at NAS Corpus Christi, Texas, and this set the stage for the huge expansion of World War II which was to follow.

Civilian Pilot Training Program. In late 1938, as a new war threatened to engulf Europe, U.S. policy-makers moved to establish the Civilian Pilot Training Program (CPTP), providing free pilot training to college students. The stated objectives of CPTP were to generate 20,000 new pilots per year and also stimulate the general aviation industry by creating demand for more flight instructors and more new training aircraft. When activated in 1939, CPTP prescribed a training syllabus of a 72-hour ground school followed by 35 to 50 hours of flight instruction, and for the military branches, CPTP provided a ready-made "pilot elimination" course for cadet applicants.

By late 1941, CPTP had grown to 35 flying schools nationwide, and after America's entry into World War II in December, the program was renamed War Training Service (WTS). Another side benefit of CPTP and WTS was extension of flight training to many African Americans (e.g., at Tuskeegee Institute, Hampton Institute, Howard University, and Virginia State University) and to women who went on to form the Women Airforce Service Pilots (WASP) during the war. When the government phased-out CPTP/WTS in mid–1944, 435,165 people had been taught to fly.

World War II Era—1941–1945

Author's Note. In terms of sheer magnitude, the colossal buildup of American military air power during World War II was a one-time event, never likely to be repeated. Since it is not within the scope of this book to explain all of the myriad details of how military flying training expanded during this era, readers are encouraged to refer to Martin W. Bowman, *USAAF Handbook 1939–1945* (Mechanicsburg, PA: Stackpole, 1997), and Naval Aviation News, eds., *Naval Aviation Training,* vols. 1–11 (Washington, D.C.: Deputy Chief of Naval Operations ([Air Warfare] and Naval Air Systems Command, 1990).

U.S. Army Air Corps/Army Air Forces. The USAAC commenced 1941 with a goal of training 12,000 new pilots plus a proportionate level of new navigators, bombardiers, flight engineers, radio operators, and air gunners, and by the end of the year, was operating 7,340 aircraft out of three training centers, which included 41 primary contract flying schools. In 1942, after war had been declared, the pilot goal was upped to 75,000 and all flying training reconstituted as Air Corps Flying Training Command in January, then as Army Air Forces Flying Training Command (AAFTC) in March, further organized into three major subdivisions: Eastern Flying Training Command, headquartered at Maxwell Field, Alabama; Central Flying Training Command, headquartered at Randolph Field, Texas; and Western Flying Training Command, headquartered at Santa Ann AAF, California. Through the course of the war, it is an amazing fact that AAFTC facilities grew to approximately 158 primary, basic, advanced, and specialized aircrew training bases, and at its peak in mid–1944, operated an inventory of 27,923 training aircraft. From start to finish, between 1939 and 1945, the USAAC/AAF trained a total of 193,000 pilots plus 408,000 in other aircrew positions.

As pilot and aircrew demand increased during the war, Flying Cadet prerequisites changed to waive the college requirement and enable graduates to be commissioned as Flight Officers (warrant rank) rather than enlisted sergeants. For pilot candidates, training was typically a 36- to 40-week course in four segments: (1) Preflight (seven to 10 weeks) that included close order military drill, physical training, pistol shooting, water training, and classes on ship and aircraft recognition, codes, aviation physics, mathematics, and map and chart reading; Primary (12 weeks) that included 60 hours of flight, 94 hours of ground school studies, and 54 hours of military training; Basic (10 weeks) that included 70 hours of flight with emphasis upon formation, nav-

Multi-engine advanced training during World War II showing a lineup of Curtiss AT-9s.

igation, instrument, night, and cross-country flying, 94 hours of ground school, and 47 hours of military training; (4) Advanced, either single-engine or twin-engine, (10 weeks) that included 70 hours of flight, 60 hours of ground school, and 19 hours of military training. Upon graduation, new pilots went into Transition Training for the type of aircraft they would fly in combat, e.g., the B-17, B-24, B-25, B-26, C-47, P-40, P-38, etc., after which they would form into Operational Training Units (OTU) earmarked for overseas deployment. All navigators and bombardiers attended a similar preflight school followed by a six-week gunnery course. Navigator school was a 20-week course consisting of 104 hours of flight and 782 hours of ground school; and bombardier school 20 weeks that included 120 hours of flight and 718 hours of ground school. Dual-role officers needed for medium bomber lead crews and B-29s attended a 12-week bombardier-navigator course in addition to navigator school. Besides designated air gunners themselves, other aircrew (e.g., radio operators, flight engineers, etc.) completed the six-week gunnery course.

In addition to the USAAF flying training regimen, several other programs were created to augment demand for new pilots in more specialized areas:

Glider Pilot Training. A separate and distinct plan for training glider pilots was set up in late 1941, with all but the advanced phase being accomplished by civilian contract schools. The course started with about 40 flight hours of instruction in two-place, powered lightplanes (e.g., Piper J-3) with emphasis upon the ability to make dead stick landings, followed by a course of

New pilots in fighter transition training. Note Curtiss P-40B in background.

unspecified length in towed training gliders, originally militarized sport sailplanes like the Schweitzer TG-4 but later modified lightplanes like the Taylorcraft TG-6 in which the engine had been removed to accommodate a third seat forward. The course ended with an advanced phase in operational troop gliders like the Waco GC-4, upon the completion of which, the graduate received a Glider Pilot Rating and a commission as a Flight Officer. In late 1944 AAFTC discontinued all glider pilot training in favor of drafting regular pilots from Troop Carrier Command and running them through the advanced glider pilot course.

Army Ground Forces Pilot Training. In early 1942, under the supervision and control of the Field Artillery School at Fort Sill, Oklahoma, the Army established a school for training liaison pilots (e.g., artillery spotting, battlefield reconnaissance, staff/courier movement, and medical evacuation) who would be attached directly to Army Ground Forces (AGF). Pilot candidates typically came from Army enlisted ranks and USAAF cadets who had washed out of basic or advanced training. The course comprised approximately 60 hours of flight with emphasis on taking off over obstacles, short field landings, and low-altitude navigation, plus a ground school curriculum. Graduates received a Liaison Pilot Rating, initially as Flight Staff Sergeants but later as Flight Officers. Most aviation historians regard the AGF school as the origin of what later became known as Organic Army Aviation.

Women Airforce Service Pilots. The first female military pilots were experienced civilian pilots who, in September 1942, following a 30-day orientation course, began flying under the supervision of Air Transport Command as the Women's Auxiliary Ferrying Squadron (WAFS).

Crew of WASP pilots seen at Laredo AAF, Texas, delivering a Martin B-26.

Independently, the Women's Flying Training Detachment (WATD), known as the "guinea pigs," was organized in November 1942 as the first program designed to actually train new female pilots for eventual service in a USAAF auxiliary support role. WATD applicants were required to complete a preflight, primary, basic, and advanced training syllabus identical to that of their USAAF male counterparts, and the first class of 23 women pilots graduated on April 24, 1943. Both WAFS and WATD were merged into a single entity as the Women Airforce Service Pilots (WASP) on August 5, 1943, and while rated as pilots, WAFS/WASP graduates were classified as Federal Civil Service employees rather than active or reserve USAAF officers. In operational service, WASP pilots ferried aircraft from factories to embarkation points, towed targets at gunnery schools, and provided flight instruction at some USAAF primary schools. In late 1944, when the USAAF had achieved a large surplus of male pilots, the WASP program was officially disbanded. Altogether, out of about 25,000 applicants, WASP accepted 1,830 trainees and graduated 1,074 new pilots.

U.S. Navy, Marine Corps and Coast Guard. Passage of the Two-Ocean Naval Expansion Act prompted a rapid buildup of Naval Air Training during 1941, with 3,112 new Navy, Marine, and Coast Guard pilots added by the end of the year (versus 708 in 1940). Even so, training quotas were raised from 800 to 2,500 entering students per month, with the aim of producing 20,000 new pilots a year by 1943. Toward this end, 10,869 new Naval Aviators graduated in 1942, 20,842 in 1943, 21,069 in 1944, and 8,880 in 1945, for a total of 65,478 between 1940

Group of Naval Aviation Cadets seen in Ready Room at NAS Corpus Christi, Texas, in 1942.

and 1945. During the same timeframe, 2,208 new enlisted Naval Aviation Pilots also completed the program. Pilot training facilities expanded from five training centers and 13 reserve bases in 1941 to 20 training centers and 20 reserve bases in 1945. Flight Preparatory Schools were opened at 20 colleges where cadets began two months of elementary flight training (referred to as "elimination training") and ground training in a CPTP/WTS school, and upon successful completion, attended a three-month Preflight before proceeding on to Primary at one of the flight training centers.

Through the course of the war, in order to shorten and accelerate naval pilot training, the curriculum became more specialized. Primary training was a 10- to 14-week course consisting of 90 to 100 hours of flight plus ground school. A 14- to 18-week Intermediate course came next, consisting of 160 hours of flight in three phases: formation, instrument and night, and cross-country instruction in basic and advanced trainers that included the North American NJ, Vultee SNV, Curtiss SNC, North American SNJ, Vought OS2U, Beech SNB, and Consolidated PBY, depending upon which type of aircraft the student aviator was destined to fly, while students selected as carrier pilots completed their training with precision takeoffs and landings in SNJs. Graduation and commissioning were followed by eight weeks of operational training at one of 17 naval stations in the type of aircraft to be flown in combat (i.e., fighter, scout-bomber, torpedo-bomber, floatplane scout, flying patrol boat, or land-based bomber). Finally, pilots selected for carrier duty underwent qualification on one of two training carriers (USS *Wolverine* and *Sable*) or an escort carrier (CVE).

Cold War Era 1946–1990

Reorganization of Armed Services. The National Security and National Defense Acts, both signed into law in July 1947, initiated an across-the-board restructuring of the U.S. military establishment. Taken together, they merged the War (Army, including AAF) and Navy (including USMC) Departments into an overall Department of Defense (DoD) and established the U.S. Air Force (USAF) as a separate service. A policy statement issued by the Secretary of Defense in April 1948, popularly known as the "Key West Agreement," defined the role of aviation within the respective services, namely: (1) the Navy would retain the strategic and tactical aircraft needed to support its operations at sea; (2) the Army would retain only aircraft and related aviation units needed for reconnaissance and medical evacuation; and (3) the USAF would control all other land-based strategic and tactical aircraft. The single greatest external influence shaping American military policy during the second half of the 20th century was the Cold War and with it, the threat of an all-out nuclear exchange (i.e., mutual assured destruction) between the United States and the Soviet Union. The massed forces of the Soviet Bloc, which attained nuclear weapons capability in 1949, compelled the United States and its allies to maintain a strong military presence, in effect, a war footing, in both hemispheres of the world. Although the Cold War served to create a nuclear stalemate between the major powers (which included Communist China after late 1949), it did not prevent limited military conflicts between conventional forces in contested spheres of influence like Korea (1950–1953) and Southeast Asia (1961–1973).

U.S. Army Air Forces/U.S. Air Force. Flying training activities during this era underwent many fluctuations in terms of both size and organization. In mid–1946, when Air Training Command (ATC) replaced AAFTC, flying training shrank to only seven operational bases. Elimination training and civilian contract schools were discontinued, with all instruction being combined into a one-year basic and advanced course for pilot trainees and a similar consolidation for student navigators and bombardiers. But mounting Cold War tensions and the start of the Korean War between 1949 and 1953 triggered a surge in flying training in which ATC expanded to 10 flying training wings and nine civilian contract primary schools. Pilot training shifted to a 16-month, four-phase program: a 12-week preflight that included elimination training, an 18-week primary course, a 16-week basic course, and a 12-week advanced course. New aircraft were introduced to all phases: T-33s to advanced (1949), T-28s to basic (1951), T-34s to primary (1953), and T-29s (1949) to navigator and bombardier schools. Although pilot and aircrew production inevitably declined following the Korean Armistice, Cold War force levels allowed ATC to maintain a core of eight flying training wings and six contract primary schools throughout the 1950s, and by the end of the decade, ATC realized its goal of eliminating propeller-driven aircraft from the primary phase as new T-37s entered service to replace T-34s.

The next decade saw many changes in ATC organization. As the first supersonic T-38s became operational during the early 1960s, ATC discontinued the four-phase program in favor of a generalized undergraduate pilot training (UPT) syllabus. UPT eliminated civilian contract primary schools and established a primary/basic course in T-37s followed by an advanced phase in T-38s. Equally significant, the Aviation Cadet program ended for pilots in late 1961 and for navigators in early 1965, placing complete reliance upon the Air Force Academy (AFA), Air Force Reserve Officer Training Corps (AFROTC), and Officer Training School (OTS) to recruit new pilot and aircrew trainees, and in 1965, to lower pilot attrition rates, flight screening in lightplanes (elimination training) was instituted in all three programs. While the escalation of U.S. military involvement in Southeast Asia during the mid–1960s generated small increases in

Four student pilots in advanced jet training pose in front of a T-33A at Williams AFB, Arizona, in the late 1950s.

pilot and aircrew training, ATC ended the decade with eleven flying training wings composed of nine UPT wings, one navigator training wing, and one combat crew training wing.

ATC continued the UPT syllabus into the 1970s but trimmed its operations to eight flying training wings (seven UPT and one navigator) and one combat crew training wing. Increasing reliance on flight simulators during this period enabled ATC to reduce costs by shortening the required flying hours within the UPT syllabus. The 1970s also established the trend toward inter-service pilot and aircrew training programs, starting in 1971 when USAF student helicopter pilots commenced UPT at the Army Aviation School, followed in 1976 by the USAF-sponsored inter-service undergraduate navigator training (i.e., IUNT) of Navy and Marine flight officer trainees. Another significant milestone came in 1975 as ATC admitted the first classes of female trainees to UPT and Navigator School. In 1978 navigator training changed from a generalized to a specialized syllabus in which the advanced phase was split between trainees on the transport-tanker-bomber track and those on the weapons system track in tactical aircraft. ATC opened the 1980s with the eight flying training wing structure and lengthened UPT by adding more flying hours to the syllabus; and operations expanded even further in 1981 with the start of Euro-NATO Joint Jet Pilot Training (ENJJPT) at Shepherd AFB, Texas. But efforts to develop a New Generation Trainer (NGT) to replace ATC's aging fleet of Cessna T-37s ultimately proved to be unsuccessful, and by the end of the decade, as a portent of things to come, funding cuts were already forcing ATC to make manpower reductions due to the winding down of the Cold War.

U.S. Army. In 1946, as part of the huge post–World War II military demobilization, the Army Ground Forces School for liaison pilots at Fort Sill, Oklahoma, was deactivated. And when separated from the newly-established U.S. Air Force in July 1947, the Army was left with a force of approximately 650 two-seat light aircraft (e.g., L-4s, L-5s, and L-16s) and a handful of H-13 helicopters. However, responsibility for training new Army Aviators, maintaining training aircraft, and providing bases was placed under overall USAF jurisdiction within Air Training Command (ATC). Army pilot training, fixed and rotary-wing, was carried out by ATC's 3585th Pilot Training Wing (Liaison-Helicopter), initially at San Marcos Field, Texas until 1949, when operations moved to Waco AFB, Texas then in 1951, back to San Marcos (renamed Gary AFB in 1953). In 1949, in a departure from the USAF system of obtaining pilot candidates through the Aviation Cadet and Officer Training School programs, the Army created the Warrant Officer Helicopter Pilot military occupational specialty (MOS) as a means of recruiting candidates from the enlisted ranks and qualified high schools graduates, with the very first class of Warrant Officer Candidates (WOC) being activated in April 1951.

As a consequence of the rapid growth of Organic Army Aviation during the Korean War, the Department of Defense (DoD) began taking active steps to move the task of training new Army pilots (and aircraft mechanics, too) from USAF to Army control. Towards this end, in 1954, DoD authorized the Army to move the Army Aviation Center (created in 1953) from Fort Sill to the site of Camp Rucker, Alabama, together with the adjacent Ozark airfield, and in late 1955, the facility gained permanent status as Fort Rucker. Although helicopter training actually commenced at Rucker in late 1954, the en mass change from USAF control did not take place until mid–1956 when the Army activated the Primary Helicopter School at Camp Wolters, Texas (formerly Wolters AFB) and assumed control of the fixed-wing school at Gary AFB, which was renamed Camp Gary. As before, civilian contract instructors accomplished primary pilot training at both bases, while advanced training, fixed and rotary-wing, was carried out by Army instructor-pilots at Fort Rucker, then in 1959, primary training of fixed-wing student aviators at Camp Gary ended and moved to Fort Rucker.

The 1960s proved to be a decade of considerable expansion and reorganization for Army Aviation, especially in the production of the thousands of new rotary-wing pilots needed to support the Army's "Airmobile" doctrine of using helicopters to transport large numbers of assault troops and supplies. As U.S. military involvement in Vietnam gained momentum, training facilities at Fort Wolters (renamed in 1963) grew to an inventory of over 1,300 training helicopters (TH-13s and TH-55s) operating out of three heliports and 25 stage fields, and by the late 1960s, boasted an output of 600 new helicopter pilots per month, with graduates continuing on to advanced and transition training at Rucker in the type of helicopter to be flown (e.g., UH-1, OH-6, CH-47, CH-54). Meanwhile, primary training of fixed-wing pilots, using an all–Army instructor force, moved from Fort Rucker to Hunter Army Air Field at Fort Stewart, Georgia in 1967, where students received initial flight training in T-41s, followed by an instrument course in T-42s. After graduation, fixed-wing pilots went to Rucker for advanced and transition training in the type of aircraft to be flown (e.g., L-19 [O-1], L-20 [TU-6], U-1, CV-2, U-21, OV-1).

During the early 1970s, with the ending of hostilities in Vietnam and the consequent force reductions, the Army ceased operations at Wolters and Steward and consolidated all of its aviator training at Rucker. Even so, the tactical mobility required of the post–Vietnam Army caused helicopter pilot training to be maintained at a high level, and because the Army emerged better equipped to train high volumes of helicopter pilots, it has been sponsoring rotary-wing UPT for USAF student pilots since 1971. As it moved through the 1970s, organic Army aviation evolved its training to refocus on a higher-intensity conflict in Europe (i.e., Cold War), with greater emphasis placed upon using helicopters in expanded roles such as close air support, battlefield interdiction, and intra-theater logistics and with less reliance on fixed-wing aircraft.

Organic Army aviation achieved a milestone in April 1983 when it officially became the Army Aviation Branch, enabling it to consolidate many aviation-related functions (e.g., transportation, logistics, training) under one command. In 1988 approval of the Army Aviation Modernization Plan provided funding for many new helicopters, which included, among other things, procurement of 181 Bell TH-67s for the Primary Helicopter School.

U.S. Navy, Marine Corps and Coast Guard. In the wake of the post–World War II demobilization, most of the Navy's wartime fleet went into storage ("mothballs"), leaving seven active carrier air groups and one replacement group, and Marine aviation, likewise, dwindled to two under-strength air wings, while the Coast Guard reverted to Treasury Department control at prewar force levels. Naval Air Training downsized to just two training centers—NAS Pensacola-Whiting for primary/basic and NAS Corpus Christi for advanced—and like the USAF, discontinued the preparatory and elimination phases and consolidated pilot training into a primary-basic course in SNJs and an advanced course in either F6Fs (carrier track), SNBs (patrol, transport and navigator track), or helicopters (at the Coast Guard-run school). Although the Naval Aviation Cadet (Nav-Cad) program remained in effect, Naval Aviation Pilot (NAP) training ended in 1948. When, in mid–1950, the Cold War suddenly erupted into a full-blown conflict in Korea, the Navy and the Marines both were forced to draw heavily on recalled reservists, active and inactive, to meet quotas for pilots and aircrews needed.

As the 1950s dawned, the Naval Air Training establishment faced the difficult task of completely modernizing its infrastructure, not only in terms of boosting pilot and aircrew production to match the needs of a significantly larger Navy and Marine post–Korea force but also to prepare a new generation of pilots to operate high-performance jets from carriers. To handle the overflow, new training centers opened at NAS Memphis, Tennessee and NAS Chase Field, Texas. Procurement programs extending all the way into the early 1960s accounted for introduction of over 2,600 new training aircraft to service: 423 Beech T-34Bs (primary), 489 North American T-28Bs (basic), and 699 Lockheed TV-2s (advanced jet); and for the critical carrier qualification phase, 299 North American T-28Cs, 150 Lockheed T2V-1s, 400 Grumman F9F-8Ts, and 217 North American T2J-1s. Pilot production increased by 250 percent, from 900 per year in late 1940s to 2,200 per year in the late 1950s. In terms of reorganization, primary and basic pilot training reverted back to distinct phases, carrier qualification became part of the undergraduate curriculum, helicopter training, now under Navy control, expanded operations with 214 new training helicopters acquired between 1949 and 1958, and last, Advanced Training Units (ATU) were established to transition pilot and aircrew graduates into operational aircraft. Finally, in order to attract college graduates who would undergo pilot or aircrew training as commissioned officers, the Navy established Aviation Officer Candidate School (AOCS) in 1955 to augment the Nav-Cad program.

Although pilot and aircrew production dropped slightly during the early 1960s, the Naval Air Training establishment was well positioned for the surge brought on by the Vietnam War, increasing pilot production from an average of about 1,600 per year in the mid–1960s to 2,400 per year by the late 1960s. With sufficient numbers of pilot and aircrew candidates being recruited through AOCS, the Nav-Cad program terminated in 1968. Once the Vietnam War ended in 1972, pilot and aircrew production returned to levels comparable with the early 1960s, while long-term procurement programs for new training aircraft continued uninterrupted: 97 North American T-2Bs (old designation T2J-2) in 1964 and 1965, 281 McDonnell Douglas TA-4Js from 1969 to 1974, 231 North American T-2Cs from 1970 to 1973, 352 turboprop Beech T-34Cs starting in 1975, 61 Beech T-44As (advanced multi-engine) in 1977 and 1978; for helicopter training, 40 Bell TH-57As and 45 Bell TH-1Ls in 1968 and 1969; and for Naval Flight Officer (NFO) training, 42 North American T-39Ds in 1962 and 1963 and nine Grumman TC-4As in

1967. The Navy blazed the trail for the other armed services when it admitted the first group of female candidates to pilot training in March 1973. As a result of an inter-service agreement implemented in 1976, navigator training of NFOs earmarked for land-based aircraft—known thereafter as Inter-Service Undergraduate Navigator Training (IUNT)—was merged into the USAF training in Boeing T-43s. During the fleet buildup of the Reagan era in the mid–1980s, shortfalls in pilot and NFO production led to a temporary revival of the Nav-Cad Program in 1986, accepting candidates with an associate's degree or 60 hours of college credit who would be commissioned upon graduation with the understanding that they would be required to complete their college education during their first shore duty assignment. But because of previous procurement programs, only a few types of new aircraft were added to the training fleet: 15 Cessna T-47As procured for NFO training in 1983 and 1984 to replace North American T-39Ds with high-time airframes and 140 Bell TH-57As and Bs for helicopter training between 1986 and 1989 in a across-the-board replacement of Bell TH-57As and TH-1Ls.

Modern Era 1990–Present

Post-Cold War Force Reductions. The Cold War had been the most important factor in the shaping of U.S. military policy for over 40 years. Its demise actually began in the mid–1980s, but it was not officially and finally over until the formal dissolution of the U.S.S.R. in December 1991. While the process of reducing manpower, restricting procurement, and closing bases within the military branches started in the late 1980s, its full effect was temporarily delayed by the Iraq's invasion of Kuwait in August 1990 and the consequent mobilization for Operation Desert Storm. But once the Gulf War ended in April 1991, military reductions began in earnest, and within a few years, manpower diminished by a factor of 35 percent, 93 bases and installations closed, 200 ships were decommissioned, and thousands of aircraft were either placed in storage or scrapped.

U.S. Air Force. In a marked departure from past practices, the USAF and the Navy agreed in 1990 to a program entitled Joint Primary Aircraft Training System (JPATS) with the goal of selecting and procuring one type of primary trainer to replace both the T-37 and the T-34, though actual service entry would be gradual, over a decade or more. As part of a post–Cold War realignment of major USAF commands, Air Training Command merged with the Air University and Air Force Recruiting Service to form Air Education and Training Command (AETC) on July 1, 1993. With respect to pilot and aircrew training, AETC commenced with 1,377 active aircraft and a structure of seven flying training wings plus one air mobility wing (strategic airlift and air refueling post–UPT pilot training). AETC retained the UPT curriculum until the mid–1990s, when it was phased-out in favor of Specialized Undergraduate Pilot Training (SUPT) in which pilot students completing the primary-basic phase in T-37s proceeded on to a bomber-fighter track in T-38s or an airlift-tanker track in newly procured Beech T-1As. Likewise, navigator training was renamed Specialized Undergraduate Navigator Training (SUNT) with tactical and airlift-tanker tracks in conjunction with inter-service training of Navy and Marine NFOs under the heading Inter-Service Undergraduate Navigator Training (IUNT). Another change, termed Joint Specialized Undergraduate Pilot Training (JSUPT), emerged in 1996 as a result of an inter-service agreement in which the Navy would provide advanced multi-engine training to USAF student pilots on the C-130 airlift track in Beech T-44s. On the negative side, the pre–SUPT flight-screening program (FSP) was suspended mid–1997 when all Slingsby T-3s were grounded.

As it entered the new century, AETC possessed an inventory of 1,571 aircraft organized into six flying training wings and one flying training group (SUPT/SUNT training of student

Instructor and student seen in front of their T-6A Texan II at Sheppard AFB, Texas, in 2011.

pilots, weapons system officers [WSO], and combat systems officers [CSO] on the tactical fighter track), plus five post–SUPT training wings for Special Operations (MC-130 or helicopter), Tactical Airlift (C-130), Air Mobility (C-5, C-141, C-17, KC-135, and KC-10), Aircrew Survival, and Fighter (F-15 and F-22). In response to the War on Terrorism during the early 2000s (i.e., Operations Enduring Freedom and Iraqi Freedom), AETC generated a 25 percent increase in SUPT and SUNT graduates, an average of over 1,200 per year compared to 800 per year in the late 1990s. The first JPATS Raytheon T-6As began entering AETC service on a slow-rate basis in May 2000, with the process of fully integrating them into primary-basic SUPT continuing until July 2009, when the very last T-37s were retired. Pre-SUPT IFS was reinstituted in late 2006 with training in light aircraft provided by a civilian contractor, and since 2009, has been conducted at the Air Force Academy in USAF-leased or owned aircraft. As of the end of 2013, AETC operated a fleet of 1,361 aircraft organized into five flying training wings for SUPT/SUNT and five wings for post–SUPT/SUNT, two for fighters, plus one each for special operations, tactical airlift, and air mobility. Helicopter SUPT for USAF student pilots, now in the TH-67A, is still performed per inter-service agreement at the Army Aviation Helicopter School at Fort Rucker, followed by an advanced course in the TH-1H, either at Fort Rucker or Lackland AFB.

U.S. Army. During the early to mid–1990s, the Army underwent a post–Cold War-Gulf War downsizing process comparable to the other services (-35 percent overall), however, new helicopter procurement under the Army Aviation Modernization Plan remained more or less intact, so that the Army Aviation School at Fort Rucker began receiving the first of 183 Bell TH-67As for primary training in 1993. Upon completion of a 20-week primary phase in the TH-67, pilot trainees moved on to a 12-week intermediate phase in Bell OH-58s or UH-1s, then to an advanced phase in the type of helicopter to be flown (e.g., CH-47, OH-58. UH-1, UH-60,

Army student conducting preflight of TH-67A in primary helicopter training at Fort Rucker, Alabama, in 2009.

or AH-64). With less emphasis on fixed-wing aircraft, all pilot trainees were required to complete the entire helicopter curriculum before becoming eligible for a fixed-wing transition. In 1994 the Army began implementing the Aviation Restructure Initiative (ARI), which contemplated a smaller but better organized and equipped Aviation Branch as it moved towards the 21st Century. While exact numbers are not reported anywhere, it appears that Army Aviation new pilot production averaged an estimated 1,000–1,100 graduates per year through most of the 1990s.

As Army Aviation moved into the new century, in response to the post–9/11 War on Terror, new pilot production increased by approximately 25 percent, but more importantly, pilot training was completely revamped under a program entitled Flight School Twenty-First (FSXXI). Under FSXXI, all primary, including both pilot instruction and maintenance support, is carried out by a civilian contractor, and the training syllabus shortened and condensed to comprise a two-week prelight, a 20-week primary, instrument, and terrain-following navigation phase in the TH-67, followed by 46 to 50 weeks of advanced instruction in a UH-60, CH-47, or AH-64. According to recent internet information, the Army currently operates a fleet of 172 TH-67As at its Fort Rucker Aviation School which, since FSXXI, are carried under civil registrations as "public use" aircraft.

U.S. Navy, Marine Corps and Coast Guard. In the five years following the end of the Cold War, the Navy and Marines experienced force reductions of 36 percent and 12 percent, respectively, with average new pilot/NFO production dropping proportionately to about 1,100/500 per year. In 1993, the Nav-Cad program ended for the last time, and in 1994, AOCS was merged into OCS, with aviation candidates being classified as either Student Naval Aviators (SNA) or Student Naval Flight Officers (SNFO). In terms of new aircraft procured, McDonnell Douglas T-45s began entering service as tailhook/advanced jet trainers in 1993, but due to slow-

Student pilots prepare to exit a T-45C Goshawk during flight operations aboard the USS *Abraham Lincoln* CVN-72 in 2004.

rate production, the process extended to 1999 for replacement of TA-4s and to 2004 for T-2s. For NFOs, 17 rebuilt Rockwell Sabreliners Joined the inventory in 1993 as T-39Ns to replace T-47As. To augment T-44As on the advanced multi-engine track, approximately 14 Beech Huron transports were redirected to training duties in 1996 as the TC-12B. The process of replacing T-34s with the first of 49 Raytheon T-6As (JPATS) began in early 2003; however; the Navy has since elected to shift balance of the order for 262 aircraft to the T-6B, the first example of which became operational in mid–2010. At present, only one squadron (VT-28) still conducts primary in the T-34. Since the termination of USAF-sponsored IUNT in T-43s in 2010, navigator training of NFOs has been performed in T-6As of VT-10.

No exact, year-by-year figures on pilot and NFO production after 1995 are readily available, although official sources suggest that current levels are only slightly below those reported above for 1990–1995. Under the current Naval Air Training curriculum, all SNA candidates must complete a four-phase training syllabus: (1) An Introductory Flight Screening phase conducted at one of three civilian schools, consisting of 14 hours of flight and ground school in preparation for an FAA Private Pilot Knowledge Test; (2) A 15-week Preflight Indoctrination phase; (3) A six-month Primary Flight Training phase that includes a ground school plus contact, basic instrument, precision aerobatics, formation, night, and navigation instruction in a T-34C or T-6B; and (4) A 43- to 50-week advanced phase according to four different "pipelines": (a) Tailhook-Advanced Strike [tactical jets]; (b) Tailhook-Multi-Engine [E-2/C-2]; (c) Maritime [P-3/P-8]; and (d) Rotary-Wing [in TH-57s in preparation for training in operational helicopters]. Marine SNAs destined for MV-22s complete a hybrid syllabus in both T-44s and TH-57s.

Series I

Fixed Wing Trainers, 1917 to 1947

A. USAAS, USAAC and USAAF

Curtiss JN-4, JN-6 and JNS—1916

Technical Specifications (JN-4D [JN-4HG])

Type: Two-place primary trainer.
Manufacturer: Curtiss Aeroplane & Motor Co., Hammondsport, Garden City, and Buffalo, New York; Canadian Aeroplanes, Ltd., Ontario, Canada.
Total U.S. military versions: 6,409 (6,148 USAAS; 261 USN/USMC).
Unit Cost: Varied from $7,500 to $8,100.
Powerplant: One 90-hp Curtiss OX-5 (V-503) [150-hp Wright-Hispano E (V-718)] 8-cylinder water-cooled inline engine driving a two-bladed, fixed-pitch wooden propeller.
Armament (JN-4HG): One .30-calibre machine gun on scarf ring in rear cockpit.
Performance: Max. speed 75 mph. [91 mph], cruise 60 mph [75 mph]; ceiling 6,500 ft. [7,500 ft.]; range 155 mi.
Weights: 1,390 lbs. [1,625 lbs.] empty, 1,920 lbs. [2,269 lbs.] loaded.
Dimensions: Span 43 ft. 7 in., length 27 ft. 4 in., wing area 352 sq. ft.

On the topic of American military training aircraft, there is perhaps no better place to begin than with the Curtiss JN, best remembered as the "Jenny." Jennies not only trained 90 percent of U.S. Army pilots during the First World War but played an equally significant role in establishing the United States as a mass-producer of military aircraft. The JN series emerged in 1914 as a hybrid design for a tractor-type military trainer that combined the best features of the model N, designed by Curtiss, and the model J, designed in Great Britain by former Sopwith engineer B. Douglas Thomas. After two model J prototypes underwent official Army trials during late 1914, Curtiss received an order in December for eight modified model Js. Since the J incorporated major elements of the N's design, the two prototypes became the JN-1, and the eight modified aircraft delivered to the Army in early 1915, the JN-2. In their original configuration, the JN-1 and JN-2 were equal-span, two-bay biplanes with a one-piece, all-moving rudder and the shoulder-yoke aileron control system similar to that seen on earlier Curtiss pushers. Construction of all major airframe components consisted of a built-up spruce framework covered in a linen fabric. When the Army expressed official disappointment with the handling qualities of the JN-2, Curtiss

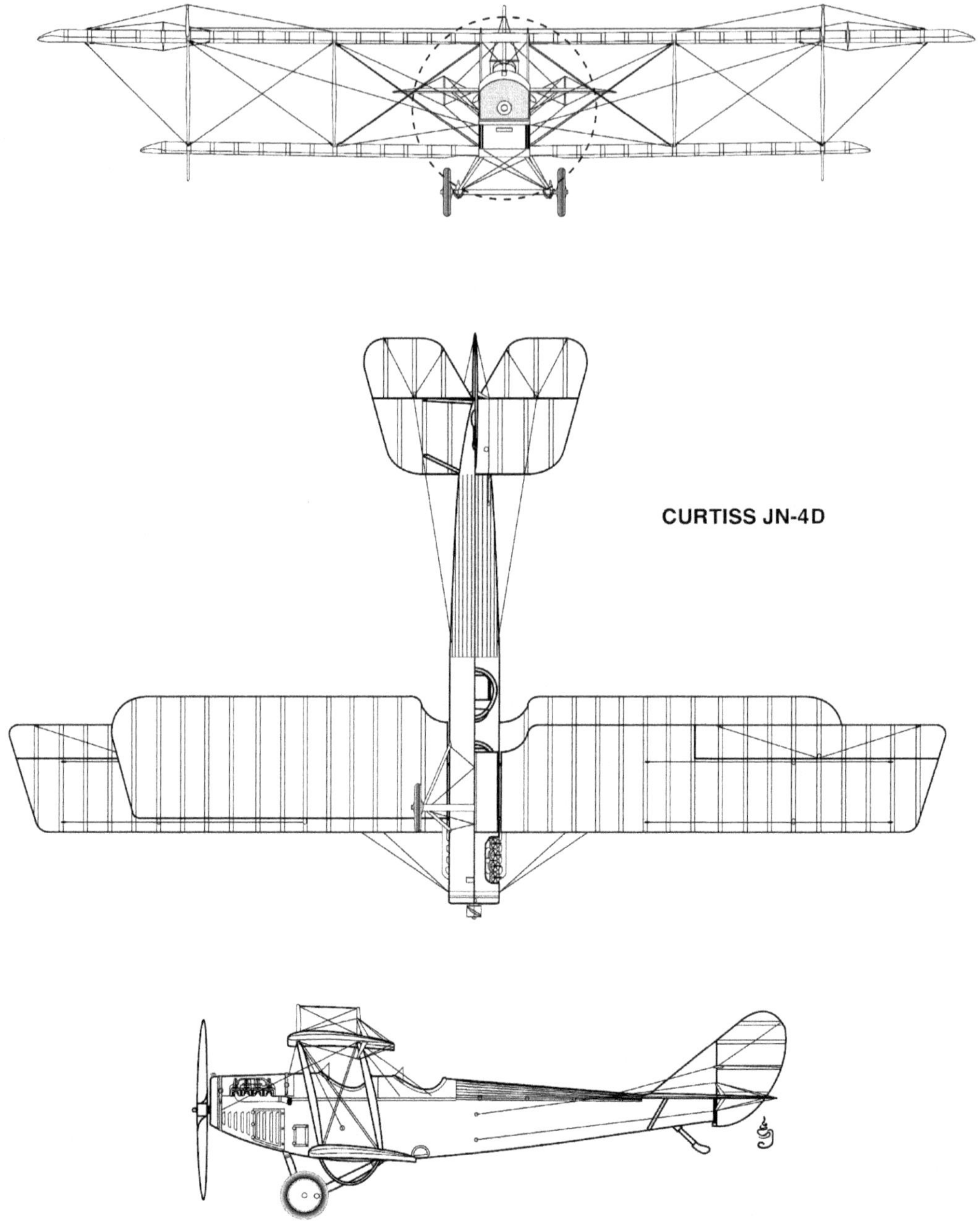

CURTISS JN-4D

responded in late 1915 with the much improved JN-3, which differed in having increased upper wingspan, enlarged vertical tail surfaces with a separate fin and rudder, and a Deperdussin control system using a wheel to move the ailerons on the upper wing. The Army took delivery of two JN-3s and subsequently modified five JN-2s to the -3 standard.

What became the first true Jenny, the JN-4, appeared in 1916 with upper wingspan increased to 43 feet 7 inches and Deperdussin controls. When the first of 21 JN-4s began arriving for Army service in late 1916, they joined JN-2s and JN-3s of the 1st Aero Squadron in the Mexican Border War against Pancho Villa, becoming the first aircraft to be used in combat by the U.S. Army.

During the same timeframe, British interest in acquiring the Jenny for pilot training led to formation of Canadian Aeroplane, Ltd. in Toronto, Ontario (originally as a Curtiss subsidiary but later taken over by the Canadian government) to build the JN-4 (Can) or "Canuck" for the Royal Flying Corps as well as the U.S. Army. Differences between the Canuck and the standard JN-4 version included a lighter structure, revised lower wing outline, and stick controls. Of the 1,269 Canucks built, 869 went to the Royal Flying Corps and 400 to the Army Air Service. Meanwhile, in early 1917, Curtiss switched American production to the JN-4A, which came with enlarged tail surfaces for better stability, a stronger fuselage structure, upper and lower wing ailerons, and down-thrust added to the motor, with 601 being delivered to the Army Air Service before the end of the year. In parallel, the Army also took delivery of 76 JN-4Bs without motor down-thrust and tested one JN-4C with experimental wings.

The definitive and most numerous Jenny variant, the JN-4D, appeared in June 1917 with stick controls, upper wing ailerons only, and semi-circular cutouts on the inner trailing edges of both wings for improved visibility. Although Curtiss's new plant in Buffalo was the largest aircraft production facility in the world by the time the U.S. entered the World War in April 1917, War Department demand for JN-4Ds outstripped its capacity to such extent that production contracts were awarded to six different American manufacturers: Fowler Airplane Corp., Howell & Lesser, Liberty Iron Works, St. Louis Aircraft Corp., Springfield Aircraft Corp., and U.S. Aircraft Corp. Between mid–1917 and late 1918, the Army Air Service accepted delivery of an estimated 3,394 JN-4Ds, with 1,404 completed by Curtiss, 1,310 by other American companies, and 680 by Canadian Aero.

Installation of a 150-hp Wright-Hispano engine in the JN-4D airframe in place of the OX-5 resulted in the introduction of the JN-4H advanced trainer version in early 1918. Of the 839 delivered to the Army Air Service, 372 were completed as JN-4H dual-control trainers, 100 as JN-4HB bombing trainers, and 367 as JN-4HG gunnery trainers. Later in 1918, the JN-4H was succeeded in production by the JN-6H, differing mainly in having strut-connected upper and lower wing ailerons. An estimated 1,119 JN-6H variants had been delivered to the Army Air Service by the time hostilities ceased in November 1918, which included 154 JN-6HB bombing trainers, 584 JN-6HG gunnery trainers, 256 JN-6HO observation trainers, and 125 single-control JN-6HP pursuit trainers. Although the vast majority of Army Air service Jennies were sold surplus on the civilian market after the war ended, an unspecified number of Wright-Hispano-

JN-4D, the definitive and most numerous Jenny variant, with 3,394 produced by six manufacturers. The majority were sold surplus on the civilian market following the armistice.

One of 154 JN-6BH bombing trainers equipped with 150-hp Wright-Hispano engines. Note increased rudder area and bomb racks below the bottom wing center section.

powered variants underwent postwar rebuilding programs, including some with welded, steel tube fuselages, and remerged as the JNS. The final examples of the JNS are thought to have been withdrawn from National Guard units during the late 1920s. After leaving military service, surplus Jennies in private hands trained hundreds of new civilian pilots, and as barnstormers, introduced aviation to small communities all over the nation.

Naval Variants: Although the Navy acquired small numbers of early Jenny variants, one JN-1 and two JN-1Ws (float-equipped) in 1916 and approximately 15 JN-4As and Bs in 1917, the Wright-Hispano-powered JN-4H and -4HG became the principal naval versions. Thirty JN-4Hs were procured in 1918 for advanced pilot training, followed by 90 JN-4HGs outfitted as gunnery trainers. A final batch, 113 JN-4H and 10 JN-6H advanced trainers, were accepted by the Navy during the early post-war era. A number of JN-4Hs/HGs were transferred to the Marines, where they were used for training and observation duties. The Navy and Marines began phasing out their Jennies in the mid–1920s as newer types of trainers became available, the final examples being withdrawn in 1926.

Standard J-1 and JR-1–1917

Technical Specifications (J-1)

Type: Two-place primary trainer.
Manufacturer: Standard Aero Corp., Plainfield, New Jersey.
Total U.S. military versions: 1,607 (USAAS).
Unit Cost: $8,000.
Powerplant: One 100 hp Hall-Scott L-4A (L-550) 4-cylinder water-cooled inline engine driving a two-bladed, fixed-pitch wooden propeller.
Armament: None.

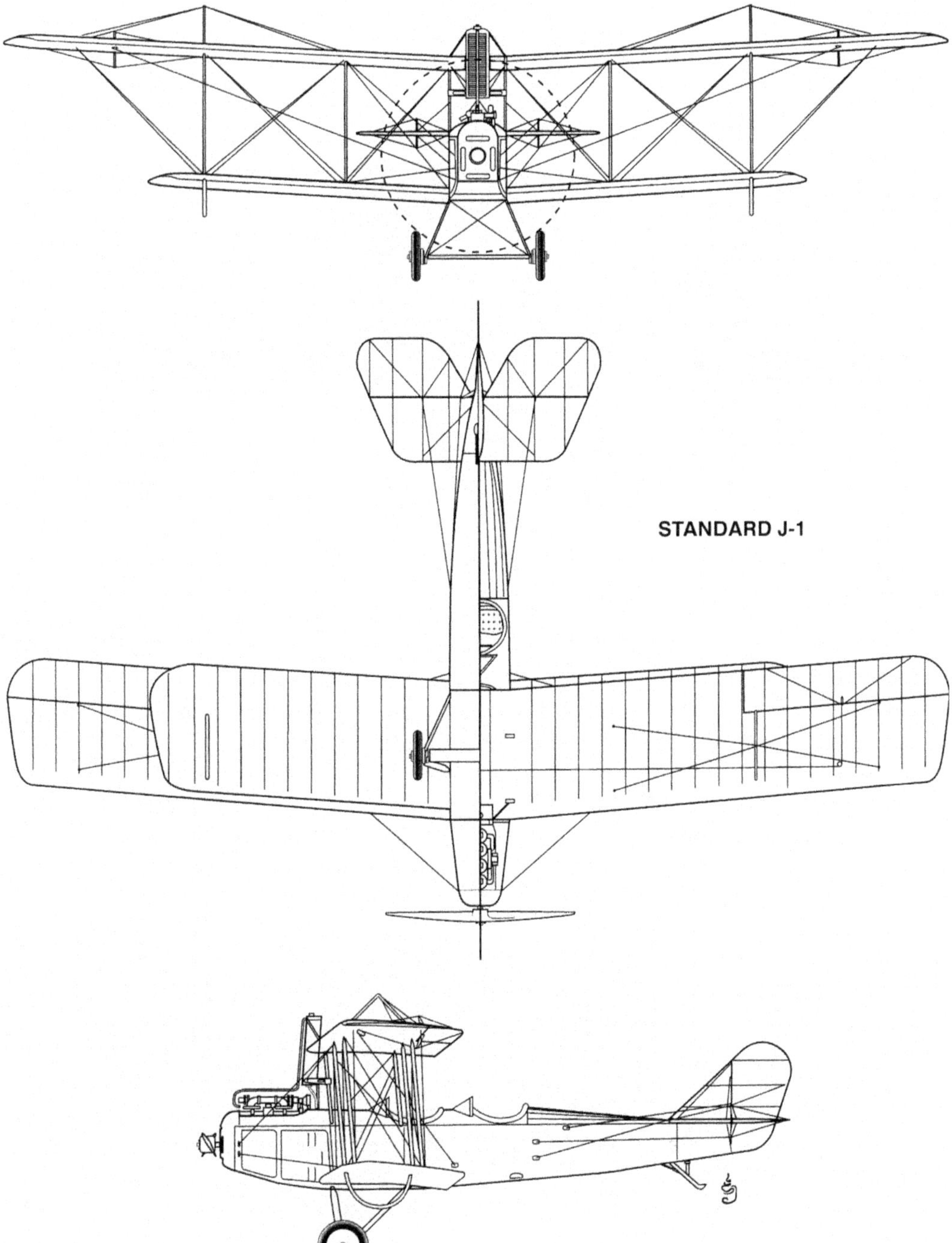

Performance: Max. speed 72 mph, cruise 64 mph; ceiling 5,800 ft.; range 235 mi.
Weights: 1,512 lbs. empty, 2,025 lbs. loaded.
Dimensions: Span 43 ft. 10 in., length 26 ft. 2 in., wing area (not reported).

Although distinguished as the second most numerous type of primary trainer used by the Army during the First World War, the Standard model J never achieved the acclaim of the con-

The second most numerous World War I primary trainer, the Standard J-1s were unpopular with instructors and students because of excessive airframe vibration produced by their Hall-Scott engines.

temporaneous Curtiss Jenny. Much of its design had been derived from the Sloan H-2 of 1915, a two-bay, tractor-type biplane having swept-back, equal span wings, and like most aircraft of the period, constructed chiefly of a fabric-covered, built-up wooden framework. Following Sloan's reorganization as Standard Aero in 1916, the design reemerged as the model J, which differed in having an un-swept layout with a noticeably larger upper wing. Army trials in early 1917 resulted in a contract for 100 production models as the J-1, then following U.S. entry into the World War, the order was increased over time to about 2,700 aircraft which, in addition to Standard, included subcontracts awarded to Dayton-Wright, Fisher Body, and Wright-Martin. However, as events turned out, the War Department cancelled Standard's contract in June 1918 when J-1 deliveries to the Army Air Service stood at 1,601 aircraft.

In service, despite having a somewhat better safety record than the Jenny, J-1s were unpopular with students and instructors alike, due reportedly to the excessive airframe vibration generated by the Hall-Scott engines. The Army Air Service studied the possibility of converting existing J-1s to smoother-running OX-5 engines, but at a cost of $2,000 per conversion, the scheme was deemed impractical. With sufficient Jennies on hand to meet training needs, the Army Air Service withdrew all J-1s from operational service in June 1918, the same date the contract was terminated, and most J-1s delivered afterward were never removed from their packing crates. J-1s, like Jennies, were sold surplus on the civil market and used for barnstorming and pilot training, and a number of them converted to OX-5 or Hispano engines reappeared as the SJ-1.

In early 1918 the Army Air Service evaluated an extensively modified variant of the J as an advanced trainer under the designation JR-1. Differences from the basic J included equal-span wings with slight sweepback, revised vertical tail surfaces, and a 175-hp Hall-Scott A-5 engine. Six examples were tested but no production was ordered. In late 1918, six equipped with 150-hp Wright-Hispano engines were acquired for the U.S. Mail Service as the JR-1B.

Thomas-Morse S-4 (S-5)—1917

Technical Specifications (S-4C)

Type: One-place pursuit trainer.
Manufacturer: Thomas-Morse Aircraft Corp., Ithaca, New York.
Total U.S. military versions: 569 (559 USAAS; 10 USN).
Unit Cost: $5,400.
Powerplant: One 80-hp Le Rhône 9C 9-cylinder air-cooled rotary engine driving a two-bladed, fixed-pitch wooden propeller.
Armament: One fixed .30-calibre machine gun.
Performance: Max. speed 95 mph, cruise 85 mph; ceiling 15,000 ft.; range 225 mi.
Weights: 978 lbs. empty, 1,373 lbs. loaded.
Dimensions: Span 26 ft. 6 in., length 19 ft. 10 in., wing area (not reported).

Designed by Englishman B. Douglas Thomas (same designer as Jenny but no relation to the company founders), the S-4 prototype flew for the first time in June 1917 and was subsequently

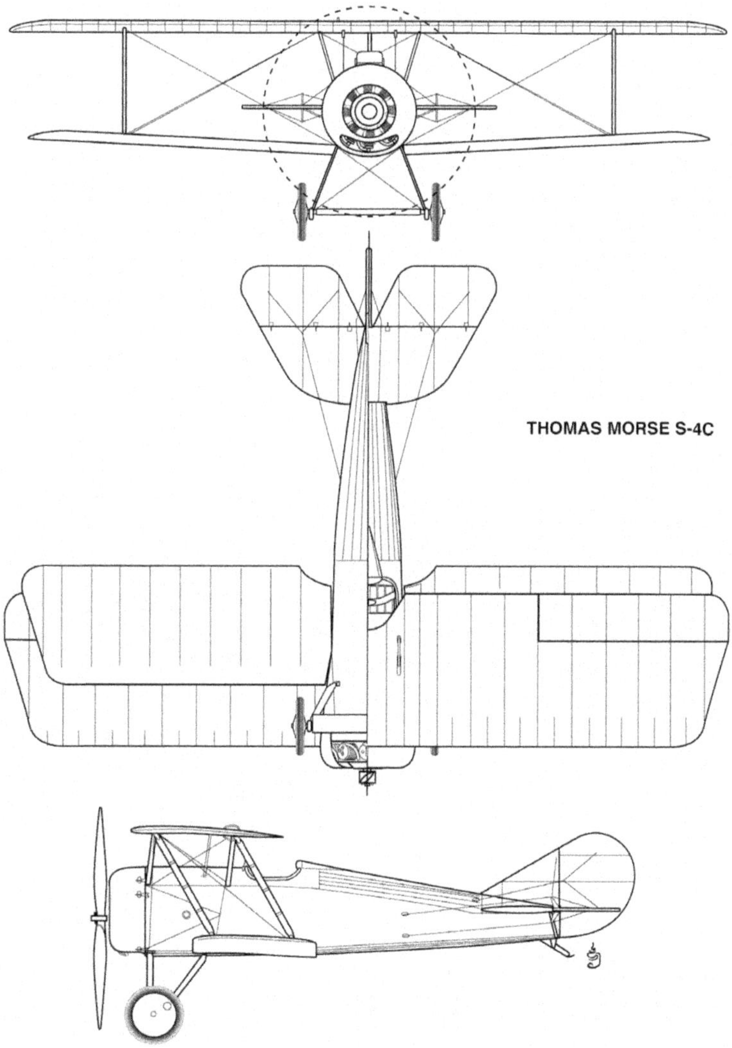

Nearly new Thomas-Morse S-4C seen on snow covered field in late 1918. Most equipped Air Service pursuit schools. Many were sold surplus after the war ended.

delivered to the Army Air Service for evaluation as a fighter. The simple design appeared as a tractor-type, single-bay biplane of conventional fabric-covered wooden construction. Though considered too lightly armed to be a fighter, the S-4 demonstrated sufficiently good flying characteristics to be ordered into production as a single-seat pursuit trainer. Deliveries of 97 S-4B production models with armament commenced in late 1917, followed in 1918 by deliveries of 461 improved S-4Cs, which differed from Bs in having slightly reduced wingspan and changes to the ailerons and elevator. In late 1918, the Army tested an S-4E with a 110-hp Le Rhône rotary and tapered wings of reduced span but did not order the type into production.

Dubbed "Tommy" by the pilots who flew them, S-4B and Cs equipped most of the Army Air Service pursuit training schools during the First World War; however, once hostilities ended in November 1918, virtually all of them were offered as surplus on the civil market. In private hands, Tommies were popular with flying schools and sportsman pilots, with many surviving well into the 1930s.

Naval Variants: The Navy acquired six S-4Bs in late 1917 which were placed on the inventory as the S-5 after being fitted with floats similar to those of the Sopwith Baby (i.e., two short main floats and a tail float), and in 1918, took delivery of four S-4Cs that were likewise fitted with floats but apparently retained their original designation. The S-5s and S-4Cs saw only brief service and were withdrawn once the Navy judged single-seat scouts to be impractical.

Standard E-1–1918

Technical Specifications (E-1)

Type: One-place pursuit trainer.
Manufacturer: Standard Aero Corp., Plainfield, New Jersey.
Total U.S. military versions: 128 (USAAS).
Unit Cost: $7,000.
Powerplant: One 100-hp Gnôme 9B or 80-hp Le Rhône 9C 9-cylinder air-cooled rotary engine driving a two-bladed, fixed-pitch wooden propeller.

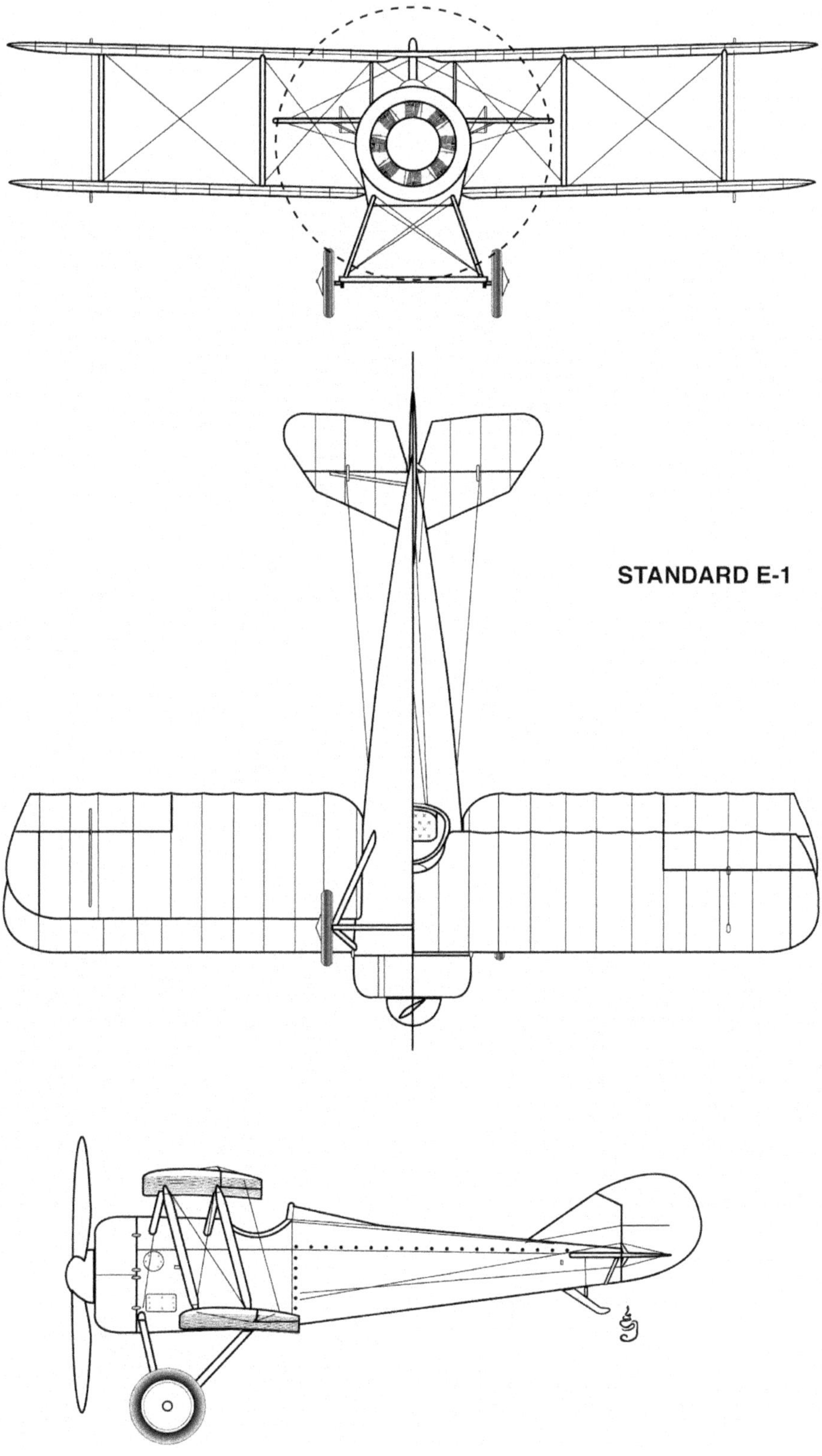

STANDARD E-1

The Standard E-1 was originally conceived as a small, airfield defense fighter, and most served as pursuit trainers and some continued after hostilities ended. Three were used as radio-controlled drones in torpedo experiments.

Armament: None, but equipped with a gun camera.
Performance: Max. speed 100 mph, cruise 90 mph; ceiling 14,500 ft.; range 225 mi.
Weights: 815 lbs. empty, 1,144 lbs. loaded.
Dimensions: Span 24 ft., length 18 ft. 10 in., wing area (not reported).

Like the S-4 reported above, the Standard E-1 or M-Defense came into being in late 1917 as a proposed pursuit aircraft, specifically to serve as an airfield defense fighter. It emerged as a very compact, equal-span biplane powered by a 100-hp Gnôme rotary and retained the two-bay strut layout seen on the company's earlier J series. Primary construction consisted of a wooden airframe covered in fabric. The War Department gave Standard an order for four E-1 prototypes, with the first arriving for evaluations in January 1918. Although the type demonstrated good maneuverability and stability, Army officials concluded the E-1 was too underpowered and lightly armed (one machine gun) for frontline combat but would be useful a pursuit trainer. Standard reportedly received an initial order for 460 aircraft to be powered by Gnôme rotaries, however, serial number records indicate the contact was limited to 53 Gnôme-powered E-1s accepted as of mid 1918 and 75 Le Rhône-powered examples delivered at the time of the Armistice in November 1918.

E-1s began equipping flying schools during the last year of the war, and some data suggests that a small number continued in service as pursuit trainers after hostilities ended. Three E-1s modified by Sperry in 1919 were used as radio-controlled drones in torpedo experiments.

Lewis & Vought VE-7, VE-8 and VE-9–1918

Technical Specifications (VE-7)

Type: Two-place landplane/floatplane trainer, observation, and fighter.
Manufacturer: Lewis & Vought/Chance Vought Corp., Long Island City, New York; Army Air Service Engineering Div., Dayton, Ohio; and Springfield Aircraft Co., Springfield, Massachusetts.

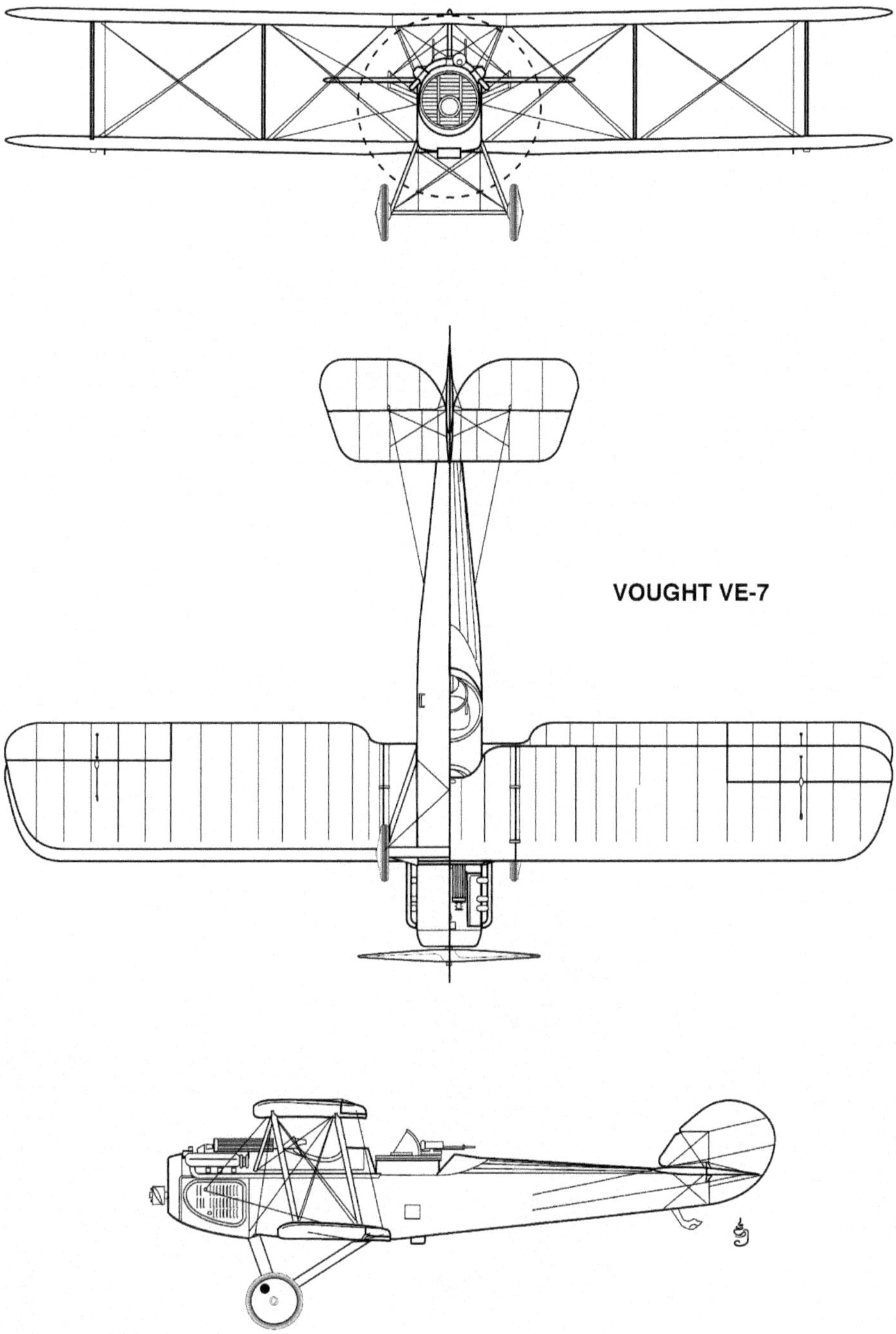

Total U.S. military (trainer) versions: 85 (45 USAAS; 40 USN/USMC).
Unit cost: Unknown.
Powerplant: One 180-hp Wright-Hispano E-2 (V-719) 8-cylinder water-cooled inline engine driving a two-bladed fixed-pitch wooden propeller.
Performance: Max. speed 106 mph; ceiling 14,800 ft.; range 290 mi.
Weights: 1,392 lbs. empty, 1,937 lbs. loaded.
Dimensions: Span 34 ft. 4 in., length 24 ft. 5 in., wing area 285 sq. ft.

A factory fresh Lewis & Vought VE-7, probably seen in mid–1918. Some served after the war as advanced trainers at Kelly Field, Texas.

One of nine Vought VE-7Hs completed by the Naval Aircraft Factory in 1922. This float-equipped example was stressed for catapulting.

The two-seat Lewis & Vought VE-7 "Bluebird," designed, built, and tested sometime during 1917, was delivered to the Army Air Service at McCook Field, Ohio for testing as a trainer sometime in early 1918. It appeared as a two-bay biplane of conventional fabric-covered, wooden construction, powered initially by a license-built 150-hp Hispano-Suiza E-1 engine. After Army officials graded the VE-7's overall performance as superior that of the JN-4H advanced trainer, the War Department awarded Lewis & Vought a preliminary contract for 14 VE-7s, with deliveries commencing in mid–1918. A second order followed for 1,000 VE-8s upgraded to 180-hp E-2 engines; however, due to the Armistice, the contract was cancelled prior to any aircraft being delivered. Later, in 1919, the Army Air Service procured six VE-7s with E-2 engines, two built by the Engineering Division and four by Springfield Aircraft. A final order came in mid–1922 when Chance Vought (Lewis & Vought's successor) received an order for 27 aircraft with revised fuel systems and detail improvements, to be delivered as the VE-9. Though no operational details exist, the VE-7s and VE-9s most likely served out their careers as advanced trainers at Kelly Field, Texas.

Naval Variants: A VE-7 piloted by Lt. Virgil Griffin holds the distinction of having been the first aircraft to takeoff from the deck of the U.S.S. Langley in October 1922. Between 1919 and 1923, the Navy Department procured 140 VE-7s and VE-9 in different variants, of which 40 were unarmed VE-7 trainers. Some of the trainers were assigned to NAS Pensacola for landplane training while others served at various bases for training and utility purposes. The last active VE-7 was withdrawn from Marine squadron VO-8M in 1928.

Airco DH-4B/4M—1918

Technical Specifications (DH-4B)

Type: Two-place observation aircraft and bomber; advanced trainer.
Manufacturer: Dayton-Wright Co., Dayton, Ohio; Standard Aircraft Corp., Patterson, New Jersey; Fisher Body Div. of General Motors, Cleveland, Ohio; and Boeing Airplane Co., Seattle, Washington.
Total U.S. military versions: 4,486 (4,264 USAAS/USAAC; 222 USN/USMC).
Unit Cost: $11,250.
Powerplant: One 400-hp (various auto manufacturers) Liberty (V-1649) 12-cylinder water-cooled inline engine driving a two-bladed fixed-pitch wooden propeller.
Armament: Two fixed forward-firing .30-caliber machine guns, two flexible .30-caliber machine guns in rear cockpit, and up 322 lbs. of bombs carried on wing racks.
Performance: Max. speed 123 mph, cruise 90 mph; ceiling 14,000 ft.; range 400 mi.
Weights: 2,392 lbs. empty, 3,557 lbs. loaded.
Dimensions: Span 42 ft. 51/2 in., length 29 ft. 11 in., wing area 440 sq. ft.

Based on Geoffrey de Havilland's British design for Airco, a total of 4,846 DH-4s powered by American-designed and made Liberty engines were license-built in the U.S. during 1918 and 1919. From 1919 to 1923, to improve crash protection and crew communication, 1,538 DH-4s were completed or converted to DH-4B standard, which entailed moving the main fuel tank forward and the cockpit aft, repositioning the landing gear, and skinning the entire fuselage with plywood. Then starting in 1923, another 163 DH-4s were remanufactured by Boeing with new welded, steel tube fuselages, with 133 returning to Army service as the DH-4M. An unspecified number of DH-4Bs were converted into dual-control trainer variants as the DH-4BM-1, DH-4BM-2, and DH-4BT and presumably used afterward for advanced bombing and attack training; and at least 22 DH-4Ms underwent conversion into dual-control trainers as the DH-4M-1T. The last DH-4Ms were retired from Army Air Corps service in 1932.

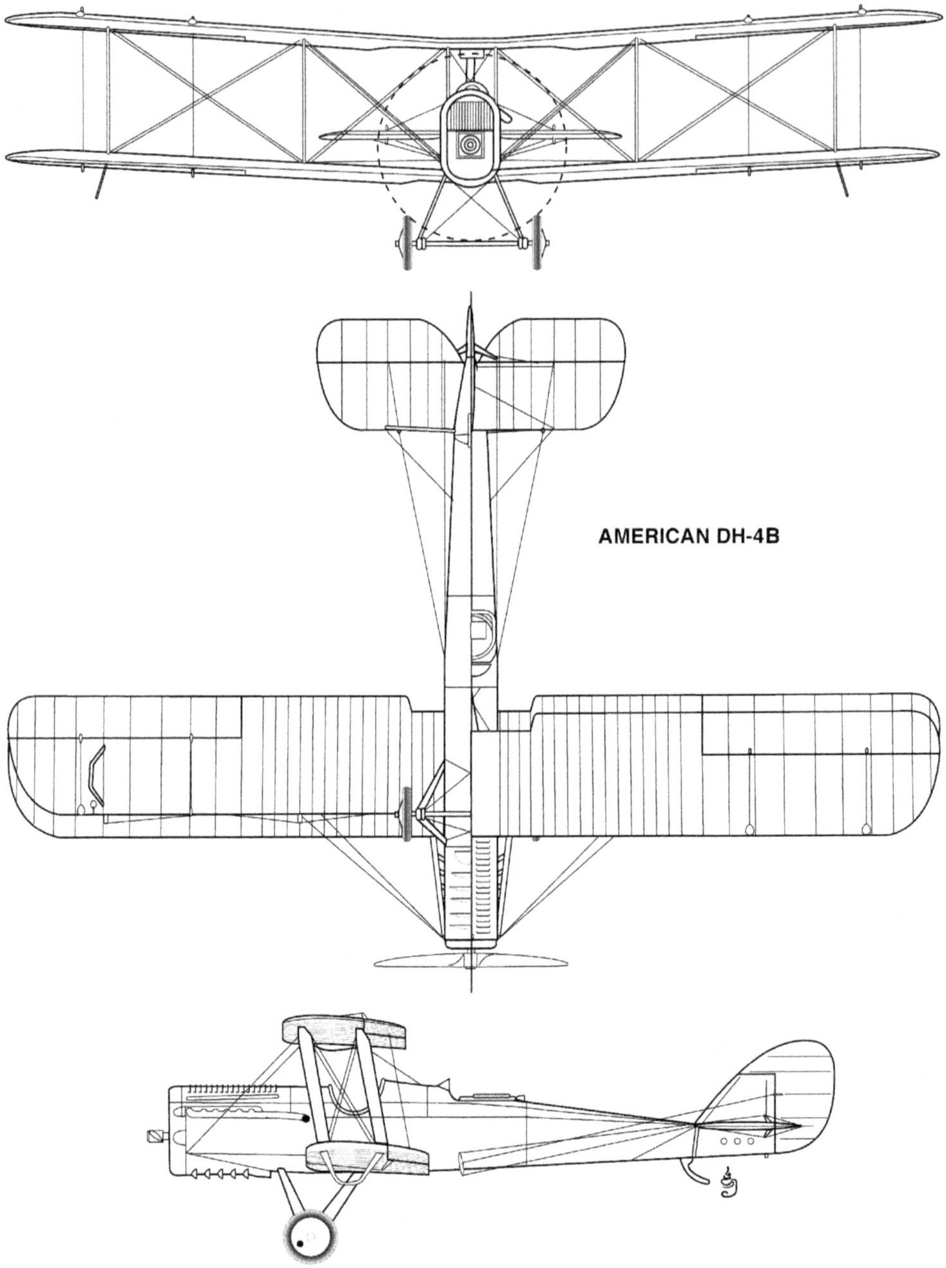

Naval Variants: Although the naval service received some 222 DH-4 variants, including 192 Navy DH-4Bs and 30 Marine DH-4Ms (designated O2B-1), none were reported to have been used for training duties.

Synopsis of USAAS Trainer Procurement, 1920 to 1924

Even though the Army Air Service possessed large stocks of training aircraft at the time of the Armistice in late 1918—primarily Curtiss JN-4s and JN-6s—the aircraft were completely

One of 188 DH-4s remanufactured by Boeing in 1923 with welded steel tube fuselages that reappeared as the DH-4M. Most were used for bomber training, though 22 were converted to dual-control advanced trainers.

obsolete and rapidly deteriorated with use as a consequence of their wooden-framed construction. In order to find a long-term solution, the Army tested 12 different types of trainers between 1920 and 1924, ultimately selecting the Consolidated PT-1 for quantity production. During this era the Engineering Division at McCook Field (later Wright Field) in Dayton, Ohio became the Army's focal point for testing state-of-the-art aeronautical concepts in terms of construction methods, powerplants, and aerodynamic design. At the same time, however, out of an effort to conserve costs, new aircraft requirements often specified the use of outmoded engines and other components left over from War Department surplus stocks. Under the Army's 1919 designation scheme, the prefix "TW" identified a trainer equipped with a water-cooled engine and "TA" one with an air-cooled engine; then under a new system adopted in 1924, "PT" identified a primary trainer, "BT" a basic trainer, and "AT" an advanced trainer.

Because of the very limited historical data available on some of the prototypes, complete technical specifications are unknown and will go unreported.

Engineering Division TW-1–1920

Technical Specifications (TW-1)

Type: Two-place water-cooled primary trainer.
Manufacturer: Army Air Service Engineering Div., McCook Field, Dayton, Ohio.

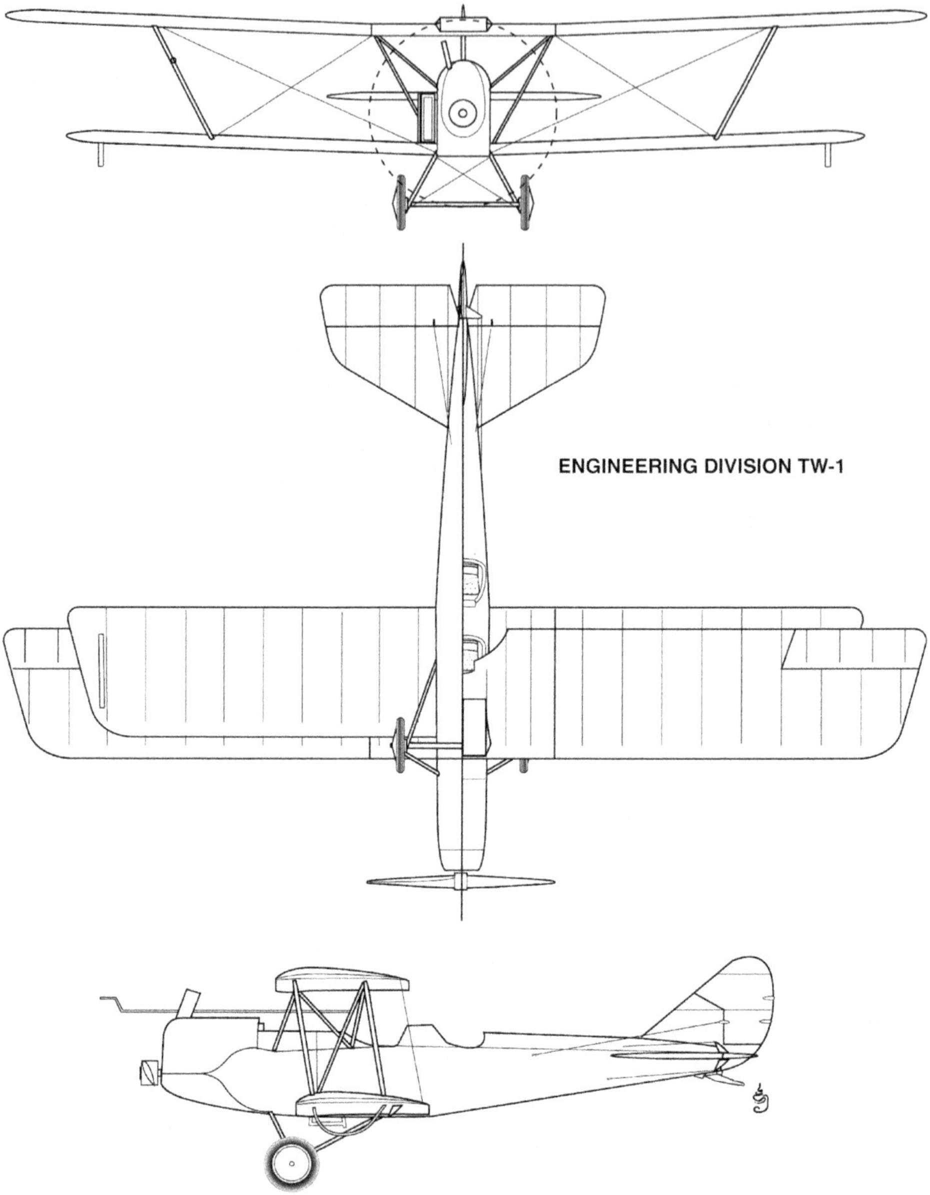

Total U.S. military versions: 2 (USAAS).
Unit cost: Unknown.
Powerplant: One 230-hp Liberty 6 (L-825) 6-cylinder or 315-hp Packard 1A-1237 (V-734) 12-cylinder water-cooled inline engine driving a two-bladed fixed-pitch wooden propeller.
Armament: None.
Performance: Max. speed 97 mph; ceiling and range not reported.
Weights: Empty (not reported), 3,225 lbs. loaded.
Dimensions: Span 41 ft., length 29 ft. 11 in., wing area (not reported).

In 1920 and 1921 the Army's Airplane Engineering Division at McCook Field built and tested two trainer prototypes under the designation TW-1. Photographs seem to indicate that the wings, fuselage, and tail group were adapted from a Curtiss JN-4 or JN-6. Innovative features

Built by the USAAS Engineering Division at McCook Field, Ohio, the TW-1 shown was one of two prototypes tested but never placed in production.

included a single-bay biplane layout that utilized a Fokker-inspired system of metal N-type cabane and interplane struts. Though designated as trainers, these aircraft were apparently never used as such and functioned primarily as flying testbeds.

Elias TA-1–1920

Technical Specifications (TA-1)

Type: Two-place air-cooled primary trainer.
Manufacturer: G. Elias and Brother, Buffalo, New York.
Total U.S. military versions: 3 (USAAS).
Unit cost: Unknown.
Powerplant: One 170-hp ABC Wasp (R-667) 7-cylinder or 200-hp Lawrance J-1 (R-787) 9-cylinder air-cooled radial engine driving a two-bladed fixed-pitch wooden propeller.
Armament: None.
Performance: Max. speed 90 mph; ceiling and range not reported.
Weights: Empty (not reported), 2,235 lbs. loaded.
Dimensions: Span 32 ft. 7 in., length 24 ft. 9 in., wing area (not reported).

The three aircraft purchased from Elias in 1922 under the designation TA-1 were equal-span, single-bay biplanes of conventional built-up wooden construction and fabric covering, characterized by a boxy fuselage cross-section. Other than being equipped with new types of radial engines, the three aircraft did not exhibit any advanced features of note. During competitive flight trials conducted from 1920 to 1923, Army Air Service officials rated the TA-1's graded performance as inferior to that of other trainer prototypes being tested (e.g., the Dayton Wright TA-3 and Huff-Daland TW-5), and further development was discontinued. All three aircraft were scrapped in 1926.

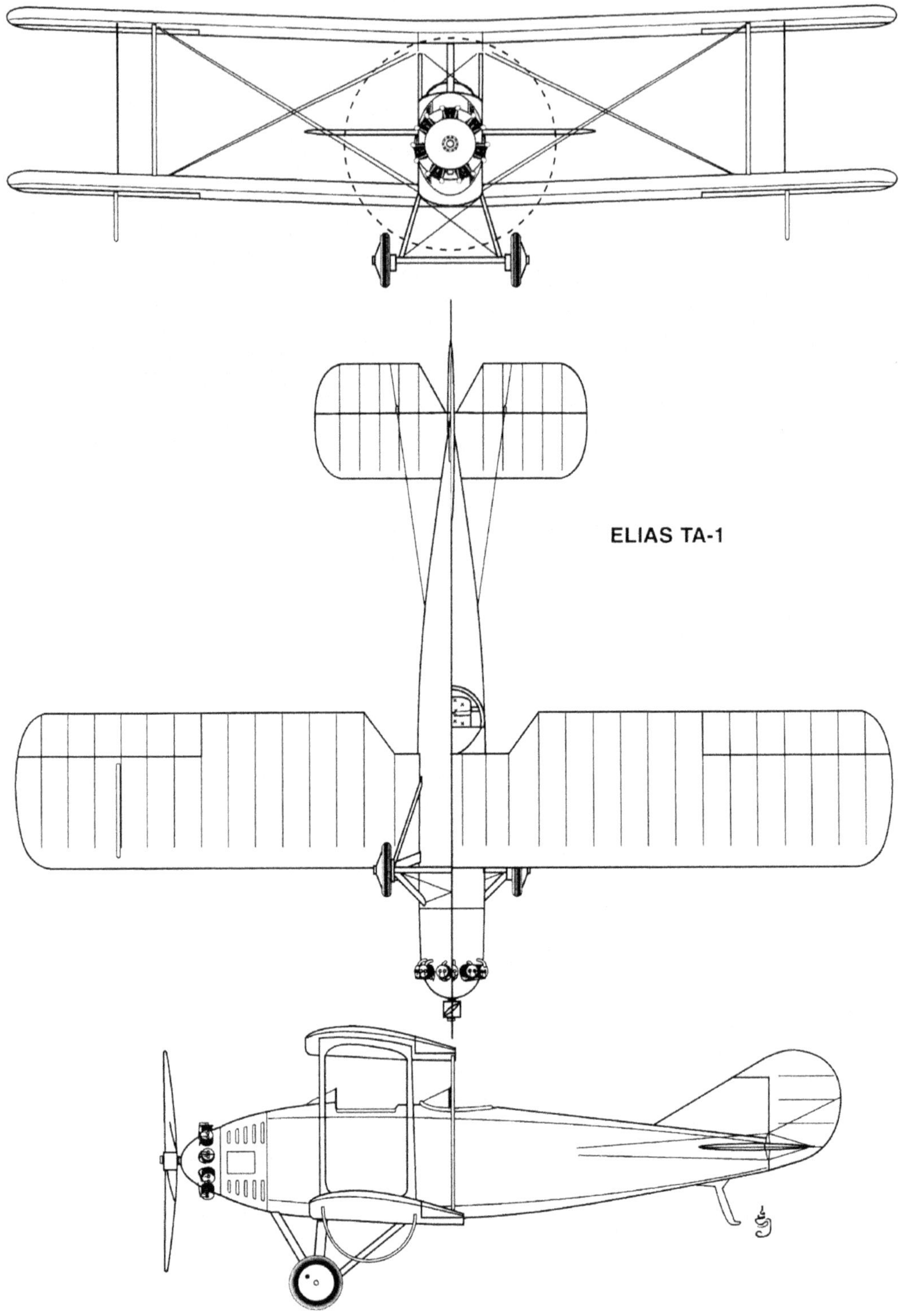

ELIAS TA-1

TA-1, one of three radial-powered trainer prototypes built by the short-lived Elias firm, did not perform well when tested against other competing prototypes evaluated from 1920 to 1923.

Huff-Daland TA-2–1920

Technical Specifications (TA-2)

Type: Two-place air-cooled primary trainer.
Manufacturer: Huff-Daland Airplane Co., Bristol, Pennsylvania.
Total U.S. military versions: 3 (USAAS).
Unit cost: $4,800.
Powerplant: One 170-hp ABC Wasp (R-667) 7-cylinder, 100-hp Anzani (displ. unknown) 10-cylinder, or 200-hp Lawrance J-1 (R-787) 9-cylinder air-cooled radial engine driving a two-bladed fixed-pitch wooden propeller.
Armament: None.
Performance: Max. speed 90 mph; ceiling and range (not reported).
Weights: Empty (not reported), 1,765 lbs. loaded.
Dimensions: Span 30 ft. 10 in., length 24 ft. 2 in., wing area (not reported).

In 1920 the Army acquired three civil Huff-Daland model HD.4s that were evaluated under the designation TA-2. The first example was reportedly used only for static (non-flying) tests and the other two to evaluate new types of radial engines and serve as experimental testbeds in other projects. As built, the TA-2's equal span wings came with two bays of metal N struts braced by unusual truss-type diagonal girders instead of the usual rigging wires. Another unusual feature was longitudinal braces running from the cabane struts to the back of the fuselage. The fabric-covered structure consisted of a steel tube fuselage frame and built-up wooden wings. The two flying TA-2s were later rebuilt with modified fuselages, wings of less area, balanced rudders, and 90-hp OX-5 engines, and one was later modified for installation of a Lawrance J-1 engine. Both were scrapped in 1926.

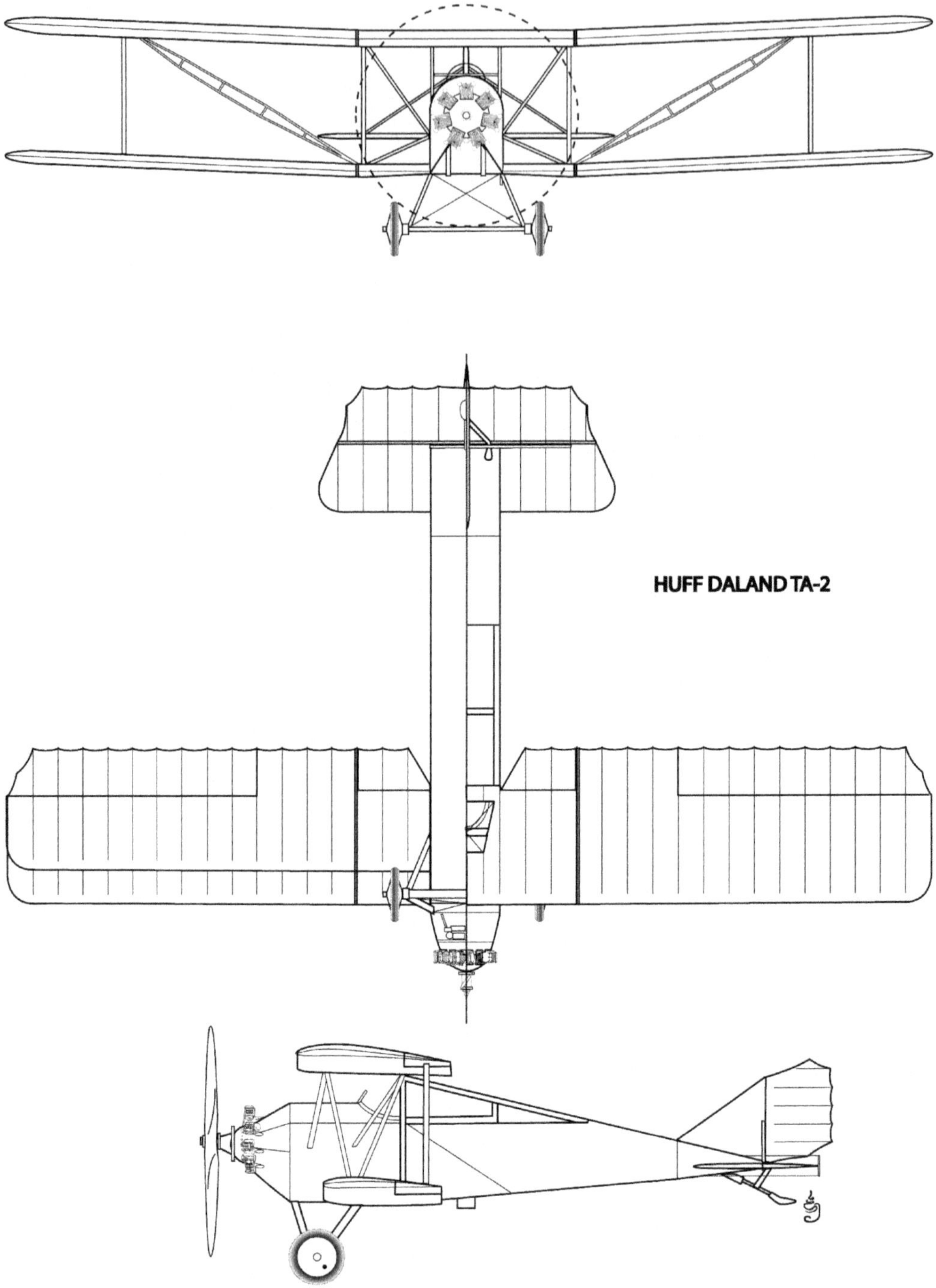
HUFF DALAND TA-2

One of three civil Huff-Daland HD.4s acquired for testing as the TA-2. Note unusual interplane girders used in place of rigging wires. Two were later used as engine testbeds.

Cox-Klemin TW-2—1921

Technical Specifications (TW-2)

Type: Two-place air-cooled primary trainer.
Manufacturer: Cox-Klemin Aircraft Corp., College Point, New York.
Total U.S. military versions: 3 (USAAS).
Unit cost: Unknown.
Powerplant: One 180-hp Wright-Hispano E (V-718) 8-cylinder water-cooled inline engine driving a two-bladed fixed-pitch wooden propeller.
Armament: None.
Performance: Max. speed 99 mph; ceiling and range not reported.
Weights: Empty (not reported), 2,505 lbs. loaded.
Dimensions: Span 29 ft., length 23 ft. 9 in., wing area (not reported).

Very little data survives regarding the three Cox-Klemin TW-2s delivered to McCook Field in 1921 and 1922, except that one was apparently used for static tests and the two flying examples used to test different engines. One photograph indicates the TW-2 was a single-bay, equal-span biplane of conventional design that employed a Fokker scheme of metal cabane and N-type interplane struts with no wire bracing between them. Cox-Klemin reportedly delivered an improved trainer design to McCook Field in 1924, the model CK-2A with a steel tube fuselage and interchangeable wings, but the aircraft was never fully evaluated nor assigned an Army serial number. In any event, Cox-Klemin ceased all operations in 1925.

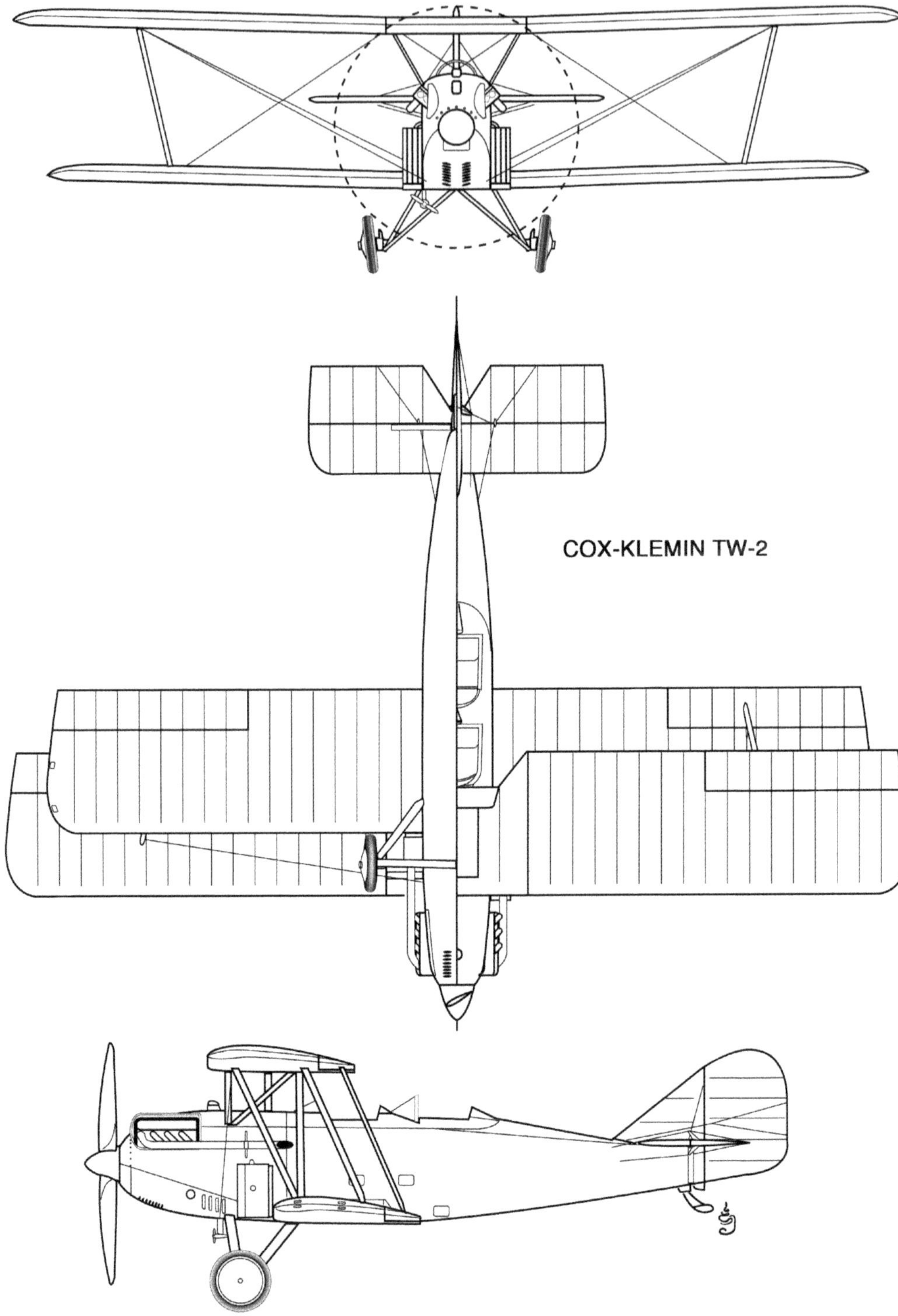

COX-KLEMIN TW-2

A Cox-Klemin TW-2 prototype seen at McCook Field, Ohio, in 1922 or 1923. Like many other designs of the early 1920s, the TW-2 employed the Fokker system of cabane and interplane struts without wire rigging.

Engineering Division TA-4–1921 (cancelled)

Dayton-Wright TA-3, TW-3 and TA-5–1922

Technical Specifications (TA-3 [TW-3])

Type: Two-place air-cooled [water-cooled] primary trainer.
Manufacturer: Dayton-Wright Airplane Co., Dayton, Ohio; design rights sold in 1923 to Consolidated Aircraft Corp., Buffalo, New York.
Total U.S. military versions: 34 (USAAS).
Unit cost: Unknown.
Powerplant: One 80-hp or 110-hp Le Rhône 9C/B 9-cylinder air-cooled rotary engine [150-hp Wright-Hispano E (V-718) 8-cylinder water-cooled inline engine] driving a two-bladed, fixed-pitch wooden propeller.
Armament: None.
Performance: Max. speed 85 mph [103 mph]; ceiling (not reported); range 200 [370] mi.
Weights: 1,240 lbs. [1,700 lbs.] empty, 1,753 lbs. [2,407 lbs.] loaded.
Dimensions: Span 30 ft. 11 in. [34 ft. 9 in.], length 22 ft. 7 in. [25 ft. 7 in.], wing area (not reported).

The TA-3, TW-3, and TA-5 all originated from the "Chummy" trainer, a rotary-powered single-bay biplane with side-by-side seating which had been test-flown in 1921. Designed by Virginius Clark, Dayton-Wright's chief engineer, the Chummy incorporated many advanced features such as a new Clark Y thick-section airfoil, a welded fuselage framework of chrome-molybdenum steel tubing, and metal wing and landing gear struts. The Army ordered three Le Rhône-powered examples which were delivered to McCook Field in 1922 to be evaluated as trainers under the designation TA-3. Acceptance trials indicated a very rugged airframe with excellent flight characteristics, however, the aircraft was considered underpowered in terms of speed and climb. After

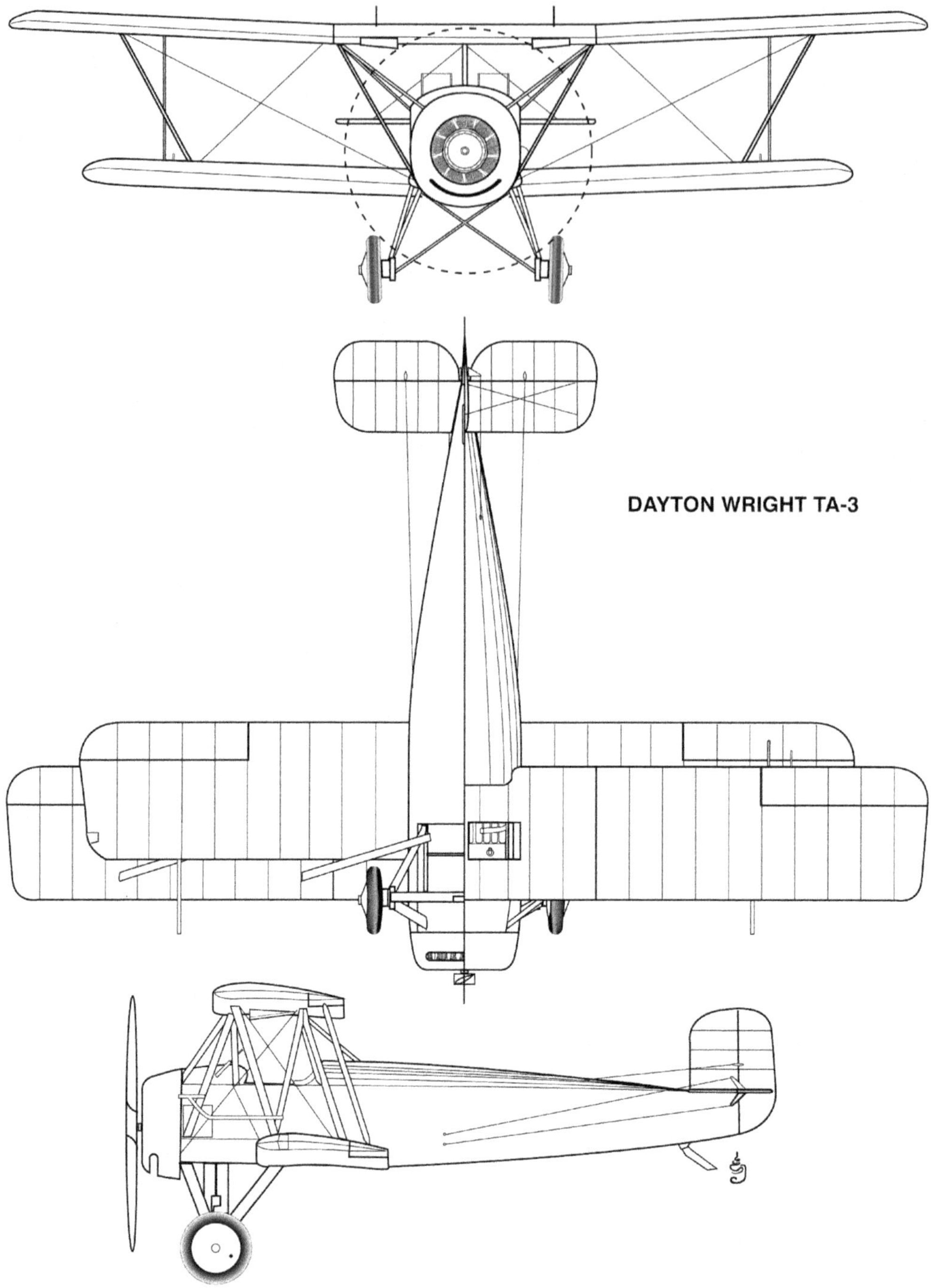
DAYTON WRIGHT TA-3

Based on the civil "Chummy," the rotary-powered TA-3 as it appeared in its original configuration in 1923. Air Service officials liked the design enough to order 20 more but wanted a more powerful engine.

Modified TA-3 with a Lawrance radial and enlarged vertical tail surfaces as the TA-5. Further development of the design resulted in mass production as the Consolidated PT-1.

Dayton-Wright re-engined one TA-3 with a 110-hp Le Rhône, the Army awarded a contract for ten new aircraft delivered in 1923 (note: these became the last new Army aircraft to be powered with rotary engines).

Although Army officials liked the TA-3, they believed the design would achieve better long-term potential as a trainer if equipped with a more powerful engine; thus, in early 1923, Dayton-Wright received a contract for three new airframes to be powered with 150-hp Wright-Hispano engines under the new designation TW-3. To offset the added weight, wingspan was increased four feet nine inches. Then in June 1923, following successful trials, Dayton-Wright received a second contract for 20 production TW-3s, which, after a transfer of production rights, was completed by the newly organized Consolidated Aircraft Corp. In operational service, the upper cowlings were removed from TW-3s to improve over the nose visibility. The one-of-a-kind TA-5, essentially a TW-3 modified to accept a 220-hp Lawrance J-1 engine, was delivered to McCook Field in 1923 to test the feasibility of a single-wheel, centerline landing gear configuration. Further efforts by Consolidated to improve upon the TW-3 design resulted in the tandem cockpit TW-8 of 1924, which is reported separately as the PT-1.

Fokker TW-4–1922

Technical Specifications (TW-4)

Type: Two-place air-cooled primary trainer.
Manufacturer: Nederlandse Vliegtuigenfabriek (Dutch Fokker), Amsterdam, Netherlands.
Total U.S. military versions: 1 (USAAS).

The Fokker TW-4, pictured in 1922, was the only type of monoplane trainer to be tested by the Air Service. Testing indicated that the restricted upward visibility would make training operations too hazardous.

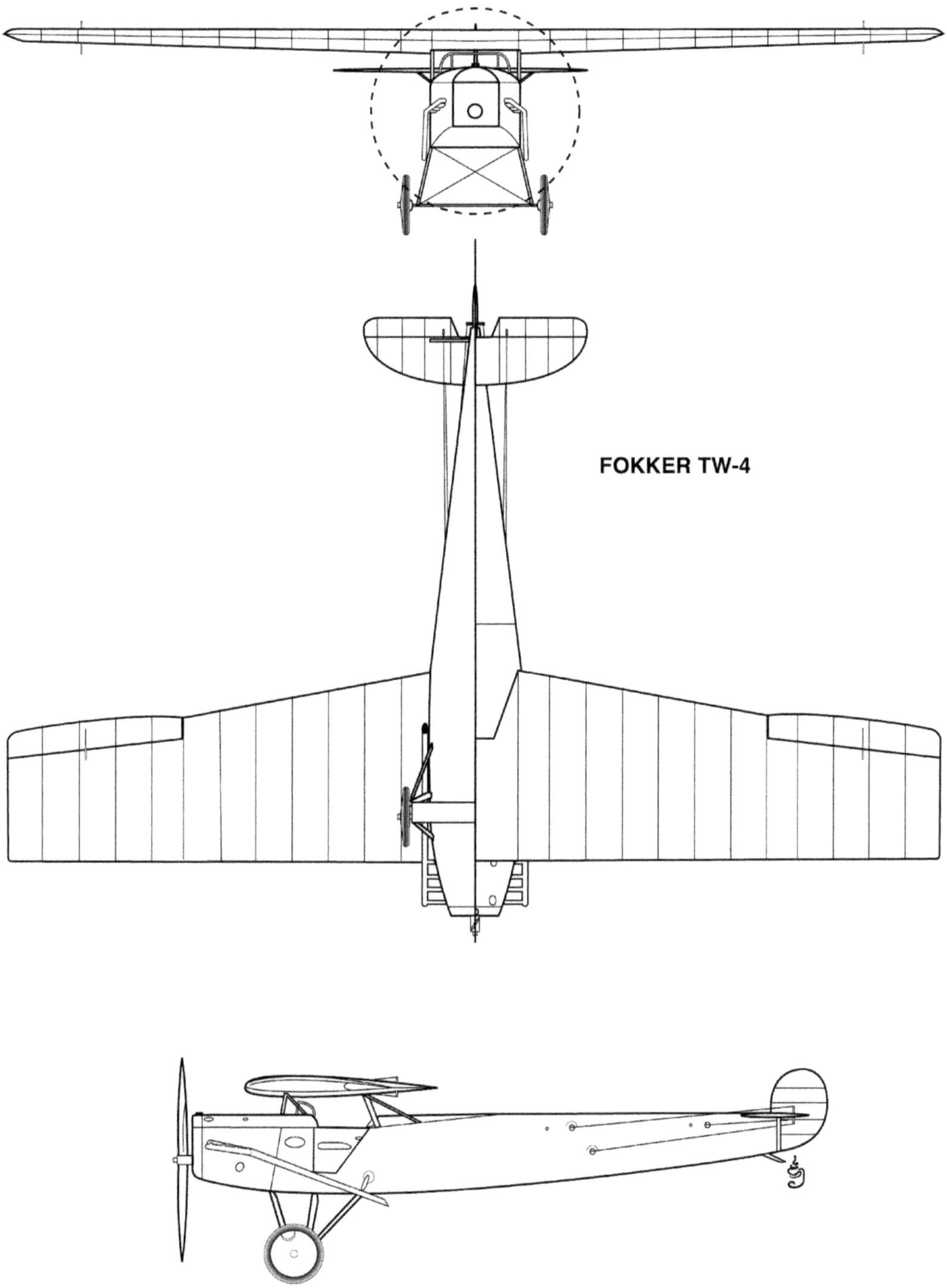

FOKKER TW-4

Unit cost: Unknown.
Powerplant: One 90-hp Curtiss OX-5 (V-503) 8-cylinder water-cooled inline engine driving a two-bladed fixed-pitch wooden propeller.
Armament: None.
Performance: Max. speed 87 mph; ceiling and range not reported.
Weights: 1,342 lbs. empty, 1,947 lbs. loaded.
Dimensions: Span 41 ft. 8 in., length 28 ft., wing area (not reported).

The Fokker TW-4 is unique in having been the only type of monoplane trainer to be evaluated by the Army Air Service during the 1920s. Designed by Reinhold Platz and initially flown in Germany in 1919 as the model S.1, Fokker shipped one airframe to the U.S. to be modified for installation of an OX-5 engine, and following delivery to McCook Field for evaluations in 1922, it was accepted under the assigned designation TW-4. Like many Fokker designs of this era, the TW-4 featured a welded, steel tube fuselage and tail group and a fully-cantilevered, built-up wooden wing. The instructor and student sat side-by-side beneath the parasol-mounted wing. Testing of the TW-4 was discontinued in 1923 after test pilots reported that the restricted visibility under the wing would make normal training operations too hazardous.

Huff-Daland TA-6 and TW-5/AT-1 and AT-2 (HN)—1923

Technical Specifications (TW-5/AT-1)

Type: Two-place water-cooled advanced trainer.
Manufacturer: Huff-Daland Airplane Co., Bristol, Pennsylvania.
Total U.S. military versions: 19 (16 USAAS; 3 USN).
Unit cost: Unknown.
Powerplant: One 180-hp Wright-Hispano E (V-718) 8-cylinder water-cooled inline engine driving a two-bladed fixed-pitch wooden propeller.
Armament: None.
Performance: Max. speed 114 mph; ceiling 13,700 ft.; range (not reported).
Weights: 1,687 lbs. empty, 2,212 lbs. loaded.
Dimensions: Span 31 ft. 1 in., length 24 ft. 8 in., wing area (not reported).

Huff-Daland, organized in 1920, is noteworthy in having been one of the first American aircraft manufacturers to adopt the Fokker-inspired construction techniques of welded, steel tube fuselages and thick airfoil section semi- or fully-cantilevered wings. Huff-Daland received an Army contract in late 1922 to build one trainer prototype, designated TA-6, to be powered by a Lawrance J-1 radial engine. The TA-6, as delivered in April 1923, emerged with semi-cantilevered, tapered wings of unequal span, a new type of cross-axle landing gear, and tandem cockpit seating that used a Gosport system for instructor-student communications. The Army liked the design, but as a cost-savings measure, ordered five more airframes with surplus Wright-Hispano engines and upper wingspan increased by nine inches (to offset the 250-lb. weight increase), which were delivered in late 1923 as the TW-5. In 1925 Huff-Daland received a follow-up order for 10 TW-5s that joined the Army Air service inventory under the new advanced trainer designation AT-1, and one AT-1 became the AT-2 after being converted to single-seat configuration and tested during 1926. Huff-Daland was merged into Keystone Aircraft Corp. in 1930.

Naval Variants: In 1923, the Navy procured a total of six essentially identical aircraft, three as trainers under the designation HN-1 and three for observation as the HO-1, the latter having provision for a .30-calibre Lewis gun mounted in the rear cockpit. The HN/HO-1s differed

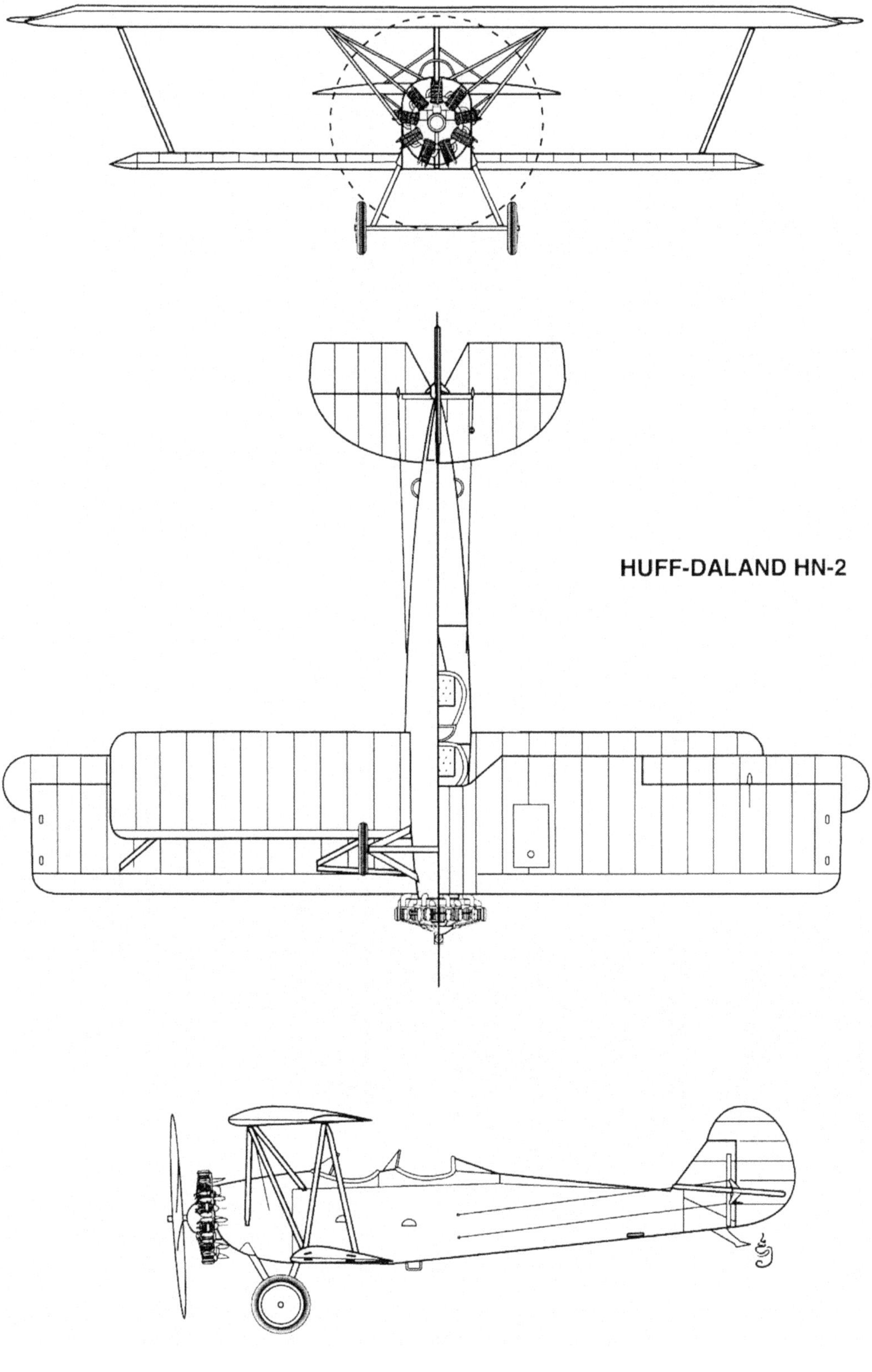

HUFF-DALAND HN-2

Huff-Daland TA-6 as delivered in 1923 with a Lawrance radial. To reduce costs, the Air Service ordered 10 more with surplus Wright-Hispano engines which entered under the designation TW-5.

Wright-Hispano-powered Navy HN-1 seen in 1922 or 1923 taking off from the Washington Monument reflecting pool. Fokker wing layout is evident in this photograph.

from the Army TW-5s in being convertible to floats and having constant chord upper and lower wings very similar in outline to those of the Fokker D.VII. Later than same year, the Navy ordered three additional trainers powered with 200-hp Lawrance J-1 radial engines that joined the inventory as the HN-2. The HNs served at NAS Pensacola as trainers until withdrawn in the mid–1920s.

Consolidated PT-1, PT-2, PT-3, PT-4, PT-5 and PT-8 (NY and N3Y)—1925

Technical Specifications (PT-1 [NY-2])

Type: Two-place primary trainer.
Manufacturer: Consolidated Aircraft Corp., Buffalo, New York.
Total U.S. military versions: 778 (471 USAAS/AAC; 307 USN/USMC).
Unit cost (PT-1): $8,000.
Powerplant: One 180-hp Wright-Hispano E (V-718) 8-cylinder water-cooled inline engine [220-hp Wright J-5 *Whirlwind* (R-790–8) 9-cylinder air-cooled radial engine] driving a two-bladed fixed-pitch wooden [metal] propeller.
Armament (NY-1A/2A): One flexible .30 calibre machine gun in the rear cockpit.
Performance: Max. speed 99 mph [98 mph], cruise 79 mph [75 mph]; ceiling 13,450 ft. [15,200 ft.]; range 270 mi. [300 mi].
Weights: 1,805 lbs. [1,801 lbs.] empty, 2,577 lbs. [2,627 lbs.] loaded.
Dimensions: Span 34 ft. 9 in. [40 ft.], length 27 ft. 8 in. [27 ft. 11 in.], wing area 284 sq. ft. [370 sq. ft.].

The Consolidated PT-1, together with the follow-on PT-3, formed the nucleus of the Army Air Service's (became Army Air Corps in July 1926) primary training fleet from 1925 until the mid–1930s. To address the inherent visibility limitations of the TW-3's side-by-side seating arrangement, Consolidated received official approval in late 1923 to redesign the fuselage and deliver a new prototype. Under the provisional designation TW-8, the aircraft arrived at McCook Field in June 1924 with a narrow-section fuselage that had been lengthened five feet one inch to accommodate tandem cockpit seating. Following successful trials, the Army awarded Consolidated a series of contracts in 1924 and 1925 for 220 production aircraft to be delivered under the new designation PT-1. The first production batch featured a flat turtleback behind the rear cockpit and a streamlined cowling over the nose, but the second run of 170 PT-1s came with rounded turtlebacks and lacked cowlings to improve downward visibility and maintenance access.

Influenced by the Navy's recent success with the NY-1, a radial-powered PT-1 derivative, the Army directed Consolidated to modify one PT-1 for installation of a 220-hp Wright J-5 (R-790) radial engine, and the aircraft was returned to Wright Field (formerly McCook Field) for tests in the fall of 1927 as the XPT-2. (Note: the "X" prefix identifying experimental prototypes was adopted sometime in 1925.) A second PT-1 airframe, having the same J-5 radial engine but fitted with the NY-1 wings (34 ft. 6 in. span with Clark "Y" airfoil) and reshaped vertical tail surfaces, arrived for testing in late 1927 as the XPT-3. Upon completion of official trials in 1928, the Army awarded Consolidated a contract to produce 130 essentially identical aircraft as the PT-3, followed in 1929 by a second contract to deliver 120 more aircraft with small detail changes as the PT-3A. The designation XPT-4 was reserved in 1928 for a PT-3 to be fitted with an experimental 125-hp Fairchild-Caminez crankless radial engine, but the Army cancelled the project before an aircraft was converted. The XPT-3 became the XPT-5 after conversion to a 175-hp Curtiss R-600 Challenger radial engine in 1929 but later returned to the XPT-3 standard. One PT-3 airframe modified by Consolidated for foreign export in the observation role became the

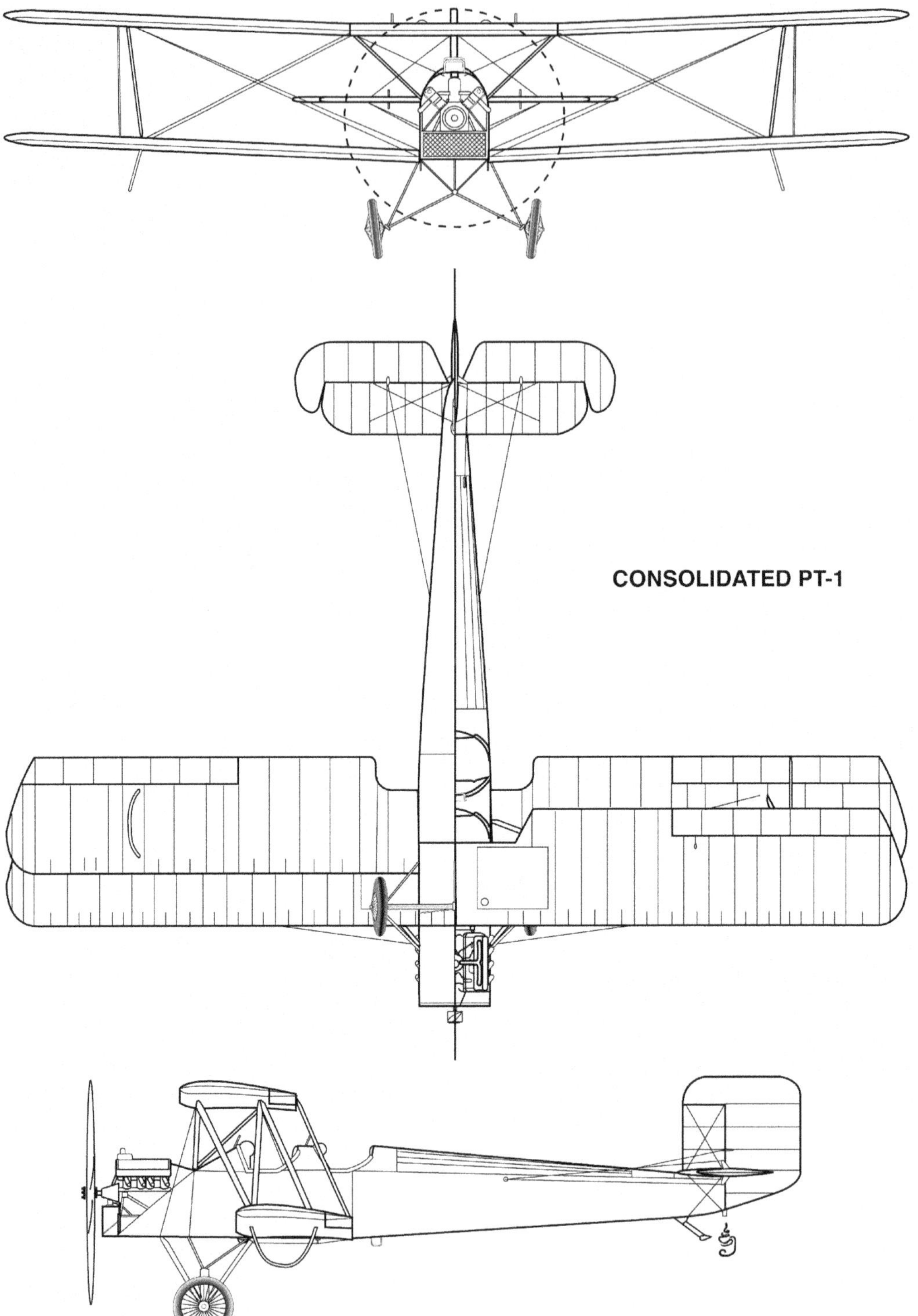

CONSOLIDATED PT-1

An early production Consolidated PT-1 seen at Brooks Field, Texas, during in the mid–1920s. The order for 220 PT-1s represented an across-the-board replacement of the Army's World War I-era primary training fleet.

Continuing refinements of the PT-1 design resulted in the PT-3A. The aircraft depicted appears in 1930s markings with large buzz numbers on either side.

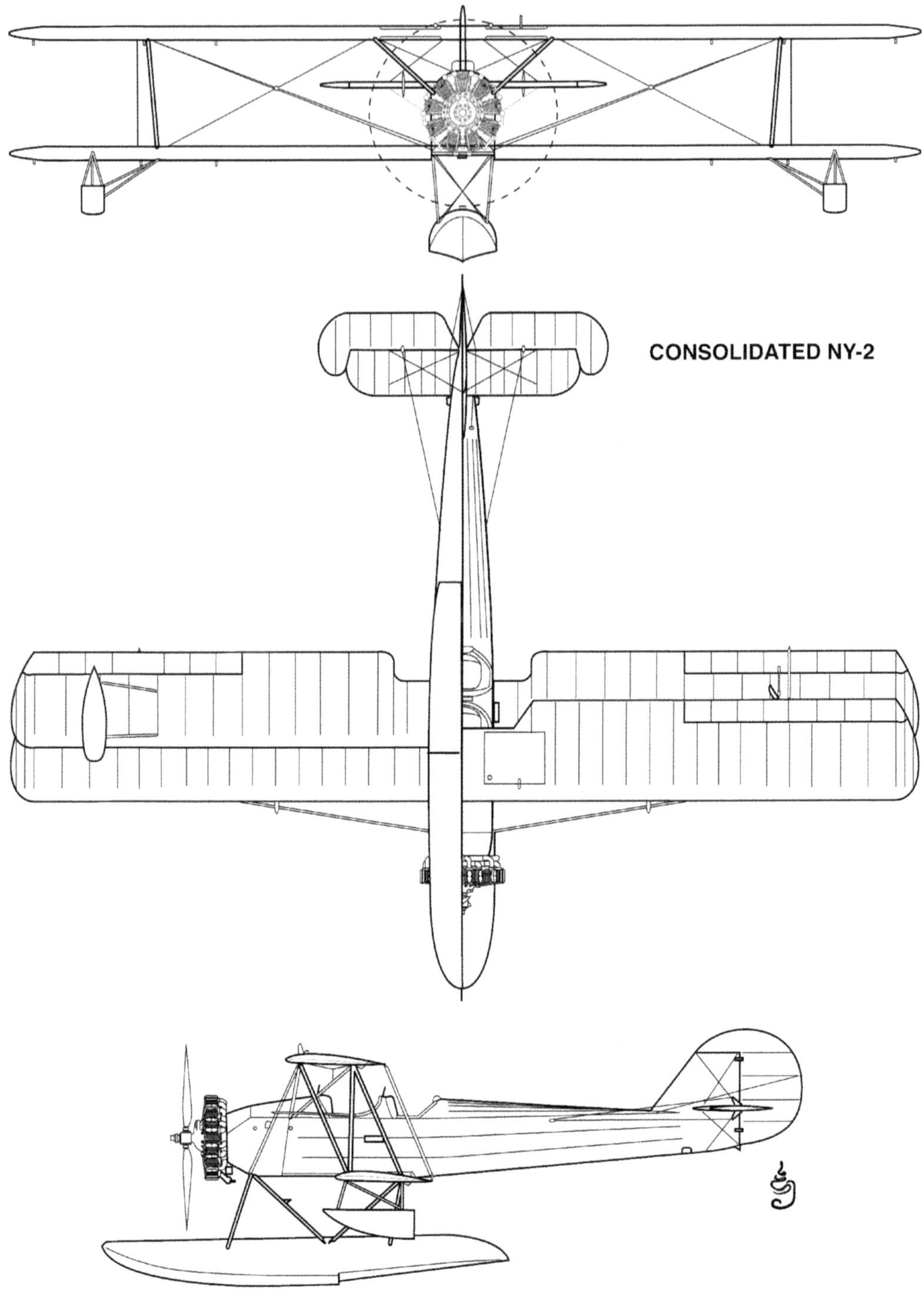

CONSOLIDATED NY-2

The Navy's most important trainer from the mid–1920s to the mid–1930s, this float-equipped NY-2 operated out of NAS Pensacola, Florida. Note Gosport tube on face of instructor in front cockpit.

XO-17A demonstrator, then in 1931, after being refitted with a 225-hp Packard DR-980 diesel radial engine, was briefly tested as the XPT-8.

Whereas the TA-3s and TW-3s served only to augment older aircraft, the quantity procurement of PT-1s in 1924 and 1925 represented an across-the-board replacement of the Army's entire primary training fleet. As new PT-3s arrived for duty during 1928 and 1929, PT-1s were phased out of frontline service and transferred to the National Guard flight program where they were allocated to 18 different squadrons to replace JNSs in the training role. The PT-3 continued as the Army Air Corps' most important primary trainer until deliveries of Boeing-Stearman PT-13s started in 1936 and 1937.

Naval Variants: In 1925 the U.S. Naval Bureau of Aeronautics (BuAer) asked Consolidated to adapt the PT-1 to naval specifications that included an air-cooled Wright J-4 powerplant and the ability to convert to either a wheeled or centerline float undercarriage. Fin and rudder area were increased accordingly to offset the side area of the floats. After the Navy awarded a contract for 76 aircraft as the NY-1, the first production model (no experimental prototype) was flown on November 12, 1925, and deliveries commenced in May 1926. In service, a number of these were re-designated NY-1A when retrofitted with a flexible .30-calibre machine gun mount and used for gunnery training. Early operational experience with the NY-1 revealed that in a float configuration, at gross weight, the aircraft suffered from excessive wing-loading. In late 1926, to overcome these shortcomings, Consolidated introduced the NY-2 having a 5-foot 6-inch increase in wingspan and more power from a 220-hp J-5 engine. Following completion of successful trials in 1927, the Navy ordered 186 NY-2 production models (the largest Navy trainer order since 1918), all having been delivered by the end of 1928.

A further 25 NY-2s specifically outfitted for gunnery training were ordered in 1928 as the NY-2A. The final 20 variants built, NY-3s with 240-hp Wright R-760–94 7-cylinder engines, went directly to various Navy and Marine reserve units as they were delivered in 1929. In later service, NY-1s receiving the larger wings of the -2 and 220-hp J-5 engines were returned to duty as the NY-1B. A single NY-2 airframe fitted with a Wright R-790-A engine was tested in 1929 as the XN3Y-1 but not placed in production. From the late 1920s onward, NY variants not only established themselves as the Navy's most important trainer for primary flight instruction and gunnery, but also formed a vital training component within the ten Navy and Marine reserve units organized during 1928. NY-1Bs, -2s, and -2Bs continued in service as the Navy's most numerous type of primary trainer until Naval Aircraft Factory N3N-1s began arriving in quantity during 1936. All NY variants had been replaced at training centers by the end of 1937, but the very last NY-3s were not withdrawn from the reserves until 1939.

Boeing AT-3–1926

Technical Specifications (AT-3)
Type: One-place advanced trainer.
Manufacturer: Boeing Airplane Co., Seattle, Washington.
Total U.S. military versions: 1 (USAAS/AAC).
Unit cost: Unknown.
Powerplant: One 180-hp Wright-Hispano E (V-718) 8-cylinder water-cooled inline engine driving a two-bladed fixed-pitch metal propeller.
Armament: None.
Performance: Max. speed 129 mph; ceiling and range not reported.
Weights: Empty (not reported), 2,476 lbs. loaded.
Dimensions: Span 32 ft., length 23 ft. 5 in., wing area 241 sq. ft.

The Boeing model 15, originally developed in 1923 as a fighter for the Army (113 built as the PW-9, -9A, -9B, -9C, and -9D) and the Navy (40 built as the FB-1, -2, -3, -4, and -5) was

The sole AT-3, a PW-9A fighter modified for installation of a less powerful Wright-Hispano engine, underwent evaluations as an advanced pursuit trainer at McCook Field in 1926.

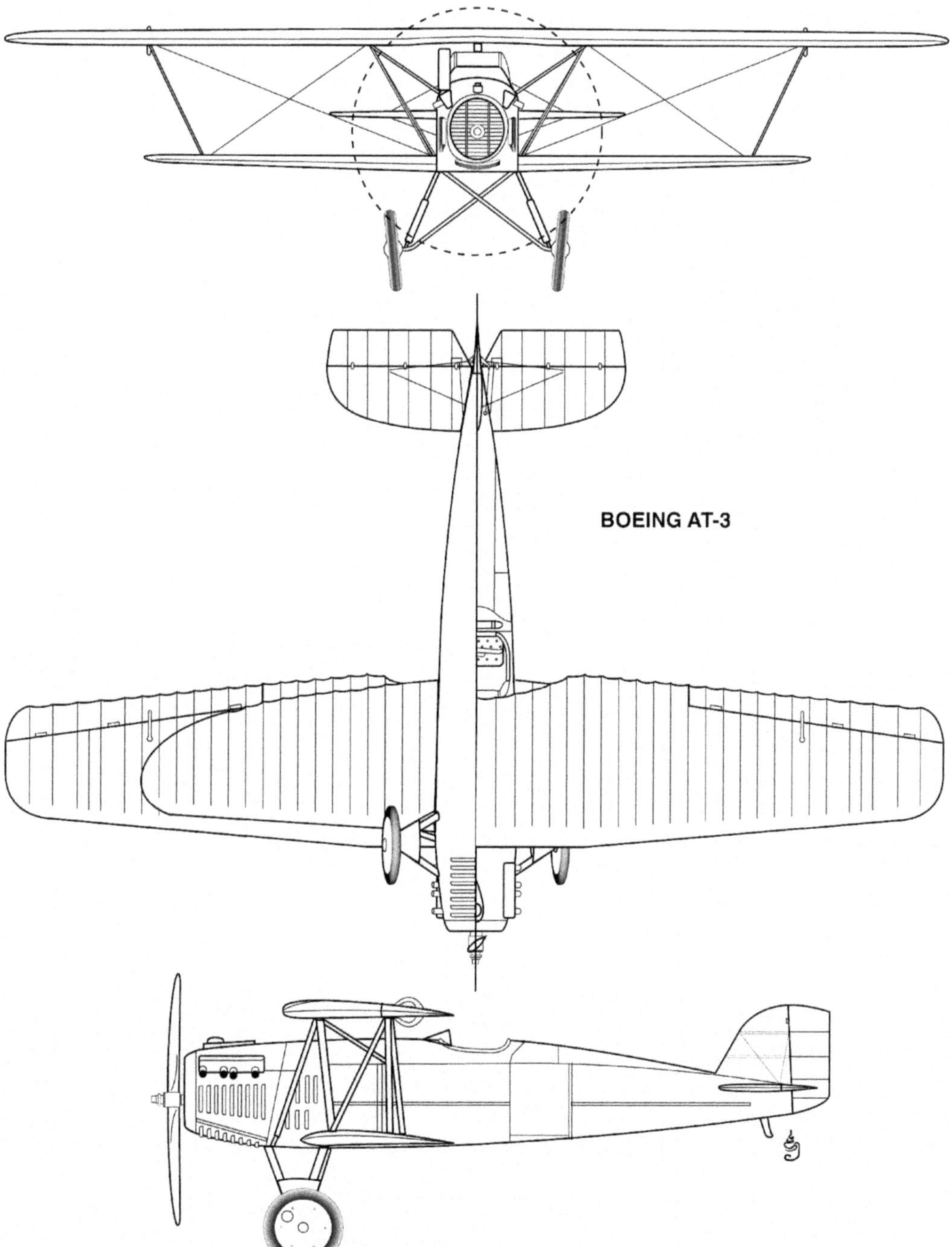
BOEING AT-3

itself based upon the structural design of the Fokker D.VII and powered by a 435-hp Curtiss D-12 (V-1150) water-cooled inline engine. In 1926 one PW-9A modified for installation of a Wright-Hispano engine was re-designated AT-3 and evaluated at McCook Field for potential use as an advanced trainer, however, no further conversions were ordered.

Curtiss AT-4 and AT-5—1926

Technical Specifications (AT-5)

Type: One-place advanced trainer.
Manufacturer: Curtiss Aeroplane & Motor Co., Buffalo, New York.
Total U.S. military versions: 71 (USAAC).
Unit cost: $9,800 (as P-1A/B).
Powerplant: One 220-hp Wright J-5 (R-790–1) Whirlwind 9-cylinder air-cooled radial engine driving a two-bladed fixed-pitch metal propeller.
Armament: Two fixed .30-calibre machine guns.
Performance: Max. speed 125 mph; ceiling and range not reported.
Weights: 1,800 lbs. empty (est.), 2,471 lbs. loaded.
Dimensions: Span 31 ft. 6 in., length 23 ft., wing area 252 sq. ft.

The Curtiss P-1 Hawk series first made its appearance as the PW-8B fighter prototype in 1924. Originally equipped with 435-hp Curtiss D-12 (V-1150) inline engines, 71 P-1As, Bs, and Cs subsequently served as advanced trainers after being converted to less powerful engines. The XAT-4, a P-1A fitted with a 180-hp Wright-Hispano E inline engine, arrived at McCook Field in 1926 for trials, and over the next two years, 35 more P-1As and Bs underwent the same conversion and returned to service as the AT-4. Five P-1Bs modified for installation of 220-hp Wright J-5 radial engines were redelivered in 1927 under the designation AT-5, and then in 1928, 31 P-1Cs became AT-5As that, in addition to J-5 radial engines, featured an extended nose and a tight-fitting NACA-type cowling. All AT-4 and AT-5/5A trainer variants were assigned to the 43rd Pursuit Squadron (School) at Kelly Field, Texas. During 1929 most surviving Hawk trainers were restored to their original fighter configuration, 24 AT-4s as the P-1D, four AT-5s as the P-1E, and 24 AT-5As and the P-1F.

Synopsis of USAAC Trainer Procurement, 1930 to 1940

Primary Trainers: The Army Air Corps began seeking a replacement for the PT-3 as early as 1930 and starting with the Fleet XPT-6, tested six different designs until arriving at the selection of the Boeing-Stearman PT-13 in 1936. Then in April 1939, in response to increasing military tensions in Europe and the Far East, the U.S. Congress approved an expansion plan authorizing the Army Air Corps to procure 3,000 new aircraft. As a component of this plan, Air Corps Material Division circulated new primary trainer requirements that set in motion the development and eventual mass production of such well-known types as the Boeing-Stearman PT-17 (PT-13 powered by a Continental R-670 engine), Fairchild PT-19, and Ryan PT-22.

Basic Trainers: The intermediate step of employing a basic trainer to prepare pilots for the transition to larger and heavier aircraft was not put into practice until 40 converted Douglas O-2Ks began entering service as the BT-1 in 1930, followed by 176 BT-2s (a similar conversion of the Douglas O-32) from 1930 to 1932. Efforts in the mid- 1930s to replace the BT-1 and -2 with a modern monoplane resulted in procurement of 30 Seversky BT-8s and 266 North American BT-9s between 1935 and 1939. The Army Army Air Corps expansion plan of 1939 prompted,

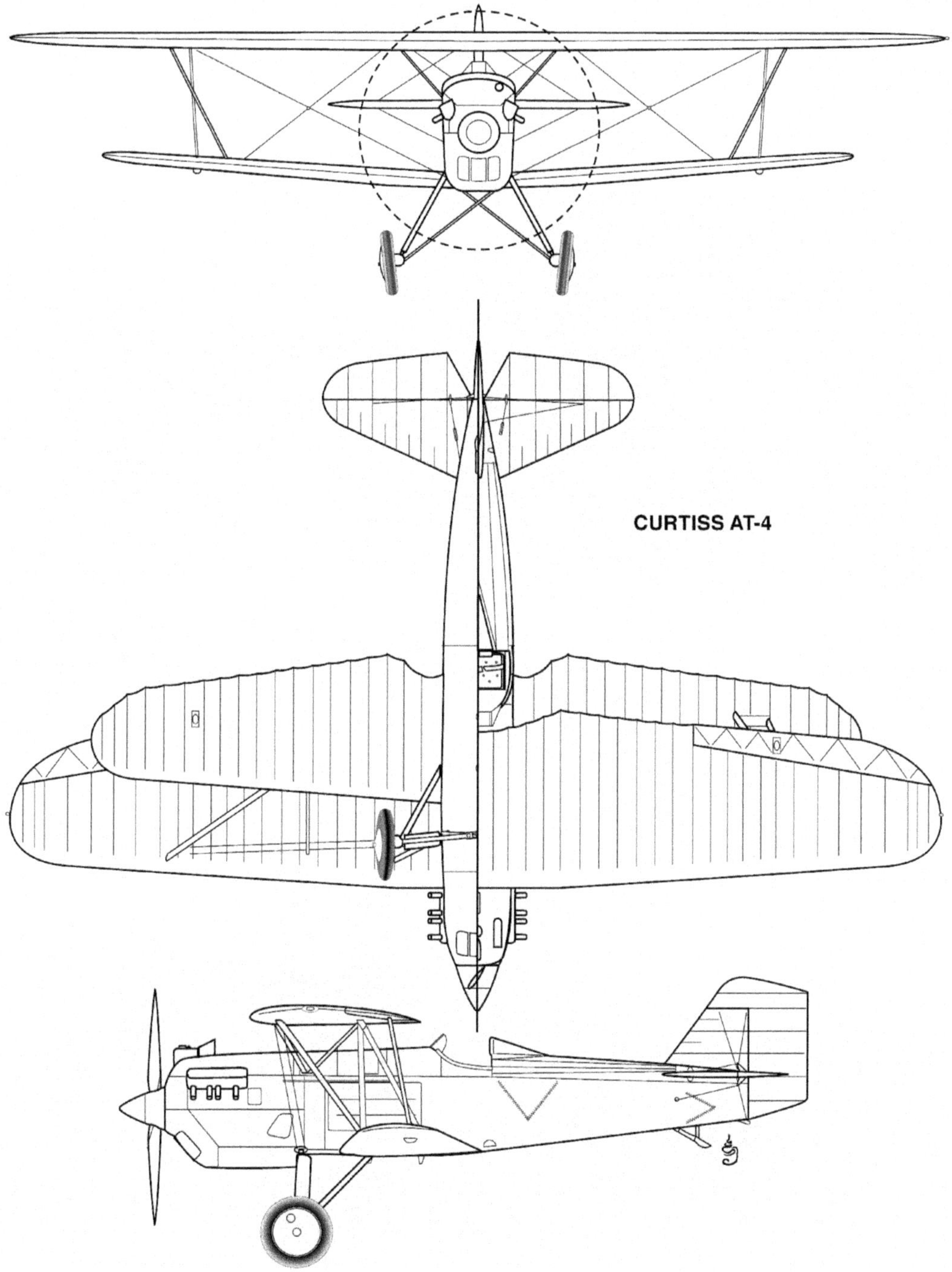

CURTISS AT-4

One of 36 Curtiss P-1A/Bs refitted with Wright-Hispano engines between 1926 and 1928 and returned to service under the designation AT-4 as advanced trainers at Kelly Field, Texas.

A second batch of P-1B/Cs modified for installation of Wright J-5 radial engines returned to service as AT-5 advanced trainers, all with the 43rd Pursuit Squadron School at Kelly Field, Texas.

among other things, the testing and procurement in 1940 of 300 Vultee BT-13s, a development of the BC-3 that became the most important basic trainer of Army Air Forces during the wartime era, and 251 North American BT-14s, an improvement of the BT-9.

Advanced Trainers: The Army Air Corps scheme of using re-engined fighters (e.g., Curtiss AT-4 and -5) to train future pursuit pilots was discontinued in 1929; however, the advent of much higher performance monoplanes during the mid and late 1930s (e.g., Curtiss P-36, Boeing B-17, Northrop A-17, etc.) dictated the need for a trainer capable of transitioning pilots to this new breed of aircraft. To this end, Air Corps Material Division issued a new Basic Combat (BC) requirement in 1936 for a 200 mph, two-place trainer equipped with retractable landing gear. Responding with an extensive redesign of its BT-9, North American delivered the first BC prototype for evaluations in March 1937 and was rewarded with a contract for 180 BC-1 production models. Eighty-two improved BC-1As, ordered in 1939, began entering service in 1940 under the new designation AT-6. The BC-2 of 1938 was simply a modified BC-1 and the Vultee BC-3, tested in 1939, ultimately metamorphosed into the BT-13. When the U.S. Congress approved the Air Corps Expansion Plan in April 1939, it not only set the stage for a process that generated an exponential increase in AT-6 orders but also fostered development and eventual production of new types of aircraft designed to train navigators, bombardiers, and air gunners as well as multi-engine pilots (e.g., Beech AT-7 and -11, Cessna AT-8 and -17, and Curtiss AT-9).

Fleet PT-6 (N2Y)—1930

Technical Specifications (PT-6/N2Y-1)

Type: Two-place primary trainer.
Manufacturer: Fleet Div. of Consolidated Aircraft Corp., Buffalo, New York.
Total U.S. military versions: 22 (16 USAAC; 6 USN).
Unit cost: $5,500.
Powerplant: One 100-hp Kinner K-5 (R-370–1) 5-cylinder air-cooled radial engine driving a two-bladed fixed-pitch metal propeller.
Armament: None.
Performance: Max. speed 105 mph; ceiling 12,200 ft., range 350 mi.
Weights: 1,072 lbs. empty, 1,637 lbs. loaded.
Dimensions: Span 28 ft., length 21 ft. 5 in., wing area 194 sq. ft.

Consolidated established Fleet as a separate division in 1929 to produce smaller, less complex aircraft for civilian and military markets. The Fleet model 1, a single-bay biplane derived from the design of the earlier Consolidated model 14, was introduced the same year with a 110-hp Warner *Scarab* 7-cylinder radial engine and an usual "bathtub" cockpit having no structure between the front and rear seats. Though smaller and lighter than the PT-1/PT-3, it followed the same proven formula of a simple and rugged airframe combined with docile handling qualities. The Air Corps Material Division (succeeded the Engineering Division in 1926) acquired a Fleet model 2 for test purposes in 1930 as the XPT-6, which differed from the model 1 in having a Kinner K-5 engine and a conventional two-hole cockpit enclosure. Following military trials, Fleet received a contract to deliver ten more model 2s as the YPT-6. (Note: the "Y" prefix, adopted in 1928, denoted a pre-production service test model.) A final batch of five model 2s with enlarged cockpit openings were delivered in 1931 as the PT-6A. Although Army officials ultimately deemed the PT-6 to be too light and small, a new Fleet facility located in Canada went on to produce over 300 derivatives of the design for the civil market and some 600 for the Royal Canadian Air Force.

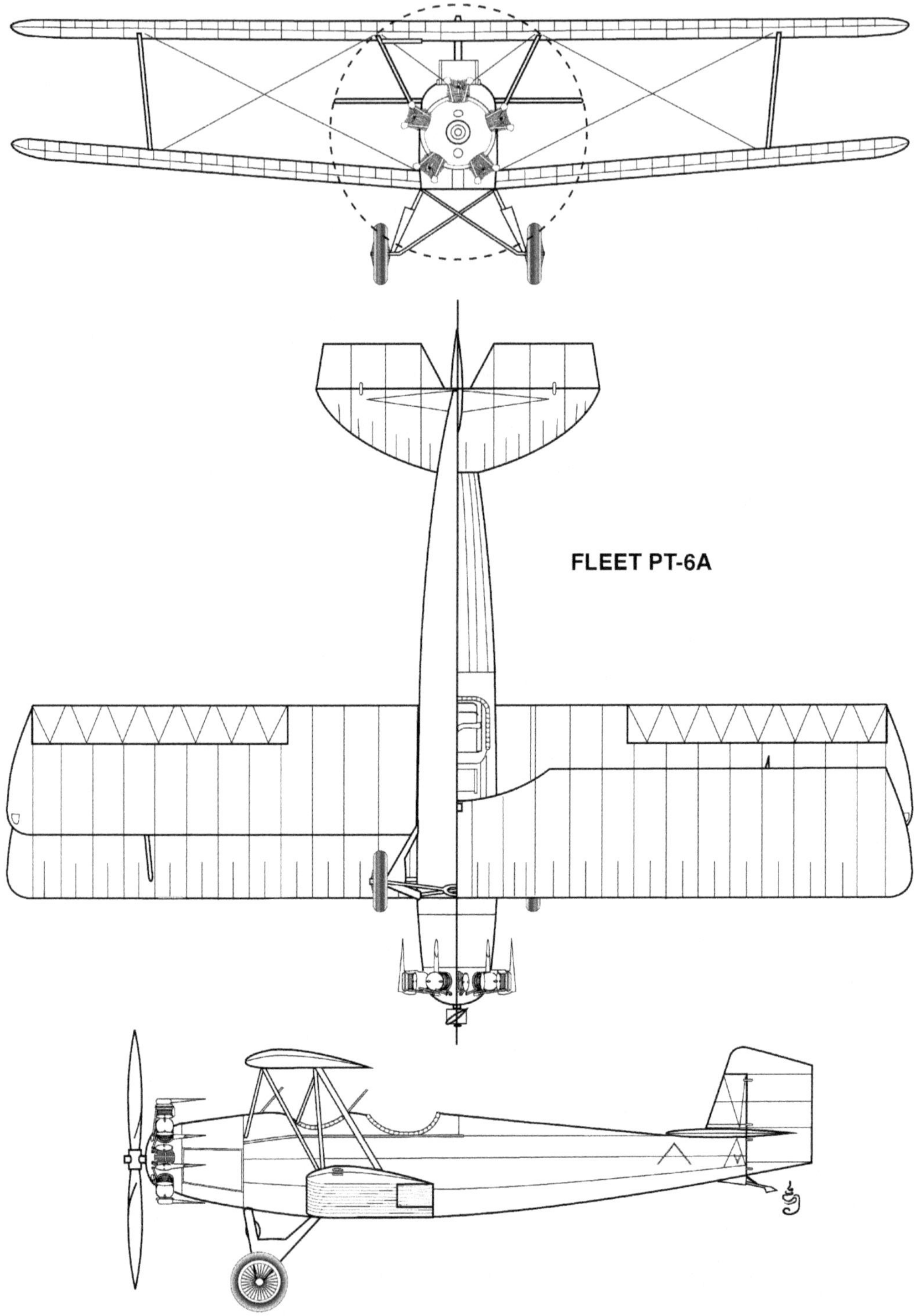
FLEET PT-6A

One of five Fleet PT-6As with enlarged cockpit openings delivered to the Air Corps in 1931. This was not procured in quantity after being deemed too light and small to be a suitable primary trainer.

Naval Variants: As part of its "lightweight" trainer competition, the Navy acquired one Fleet model 1 for testing in 1929 under the designation XN2Y-1. Though not selected to replace the NY as a primary trainer, BuAer nevertheless ordered five Kinner-powered Fleet Model 2s as the N2Y-1 to serve as familiarization trainers for the "skyhook" pilots assigned to the rigid airships *Akron* and *Macon*. One of the N2Y-1s, after being fitted with a skyhook to engage the airships' "trapeze" systems, was re-designated XN2Y-2. After *Akron* and *Macon* were lost, two N2Y-1s were placed in storage at NAF and two others fitted with tailhooks served with the utility units aboard the carriers *Saratoga* and *Ranger* until 1937. The XN2Y-2 was transferred to the Pennsylvania Aircraft Syndicate in 1934 to become the basis for the XOZ-1 autogiro.

Mohawk PT-7–1930

Technical Specifications (XPT-7)

Type: Two-place primary trainer.
Manufacturer: Mohawk Aero Corp., Minneapolis, Minnesota.
Total U.S. military versions: 1 (USAAC).
Unit cost: $4,800.
Powerplant: One 100-hp Kinner K-5 (YR-370–1) 5-cylinder air-cooled radial engine driving a two-bladed fixed-pitch wooden propeller.
Armament: None.
Performance: Max. speed 115 mph, cruise 98 mph; ceiling 12,100 ft., range 500 mi.
Weights: 1,115 lbs. empty, 1,773 lbs. loaded.
Dimensions: Span 34 ft. 9 in., length 24 ft. 5 in., wing area (not reported).

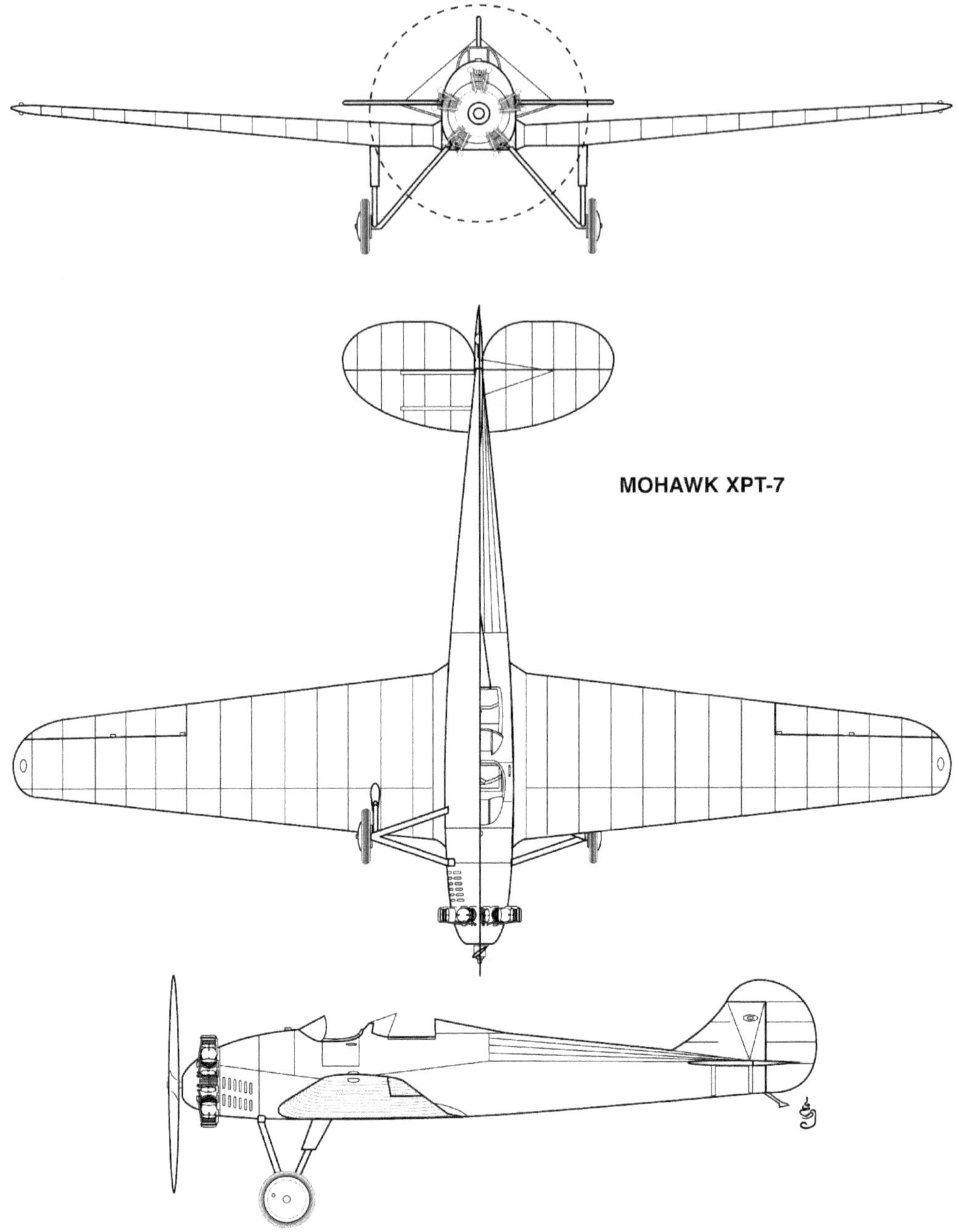

MOHAWK XPT-7

Of the five M-1C "Pintos" completed by Mohawk Aero in 1929 and 1930, a single example purchased by Air Corps Material Division was tested at Wright Field in 1930 as the XPT-7. It is distinguished as the only type of monoplane to be evaluated as a trainer since the Fokker TW-2 of 1922. Few details exist about the XPT-7's structural composition, however, photos suggest a fabric-covered, steel tube fuselage and a fully-cantilevered, built-up wooden wing. Though not mentioned anywhere, the wide track landing gear would likely have given the aircraft good landing and ground handling characteristics. In any case, Mohawk Aero ceased operations in 1930, rendering impractical any plan to acquire more of these aircraft.

One of five M-1C "Pintos" built by Mohawk Aero Corp. The aircraft depicted was tested at Wright Field, Ohio, in 1930 and is the XPT-7. The company went out of business before more aircraft could be built.

Douglas BT-1 and BT-2–1930

Technical Specifications (BT-1 [BT-2B])

Type: Two-place basic trainer.
Manufacturer: Douglas Aircraft Co., Santa Monica, California.
Total U.S. military versions: 237 (USAAC).
Unit cost: $8,000-$9,000 (est.).
Powerplant: One 420-hp Liberty (V-1650–1) 12-cylinder water-cooled inline engine [450-hp Pratt & Whitney R-1340–11 *Wasp* 9-cylinder air-cooled radial engine] driving a two-bladed fixed-pitch metal propeller.
Armament: Provision for one fixed .30-calibre machine gun and one flexible .30 calibre machine gun in the rear cockpit, plus 400 lbs. of bombs carried on wing racks.
Performance: Max. speed 128 mph [134 mph], cruise 103 mph [117 mph]; ceiling 16,279 ft. [19,200 ft.], range 360 mi. [320 mi.].
Weights: 3,032 lbs. [2,919 lbs.] empty, 4,785 lbs. [3,948 lbs.] loaded.
Dimensions: Span 39 ft. 8 in. [40 ft.], length 28 ft. 9 in. [31 ft. 2 in.], wing area 411 [362] sq. ft.

The Douglas BT-1 and BT-2, both originating from O-2 and O-32 observation designs, served as the most numerous type of basic trainer in the Army Air Corps inventory until the development of monoplane replacements during the mid- and late 1930s. Despite being powered by surplus Liberty engines, the 312 O-2 variants procured by the Army between 1924 and 1929 embodied the important structural and aerodynamic advances of the 1920s. Of 59 dual-control O-2Ks delivered during the late 1920s, 40 were subsequently reassigned to training duties under the new designation BT-1. The O-32, which appeared in 1930, was basically a dual-control O-2 airframe modified to accept an R-1340 radial engine and fitted with new wings of reduced chord.

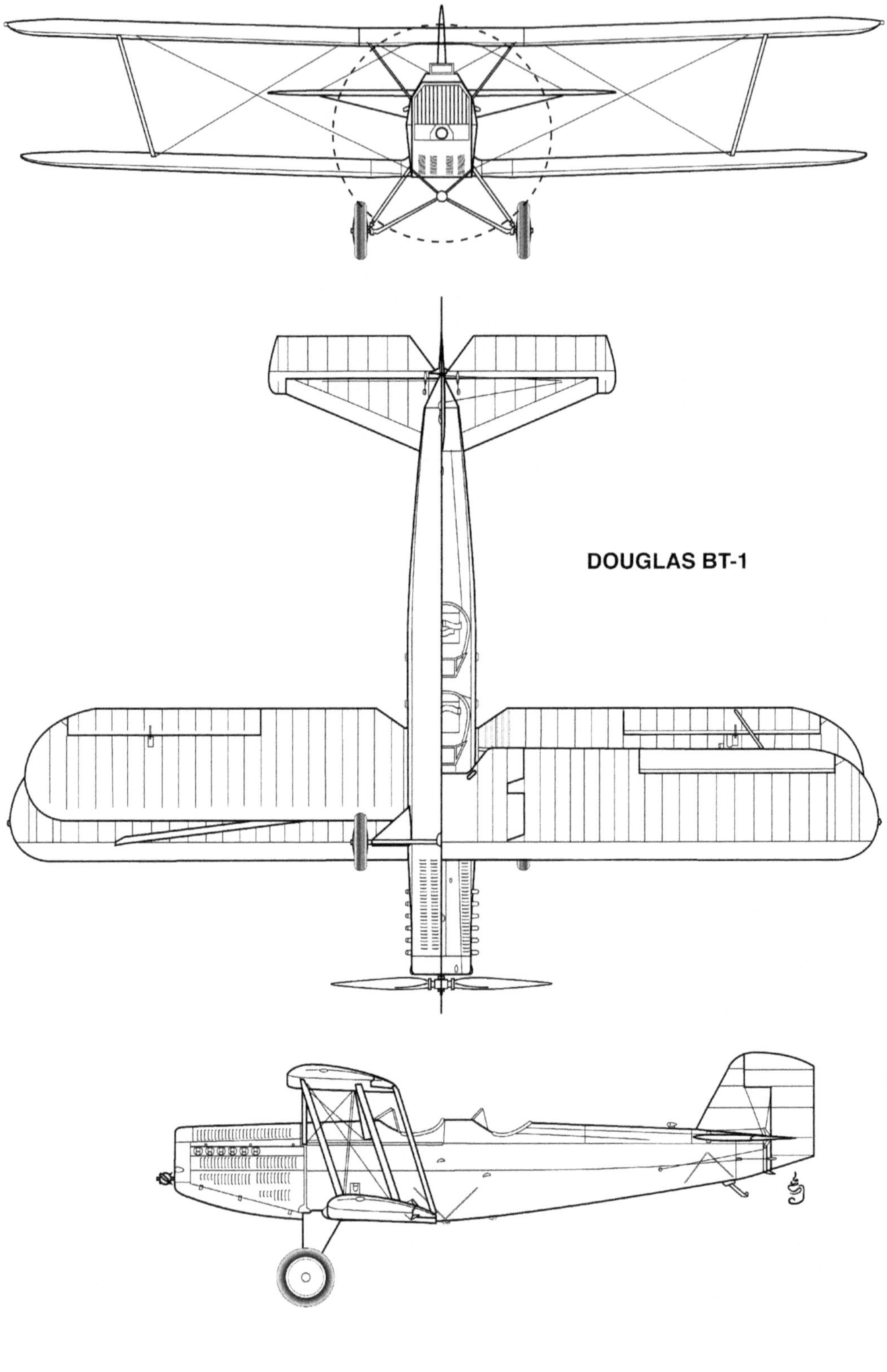

DOUGLAS BT-1

One of 40 dual-control Douglas O-2Ks reassigned to training duties as the BT-1. Although powered by older Liberty engines, the O-2Ks/BT-1s possessed comparatively modern airframes. Note instrument hood in back cockpit.

The BT-2B was derived from the Douglas O-32 airframe and engine. Many, like this example, served with National Guard training units through most of the 1930s.

After Douglas received an order for 30 production aircraft as the O-32A, the O-32 prototype was re-designated as the BT-2 trainer, and all O-32As delivered in 1931 actually entered service as the BT-2A.

In 1931, in order to reequip both frontline and National Guard training units, the Army ordered 146 BT-2Bs, which differed from BT-2As mainly in having steerable tailwheels in place of skids, followed in 1932 by 20 BT-2Cs with revisions to the landing gear. Fifty-eight BT-2Bs specially equipped for instrument training returned to service as the BT-2BI and 13 BT-2Cs as the BT-2CI. In 1940, at the end of their careers, 15 BT-4Bs and seven BT-4Cs became BT-2BR/BT-2CR target drone controllers, with another 17 BT-2Bs becoming A-4 radio-controlled aerial targets.

Stearman PT-9, BT-3 and BT-5–1931

Technical Specifications (YPT-9B)

Type: Two-place primary or basic trainer.
Manufacturer: Stearman Aircraft Corp., Wichita, Kansas.
Total U.S. military versions: 4 (USAAC).
Unit cost: $8,500.
Powerplant: One 200-hp Lycoming R-680–3 9-cylinder air-cooled radial engine driving a two-bladed fixed-pitch metal propeller.
Armament: None.
Performance: Max. speed 120 mph, cruise 100 mph; ceiling 17,000 ft., range 490 mi.
Weights: 2,084 lbs. empty, 2,755 lbs. loaded.
Dimensions: Span 32 ft., length 24 ft. 8 in., wing area 272 sq. ft.

The YPT-9 in 1931. All of the testing and various engine combinations ultimately led Stearman to develop the model 70, which was subsequently mass-produced by the thousands as the PT-13 and PT-17.

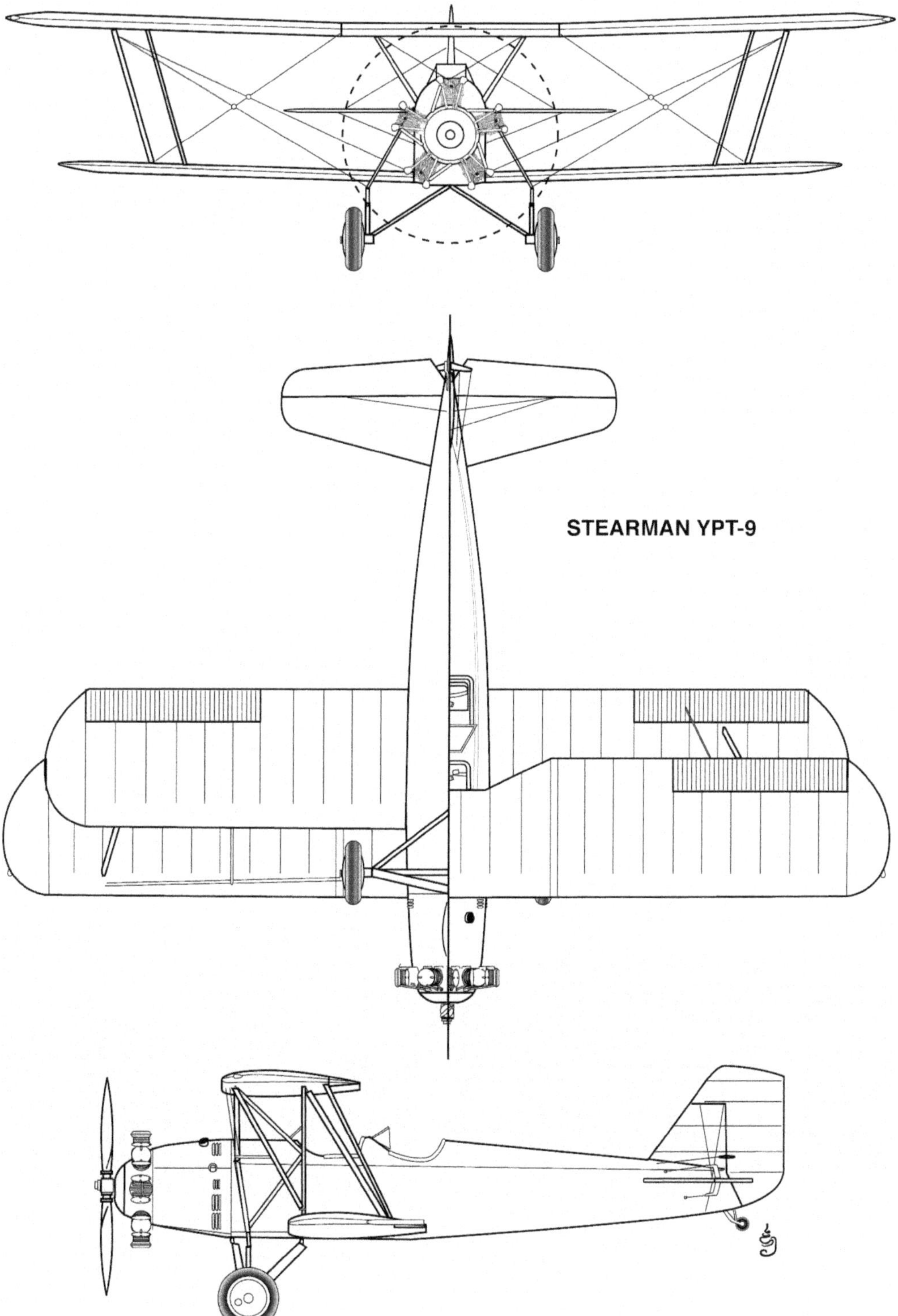
STEARMAN YPT-9

The PT-9, BT-3, and BT-5 were all the same airframe with different engine combinations. Air Corps Material Division purchased four Stearman model 6 "Cloudboys," all equipped with 165-hp Wright J-6–5 (R-540–1) engines, which joined the inventory in 1931 as the YPT-9. In general design, the YPT-9 was a conventional single-bay biplane of fabric-covered, mixed metal and wood construction, characterized by a heavily-braced landing gear structure. Engine changes resulted in a series of YPT-9 sub-variants: one to YPT-9A after substitution of a 165-hp Continental YR-545–1 engine; two to YPT-9B after a switch to 200-hp Lycoming R-680–3 engines; one to YPT-9C after a change to a 170-hp Kinner YR-720 engine. In 1932 the YPT-9C became the YBT-3 when converted to a 300-hp Wright R-975–1 engine and one YPT-9 refitted with a 300-hp Pratt & Whitney R-985–1 engine was re-designated YBT-5. The data derived from all of these tests ultimately prompted development of the Stearman model 70 in 1933, which evolved into the highly successful PT-13 and -17 generation of primary trainers.

Verville PT-10–1931

Technical Specifications (YPT-10)

Type: Two-place primary trainer.
Manufacturer: Verville Aircraft Co., Detroit, Michigan.
Total U.S. military versions: 4 (USAAC).
Unit cost: $5,250.
Powerplant: One 165-hp Wright J-6–5 (R-540–1) 5-cylinder air-cooled radial engine driving a two-bladed fixed-pitch metal propeller.
Armament: None.
Performance: Max. speed 108 mph, cruise 90 mph; ceiling 12,000 ft.; range 370 mi.
Weights: 1,562 lbs. empty, 2,204 lbs. loaded.
Dimensions: Span 33 ft., length 25 ft., wing area (not reported).

One of four YPT-10s accepted in 1931 with its original Wright R-540 engine. The demise of the Verville company later the same year precluded any potential orders for additional aircraft.

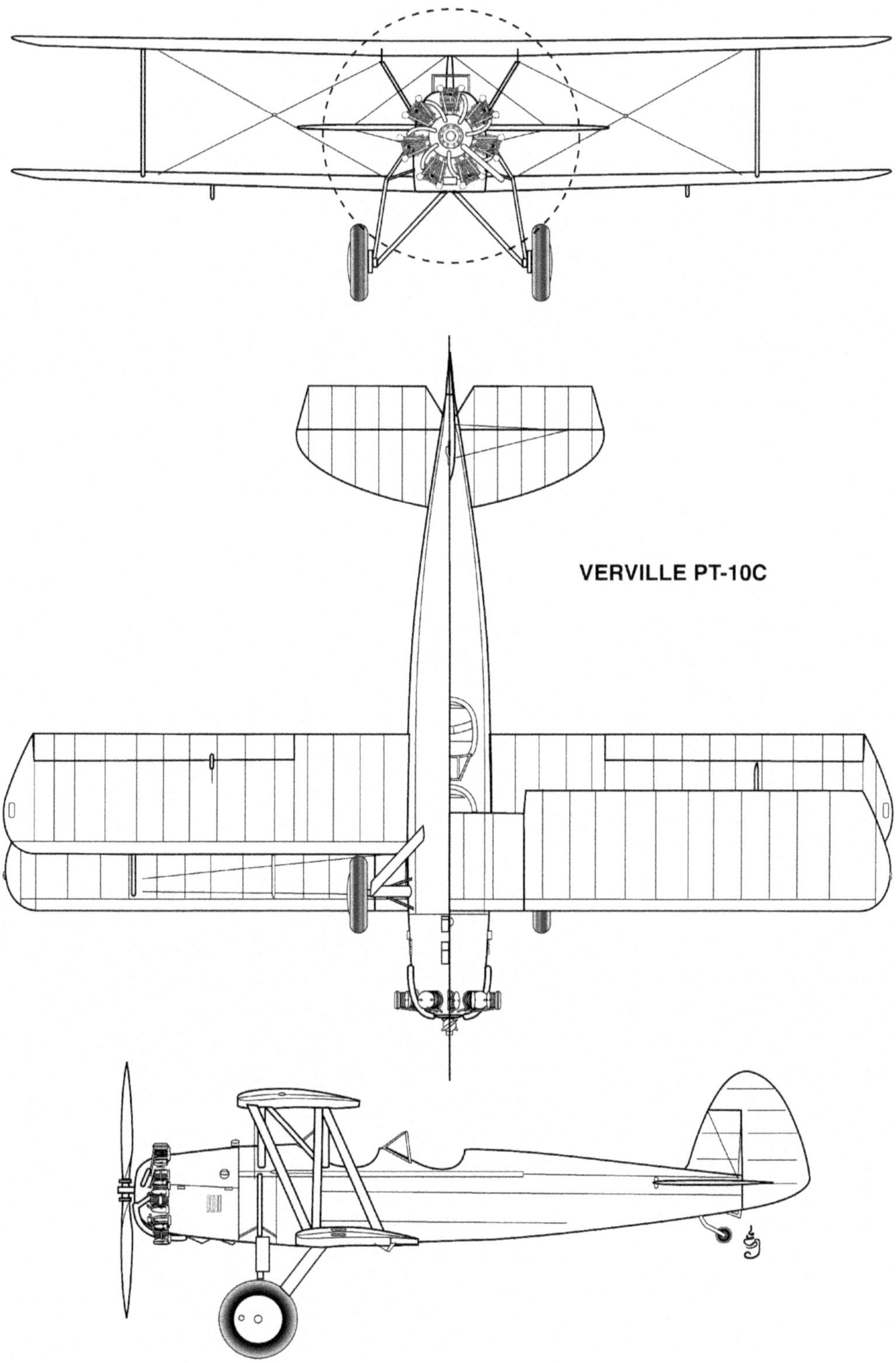

VERVILLE PT-10C

Alfred V. Verville, a well-known aircraft designer, started his own company in 1928 and introduced its second product, the AT Sportsman, to the civil market in 1930. The above-average stability shown by the AT came in part from a long tail moment and nearly full-span ailerons on lower wings rigged at less incidence, which gave positive control at all airspeeds and made the aircraft more resistant to spins. Another notable feature, which rendered single-pilot operations much easier, was an electrical starter powered by a battery. In 1931, after the AT had received civil certification, Air Corps Material Division purchased four examples powered by 165-hp Wright R-540 engines under the designation YPT-10. Like other primary trainers being tested at that time, the YPT-10s were subjected to a number of engine changes that generated subvariants: one became the YPT-10A with a 165-hp Continental YR-545–1 engine; another the YPT-10B with a 165-hp Wright YGR-540 geared engine; and a third the YPT-10C with a 170-hp Kinner YR-720 engine. In 1932, three of these became the YPT-10C when fitted with 210-hp Lycoming YR-680–1 engines. Procurement of additional PT-10s became moot when Verville's company went out of business in 1931.

Consolidated PT-11, PT-12, BT-6 and BT-7 (N4Y)—1931

Technical Specifications (PT-11D)

Type: Two-place primary trainer.
Manufacturer: Consolidated Aircraft Corp., Buffalo, New York.
Total U.S. military versions: 44 (40 USAAC; 4 USN/USCG).
Unit cost: Unknown.
Powerplant: One 215-hp Lycoming R-680–3 9-cylinder air-cooled radial engine driving a two-bladed fixed-pitch metal propeller.
Armament: None.
Performance: Max. speed 118 mph, cruise 100 mph; ceiling 13,700 ft.; range (not reported).
Weights: 1,918 lbs. empty, 2,585 lbs. loaded.
Dimensions: Span 31 ft. 7 in., length 26 ft. 11 in., wing area 280 sq. ft.

The PT-11 represented a determined but ultimately unsuccessful effort on the part of Consolidated to supply the Army Air Corps with a modernized primary trainer that would replace the PT-3. The prototype model 21, known internally as the XP-933, flew for the first time in February 1931 with a 300-hp Pratt & Whitney R-985 engine. In addition to a general aerodynamic cleanup of PT-3, the XP-933 was designed so that the position of the wings could be adjusted to compensate for powerplants of differing horsepower and weight. In May 1931, Air Corps Material Division ordered four model 21s powered by 165-hp Continental R-540–1 engines that were accepted later the same year under the designation Y1PT-11. (Note: the "Y1" prefix, adopted in 1931, denoted funding outside normal fiscal year procurement.)

The PT-11 underwent many variations before reaching final form. During 1931 all four Y1PT-11s acquired designation suffixes after engine changes: one to Y1PT-11A with a 185-hp Curtiss R-600 engine; another to Y1PT-11B with a 210-hp Kinner YR-720 (C-5) engine; and two to Y1PT-11C with 200-hp Lycoming YR-680 engines. In 1932 Consolidated delivered three new Kinner-powered model 21s as the PT-11B; then the same year, the Y1PT-1A, both Y1PT-1Bs, and all three PT-11Bs were converted to 210-hp Lycoming R-680–3 engines and redesignated Y1PT-11D. Once Army officials decided that the R-680 would be the engine of choice for a production version, Consolidated received an initial production contract in 1932 for 21 aircraft to be completed as the PT-11D. Yet, engine testing continued, and a PT-11D converted

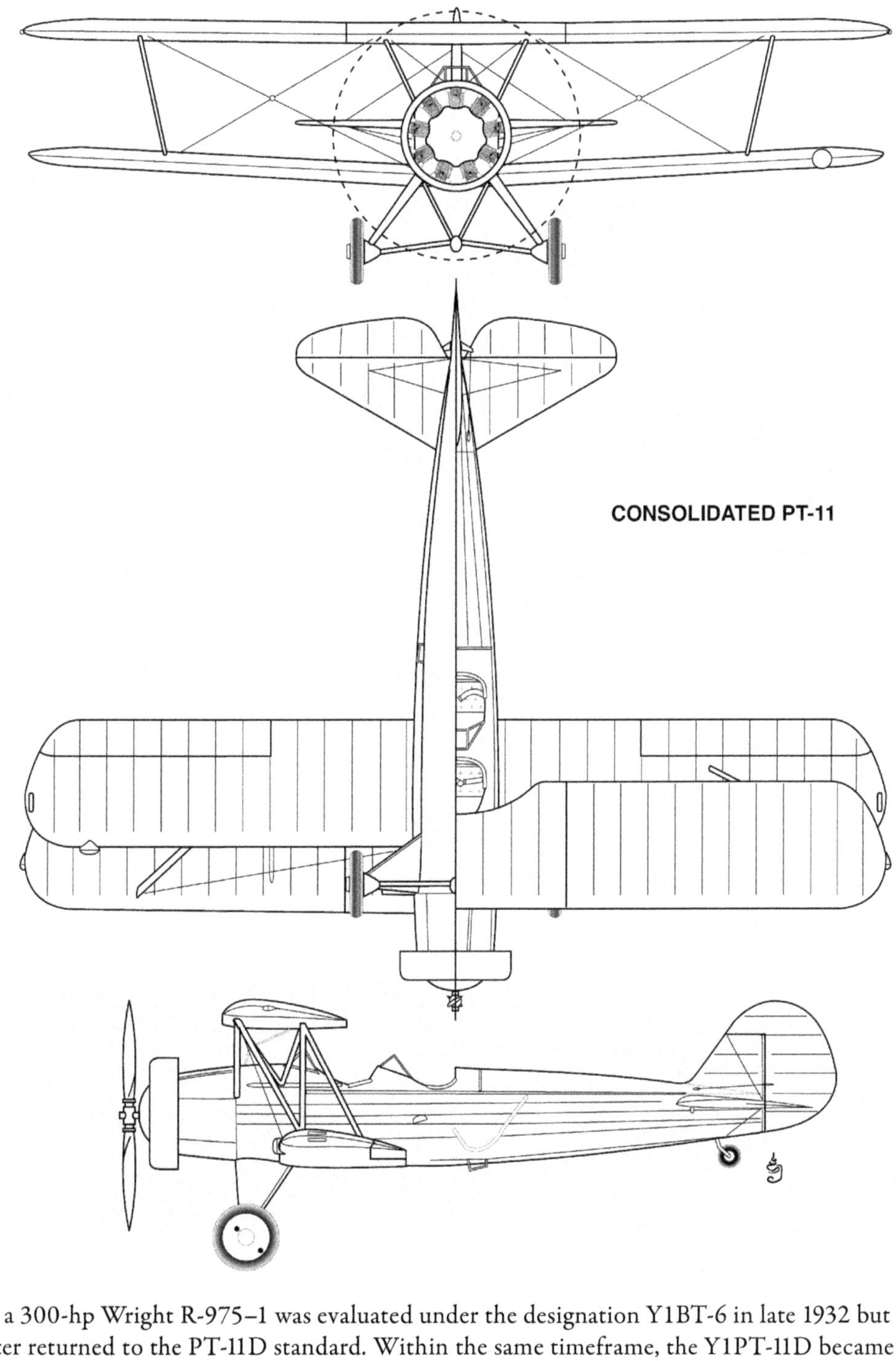

to a 300-hp Wright R-975–1 was evaluated under the designation Y1BT-6 in late 1932 but was later returned to the PT-11D standard. Within the same timeframe, the Y1PT-11D became the Y1PT-12 after a switch to a 300-hp Pratt & Whitney R-985–1 engine, with Consolidated receiving a contract to complete nine new aircraft as the PT-12; however, upon delivery in 1933, the Y1PT-12 and nine PT-12s were re-categorized as basic trainers under the designation Y1BT-7. Lack of funding conspired to delay a final decision on new primary trainer procurement until

Twenty-one of these Consolidated PT-11Ds entered operational service as primary trainers in 1932 at Randolph Field, Texas. Lack of military funding prevented further procurement at the time.

One of 10 Y1BT-6s ("Y1" denotes a service test aircraft), powered by 300-hp R-985-1 engines, that the Air Corps evaluated in 1933 as potential basic trainers.

1936, by which time the Boeing-Stearman model 70 had entered the contest and was ultimately placed in quantity production as the PT-13.

Naval Variants: A single R-540–1-powered Model 21-A, identical to the Y1PT-11, was purchased by BuAer in 1932 and allocated to the Coast Guard at Cape May, New Jersey without any assigned designation. This aircraft, used for general training duties until late 1941, was afterward refitted with a 210-hp Lycoming R-680 engine and given the designation N4Y-1. In 1934 BuAer acquired a further three Lycoming-powered variants, identical to the Army PT-11D, for evaluation as the XN4Y-1 and assigned them to the VD-2D8 training unit at NAS Pensacola. In the interval, however, naval officials had decided to re-equip the training establishment with NAF-built N3Ns, with the consequence that no further production of the N4Y-1 was ordered.

Curtiss BT-4–1932

Technical Specifications (XBT-4)

Type: Two-place basic trainer.
Manufacturer: Curtiss Aeroplane & Motor Co., Buffalo, New York.
Total U.S. military versions: 1 (USAAC).
Unit cost: Unknown.
Powerplant: One 435-hp Curtiss D-12 (V-1150–5) 12-cylinder water-cooled inline engine driving a two-bladed fixed-pitch metal propeller.
Armament: None.
Performance: Max. speed 139 mph, cruise 110 mph; ceiling 14,100 ft., range 650 mi.
Weights: 2,875 lbs. empty, 4,076 lbs. loaded.
Dimensions: Span 38 ft., length 27 ft. 2 in., wing area 353 sq. ft.

This Curtiss XBT-4, shown at Wright Field, Ohio, in 1932, was an O-1E stripped of armament and fitted with dual controls. The aircraft later returned to service as the Y1O-1G.

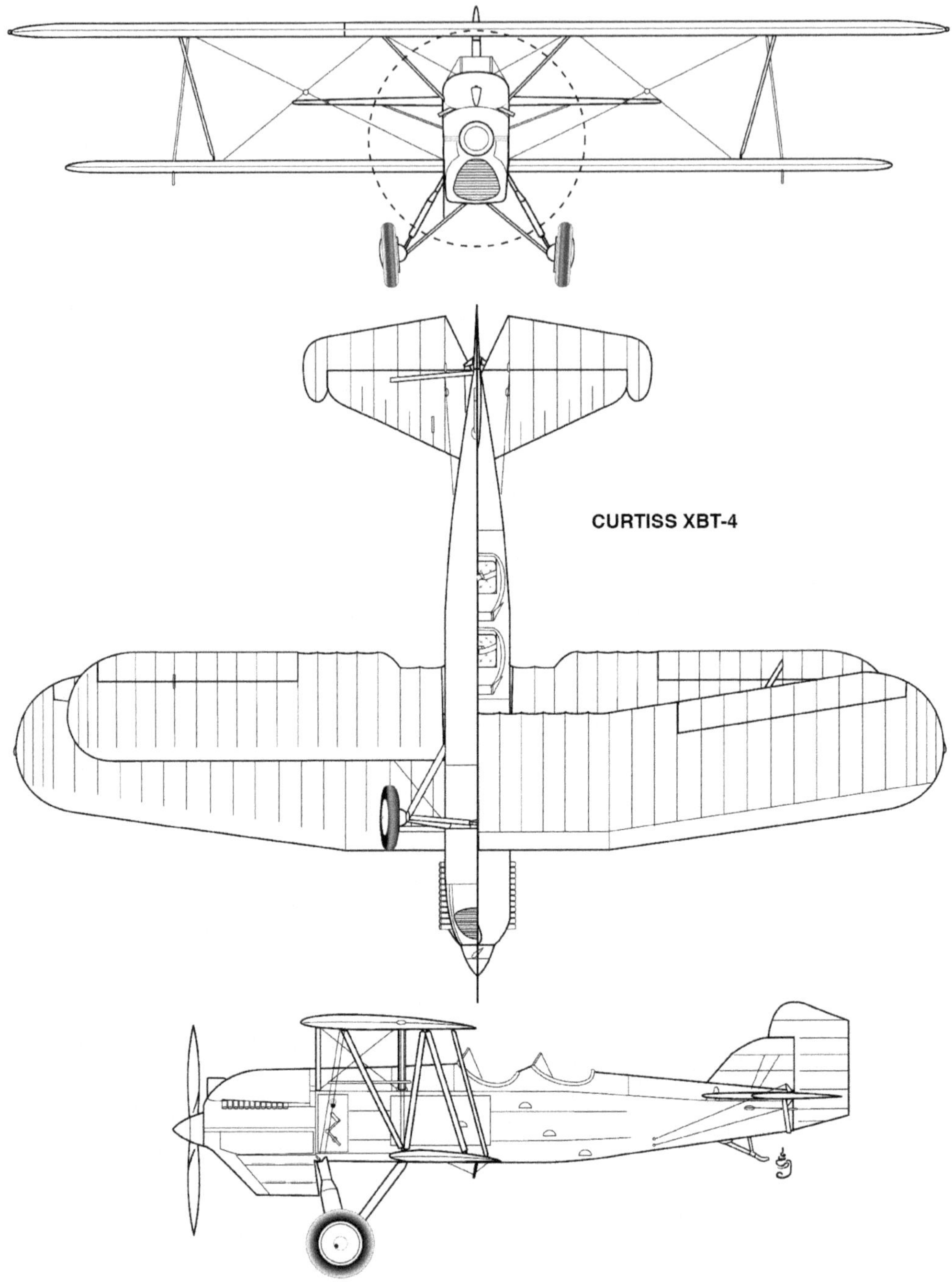

From 1927 to 1930 the Army Air Corps procured some 338 Curtiss "Falcons" that entered service in both A-3 attack (model 44) and O-1 observation (model 46) versions. One interesting feature of the A-3/O-1 was a fuselage fabricated of aluminum tubing braced by steel tie rods rather than the more common welded steel tubing. In 1932 one O-1E stripped of armament and equipped with dual controls was tested at Wright Field as the XBT-4. Once testing was concluded, no further conversions were ordered and the aircraft returned to service as the Y1O-1G.

Seversky BT-8–1934

Technical Specifications (BT-8)

Type: Two-place basic trainer.
Manufacturer: Seversky Aircraft Corp., Farmingdale, New York.
Total U.S. military versions: 30 (USAAC).
Unit cost: $22,000 (est.).
Powerplant: One 400-hp Pratt & Whitney R-985–11 9-cylinder air-cooled radial engine driving a two-bladed variable-pitch metal propeller.

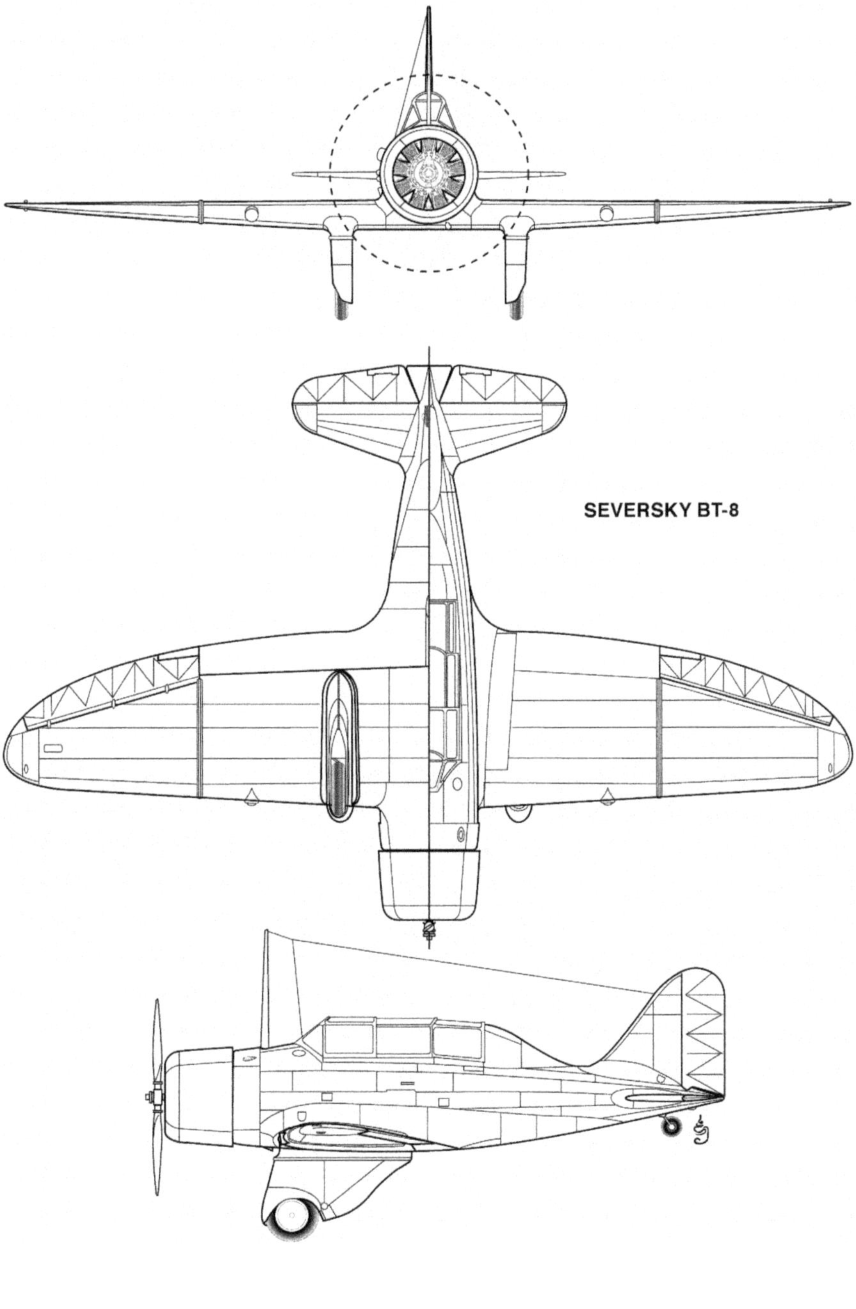

The Seversky BT-8 was the first all-metal monoplane trainer to be procured by the Air Corps. Thirty were delivered to training units in 1936 but soon withdrawn due to unforgiving stall characteristics.

Armament: None.
Performance: Max. speed 175 mph, cruise 118 mph; ceiling 14,750 ft., range 528 mi.
Weights: 2,732 lbs. empty, 4,057 lbs. loaded.
Dimensions: Span 36 ft., length 24 ft. 3 in., wing area 220 sq. ft.

Alexander de Seversky, a former ace of the Russian Imperial Air Service during World War I, came to the U.S. in 1918 and organized Seversky Aircraft Corp. in 1931. The company's first design, the SEV-3, appearing in July 1933 as a sleek, all-metal monoplane powered by a 420-hp Wright R-975-E3 supercharged engine and equipped with amphibious floats, subsequently established a new amphibian speed record of 179 mph. Soon after Air Corps Material Division announced a new monoplane basic trainer requirement in 1934, the SEV-3 underwent modifications that entailed installing a fixed landing gear encased in fairings, raising and relocating the cockpit forward, and substituting a non-supercharged (350-hp) R-975E engine. The modified prototype, now billed as the SEV-3XAR (i.e., Experimental Army), was delivered to Wright Field for evaluations later the same year. Besides state-of-the-art aerodynamics and construction methods, the prototype possessed innovations such as fully-enclosed, sliding canopies for the student and instructor and split-type flaps that kept approach and landing speeds within acceptable limits.

The SEV-3XA's closest competitor, the North American NA-16, appeared at Wright Field in mid–1935 with open cockpits, unfaired wheels, and a fabric-covered fuselage. At the conclusion of trials in late 1935, Seversky was declared winner of the fly-off and received a contract to build 30 aircraft under the designation BT-8. Changes to production models specified certain detail refinements and installation of government-supplied 400-hp Pratt & Whitney R-985–11 engines. As deliveries proceeded in 1936, two BT-8s went to Wright Field for continued testing and the remainder to training units at Randolph Field, Texas. In operational service, however, the BT-8 proved to be underpowered, and its high wing loading (e.g., 18.5 lbs./sq. ft. vs. 10.9 lbs./sq. ft. for the BT-2) produced dangerous stall characteristics in the hands of inexperienced pilots, with the result that the more conservative North American model NA-19 (an improved NA-16) succeeded it in production in 1936.

Boeing-Stearman PT-13, PT-17, PT-18 and PT-27 (NS and N2S)—1934

Technical Specifications (PT-13D [PT-17])

Type: Two-place primary trainer.
Manufacturer: Stearman Aircraft Div. of Boeing Airplane Co., Wichita, Kansas.
Total U.S. military versions: 8,832 (4,395 USAAC/AAF; 4,437 USN/USMC/USCG).
Unit cost: $7,700-$10,400.
Powerplant: One 225-hp Lycoming R-680–17 9-cylinder [220-hp Continental R-670–5 7-cylinder] air-cooled radial engine driving a two-bladed fixed-pitch metal (or wooden) propeller.
Performance: Max. speed 124 mph, cruise 108 mph; ceiling 11,200 ft., range 505 mi. (Note: Comparative performance specifications of the PT-13D and PT-17 are nearly identical.)
Weights: 2,028 lbs. [1,931 lbs.] empty, 2,717 lbs. [2,635 lbs.] loaded.
Dimensions: Span 32 ft. 2 in., length 24 ft. 10 in. [24 ft. 9 in.], wing area 298 sq. ft.

The legendary Stearman "Kaydet" stands as the most produced and certainly, best-remembered type of primary trainer to have ever served in the U.S. military establishment. Its origins are traceable to 1933, when Stearman, as a private venture, developed the model 70, also known as the XPT-943, to compete for Army and Navy trainer contracts against the front-running Consolidated PT-11/N4Y. Although derived from the design of the model 6 (see YPT-9, above), the XP-943 emerged with a revised wing layout having full-span lower ailerons only, a fully-cantilevered landing gear with oleo struts, reshaped tail surfaces, and a 215-hp Lycoming R-680–5 engine. After the XP-943 was demonstrated at Wright Field and NAS Anacostia in 1934, the Navy was the first to express interest in the type, as explained below. Two years later, in 1936, Air Corps Material Division gave Stearman (now a subsidiary of Boeing) a contract to deliver a model 75 (identical to the Navy's model 73 except for a Lycoming R-680–5 engine) under the designation PT-13. Following trials, Stearman received a contract for 25 more PT-13s that were delivered during 1936. A succession of new production orders over the next four years accounted for delivery of 92 PT-13As (220-hp R-680–7 engines) in 1937 and 255 PT-13Bs (280-hp R-680–11 engines) in 1940, making it the Air Corps' most important type of primary trainer. In 1941, after being re-powered with R-680–11 engines, six PT-13As were re-designated PT-13C. A final batch of 353 PT-13Ds (first standardized Army-Navy version with 225-hp R-680–17 engines) were delivered to Army Air Corps/Army Air Forces Training Command (AAC/AAFTC) in 1942.

In 1940, in the midst of increasing American war preparations, Stearman was authorized to develop the model 75N, a new standardized version equipped with a 220-hp Continental R-670–5 7-cylinder engine, which underwent testing the same year under the new designation PT-17, and from 1940 to 1943, a total of 3,519 PT-17s entered service at various AAFTC primary training centers. During the war, 18 PT-17s fitted with new instruments and blind-flying hoods returned to service as the PT-17A and three modified as aerials sprayers as the PT-17B. After conversion of a PT-13A to a 225-hp Jacobs R-755–7 7-cylinder engine in 1940, Stearman commenced production of a new version in 1941 under the designation PT-18, however, production

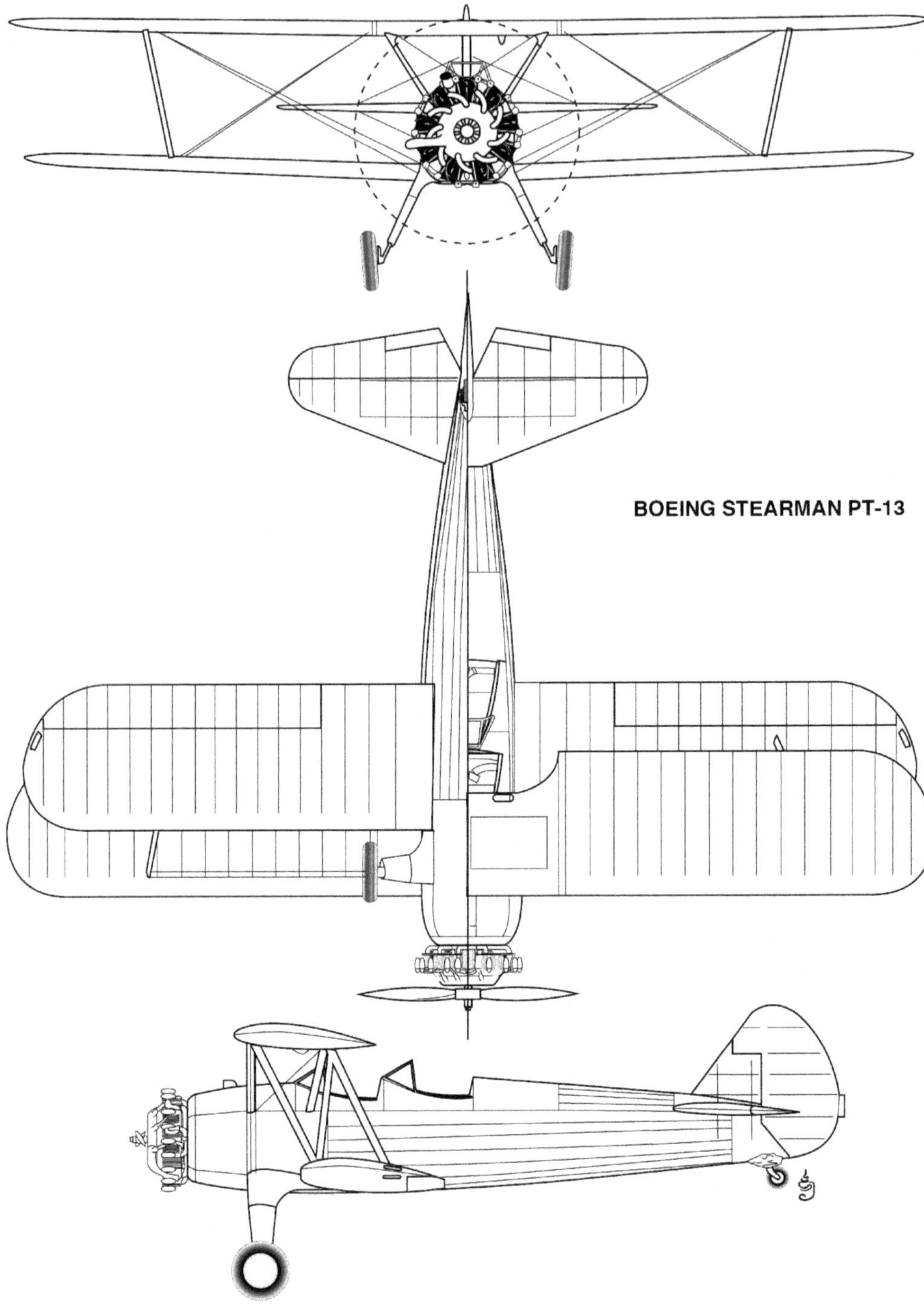

ceased after 150 examples had been delivered to the ATC, apparently due to the need for Jacobs engines elsewhere (e.g., Sherman tanks). Six PT-18s subsequently modified for instrument flying were re-designated PT-18A. A final batch of 300 Continental-powered model 75Ns with canopies and engine cowlings added for cold weather operations were built under AAF serial numbers but went to the Royal Canadian Air Force under Lend-Lease as the PT-27. In 1946, as a conse-

This photograph depicts one of the first 25 Boeing-Stearman PT-13s delivered to the Air Corps in 1936. These were followed by 92 PT-13As delivered in 1937 and 255 PT-13Bs in 1940.

As the United States went to a war footing in 1940, the standard production version became the Continental-powered PT-17 shown. More than 5,000 of the same model were built between 1940 and 1943 as the PT-17 and N2S.

Only 150 of these Jacobs-powered PT-18s had been completed in early 1942 when production terminated due to shortages of Jacobs engines that were instead needed for Sherman tanks.

The NS-1 was powered by surplus Wright J-5 engines. The Navy procured 61 of these in 1935 to supplement its primary training fleet at NAS Pensacola, Florida.

One of 1,664 N2S-3s delivered to the Navy from 1942 onwards. N2S variants became the most numerous type of primary trainer to serve with the Naval Air Training establishment during World War II.

quence of the huge demobilization and reorganization of AAFTC that followed the ending of World War II, all but small numbers of surviving PT-13s and PT-17s were withdrawn from service and declared surplus, and the very last examples were stricken from the inventory during 1947.

Naval Variants: In terms of numbers, the Stearman "Kaydet" became the Navy's most important primary trainer of the wartime era. Though not as versatile as the N3N, the Stearman Division of Boeing possessed the plant capacity to produce trainers in far greater numbers than the Naval Aircraft Factory (NAF). In 1934, the same year Stearman was acquired as a subsidiary of Boeing, the Model 70/XPT-943 demonstrator appeared at NAS Anacostia as a potential candidate to replace the Navy's aging fleet of Consolidated NY primary trainers. BuAer liked the design well enough to order one prototype, designated XNS-1, with the added specification that it be powered with a Navy-supplied 220-hp Wright R-790–8 engine. Following official acceptance trials in 1935, Stearman received a contract for 61 production NS-1s which entered service at NAS Pensacola during 1935 and 1936 to supplement N3N-1s.

In 1940, after the Army Air Corps had placed the PT-13 (model 75) and PT-17 (model 75N) in large-scale production, BuAer procured 250 PT-17 copies under the designation N2S-1, with deliveries being completed in early 1941. Also accepted during 1941 were 125 PT-13C instrument trainer variants that entered service as the N2S-2, together with 211 N2S-3s (PT-17s with R-670–4 engines), all of which were assigned to the Navy's three pilot training centers. Wartime deliveries accounted for a further 1,664 N2S-3s, 667 N2S-4s (R-670–5 engines), and 1,450 N2S-5s (PT-13Ds with Lycoming R-680–17 engines). Eleven N2S-3s were transferred to the Coast Guard in 1943 for training purposes, and small numbers of different N2S variants also served with the Marines. As part of the general downsizing that followed the end of World War II, all Navy and Marine N2S variants were withdrawn and sold surplus, and the last Coast Guard examples were retired in 1947.

North American BT-9, BT-10 and BT-14 (NJ)—1936

Technical Specifications (BT-9)

Type: Two-place basic trainer.
Manufacturer: North American Aviation, Inc., Inglewood, California.

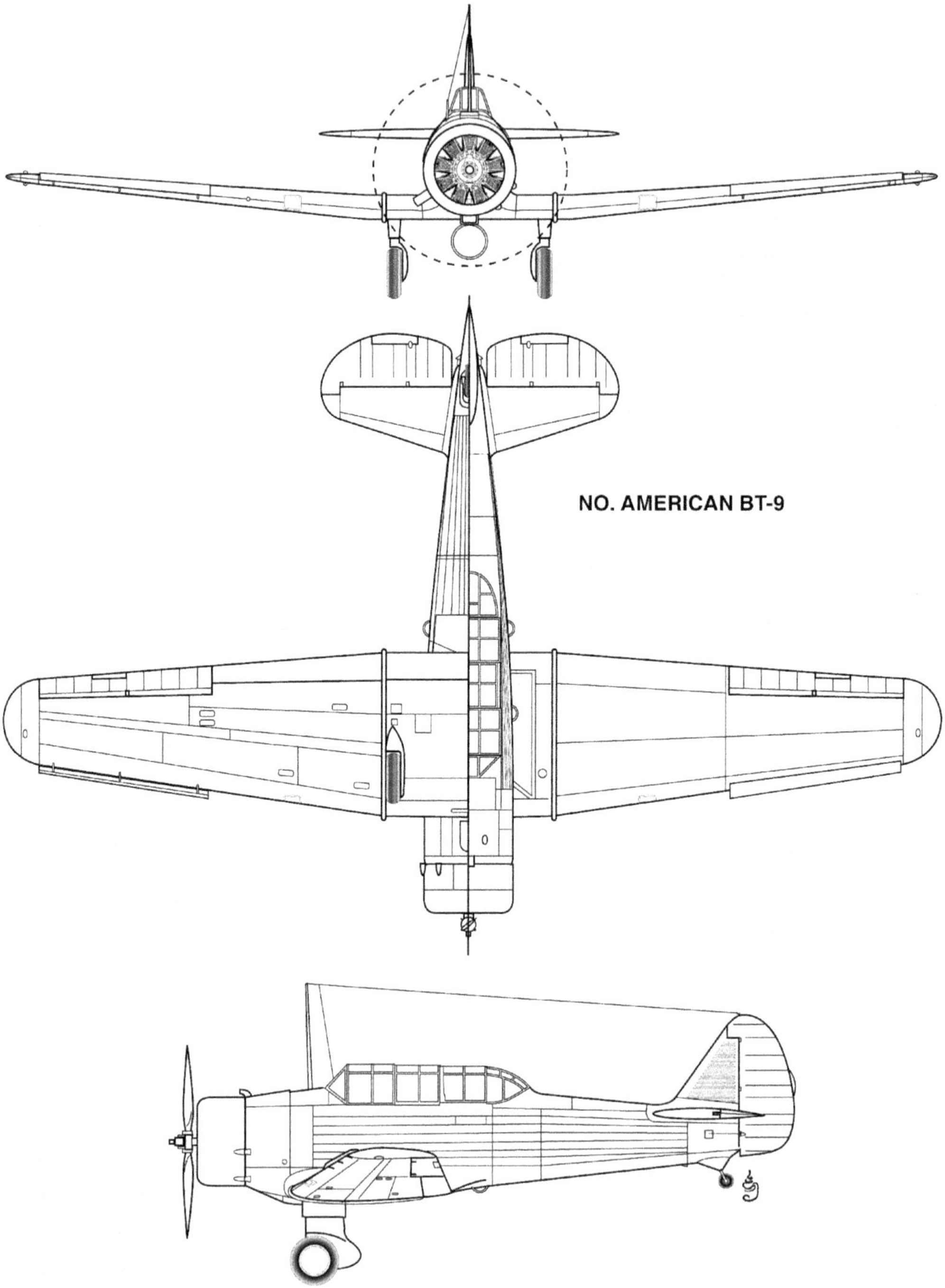

One of 60 armed North American BT-9Cs allocated to reserve units in 1938. BT-9 variants became the Air Corps most numerous type of basic trainer during the late 1930s.

The BT-14 came as an improvement of the BT-9, with the Air Corps accepting 251 during 1941. Differences from the BT-9 included an all-metal fuselage, reshaped rudder, and R-985–25 engine.

The Navy took delivery of 40 NJ-1s in 1937 that differed from the Army BT-9 in having R-1340 engines. The fuselage (dark blue) of the aircraft pictured indicates it was assigned to command duties.

Total U.S. military versions: 557 (517 USAAC/AAF; 40 USN).
Unit cost: $20,000.
Powerplant: One 400-hp Wright R-975–7 9-cylinder air-cooled radial engine driving a two-bladed variable-pitch metal propeller.
Armament (BT-9A/C): Two fixed .30-calibre machine guns.
Performance: Max. speed 172 mph, cruise 147 mph; ceiling 19,750 ft., range 877 mi.
Weights: 3,313 lbs. empty, 4,470 lbs. loaded.
Dimensions: Span 42 ft., length 28 ft., wing area 248 sq. ft.

After losing the 1935 basic trainer fly-off to Seversky, North American modified the NA-16 to incorporate an enclosed canopy and wheel fairings, and as company model NA-18, resubmitted it to Air Corps Material Division for future consideration. Though marginally slower and less modern in construction than the BT-8, the NA-18 exhibited better flying characteristics and had potential for improvement. North American's efforts were rewarded in early 1936 with a contract to manufacture 42 aircraft to a revised specification as the BT-9, the first example of which was flown in April the same year. Carried as the company model NA-19, the BT-9 retained the NA-18's R-975 engine but differed in having a taller canopy enclosure with a rollover cage, a variable-pitch propeller, and certain detail refinements. As Army Air Corps training operations grew during the late 1930s, North American received additional BT-9 orders: 40 armed BT-9A gunnery trainers delivered in 1937; 117 BT-9Bs with strengthened center-sections that entered service in 1937 and 1938; and 67 similar but armed BT-9Cs that were allocated among reserve units in 1938. In 1939, a BT-9B fitted with BC-1A-type outer wing panels and vertical tail surfaces was tested under the designation BT-9D, presumably as a prototype for the impending BT-14, and one BT-9C modified for installation of a 600-hp Pratt & Whitney R-1340 engine was tested as the Y1BT-10.

In 1940, following approval of the Air Corps Expansion Plan, North American was awarded a contract to built 251 improved BT-9 derivatives under the new designation BT-14 (company model NA-58), with deliveries being completed in 1941. The BT-14 incorporated the all-metal fuselage, wing panels, and tail group of the BC-1A/AT-6 (see below) and was re-powered with a 450-hp Pratt & Whitney R-985–25 engine. Twenty-seven BT-14 airframes completed with 400-hp R-985–11 engines entered service in 1941 as the BT-14A. Although BT-14s gave good service to Air Corps/Army Air Forces Training Command, they were rapidly overshadowed by newer Vultee BT-13s once wartime priorities dictated that North American focus its efforts on production of other aircraft (e.g., AT-6, B-25, and P-51). All surviving BT-14s were withdrawn and sold surplus soon after World War II ended.

Naval Variants: In late 1936, after North American had received an Army contract to produce the model NA-19 as the BT-9, BuAer awarded the company a contract to deliver 40 model NA-28s, identical to the BT-9 except for 500-hp Pratt & Whitney R-1340–6 engines and minor detail changes, and assigned the designation NJ-1. As deliveries of NJ-1s proceeded in 1937, 39 were assigned to training duties at NAS Pensacola, while the 40th, fitted with an experimental Ranger V-770–4 inverted V-12 engine, was briefly tested as the NJ-2 before being returned to the NJ-1 standard. During the war, NJ-1s remained in service at Pensacola as familiarization trainers for student pilots transitioning into the SNJ (AAF AT-6).

North American BC-1, BC-2, AT/T-6 and AT-16 (SNJ) Texan—1937

Technical Specifications (AT-6D/SNJ-5)

Type: Two-place landplane advanced, gunnery, and bombing trainer.
Manufacturer: North American Aviation, Inc., Inglewood, California, and Dallas, Texas; and Noorduyn Aviation, Ltd., Montreal, Quebec, Canada.
Total U.S. military versions: 15,069 (9,698 USAAF/AAF/USAF; 5,371 USN, USMC, USCG).
Unit cost: $27,000 (avg.).
Powerplant: One 600-hp Pratt & Whitney R-1340-AN1 *Wasp* 9-cylinder air-cooled radial engine driving a two-bladed, variable-pitch metal propeller.
Armament: one fixed .30-calibre machine gun in cowling (plus one fixed .30-calibre machine gun in wing on AT-6D/SNJ-5), provision for one flexible .30-calibre machine gun in the rear cockpit, and up to ten 20-lb. or four 100-lb. bombs carried on wing racks.
Performance: Max. speed 207 mph, cruise 170 mph; ceiling 24,199 ft.; range 750 mi.
Weights: 4,158 lbs. empty, 5,300 lbs. loaded.
Dimensions: Span 42 ft., length 29 ft. 6 in., wing area 254 sq. ft.

North American's celebrated AT-6 "Texan," together with its SNJ naval counterpart, is universally recognized as having been the most important American-made advanced trainer of the wartime period. Its origins stemmed from the Basic Combat (BC) Trainer Requirement of 1936, issued by Air Corps Material Division in response to the need for an aircraft that would prepare student pilots for the transition to the new high-performance monoplanes then entering Army Air Corps service (e.g., Northrop A-17, Boeing B-17, Curtiss P-36, etc.). The basic specification called for a two-seat, retractable-gear aircraft having a top speed of at least 200 mph. Using the fabric-covered fuselage, outer wing panels, and tail group of its fixed-gear BT-9 as a starting place, North American evolved the model NA-26 with a center-section that incorporated an inward-retracting main landing gear and installation of a 600-hp R-1340 geared engine. The completed NA-26 prototype, delivered to Wright Field during 1937, was subsequently purchased by the Army Air Corps and tested under the designation BC-1. At the conclusion of official trials the same year, North American received a contract to build 180 BC-1 production aircraft (company model NA-36) with minor modifications and 550-hp R-1340–37 un-geared engines. In service, 30 BC-1s modified as instrument trainers became the BC-1I, and three retrofitted

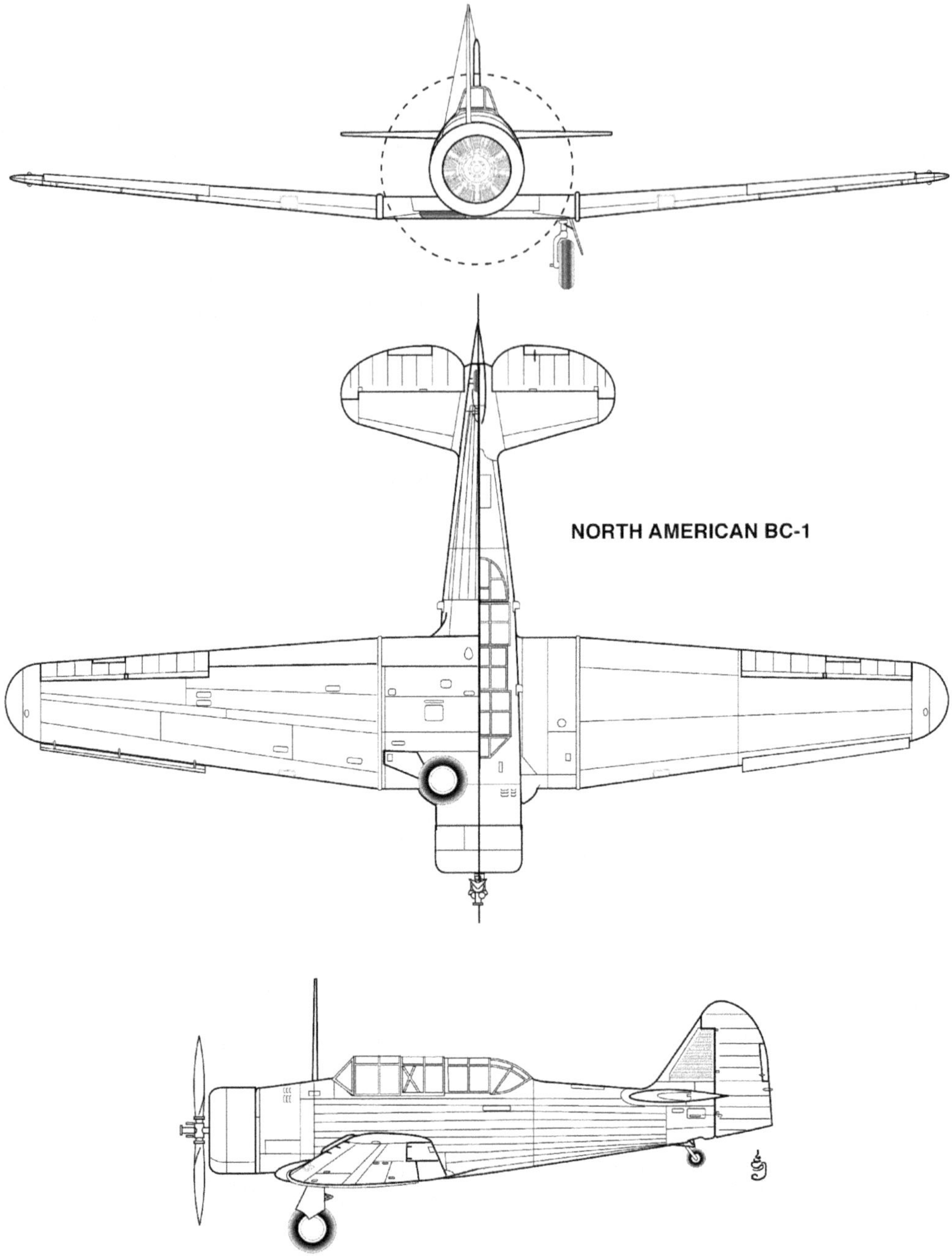

with three-bladed propellers were tested in 1938 as the BC-2. Continued refinements to the design resulted in a 1939 Army Air Corps order for 92 BC-1As (company model NA-55), which differed in having a metal-skinned fuselage, squared-off wingtips, and reshaped rudder, which became the definitive outline of the type. In the interval, the Army Air Corps resurrected the advanced trainer (AT) designation, so that the 83 BC-1As delivered in 1940 entered operational service as the AT-6 (company model NA-59). One BC-1A refitted with an AT-6A center-section was tested in 1940 as the BC-1B.

The forerunner of the legendary AT-6, the North American BC-1 won the Basic Combat competition staged at Wright Field, Ohio, in 1937 and began entering service as an advanced trainer in 1938.

In 1940, amid escalating U.S. war preparations and the anticipated procurement of thousands of training aircraft, the Army Air Corps (Army Air Forces after June 1941) and the Navy agreed upon a standardized AT-6A/SNJ-3 (company model NA-77) version powered by a 600-hp R-1340-AN1 engine, to be manufactured at a new North American plant being built in Dallas, Texas, as well as the company's existing facilities in California. As a direct result, the first of 1,549 AT-6As began entering service with Army Air Corps/Air Air Forces Training Command (AAC/AAFTC) in early 1941, shortly followed by 400 AT-6Bs (company model NA-84) gunnery trainers, identical to As except for provision of a rear-facing seat and scarf ring gun mount. Next were 2,970 AT-6Cs (company model NA-88) with 24-volt electrical systems, delivered during the balance of 1941 and into early 1942. The most numerous production variant, 4,388 AT-6Ds (company models NA-119 and -121), entered operational service with AAFTC in 1942 and 1943. One AT-6D fitted with a 575-hp Ranger V-770–9 supercharged V-12 engine was tested in 1944 as the XAT-6E, but no production was ordered. A final batch of 25 AT-6Fs, distinguished by frameless rear canopies and minor detail changes, were accepted by AAFTC during 1944 and 1945, at which point production terminated. In addition to North American's production, 1,500 AT-6As license-built by Noorduyn in Canada were delivered to AAFTC in 1942 and 1943 under the designation AT-16.

Under the new designation system adopted by the independent U.S. Air Force (USAF) in June 1948, all AT-6Ds and Fs remaining in service were re-designated T-6D and F, respectively. The postwar surge in pilot training generated by the Cold War and the Korean War led to the remanufacture of 2,068 T-6D airframes in 1949 and 1950, which reentered USAF service under new serial numbers as the T-6G. Besides a structural rebuild and a new AN1 engine and prop, refinements to the T-6G included an improved cockpit layout, updated radios, full-time hydraulic system, increased fuel capacity, reduced canopy side frames, and a steerable tailwheel. A small number of T-6Fs rebuilt to the same standard reemerged as the T-6H. Ninety-seven T-6Gs

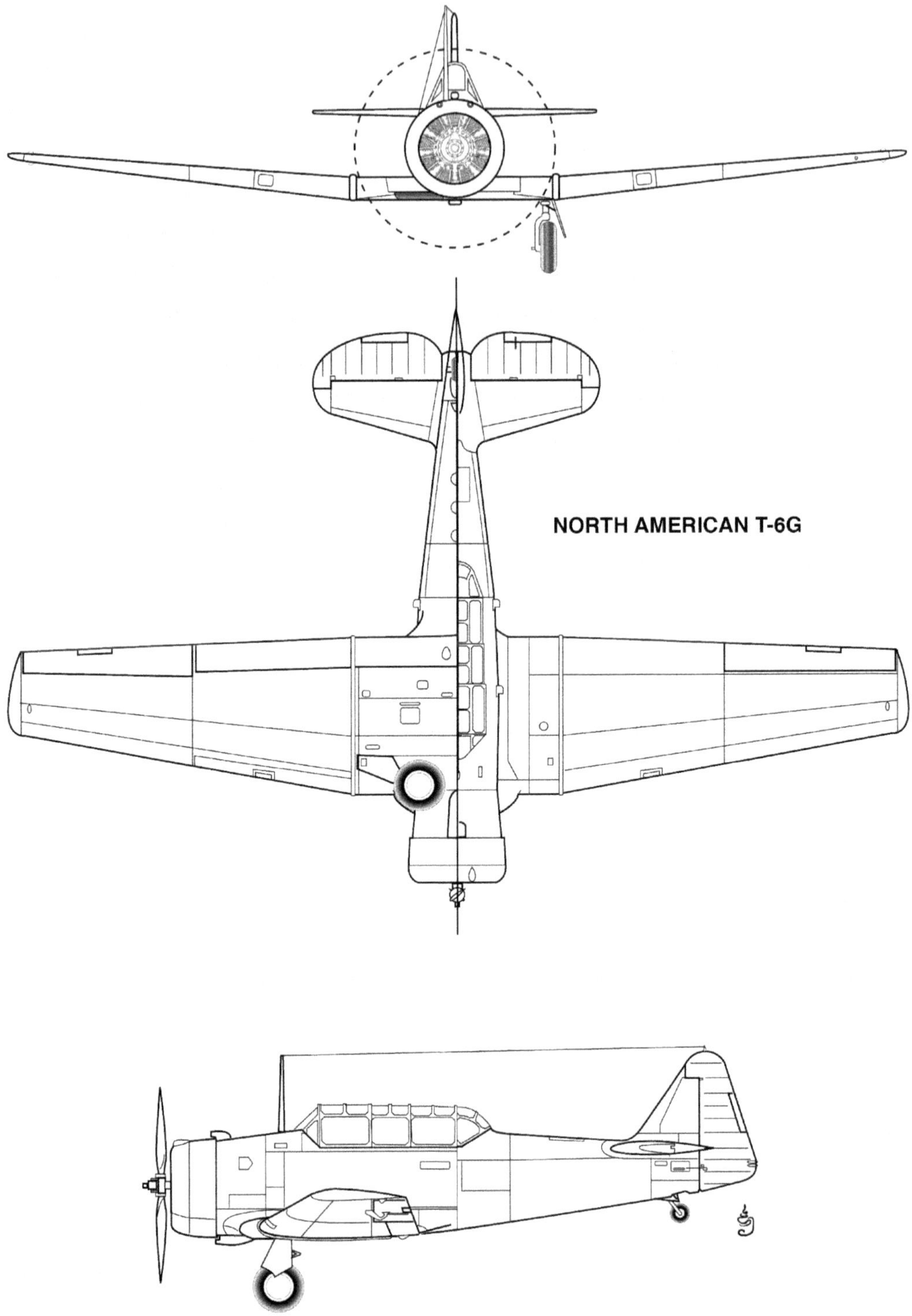

NORTH AMERICAN T-6G

The AT-6D was the most numerous AT-6 variant, and 4,388 AT-6Ds entered service with Army Air Forces Flying Training Command in 1942 and 1943. In 1949 and 1950, many were remanufactured as the T-6G.

One of 2,068 AT-6D airframes that were remanufactured in 1949 and 1950 and returned to service as the T-6G. The very last T-6Gs were retired from Air Training Command in 1955.

modified to perform forward air control missions in Korea were re-designated LT-6G. Another 285 T-6 airframes license-built in Canada by Canadian Car & Foundry during the early 1950s received the designation T-6J, however, none served in any of the U.S. military branches.

Production BC-1s were initially introduced to operational service during 1938 with the Advanced Flying School at Kelly Field, Texas, and as deliveries of BC-1A/AT-6s and AT-6As proceeded from 1939 to 1941, they were distributed among advanced schools in three newly established Air Corps Training Centers located in Alabama, Texas, and California. During World War II—as only type of single-engine advanced trainer employed by AAFTC—thousands of AT-6 variants were allocated between three Flying Training Wings (Advanced, Single-Engine) based at 12 different flying schools in Alabama, Texas, and Arizona. Then in 1945 and 1946, as a result of the massive force reductions that followed the end of World War II, all but approximately 1,000 AT-6Ds and Fs were withdrawn from AAFTC service (became Air Training Command [ATC] in mid–1946), with the remainder being allocated to reserve and ANG units as proficiency trainers, placed in storage, or sold surplus. When ATC became part of the independent USAF in September 1947, the primary, basic, and advanced phases of flight training were all combined into one "basic" course, using AT/T-6Ds and Fs as the principal aircraft type for the entire training syllabus. The expansion of ATC operations in 1949 and 1950 in response to the Cold War and the Korean War coincided with introduction of remanufactured T-6Gs to the basic course; however, the process of phasing them out of active training commenced in late 1950 as the first North American T-28As began joining ATC service. All T-6Gs had been withdrawn from ATC by the end of 1955, and the final examples were retired from reserve and ANG service during the late 1950s.

Naval Variants: Concurrent with development of the BC-1A for the Army, BuAer awarded North American a contract in 1939 for 16 essentially identical aircraft to be delivered under the designation SNJ-1, the first naval aircraft to be categorized as a scout-trainer (SN). In order to train the Navy's growing pool of reserve pilots, all SNJ-1s had originally been earmarked for reserve bases, however, as deliveries proceeded in mid–1939, they were assigned elsewhere, eventually ending up at NAS Pensacola. A second Navy order came in late 1939 for 61 SNJ-2s to be

SNJ-1 BuNo 1552 in 1939, the first of 5,371 Navy and Marine SNJs. All 16 of the SNJ-1s went to NAS Pensacola as advanced trainers and the 16 SNJ-2s delivered in 1940 went to the reserves as instrument trainers.

Two SNJ-5s out of NAS Saufley Field, Florida, seen in 1953 during carrier qualifications aboard the USS *Monterey* (CVL-26). New T-28Cs started replacing SNJs in 1956.

powered with 550-hp R-1340-56 direct drive engines, the initial batch being delivered in early 1940 to 13 different reserve bases, primarily for instrument training, and as deliveries continued, 19 were assigned to NAS Pensacola, 15 to NAS Anacostia, and a number of others to flag officers as command aircraft. In 1940, under the joint Army-Navy standardization plan mentioned above, BuAer contracted for 420 SNJ-3s (Army AT-6A) that were delivered in 1941 and 1942 and primarily assigned to Navy flight training centers at Corpus Christi, Jacksonville, Pensacola, and Miami.

During the course of World War II, the Navy accepted delivery of a further 125 SNJ-3s, 2,400 SNJ-4s, 1,568 SNJ-5s and 931 SNJ-6s. SNJs retrofitted with arresting gear for carrier qualification training carried a "C" suffix (i.e., SNJ-3C, -4C, and -5C). Between 1941 and 1945, BuAer allocated over 100 SNJs to the Marines for instrument and proficiency training in various units, and from 1943 to 1945, transferred 15 SNJ-5 and -6s to the Coast Guard for cross-country navigation training. Following World War II, the Naval Aviation Training Establishment continued to use SNJ-5s and -6s as its chief type of aircraft for primary, basic, and advanced pilot training, including carrier qualification, and many underwent a modernization program similar to the T-6G from which they emerged as the SNJ-7 or the SNJ-7B armed version. After being phased out of pilot training in the mid–1950s in favor of new Beech T-34Bs and North American T-28Bs and Cs, many SNJs went to Navy and Marine reserve units, where they functioned as instrument and proficiency trainers until being withdrawn during the 1960s, the very last examples reportedly being stricken from the naval inventory as late as 1968.

Vultee BC-3–1939

Technical Specifications (BC-3)

Type: Two-place basic combat (advanced) trainer.
Manufacturer: Vultee Aircraft, Inc., Downey, California.

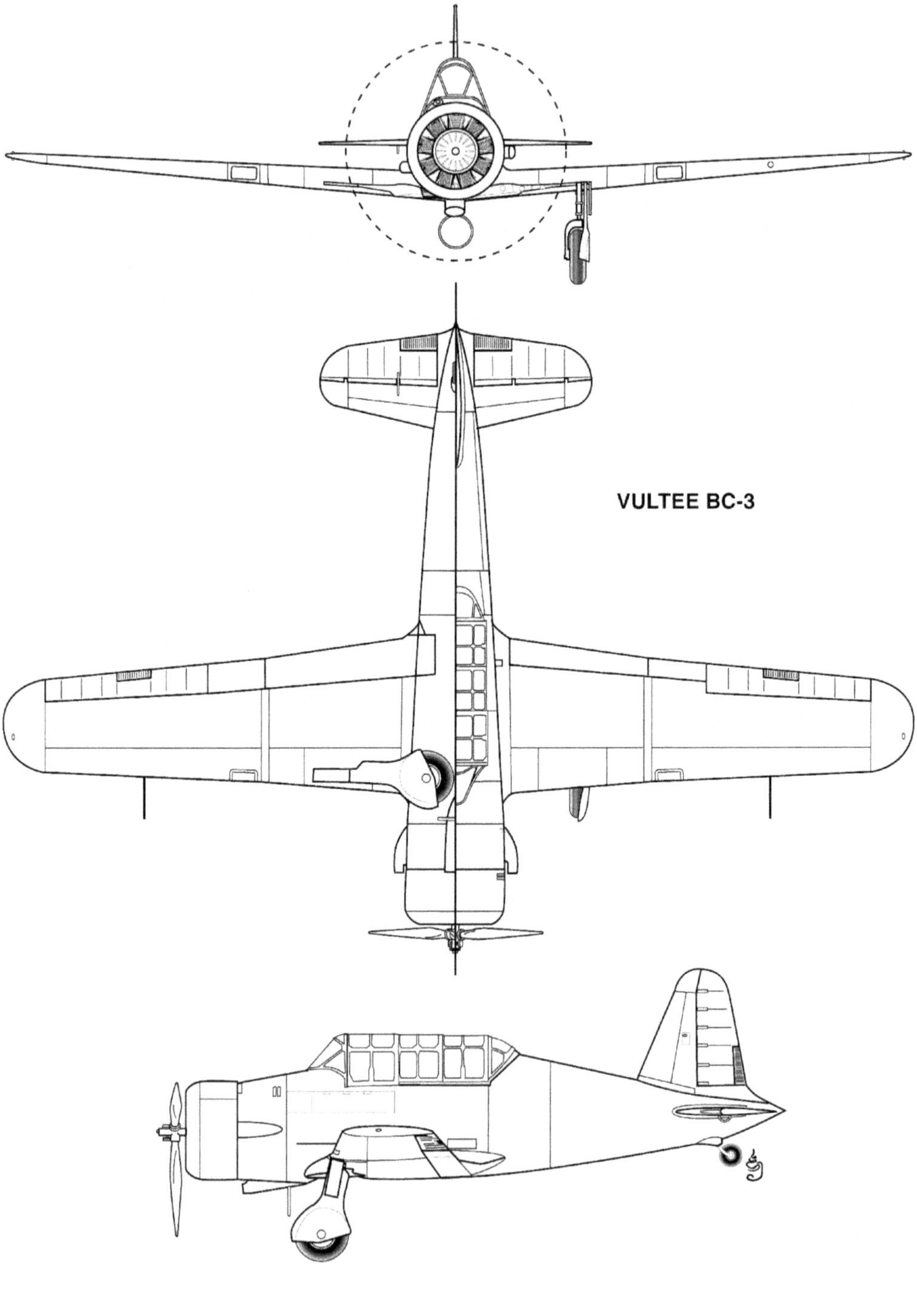

The Vultee BC-3, pictured at Wright Field, Ohio, in 1939, lost the competition against North American's improved BC-1A (redesignated AT-6 in 1940), but as a fixed gear trainer, reappeared the next year as the BT-13.

Total U.S. military versions: 1 (USAAC).
Unit cost: Unknown.
Powerplant: One 600-hp Pratt & Whitney R-1340-S3H1G *Wasp* 9-cylinder air-cooled radial engine driving a three-bladed, variable-pitch metal propeller.
Armament: None.
Performance: Max. speed 223 mph, cruise 180 mph; ceiling (not reported); range 725 mi.
Weights: 4,780 lbs. empty, 5,700 lbs. loaded.
Dimensions: Span 42 ft., length 28 ft. 10 in., wing area 239 sq. ft.

Soon after making its maiden flight in March 1939, the Vultee model 51, also known as the BC-51, was delivered to Wright Field as an unsolicited private venture to compete for the second round of Basic Combat (BC) trainer contracts, and Air Corps Material Division subsequently purchased it for testing under the designation BC-3. A progressive development of Gerard Vultee's V-1 of 1933, the design had been conceived by Richard Palmer in 1938 (chief designer following Vultee's death in 1937) as a multi-purpose airframe that could be produced in fighter (model 48) or trainer (model 51) variations. In terms of overall structural and aerodynamic design, the BC-3 represented the aeronautical state-of-the-art: an all-aluminum, semi-monocoque fuselage, a four-panel wing with NACA airfoil sections and slotted flaps, wide-track, inward-retracting main landing gear, and a greenhouse-type canopy enclosure.

Although the BC-3 met or exceeded the BC requirements, it had come too late to influence Air Corps Material Division's decision to place North American's BC-1A (AT-6) in quantity production. Undeterred by the failure to gain a military contract, Palmer reworked the design into a fixed-gear basic trainer powered by a 450-hp Pratt & Whitney R-985–25 engine, and as the model 54-A, delivered a new prototype to Wright Field in early 1940. The model 54-A is fully reported below under a separate heading as the Vultee BT-13, -15, and -16 (SNV).

Waco PT-14–1939

Technical Specifications (YPT-14)

Type: Two-place primary trainer.
Manufacturer: Waco Aircraft Co., Lorain, Ohio.

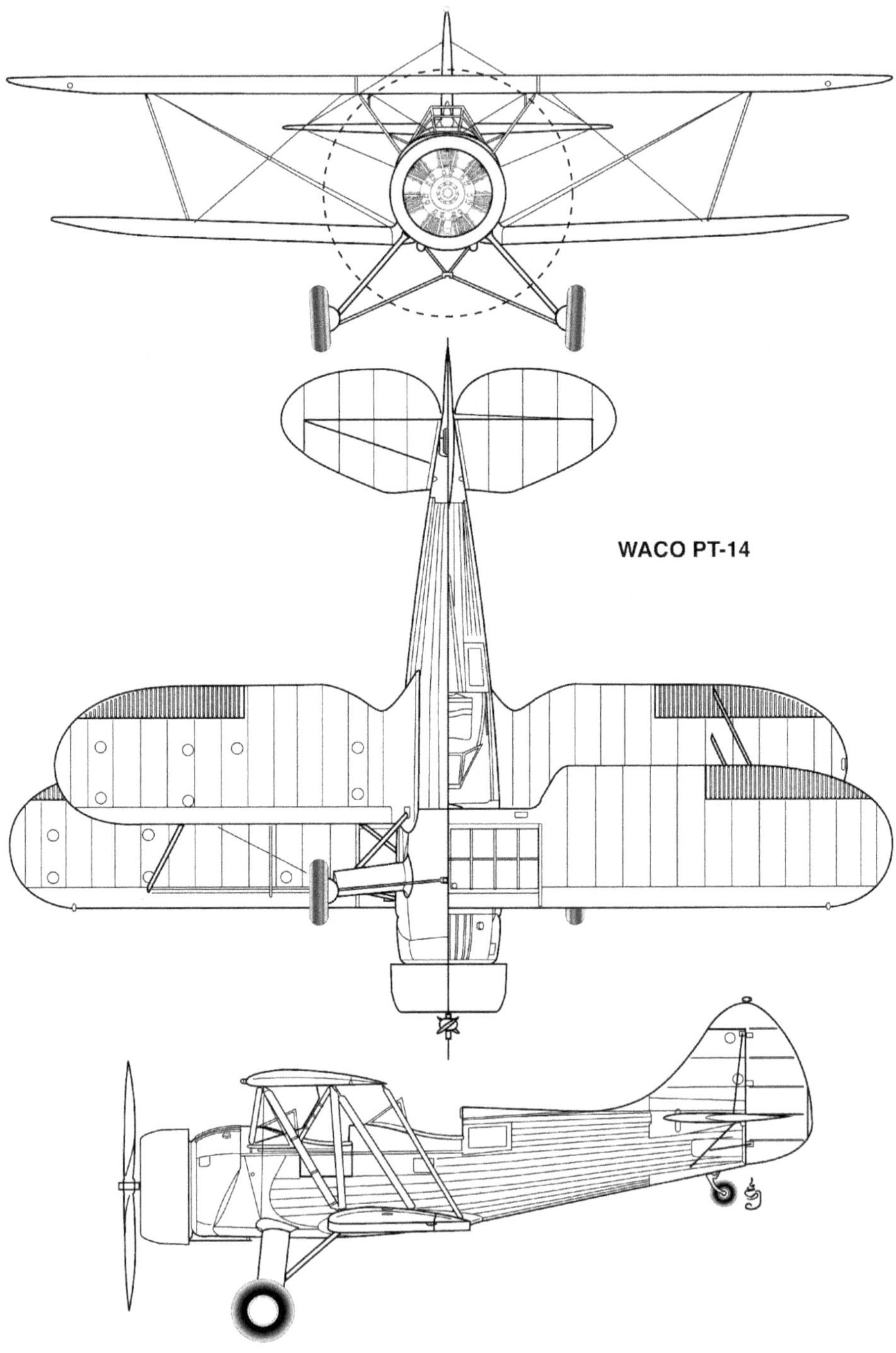

One of 13 service-test Waco Y1PT-14s delivered to the Air Corps for primary trainer evaluations in 1939. Though not ordered in quantity for the military services, more than 600 were procured for the Civilian Pilot Training Program.

Total U.S. military versions: 15 (USAAC).
Unit cost: $9,500.
Powerplant: One 220-hp Continental R-670–3 7-cylinder air-cooled radial engine driving a two-bladed, fixed-pitch metal propeller.
Armament: None.
Performance: Max. speed 128 mph, cruise 114 mph; ceiling 14,800 ft.; range 400 mi.
Weights: 1,870 lbs. empty, 2,650 lbs. loaded.
Dimensions: Span 30 ft., length 23 ft. 6 in., wing area 244 sq. ft.

The Waco F biplane series of sport aircraft and trainers first appeared in 1930 and were thereafter produced for the civil market in a variety of configurations with different powerplants. Construction was a conventional fabric-covered airframe, consisting of a welded, steel tube fuselage faired with wooden formers and stringers, wire-braced, built-up wooden wings with positive stagger, and braced main landing gear. Continuing refinements to the design led to introduction of models UPF (220-hp Continental R-680 engine), YPF (225-hp Jacobs R-755 engine), and ZPF (285-hp Jacobs R-830 engine) between 1935 and 1937. At a time when the civil market for open-cockpit biplanes had all but disappeared, Air Corps Material Division procured one UPF-7 (7 indicates a 1937 model) in 1939 to be evaluated as a primary trainer under the designation XPT-14, and in 1940, upon conclusion of trials, awarded Waco a pre-production contract to deliver 13 additional aircraft as the YPT-14. Although further military production of the type was ultimately declined in favor of other aircraft (e.g., PT-17, PT-19, and PT-22), Waco did in fact receive other government contracts which resulted in production of approximately 600 UPF-7s for the Civilian Pilot Training Program (see page 10 above) between 1940 and 1942. One civil UPF-7 impressed by the government in 1942 was carried on the Army Air Forces inventory as the PT-14A.

Fleetwings BT-12–1939

Technical Specifications (BT-12)

Type: Two-place basic trainer.
Manufacturer: Fleetwings, Inc. (Kaiser-Fleetwings, Inc. in 1941), Bristol, Pennsylvania.

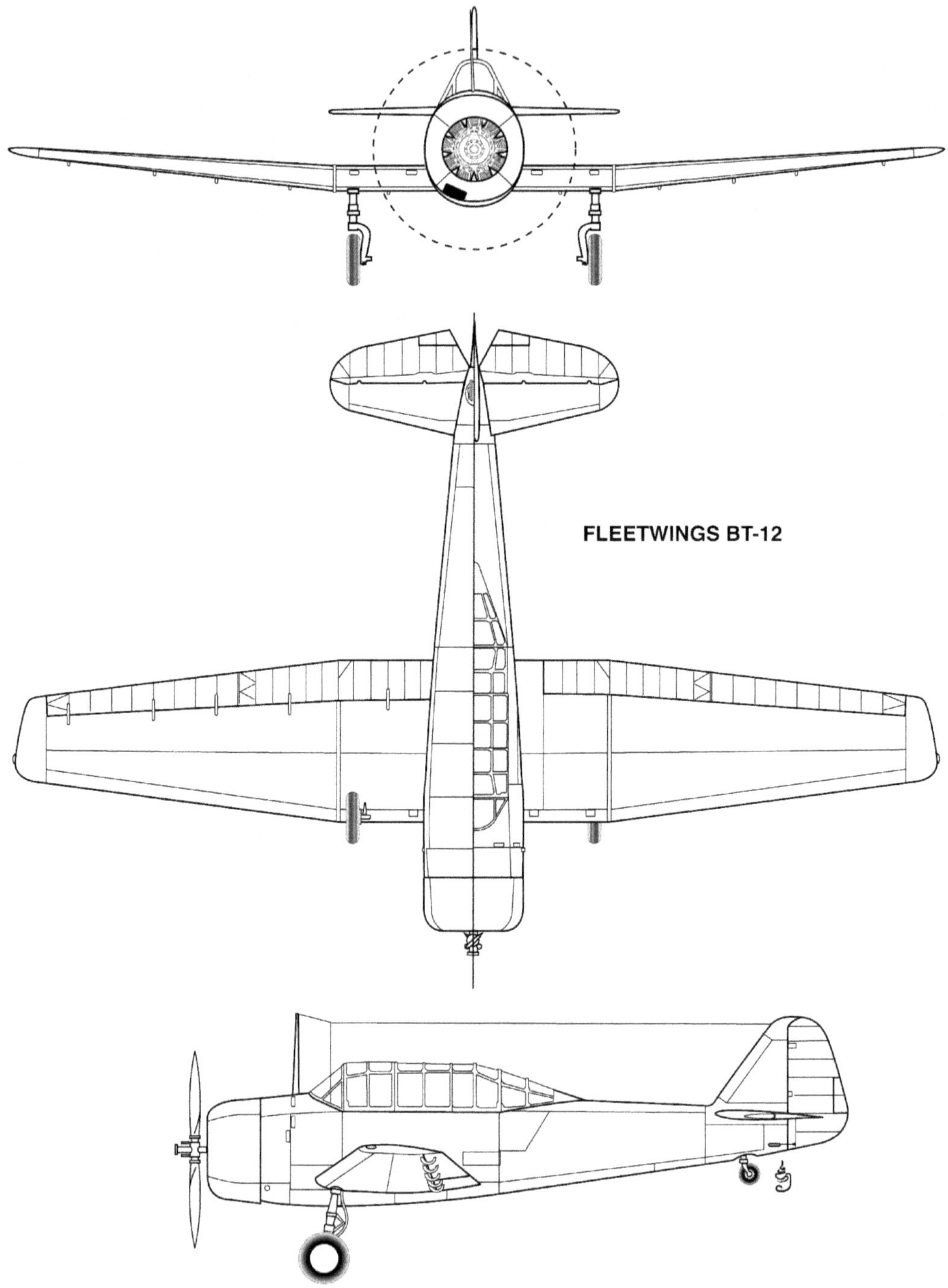

One of the 200 production Fleetwings BT-12s ordered in 1942. Only 24 had been delivered when the balance of the contract was cancelled in 1943. None are though to have served operationally.

Total U.S. military versions: 25 (USAAC/AAF).
Unit cost: Unknown.
Powerplant: One 450-hp Pratt & Whitney R-985-AN1 9-cylinder air-cooled radial engine driving a two-bladed, variable-pitch metal propeller.
Armament: None.
Performance: Max. speed 195 mph, cruise 175 mph; ceiling 23,800 ft.; range 550 mi.
Weights: 3,173 lbs. empty, 4,410 lbs. loaded.
Dimensions: Span 40 ft., length 29 ft. 2 in., wing area 240 sq. ft.

Fleetwings had been founded in 1930 for the express purpose of developing aircraft whose airframes were fabricated primarily from stainless steel. The basic technique involved a framework made up of spot-welded stainless steel components (formers, stringers, spars, ribs, etc.) to which a very thin stainless skin was riveted. In 1939, after announcing a competition for a new basic trainer production contract, Air Corps Material Division contracted with Fleetwings to deliver one prototype of its model 23 trainer for testing under the designation XBT-12. Comparable to the BT-9 in aerodynamic design, the XBT-12 appeared later the same year with an airframe comprised entirely of stainless steel with the exception of molded plywood panels on the cockpit sides, fabric-covering on the fuselage sides aft of the wing and on the flaps and control surfaces. While the XBT-12 demonstrated acceptable flying characteristics and improved performance over the BT-9, no production was ordered when trials concluded in 1940.

In 1942, amid government concerns that the tens of thousands of military aircraft needed for the war effort would potentially generate aluminum shortages, Fleetwings received an Army Air Forces contract to deliver 200 BT-12 production models. Given the factory name "Sophomore," production BT-12s differed from the prototype only in having AN1 engines, plywood panels on the fuselage in place of fabric, and minor detail changes. Only 24 BT-12s had been accepted, however, when the balance of the contract was cancelled in 1943. Some historians speculate that production may have been halted prematurely due to Fleetwings' involvement in important wartime stainless steel subcontract work. No BT-12s are known to have actually served with AAFTC.

St. Louis PT-15–1939

Technical Specifications (PT-15)

Type: Two-place primary trainer.
Manufacturer: St. Louis Aircraft Co., St. Louis, Missouri.

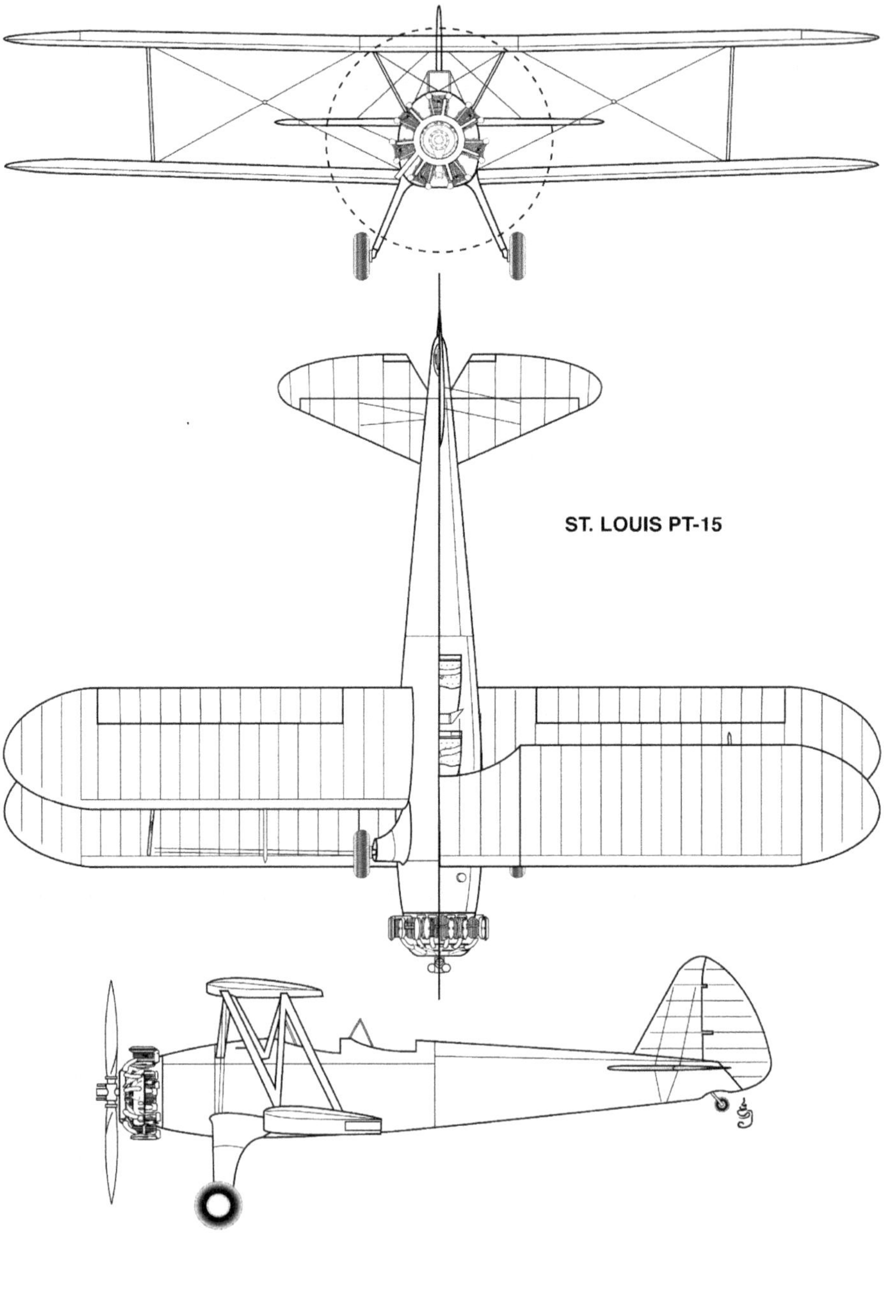

One of 13 service-test St. Louis PT-15s assigned to the Civilian Pilot Training Program at Parks Air College near St. Louis, Missouri, in 1940. Instead of PT-15s, the company license-produced PT-19s and -23s.

Total U.S. military versions: 14 (USAAC/AAF).
Unit cost: Unknown.
Powerplant: One 225-hp Wright R-760–1 7-cylinder air-cooled radial engine driving a two-bladed, fixed-pitch metal propeller.
Armament: None.
Performance: Max. speed 124 mph, cruise 105 mph; ceiling 14,500 ft.; range 355 mi.
Weights: 2,058 lbs. empty, 2,770 lbs. loaded.
Dimensions: Span 33 ft. 10 in., length 25 ft., wing area 280 sq. ft.

The origins of the PT-15 can be traced to 1935, when the St. Louis Aircraft Co. delivered its company model PT-35 to Wright Field for evaluations against the Consolidated PT-11D and the Boeing-Stearman XP-943/PT-13. Powered by a Wright R-760 engine and very similar in appearance and construction to the PT-13, Air Corps officials evinced no interest in the design and returned it to the manufacturer. Four years later, after approval of the Air Corps Expansion Plan in April 1939, St. Louis Aircraft received a contract to deliver a new trainer prototype under the designation XPT-15. Known internally as the company model P-1W, the XP-15 differed structurally from the earlier PT-35 in having an all-aluminum, semi-monocoque fuselage, aluminum wing spars, and strengthened landing gear. Upon completion of trials in 1940, St. Louis received a second order for 13 service-test aircraft as the YPT-15. But as events turned out, further production of the PT-15 was overtaken by government plans to procure other types of primary trainers by the thousands (e.g., Boeing-Stearman PT-17s, Fairchild PT-19s, and Ryan PT-22s), with St. Louis Aircraft ultimately becoming a license-contractor to build Fairchild PT-19s and -23s. All 13 of the YPT-15s (re-designated as the PT-15) were assigned to Parks Air College in southern Illinois near St. Louis, where they were employed in the school's Civilian Pilot Training Program.

Ryan PT-16, PT-20, PT-21 and PT-22 (NR)—1939

Technical Specifications (**YPT-16** [**PT-22**])

Type: Two-place primary trainer.
Manufacturer: Ryan Aeronautical Co., San Diego, California.
Total U.S. military versions: 1,278 (1,178 USAAC/AAF; 100 USN).

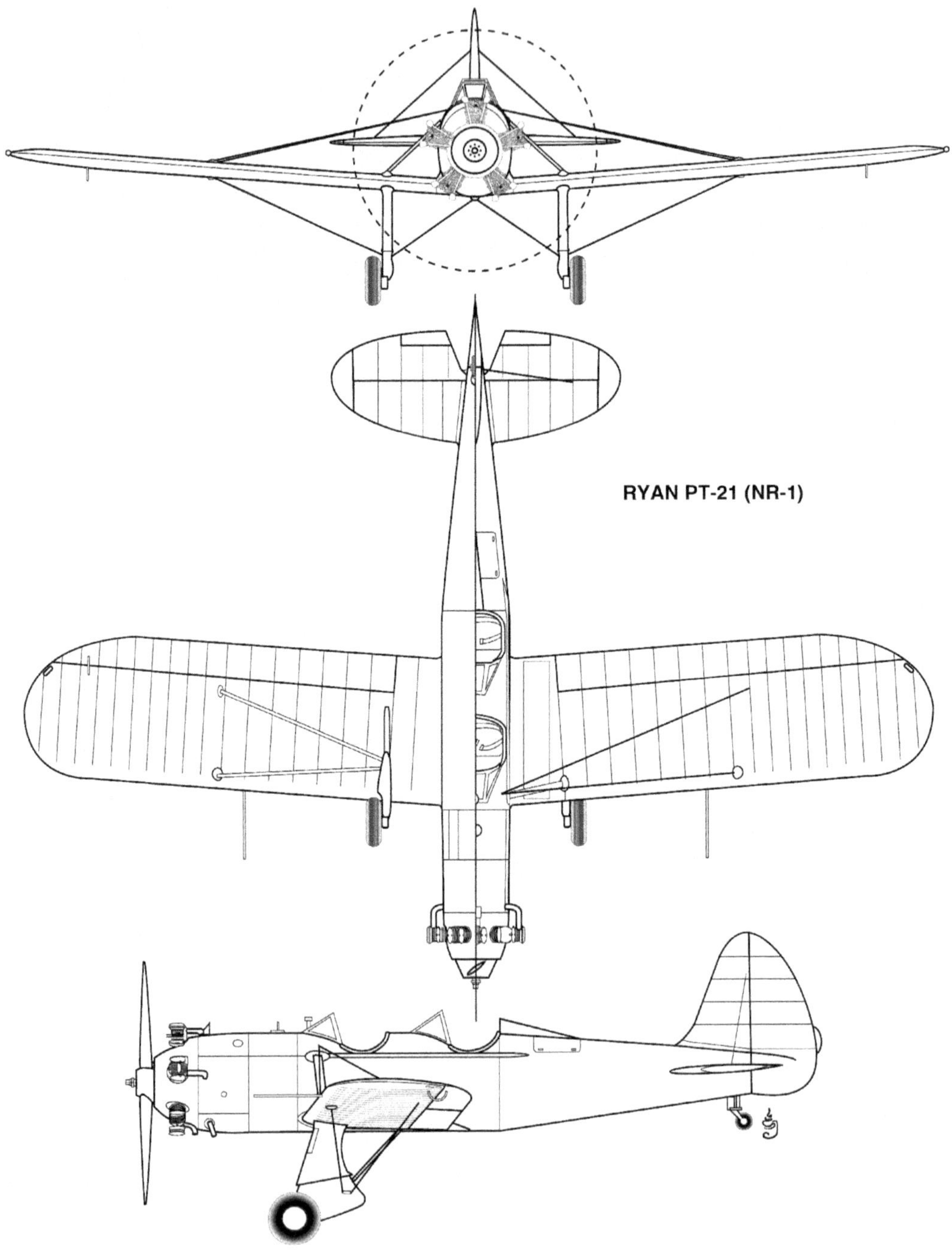

This PT-16A at the Ryan plant in 1941 was one of 14 YPT-16s returned to the factory to be modified for installation of Kinner R-440 engines.

Unit cost: $8,500.
Powerplant: One 125-hp Menasco L-365–1 4-cylinder air-cooled inline engine [160-hp Kinner R-540–1 5-cylinder air-cooled radial engine] driving a two-bladed, fixed-pitch wooden propeller.
Armament: None.
Performance: Max. speed 126 mph [125 mph], cruise 108 mph [100 mph]; ceiling 17,200 ft. [15,400 ft.]; range 366 mi. [231 mi.].
Weights: 1,081 lbs. [1,308 lbs.] empty, 1,600 lbs. [1,860 lbs.] loaded.
Dimensions: Span 30 ft. [30 ft. 1 in], length 21 ft. 5 in. [22 ft. 5 in.], wing area 124 [134] sq. ft.

The Ryan ST first appeared on the civil market in 1934 as a two-seat, open-cockpit monoplane that could be used for either sport or training. It combined very sleek aerodynamic lines with a hybrid airframe consisting of an aluminum-clad fuselage having a tubular steel framework forward and a semi-monocoque all-aluminum structure aft of the wing, an integral wing/fuselage center-section with tubular steel spars, and wire-braced, fabric-covered wing panels with aluminum ribs and wooden spars. A knee-action-type main landing gear was streamlined with fairings and wheel pants and tied into the center-section with struts between the wings and fuselage. The ST-A emerged in 1935 with a 125-hp Menasco C-4 engine and the ST-M in 1937 as a military export version with larger cockpit openings to accommodate parachutes.

In 1939 Air Corps Material Division procured an off-the-shelf ST-M for evaluations at Wright Field under the designation XPT-16, followed by a contract for 15 service-test examples, identical except for built-in starter systems, delivered in early 1940 as the YPT-16. Soon afterward, Material Division procured 30 more ST-Ms from Ryan, essentially YPT-16s with minor detail refinements, which were delivered in 1940 under the designation PT-20. As testing continued, Material Division asked Ryan to undertake a major redesign of the ST-M around a Kinner radial engine and at the same time sent 14 YPT-16s and 27 PT-20s back for installation of 125-hp

Ryan delivered 20 of these Menasco-powered PT-20s to the Air Corps in 1940. They became PT-20As in 1941 after being returned from the factory with Kinner R-440 engines.

Well-used PT-22 seen sans spinner during engine tests during World War II. PT-22s served as primary trainers in Army Air Forces Flying Training Command throughout the wartime period.

Kinner R-440–1 engines which, when they returned to service in 1941, became the PT-16A and PT-20A, respectively. The ST-3, Ryan's redesign effort, emerged as the PT-21 in 1941 with a slightly longer and wider fuselage, wider gear track, simplified wheel fairings, reshaped rudder, swept-back wings, and a 132-hp Kinner R-440–3 engine. All 100 of the PT-21s subsequently produced by Ryan were reassigned to the Navy.

The Army production version, designated PT-22 and carried as factory model ST-3KR, differed only in having a 160-hp Kinner R-540–1 engine and no wheel fairings. In 1942, after Ryan had delivered 1,023 PT-22s, the Army commandeered an additional 25 ST-3KRs built for the Netherlands and placed them in service as the PT-22A. Though not as numerous as other types of primary trainers, PT-22s served in AAFTC throughout World War II, and many survived to be sold surplus when the war ended.

Naval Variants: As mention above, the PT-21s originally ordered under Army serial numbers were transferred to the Navy in 1941 and thereafter received bureau numbers under designation NR-1. As delivered, all NR-1s were assigned to NAS Jacksonville, Florida, as primary trainers.

Fairchild PT-19, PT-23 and PT-26–1939

Technical Specifications (PT-19A)

Type: Two-place primary trainer.
Manufacturer: Fairchild Engine & Airplane Corp., Hagerstown, Maryland; Aeronca Aircraft Corp., Middletown, Ohio; Howard Aircraft Corp., Chicago, Illinois; St. Louis Aircraft Co., St. Louis, Missouri; and Fleet Aircraft of Canada, Ltd., Fort Erie, Ontario, Canada.
Total U.S. military versions: 6,684 (USAAC/AAF).
Unit cost: $8,000-$10,500.
Powerplant: One 200-hp Ranger L-440–3 6-cylinder air-cooled inline engine driving a two-bladed, fixed-pitch wooden propeller.
Armament: None.
Performance: Max. speed 124 mph, cruise 106 mph; ceiling 15,300 ft.; range 480 mi.
Weights: 2,053 lbs. empty, 2,545 lbs. loaded.
Dimensions: Span 35 ft. 11 in., length 27 ft. 8 in., wing area 200 sq. ft.

The Fairchild PT-19 and its derivatives (PT-23 and PT-26) were noteworthy in having been the most numerous type of primary trainer used by the Army Air Corps/ Army Air Forces during the World War II era (1941–1945). Its design stemmed from a 1938 Air Corps Material Division requirement calling for a primary trainer to be built largely from non-strategic materials such as wood and fabric, although the specification allowed a tubular steel fuselage frame. As a company-funded venture, Fairchild responded with the model M-62, a monoplane powered by a 175-hp Ranger L-440–1 inline engine featuring a fabric-covered, steel tube fuselage and fully-cantilevered wooden wings and tail surfaces skinned in a plywood veneer for additional strength. After the M-62 made its first flight from the factory on May 15, 1939, it was delivered to Wright Field for testing under the designation PT-19. Although the PT-19 possessed more than twice the wing-loading of contemporaneous biplane trainers, its mechanically-operated split flaps reduced stall speed to an acceptable 53 mph. Moreover, over-the nose and upward visibility were superb, and the wide track of the main gear legs made it easy to land in crosswind conditions.

In September 1939 Material Division awarded Fairchild a contract to manufacture 270 PT-19s, which differed from the original prototype in having wooden propellers, no rear headrest, and minor detail changes. As American war preparations escalated in 1940 and 1941, Fairchild received a succession of new contracts to deliver 3,181 aircraft with a 200-hp L-440–3 engine upgrade as the PT-19A; and in 1942, when demand for the type surpassed Fairchild's capacity, an additional 477 PT-19As were completed by Aeronca and 44 by St. Louis Aircraft under government license contracts. In 1942 and 1943 Fairchild and Aeronca also produced, respectively, 774 and 143 PT-19Bs equipped with cockpit hoods for instrument training. A shortage in the supply of Ranger L-440 engines during 1942 caused one PT-19A to be modified for installation of a 220-hp Continental R-670 radial engine and tested as the XPT-23. Other than being a foot

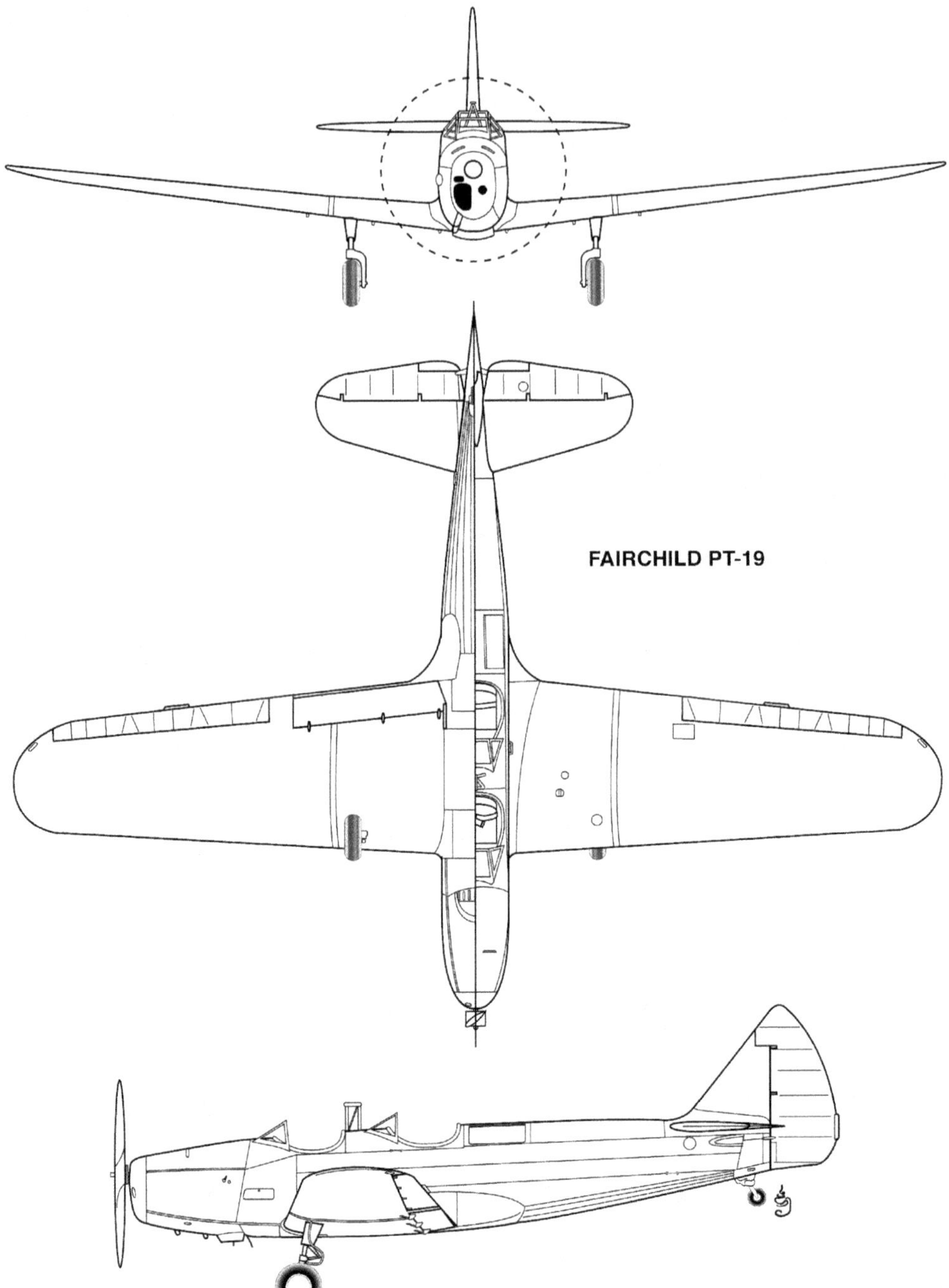

FAIRCHILD PT-19

shorter and marginally slower than a PT-19A, the XPT-23 exhibited very similar handling qualities.

Except for two more examples completed by Fairchild as templates, all PT-23 production was directed to license-contractors: 375 completed by Aeronca, 199 by Howard, 200 by St. Louis, and 93 by Fleet of Canada. In addition, Howard and St. Louis produced, respectively, 150 and 106 variants with cockpit hoods as the PT-23A. Fitting a PT-19A with a fully-enclosed canopy

All-silver PT-19B seen in 1943 markings. The 917 PT-19Bs produced by Fairchild and Aeronca in 1942 and 1943 differed from the A in having cockpit hood fittings for instrument training.

A shortage of Ranger engines in 1942 led to production of 867 PT-23s powered by Continental R-670 engines. All PT-23s were built by four license-contractors.

for cold weather operations resulted in the PT-26, the final production variant. Fairchild completed 670 PT-26s in 1943 and 1944, at which point American production ceased. Serial number records indicate that most PT-26s were Lend-Leased to the RCAF as the *Cornell I* for use in the Commonwealth Training Program. Fleet of Canada also produced 1,057 examples with L-440–7 engines as the PT-26A and PT-26B (minor detail changes) that went directly into RCAF service.

Between 1941 and 1945 PT-19s, -19As, -19Bs, -23s and -23As were distributed among the 64 Contract Flying Schools (CFS) who employed civilian flight instructors to conduct Level 1 Primary Pilot Training within the Army Air Forces Training Command (AAFTC) structure. Because AAFTC commonly used PT-19s for preflight indoctrination—introductory flights intended to determine whether or not pilot candidates possessed any aptitude for flying—the type earned the knick-name "Cradle of Heroes" for having been the first type of plane flown by most future USAAF combat pilots. As wartime training progressed, USAAF overhaul depots began reporting rapid deterioration of the wooden wing center-sections of PT-19s and -23s based in the hot and humid regions of the south and southwest, which typically entailed complete replacement of the entire component after only three or four months of service. The CFS system was disbanded as soon as the war ended in September 1945, with all remaining PT-19s and -23s being withdrawn and declared surplus.

Aircraft Research BT-11—1940 (cancelled)

Aircraft Research Corp. was apparently a trade name used by Vidal Research Corp. of Bristol, Virginia. During the late 1930s, Vidal developed a resin and wood composite material named "Weldwood," which could be pressure molded into airframe components. The BT-11 specification issued by Air Corps Material Division called for an aircraft powered by a 450-hp R-985 engine and similar in dimensions and weight to the Vultee BT-13, to be fabricated from the Weldwood composite material; however, the project was cancelled while the BT-11 was still in the mockup stage. For similar aircraft, see the Ryan YPT-25, below, and the Timm N2T in Part B, below.

Vultee BT-13, BT-15 and BT-16 (SNV) Valiant—1940

Technical Specifications (BT-13A)

Type: Two-place basic trainer.
Manufacturer: Vultee Aircraft, Inc. (later Consolidated-Vultee), Nashville, Tennessee.
Total U.S. military versions: 9,525 (7,525 AAC/AAF; 2,000 USN/USMC/USCG).
Unit cost: $23,000.
Powerplant: One 450-hp Pratt & Whitney R-985-AN1 9-cylinder air-cooled radial engine driving a two-bladed, variable-pitch metal propeller.
Armament: None.
Performance: Max. speed 180 mph, cruise 140 mph; ceiling 21,650 ft.; range 725 mi.
Weights: 3,375 lbs. empty, 4,496 lbs. loaded.
Dimensions: Span 42 ft., length 28 ft. 10 in., wing area 239 sq. ft.

Even though its career was relatively brief, the Vultee BT-13 and -15 "Valiant" deserves recognition as the preeminent U.S. military basic trainer of World War II—probably flown by more new service pilots (200,000+) between 1941 and 1945 than any other single type of training aircraft. Having originated as an advanced trainer in 1939 (see Vultee BC-3, above), when the

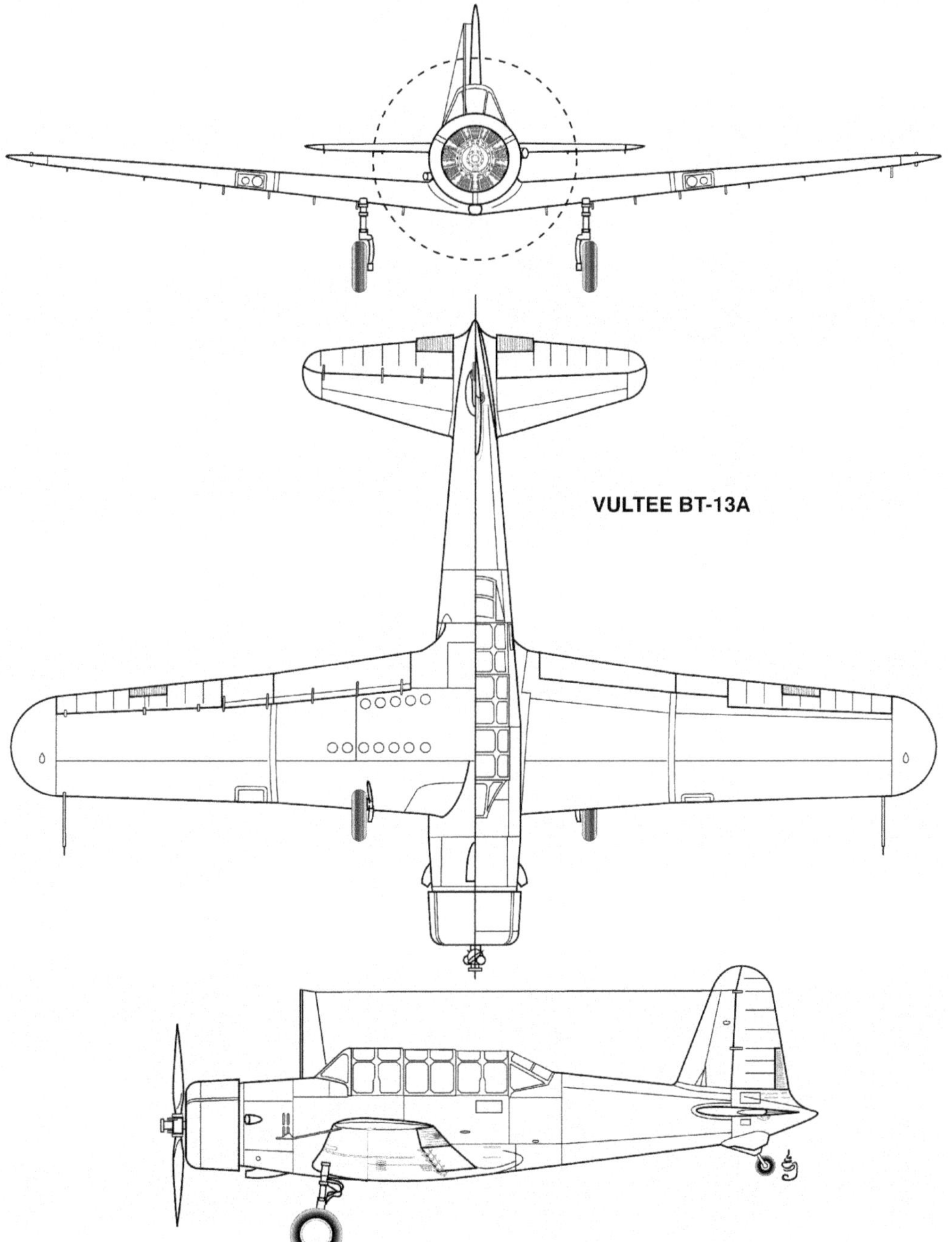

design reappeared in 1940 as the de-powered, fixed-gear Vultee model 54-A, it offered the Army Air Corps a viable alternative to the BT-9/-14 which could be mass-produced in the huge quantities needed for the planned growth of the flight training program. In August 1940, following trials, the Air Corps gave Vultee an initial contract to deliver 300 model 54-As under the designation BT-13.

Later the same year, as U.S. war preparations burgeoned, Vultee received a series of contracts

Factory new Vultee BT-13A seen in wartime natural metal finish. Over 5,300 BT-13s and -13As served with Army Air Forces Flying Training Command during World War II.

A shortage of Pratt & Whitney R-985 engines led to production of 1,693 BT-13A airframes with Wright R-975 engines as the BT-15. The example depicted is in the early 1942 blue and yellow paint scheme.

extending into the wartime era that resulted in the manufacture of 6,407 BT-13As (1,350 allocated to the Navy) that differed in having AN1 engines and no wheel fairings. During the 1941–1942 timeframe, a temporary shortage of R-985 engines led to delivery of 1,693 BT-13A airframes fitted with 450-hp Wright R-975–11 engines under the designation BT-15. Under later war contracts, Vultee completed 1,125 BT-13Bs (650 allocated to the Navy) that came with 24-volt electrical systems and some, with AN3 engines. One BT-13A, modified to incorporate a new Vidal Weldwood–type fuselage and an all-wood wing, was tested in 1942 and 1943 as the XBT-16, but no production was ordered.

As new BT-13s and -13As rolled off Vultee's production line in 1941, they were distributed among the new Basic Flying Schools being organized within the three Air Corps Training Centers at Maxwell Field, Alabama, Randolph Field, Texas, and Moffett Field, California. During the basic phase, in addition to gaining proficiency in a heavier and more complex aircraft, new pilots learned the fundamentals of formation flying, night flying, instrument flying, and cross-country navigation. In service, BT-13s soon acquired the moniker "Vultee Vibrator" due to the type's pronounced pre-stall buffet and the airframe vibration generated by the propeller in high pitch. Following the massive expansion of Army Air Forces Training Command (AAFTC) from 1942 onwards, BT-13 variants served with the 27th, 32nd, 34th, and 35th Flying Training Wings (Basic) which by 1944 consisted of 26 Basic Flying Schools located in Alabama, Arkansas, Kansas, Oklahoma, Mississippi, Texas, and California. Except for the small minority who flew BT-9s and BT-14s, most all Army Air Forces student pilots trained on BT-13s before moving on to Advanced Single Engine or Twin Engine Schools. During 1945, as training requirements diminished, AAFTC initiated the process of withdrawing BT-13s from service, and by the time the war ended, had substituted North American AT-6s to train new pilots in both basic and advanced phases. Many BT-13s were sold surplus after the war, but most were scrapped after their engines and other useable components had been removed.

One of the earliest SNV-1s (AAF BT-13A) delivered to NAS Corpus Christi in mid–1941. Note the Army BT-14, Navy JRB-1 and SNJ-3 in the background.

Naval Variants: As part of a combined Army-Navy contract awarded to Vultee in late 1940 for 3,350 BT-13As, 1,350 were allocated to the Navy under the designation SNV-1. Production SNV-1s began arriving for training duties at NAS Corpus Christi during mid–1941, and as deliveries continued, they were distributed between the Navy's major training bases at NAS Miami, NAS Jacksonville, NAS Corpus Christi, and NAS Pensacola. The Navy also accepted 650 SNV-2s which differed in having AN3 engines and 24-volt electrical systems. During the course of the war, two SNV-1s were assigned to the Coast Guard and small number also served with various Marine squadrons. Although BuAer withdrew the type from the training curriculum in May 1945, the last active SNV-2 was not actually stricken from the naval inventory until April 1946.

Synopsis of USAAF Trainer Procurement, 1941 to 1945

Primary Trainers: The most significant feature of this era was not the development of the new types of primary trainers but the sheer numbers (16,000+) in which preexisting designs (e.g., Boeing-Stearman PT-13 and -17, Fairchild PT-19 and -23, and Ryan PT-20 and -22) were manufactured from 1941 to 1945. One factor that heavily influenced wartime aircraft procurement, trainers especially, were concerns that the tens of thousands of military aircraft on order would generate shortages of aluminum and other light alloys. For this reason, PT-17s and PT-19s/-23s, which both employed wood in their airframe structures, were built in far greater numbers than the PT-22. Although the all-wood Ryan YPT-25 underwent extensive testing in 1942 and 1943, the type was not placed in production when the anticipated metal shortages failed to materialize. Another type of primary trainer procured during the war were militarized variants of civil lightplanes (e.g., Taylorcraft L-2, Aeronca, L-3, and Piper L-4) which were used to train Army Ground Forces pilots for the liaison role. From 1941 to 1944, the USAAF also procured or impressed approximately 1,300 gliders to train glider pilots for airborne operations (see Appendix 1, below, for details). In July 1946, when Air Training Command replaced AAFTC, the USAAF ceased using any type of primary trainer in its pilot training curriculum.

Basic Trainers: The Vultee BT-13 and -15 became the only type of basic trainer to be mass-produced for the USAAF after 1941. Fears over aluminum shortages led to development and testing in 1942 and 1943 of the Vultee XBT-16, an all-wood copy of a BT-13A, and the Boeing-Stearman XBT-17, structurally similar to the PT-19, but neither project progressed beyond the prototype stage. As with primary trainers, basic trainers were withdrawn from the USAAF pilot training soon after World War II ended. At least two postwar trainer projects provisionally received a BT designation (i.e., North American XBT-28 and Douglas XBT-30), however, all USAF trainer designations changed to T after 1948.

Advanced Trainers: Like the BT-13, the North American AT-6 became to only type of single-engine advanced trainer to be mass-produced for the USAAF during World War II; however, changes to the flight training program implemented in 1941 resulted in the testing and mass-production of new types of advanced trainers specialized for multi-engine—1,232 Cessna AT-8s and -17s, 791 Curtiss AT-9s, 2,371 Beech AT-10s—and crew training for navigators, bombardiers, and gunners—3,273 Beech AT-7s and -11s, 300 Lockheed AT-18s, and 175 Fairchild AT-21s. The prospect of aluminum shortages was responsible for development of new advanced trainers fabricated mostly from wood such as the Cessna AT-8 and -17, Beech AT-10, Fairchild AT-13, -14, and -21, and Boeing AT-15. During the course of the war a number of combat aircraft received AT designations after being converted to a training role: a small number of Consolidated B-24s became AT-22 flight engineer trainers for the B-29 program; 558 Martin B-26s became AT-23 crew trainers, and 660 North American B-25s became AT-24 multi-engine trainers.

Transition and Operational Trainers: During World War II, many types of combat aircraft were used for transition and operational training, and some had a T prefix added to their designation (e.g., TB-17, TB-24, TB-25, TB-26, and TB-32), though most were left unchanged. Although four types of fighters, the Lockheed TP-38, Bell-TP-39, Curtiss TP-40, and Republic TP-47, were re-designated after being converted to a two-seat, dual control configuration, fighter training was principally carried out in unmodified single-seat types that had been withdrawn from frontline service, though some received an "R" prefix to denote non-combat status. According to AAFTC records, the following types of combat aircraft served in various Combat and Transition Schools from 1942 to 1945: Northrop A-17 and -33, Douglas A-20, Douglas A-24, Curtiss A-25, Douglas A-26, Lockheed A-28 and -29, Vultee A-35, North American A-36, Boeing B-17, Douglas B-18, Consolidated B-24, North American B-25, Martin B-26, Boeing B-29, Lockheed B-34, Douglas C-32, Beech C-45, Douglas C-50 (impressed DC-3), Lockheed C-56 and -60, Waco CG-4, Douglas O-46, North American O-47, Curtiss O-52, Grumman OA-9 (Goose), Consolidated OA-10 (Catalina), Grumman OA-14 (Widgeon), Seversky P-35, Curtiss P-36, Lockheed P-38, Bell P-39, Curtiss P-40, Republic P-47, Northrop P-61, Bell P-63, Douglas UC-67, and Noorduyn UC-64. During this timeframe, AAFTC also commenced the earliest USAAF helicopter training with the Sikorsky R-4, -5, and -6.

Beech AT/T-7, AT/T-11 and TC-45 (SNB) Kansan—1941

Technical Specifications (AT-11 [TC-45G])

Type: Four or five-place navigator and bombardier crew trainer.
Manufacturer: Beech Aircraft Co., Wichita, Kansas.
Total U.S. military versions: 3,553 (2,724 USAAC/AAF/USAF; 829 USN/USMC/USCG).
Unit cost: $65,000-$70,000.
Powerplants: Two 450-hp Pratt & Whitney R-985-AN1 [AN3] 9-cylinder air-cooled radial engines driving two-bladed, variable-pitch [fully-feathering constant-speed] Hamilton Standard metal propellers.
Performance: Max. speed 215 mph [205 mph], cruise 150 mph [175 mph]; ceiling 20,000 ft. [19,000 ft.]; range 745 mi. [1,100 mi.].
Weights: 6,250 lbs. [5,900 lbs.] empty, 9,300 lbs. [9,100 lbs.] loaded.
Dimensions: Span 47 ft. 8 in., length 34 ft. 3 in. [34 ft. 8 in.], wing area 349 sq. ft.

Known over its long history as the "Twin Beech," the maiden flight of the well-known model 18 took place on January 15, 1937. Powered at first with 350-hp Wright R-760 engines (model 18A), it emerged as a very clean monoplane constructed of aluminum alloy, with metal-framed, fabric-covered control surfaces, retractable landing gear, and a distinctive twin-fin empennage. Beech initially marketed the type as either a small airliner or business transport, available in five different powerplant options. Sales were relatively modest, however, until the introduction of the Pratt & Whitney R-985-powered model 18S in 1939, which marked the point when Beech first began receiving military orders, starting with an Army Air Corps contract in late 1939 for 25 B18S models, 11 as C-45 five-passenger light transports and 14 as F-2 photo-survey aircraft. Altogether, the Army and the Navy procured 2,133 Beech 18 transport variants during the wartime era as the C/UC-45 and JRB, respectively.

In 1941 Air Corps Material Division contracted with Beech to produce the model 18S in two advanced trainer versions: AT-7 navigator trainers equipped with clear astrodomes and cabin accommodations for three student-navigator stations; and AT-11 bombardier and gunnery trainers modified to include a clear nose glazing for a student-bombardier station, a bomb bay, and provisions for top turret and lower tunnel gun positions. Early AT-11s came with a Beech-made top

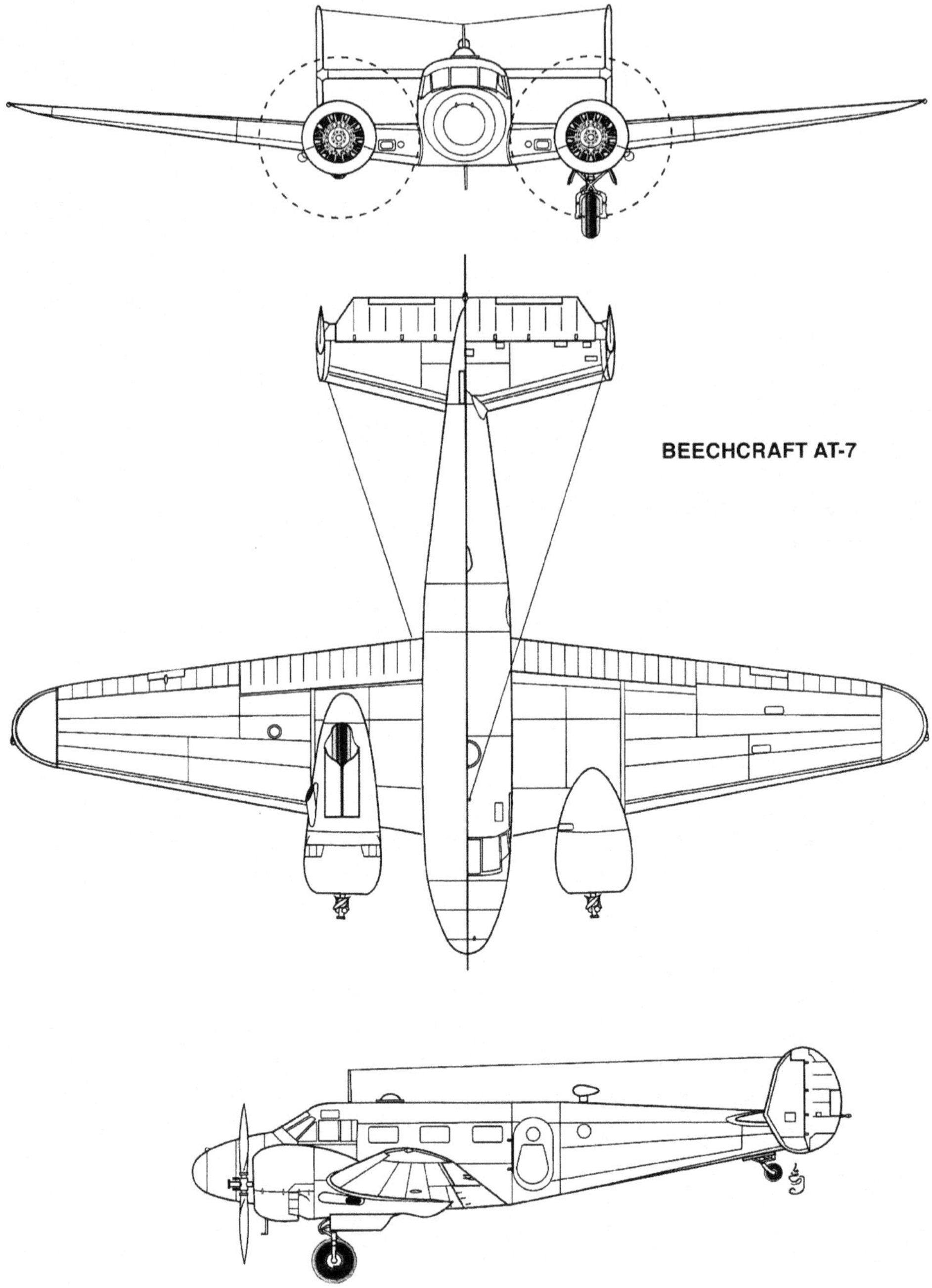

turret housing a single .30-calibre machine gun, whereas later versions possessed a twin .30-calibre turret manufactured by Crocker Wheeler. The first of 577 AT-7s and 1,582 AT-11s began arriving for service in late 1941, with deliveries of both types continuing over the next two years. During 1942 the USAAF also accepted eight navigator-trainer variants equipped with twin floats and a ventral stabilizing fin as the AT-7A. Another eight aircraft configured for cold weather operations (de-icing equipment and heaters) were delivered in 1942 as the AT-7B. The final trainer variants, 549 AT-7Cs delivered between 1943 and 1945, featured an upgrade to AN3

Early production Beech AT-7 with 1941 tail stripes, probably seen at the newly established Navigator Training School at Barkesdale Field, Louisiana.

engines and interior refinements. As training operations expanded in 1942, 36 AT-11s were converted to navigator trainers and returned to service as the AT-11A.

AT-7s were initially assigned to newly established Navigator Training Schools at Barkesdale Field and Selman Field in Louisiana. As deliveries proceeded, AT-7s and -7Cs went to one of seven navigator schools in Louisiana and Texas. The first AT-11s became operational with new Bombardier Training Schools at Lowry Field, Colorado and Barkesdale Field, Louisiana, but as AAFTC expanded and reorganized, most AT-11 served with nine different bombardier schools located in Texas, New Mexico, and Arizona where there was more territory available for bombing ranges. AT-11s also operated with nine Flexible Gunnery Schools at bases in Arizona, Florida, Nevada, and Texas. As wartime aircrew quotas peaked and then diminished, AAFTC began the process of withdrawing AT-7s and AT-11s from active training duties, and the last examples were retired from Air Training Command (ATC) during 1947. While some were reassigned to newly established reserve units as proficiency trainers, most were either placed in storage or sold surplus. In June 1948, all AT-7s and AT-11 remaining on the USAF inventory were re-designated T-7 and T-11, respectively.

During 1951 the USAF implemented a modernization program in which 372 C-45, T-7, and T-11 variants were returned to service as the C-45G under new serial numbers after being completely rebuilt and retrofitted with new wing center sections, new landing gear, constant-speed propellers, improved instruments and avionics, and autopilots. As part of the Korean War buildup, a further 96 specifically equipped for navigator training reentered service with ATC as the TC-45G. All TC-45Gs were phased out in 1955 and 1956 as more Convair T-29s arrived to replace them.

Naval Variants: Starting in 1942, the Navy took delivery of 509 AT-7s under the naval designation SNB-2 and 320 AT-11s as the SNB-1. In addition to navigator, bombardier, and aircrew training, the SNB-1s and -2s became the Navy's principal multi-engine trainer for the intermediate phase of pilot training. During the war, a number of SNB-1s and -2s were transferred to the Marine Corps to train pilots and aircrews who would be transitioning to units operating the

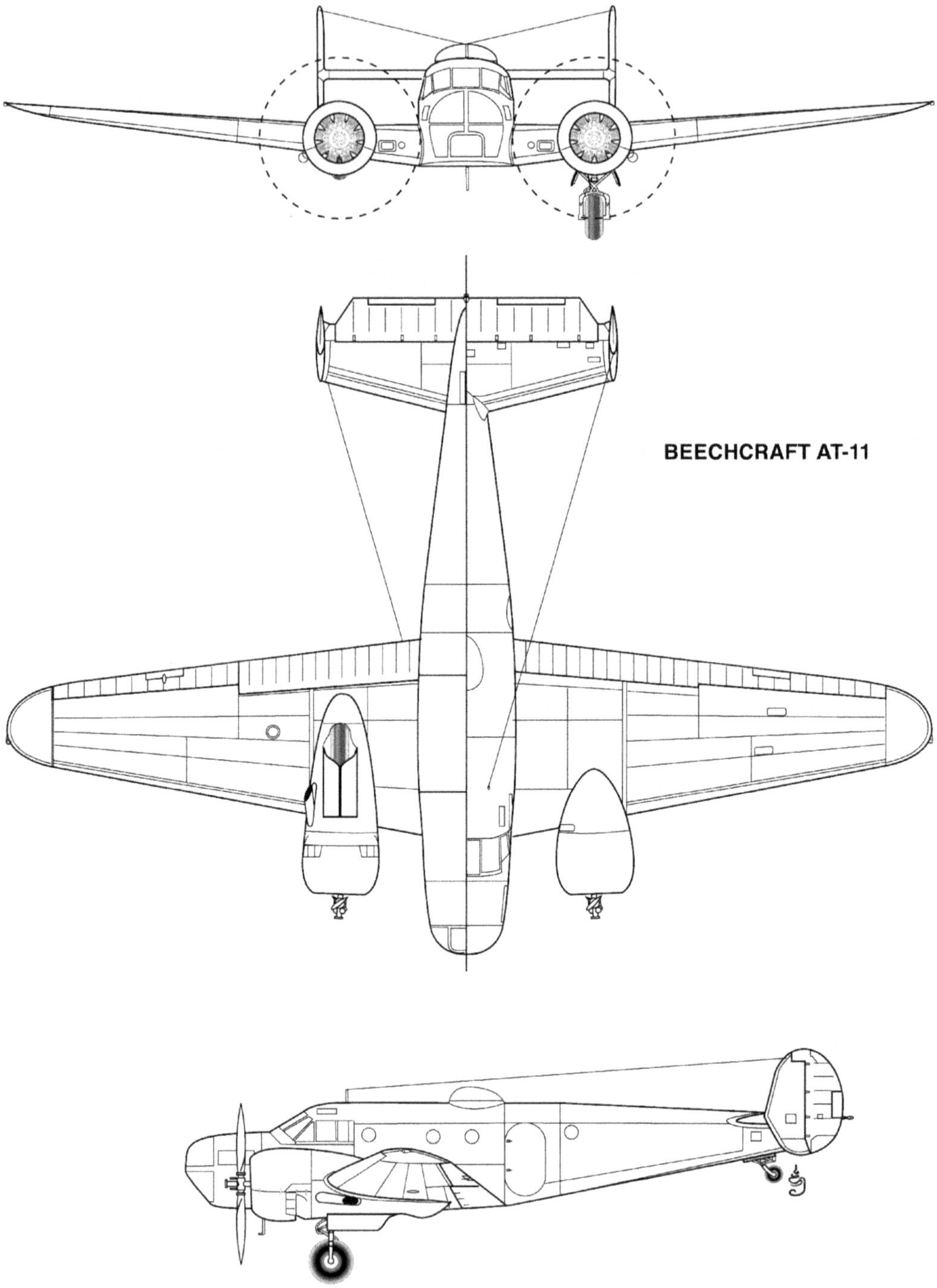

North American PBJ-1 (AAF B-25), and several ex–Navy SNB-1s were allocated to the Coast Guard for proficiency training.

Though in reduced numbers, SNB-1s and -2s continued to be utilized for naval pilot and aircrew training after World War II and others were reassigned to Navy and Marine reserve units. In 1947 BuAer contracted with Beech to rebuild and modernize 117 SNB-1s, which subsequently

Early AT-11 seen dropping practice bomb during a night bombing exercise. The Beech-made top turret is missing from this aircraft. AT-11s did double duty as bombardier and gunnery trainers.

One of 96 rebuilt TC-45Gs seen serving with the Wyoming Air National Guard in the late 1950s. All were retired from Air Training Command in 1955 and 1956.

After being refurbished from an SNB-2, this Navy SNB-5 is seen in postwar reserve markings (orange and white) while serving at NAS Philadelphia, Pennsylvania.

reentered service as SNB-4 multi-engine and instrument proficiency trainers. In 1951, as a corollary to the USAF C-45G program, an unspecified number of JRB-3s/-4s and SNB-2s reemerged as SNB-5s after being completely rebuilt and modified, and some, specialized for photographic training, became SNB-5Ps. All SNB-4s were stricken from the naval inventory during the late 1950s, while many SNB-5s and -5Ps continued to serve with Navy and Marine reserve units as flight proficiency trainer and as utility transports. In 1962, after adoption of the tri-service system, the SNB-5 was re-designated the TC-45J and the SNB-5P, the RC-45J. The very last TC-45J was withdrawn from the Navy reserve in mid–1972.

Cessna AT-8 and AT-17 Bobcat—1941

Technical Specifications (AT-17)

Type: Three or four-place multi-engine advanced trainer.
Manufacturer: Cessna Aircraft Co., Wichita, Kansas.
Total U.S. military versions: 1,232 (AAF).
Unit cost: $19,100.
Powerplants: Two 245-hp Jacobs R-755–9 7-cylinder air-cooled radial engines driving two-bladed Hamilton Standard variable-pitch metal propellers.
Performance: Max. speed 195 mph, cruise 175 mph; ceiling 22,000 ft.; range 750 mi.
Weights: 3,500 lbs. empty, 5,700 lbs. loaded.
Dimensions: Span 41 ft. 11 in., length 32 ft. 9 in., wing area 295 sq. ft.

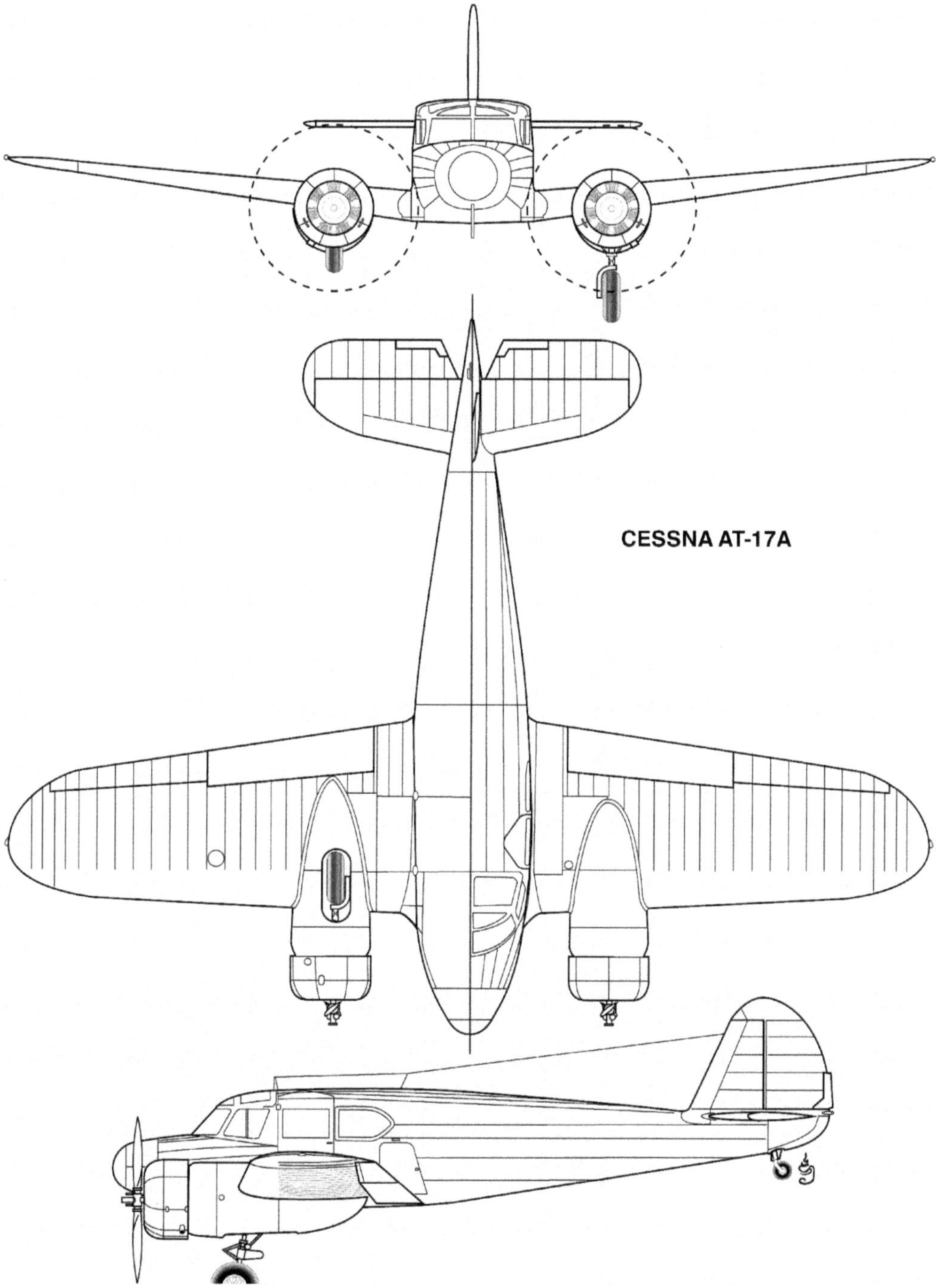

The AT-8 and -17 are notable in having been the second most numerous type of multi-engine advanced trainer employed by Army Air Forces Training Command (AAFTC) during World War II. Cessna initially offered its twin-engine Model T-50 "Bobcat" to commercial operators in early 1939 as a less expensive alternative to the Beech 18 and Lockheed 12. Most of its aerodynamic design and airframe structure—tubular steel fuselage and built-up wooden wings

One of 450 Cessna AT-17s with fixed-pitch, wooden propellers seen in wartime markings. Most served in one of 22 Advanced Twin Engine Flying Schools within Army Air Forces Flying Training Command.

and tail surfaces with fabric covering—had been derived from the company's successful single-engine Airmaster series. New features included electrically operated flaps and retractable landing gear.

After modest civilian sales, Cessna received its first military order, a contract awarded by Air Corps Material Division in 1940, to deliver 33 T-50s powered by 295-hp Lycoming R-680–9 engines, which all entered service during 1941 under the designation AT-8. Subsequent contracts awarded from 1941 onwards resulted in deliveries of 1,199 more T-50 trainer variants powered by 245-hp Jacobs R-755 engines under new designations: 450 as the AT-17 (fixed-pitch wooden propellers); 223 as the AT-17A (as -17 with metal propellers and increased payload); 466 as the AT-17B (as -17A with equipment changes and 12-volt electrical systems); 60 as the AT-17C (as -17A with upgraded radios); and 60 as the AT-17D (as -17A with equipment changes). In addition to trainer production, Cessna delivered 3,356 T-50s in a five-seat utility transport configuration to the USAAF as the C/UC-78. AT-8s were initially assigned to Barkesdale Field in Louisiana, but as the tempo of AT-17 production increased, they were assigned to one of the 22 Advanced Twin Engine Schools operated by AAFTC in ten different states. As a consequence of wooden wing spar deterioration from prolonged exposure to moisture, a number of AT-17s, -17As, and -17Bs were re-designated A-17E, -17F, and -17G, respectively, after being restricted to a maximum takeoff weight of 5,300 lbs. When World War II ended, all remaining AT-8/AT-17 variants were rapidly withdrawn from service and sold surplus.

Curtiss AT-9 Jeep—1941

Technical Specifications (AT-9)

Type: Two-place multi-engine advanced trainer.
Manufacturer: Airplane Div. of Curtiss-Wright Corp., St. Louis, Missouri.

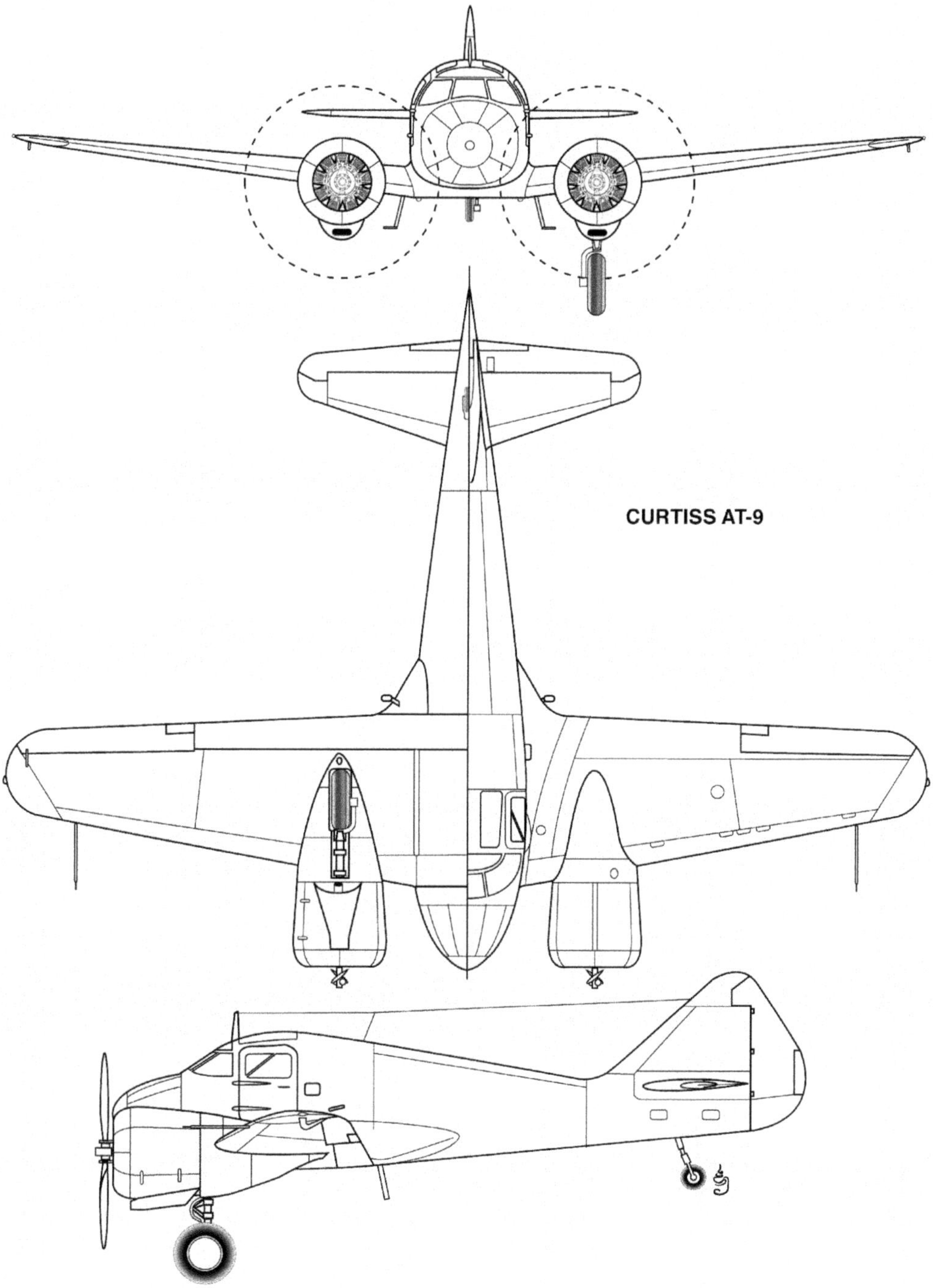

Early production Curtiss AT-9 as seen in 1942 with tail stripes. Because of their unforgiving flight characteristics, AT-9/-9As were never offered surplus to the public.

Total U.S. military versions: 792 (AAF).
Unit cost: $40,286.
Powerplants: Two 295-hp Lycoming R-680–9 9-cylinder air-cooled radial engines driving two-bladed Hamilton Standard variable-pitch metal propellers.
Performance: Max. speed 197 mph, cruise 175 mph; ceiling 19,000 ft.; range 750 mi.
Weights: 4,494 lbs. empty, 6,060 lbs. loaded.
Dimensions: Span 40 ft. 4 in., length 31 ft. 8 in., wing area 233 sq. ft.

The Curtiss model CW-25 twin-engine trainer prototype arrived at Wright Field for testing in 1941. Constructed with a fabric-covered, tubular steel fuselage and wooden wings and tail surfaces, much of the prototype's aerodynamic design had been borrowed from the company's earlier single-engine model CW-22 trainer (see Curtiss SNC in Part B, below). The military requirement for the CW-25 specified that the aircraft be designed with higher wing-loading (26 lbs./sq. ft. vs. 19 lbs./sq. ft. for the AT-8/-17) and incorporate other features (e.g., narrow gear track and heavier control forces) that would emulate the handling characteristics of newer high-performance twins like the Lockheed P-38, Martin B-26, and Douglas A-20. Following completion of trials in 1941, Curtiss received a production contract to deliver 491 aircraft to the Army Air Forces under the designation AT-9. As delivered in 1942, production AT-9s differed from the CW-25 in having all-metal airframe structures, clad entirely in stressed aluminum except for metal-framed ailerons and elevators covered in fabric. In 1942 and 1943, the AT-9s were followed in production by 300 AT-9As with R-680–11 engines and improved hydraulic systems, with all production of the type ending in February 1943.

In service as a trainer, the AT-9 lived up to its design expectations, being tricky to fly in general, and in particular, when one of the engines failed. Some P-38 pilots are reported to have said that the P-38 should have been used first to train students to fly the AT-9. The exact unit assignment of AT-9/-9As within AAFTC is not reported anywhere, however, they are likely to have been used at the various Twin Engine and Specialized Schools that prepared student pilots earmarked for transition to the higher-performance types mentioned above. All AT-9/-9As were withdrawn from AAFTC soon after the war ended in September 1945, and because of their unforgiving flight characteristics, were not offered for sale to the public but either scrapped or given to ground schools for training purposes.

Beech AT-10 Wichita—1941

Technical Specifications (AT-10)

Type: Two-place multi-engine advanced trainer.
Manufacturer: Beech Aircraft Co., Wichita, Kansas; and Globe Aircraft Corp., Fort Worth, Texas.

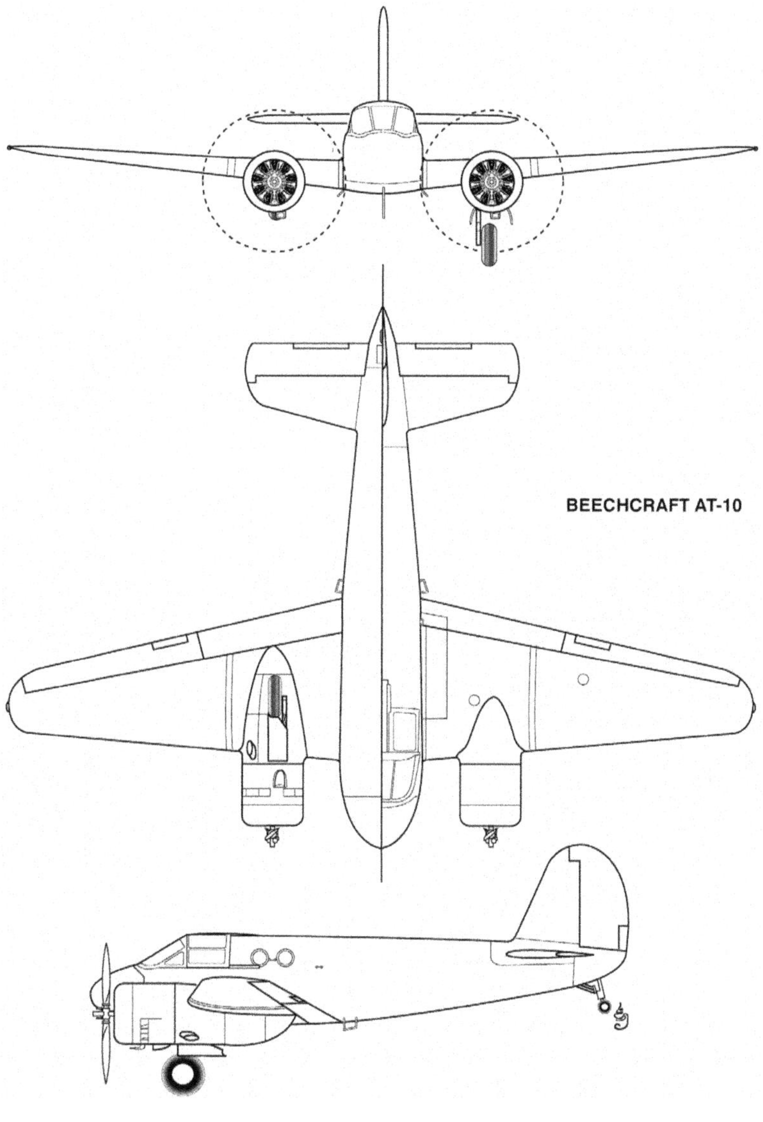

A Beech AT-10 in all-silver wartime markings. Its airframe components were fabricated from wood to save aluminum. AT-10s trained more AAF pilots during the war than any other type of twin engine trainer.

Total U.S. military versions: 2,371 (AAF).
Unit cost: Unknown (estimated between $25,000-$30,000).
Powerplants: Two 295-hp Lycoming R-680–9 9-cylinder air-cooled radial engines driving two-bladed Hamilton Standard variable-pitch metal propellers.
Performance: Max. speed 198 mph, cruise 168 mph; ceiling 16,900 ft.; range 770 mi.
Weights: 4,750 lbs. empty, 6,130 lbs. loaded.
Dimensions: Span 44 ft., length 34 ft. 4 in., wing area 298 sq. ft.

The Beech AT-10 holds the distinction of having trained more student pilots than any other type of twin-engine aircraft employed by Army Air Forces Training Command (AAFTC) during World War II. Beech commenced work on the design of the model 25 in early 1940 in response to an Air Corps Material Division requirement calling for a twin-engine, retractable-gear trainer to be constructed primarily of non-strategic materials. While using some aerodynamic features in common with the model 18 (AT-7/-11), the model 25 emerged with primary airframe structures fabricated from plywood except for a metal nose cap and engine nacelles. The model 25 prototype was destroyed in a crash shortly after delivery to Wright Field in May 1941; however, Beech managed to complete and deliver a second prototype, the model 26, within three months. Later in the year, upon completion of official trials, Material Division ordered the model 26 into production under the designation AT-10.

In order to free up some of Beech's wartime plant capacity, the government awarded Globe Aircraft a contract to license-produce AT-10s. Beech production ended in 1943 after 1,771 AT-10s had been delivered to AAFTC, and a further 600 examples had been completed by Globe when all AT-10 production terminated in 1944. The AT-10 was well regarded by instructors and students alike, and from 1942 to 1945, served in a majority of the 22 Advanced Twin Engine Schools in all three AAFTC regions (Eastern, Central, and Western Flying Training Commands). Like many other trainers procured during the wartime buildup (e.g., PT-19 and BT-13), all AT-10s were withdrawn from service once hostilities ended in September 1945. Although AT-10s were subsequently sold surplus on the civil market at bargain prices, most were acquired for their usable parts and thereafter junked.

Republic AT-12 Guardsman—1941

Technical Specifications (AT-12)

Type: Two-place advanced trainer.
Manufacturer: Republic Aviation Corp. (formerly Seversky), Farmingdale, New York.

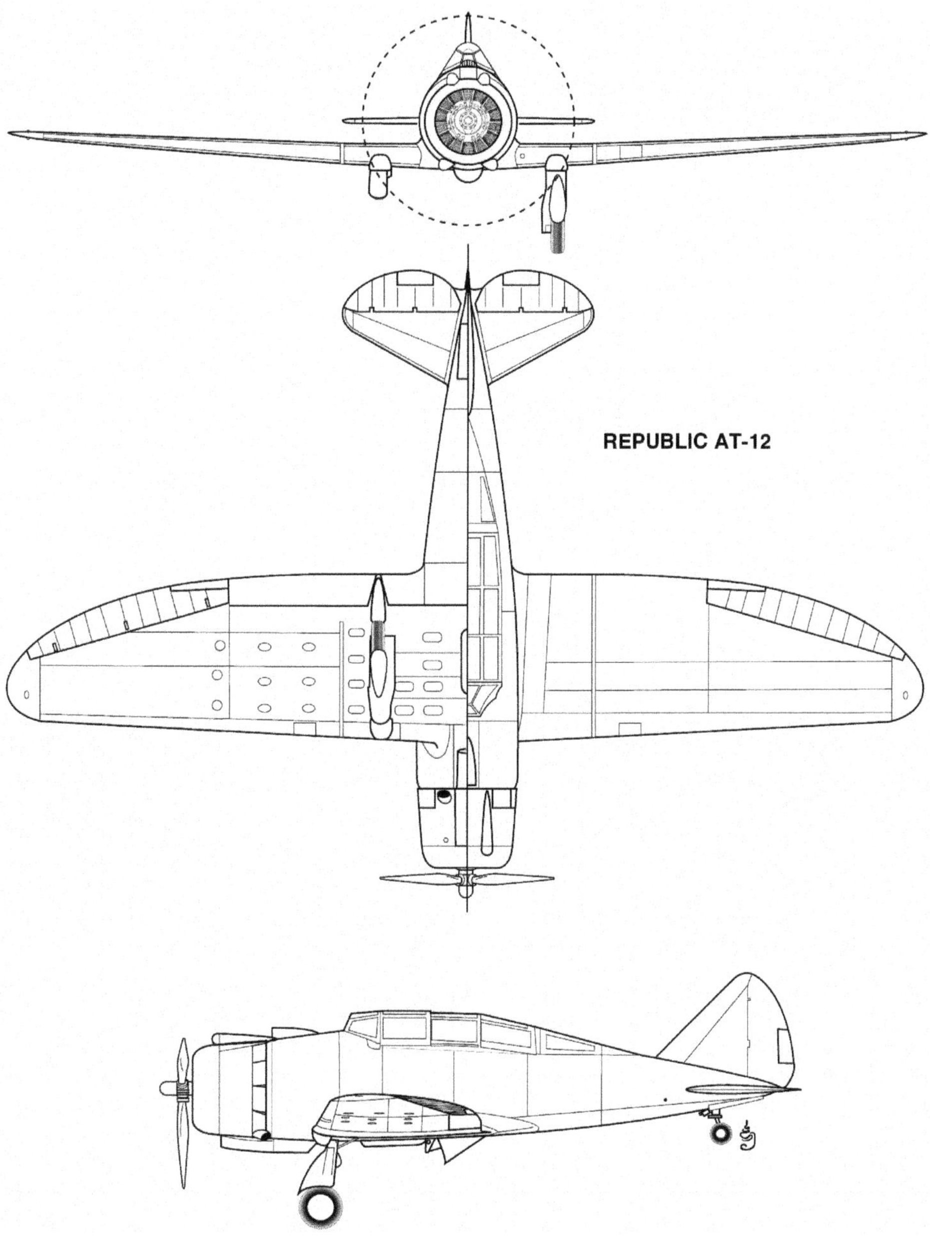

One of the 50 Republic 2PA-204As ordered by Sweden that were seized by the U.S. government in 1941 and assigned to the AAF as AT-12 advanced trainers. Note crude addition of dorsal fillet.

Total U.S. military versions: 50 (USAAC/AAF).
Unit cost: Unknown.
Powerplant: One 1,050-hp Pratt & Whitney R-1830–45 14-cylinder air-cooled radial engine driving a three-bladed variable-pitch metal propeller.
Armament: Two fixed .30-calibre machine guns, one flexible .30-calbre machine gun in the rear cockpit, and up to 1,300 lbs. of bombs carried externally.
Performance: Max. speed 285 mph; ceiling 31,400 ft.; range 950 mi.
Weights: 6,670 lbs. empty, 8,360 lbs. loaded.
Dimensions: Span 41 ft., length 27 ft. 8 in., wing area (not reported).

The AT-12 became a trainer by default in 1941 when the U.S. government seized 50 Republic model 2PA-204As destined for the Swedish Air Force. Marketed for export as a fighter-bomber or dive-bomber, the 2PA-204 had originally been evolved by Seversky Aircraft in 1938 as a two-seat development of the company's P-35, the first type of retractable-gear, monoplane fighter to enter operational service with the Army Air Corps. Except for lengthening the fuselage one foot ten inches, increasing wingspan five feet, and adding a second cockpit for the gunner, the 2PA-204A retained the structural and aerodynamic features of the P-35. Marketing and export of the design continued after Seversky was reorganized as Republic Aviation in 1939.

The government assigned the 50 2PA-204As to the Army Air Corps, however, because they were considered too obsolete for combat, Air Corps Material Division placed them on the inventory as AT-12 advanced trainers. One source indicates that the type temporarily served with Advanced Flying Schools but as AT-6 deliveries increased, many AT-12s were reassigned to airfield base commanders in the continental U.S. as fast staff transports and others are reported to have been used for test purposes. Except for two sold surplus, all remaining AT-12s were scrapped soon after World War II ended.

North American P-64–1941

Technical Specifications (P-64)

Type: Single-place advanced trainer.
Manufacturer: North American Aviation. Inc., Inglewood, California.

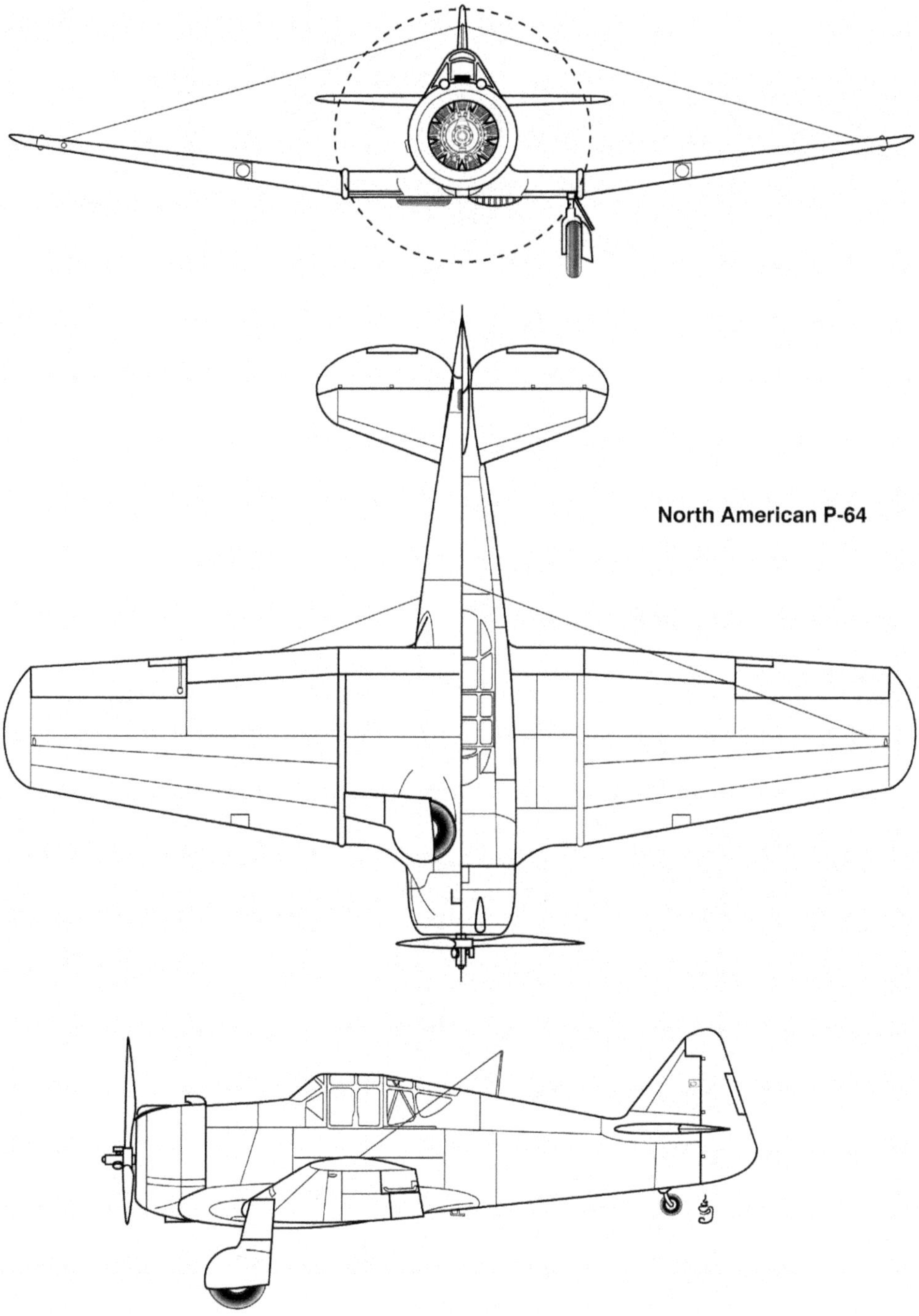

One of six commandeered North American P-64s seen at Luke Field, Arizona, probably in mid–1941, after being assigned to Advanced Flying Training. Most were used as hacks between training bases.

Total U.S. military versions: 6 (USAAC/AAF).
Unit cost: Unknown.
Powerplant: One 870-hp Wright R-1820–77 9-cylinder air-cooled radial engine driving a three-bladed variable-pitch metal propeller.
Armament: Two fixed .30-calibre machine guns in the nose and up to 400 lbs. of bombs carried externally.
Performance: Max. speed 270 mph; ceiling 27,500 ft.; range 965 mi.
Weights: 4,660 lbs. empty, 5,990 lbs. loaded.
Dimensions: Span 37 ft. 3 in., length 27 ft., wing area 228 sq. ft.

Like the AT-12 reported above, the P-64 became a trainer by an unexpected chain of events, in this case, U.S. government seizure in late 1940 of six North American model NA-68s destined for the Royal Thai Air Force. The NA-68, having originally appeared in 1939 as the model NA-50A export fighter, was an armed, single-seat offshoot of the BC-1A/AT-6 advanced trainer airframe. Besides armament, main differences from the basic AT-6 consisted of a shortened fuselage (two and a half feet) with one cockpit, a 10 percent reduction in wing area, and a more powerful (+270-hp) R-1820 engine. In early 1941, the six NA-68s received USAAC serial numbers under the designation P-64 but were deemed unsuitable for combat duty and assigned to Air Corps Flying Training Command. Although the P-64s saw brief service as fighter transition trainers with the advanced school at Luke Field, Arizona, they spent most of their wartime career as utility hacks for transportation between various training bases, and in 1943 received the noncombat designation RP-64. All were scrapped when the war ended, except for one example that survives today as part of the Experimental Aircraft Association collection.

Taylorcraft O-57/L-2–1941

Technical Specifications (L-2B)

Type: Two-place liaison and primary trainer.
Manufacturer: Taylorcraft Aviation Corp., Alliance, Ohio.

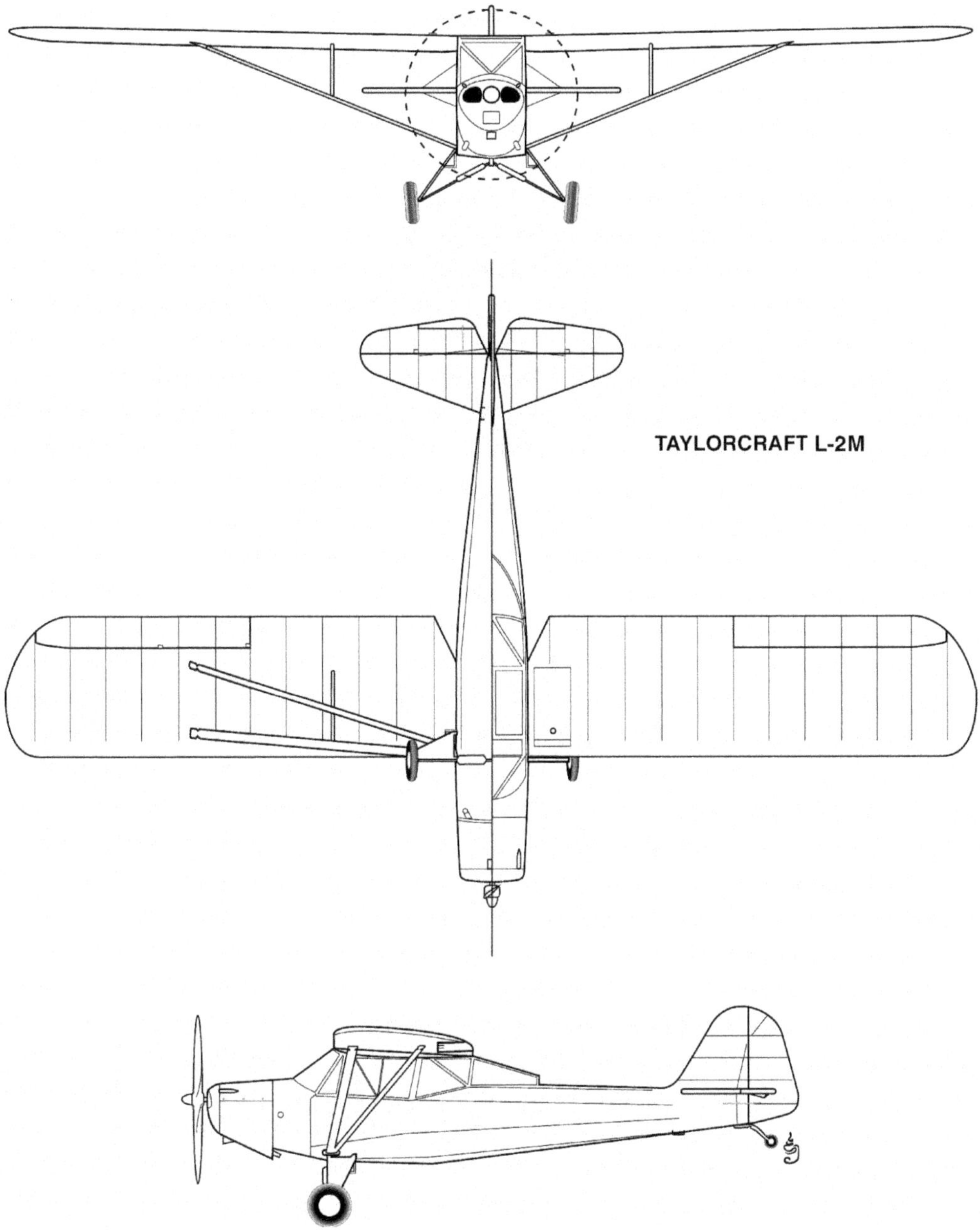

TAYLORCRAFT L-2M

Total U.S. military versions: 1,935 (AAF).
Unit cost: $1,685-$1,995.
Powerplant: One 65-hp Continental O-170–1 4-cylinder air-cooled horizontally-opposed engine driving a two-bladed fixed-pitch wooden propeller.
Performance: Max. speed 92 mph, cruise 83 mph; ceiling 12,000 ft.; range 303 mi.
Weights: 875 lbs. empty, 1,300 lbs. loaded.
Dimensions: Span 35 ft. 5 in., length 22 ft. 9 in., wing area 181 sq. ft.

In 1941 Air Corps Material Division procured four off-the-shelf Taylorcraft tandem-seat model DC65s to be evaluated as light artillery spotters under the designation YO-57. Design

One of four off-the-shelf Taylorcraft DC65s procured in 1941 as the YO-57 but later redesignated L-2. Most L-2s served with the Department of Air Training at Fort Sill, Oklahoma.

was conventional for lightplanes of the day—a strut-braced high wing layout, built with a fabric covering over a tubular steel fuselage and tail group and built-up wooden wings. Twenty production versions with minor changes followed as the O-57. Modifications to the cabin such as a plexiglas rear turtle deck and swiveling observer's seat resulted in delivery of 336 more examples in 1942 as the O-57A. The same year, the Army applied the "L" for Liaison designation to all light observation aircraft, so that the O-57 and O-57A became the L-2 and L-2A, respectively. Additional wartime contracts accounted for delivery of 490 L-2Bs with equipment changes and, finally, 900 L-2Ms which featured a close-fitting engine cowling and wing spoilers. After 1941, the government also impressed 45 civil Taylorcrafts (included some model Bs with side-by-side seating) that entered service as the L-2C through L-2L. The Army deployed no L-2 variants overseas for combat duty; instead, most served with the Department of Air Training, a subordinate unit of the Army Field Artillery School, where they were used for primary and operational training of Army Ground Forces (AGF) pilots, also known as liaison pilots. All L-2s were withdrawn and offered surplus soon after World War II ended.

Aeronca O-58/L-3 (JR)—1941

Technical Specifications (L-3C)

Type: Two-place liaison and primary trainer.
Manufacturer: Aeronca Aircraft Corp., Middletown, Ohio.
Total U.S. military versions: 1,465 (1,462 USAAC/AAF; 3 USN).
Unit cost: $1,695.
Powerplant: One 65-hp Continental O-170–1 4-cylinder air-cooled horizontally-opposed engine driving a two-bladed fixed-pitch wooden propeller.
Performance: Max. speed 87 mph, cruise 79 mph; ceiling 10,000 ft.; range 218 mi.
Weights: 835 lbs. empty, 1,260 lbs. loaded.
Dimensions: Span 35 ft. 5 in., length 22 ft. 9 in., wing area 181 sq. ft.

Analogous to the O-57/L-2 reported above, Air Corps Material Division purchased four civil Aeronca model 65TC Tandems in 1941—a high-wing lightplane of fabric-covered, mixed

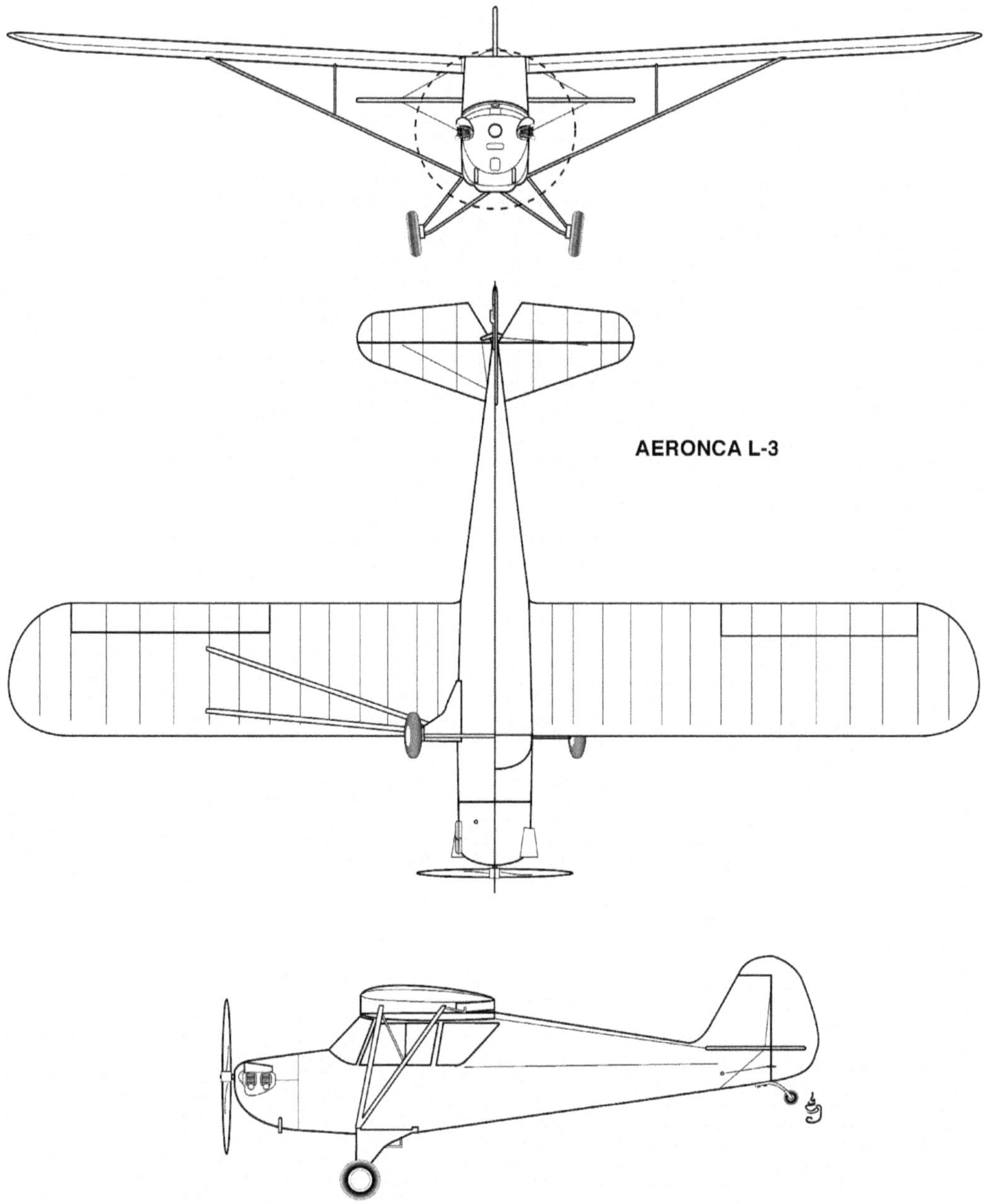

metal and wood construction—to be evaluated as light artillery spotters under the designation YO-58, and 20 more 65TCs ordered from Aeronca joined the inventory in 1942 as the O-58. The O-58A, the first of the type built to military specifications, featured a fuselage widened four inches and an extended greenhouse canopy, with a total of 20 delivered in 1942. After the switch to the "L" for Liaison designation the same year, the O-58 and O-58A became the L-3 and L-3A. Addition of radios and observation equipment resulted in the first major production variant, 875 aircraft manufactured in 1942 and 1943 as the L-3B. The final variant, 490 L-3Cs built in 1943, were dedicated trainers lacking radios. During the war, another 26 civil Aeronca Tandem and Chief (side-by-side seats) variants were impressed into service as the L-3D through

One of 20 Aeronca 65TCs that were delivered to the AAF in early 1942 as the O-58 and later redesignated L-3. Like L-2s, L-3s never saw combat but served out their careers with the Department of Air Training.

L-3L. Three L-3Cs were later transferred to the Navy as the JR-1 light utility transports. L-3s, like L-2s, were never used in combat but served out their careers with the Department of Air Training as primary and operational trainers for Army Ground Forces (AGF/Liaison) pilots. All were declared surplus at the end of World War II.

Piper O-59/L-4 (NE)—1941

Technical Specifications (L-4B)

Type: Two-place liaison and primary trainer.
Manufacturer: Piper Aircraft Corp., Lock Haven, Pennsylvania.
Total U.S. military versions: 5,911 (5,671 USAAC/AAF/U.S. Army; 240 USN/USMC).
Unit cost: $2,600.
Powerplant: One 65-hp Continental O-170–1 4-cylinder air-cooled horizontally-opposed engine driving a two-bladed fixed-pitch wooden propeller.
Performance: Max. speed 87 mph, cruise 75 mph; ceiling 11,500 ft.; range 220 mi.
Weights: 765 lbs. empty, 1,220 lbs. loaded.
Dimensions: Span 35 ft. 3 in., length 22 ft. 5 in., wing area 179 sq. ft.

Arguably the most famous lightplane in the history of aviation, the Piper J-3 Cub is certainly the most enduring, remaining in production at the time of this writing (though not by Piper), 75 years after C. G. Taylor and Walter Jamouneau conceived it. The Cub, evolving from the E-2 of 1931 to the J-3 of 1937, set the pattern for a new generation of inexpensive two-seat lightplanes: a high-wing monoplane layout braced by V struts; a steel tube fuselage/tail group and wooden wings with fabric covering; a simple conventional landing gear sprung by bungee cords; and a light, horizontally-opposed air-cooled engine. Cubs could be purchased with a variety of powerplants, ranging from 40-hp to 65-hp, and docile flight characteristics made them ideal trainers. After the government-funded Civilian Pilot Training Program (CTPT) got underway in 1939, Cubs became the preferred type of primary trainer in a majority of the new schools.

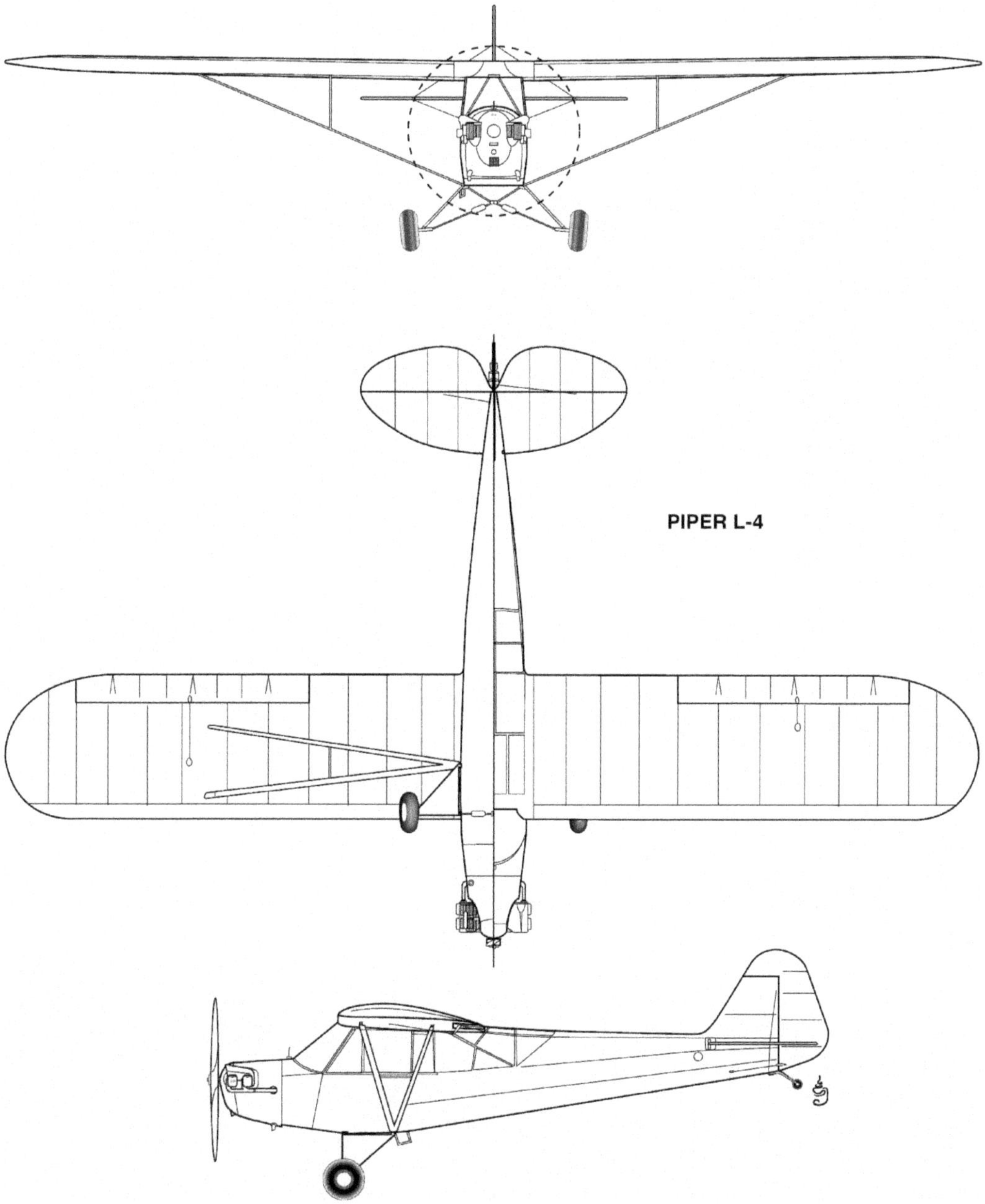

During World War II, the militarized Cub, the L-4 "Grasshopper," was produced in greater numbers than any other type of liaison aircraft and stayed active until the early 1950s. The first Cubs to be placed on the Army Air Forces inventory, four YO-58s in 1941 and 140 O-58s in late 1941 to early 1942, were essentially factory stock J-3C-65s. Military specifications to add radios and a greenhouse canopy brought about production of 948 O-58As, which from mid 1942, were delivered as the L-2A. These were followed in late 1942 to early 1943 by 981 L-4Bs earmarked for training, identical to As but lacking radios. At the same time, the government impressed some 117 civil Cubs (a mix of J-3Fs, J-3Ls, J-5As, and J-5Es) as the L-4C through L-4G. The next major production variant, 1,801 L-4Hs delivered in 1944, featured improved radios and equip-

One of the 948 L-4As delivered in 1942. L-4s saw widespread use with Army Ground Forces in every theatre of the war. In postwar organic Army Aviation, they served as fixed-wing trainers until 1953.

ment changes. Production ended in 1945 after delivery of 1,680 L-4Js with controllable propellers.

In the liaison role, L-4s attached to Army Ground Forces (AGF) became a common sight in every combat theater of World War II, where they were employed for artillery spotting, combat zone observation, and as battlefield couriers and staff transports. L-4Bs and other L-4 variants also saw wide use with the Department of Air Training as primary and operational trainers for AGF (Liaison) pilots. Unlike the L-2 and L-3, large numbers of L-4s continued in service after the war to train the first generation of organic U.S. Army aviators, the last examples not being withdrawn until 1953.

Naval Variants: During 1942 the Navy procured 230 factory stock J-3C-65s which entered operational service under the naval designation NE-1. The Navy also acquired 20 more J-3C-65s with minor equipment changes as the NE-2. The majority of NEs were assigned to various pilot training centers to serve as elimination trainers for pilot candidates in the preflight stage of training. All were withdrawn and offered surplus soon after the war ended.

Interstate O-63/L-6–1942

Technical Specifications (L-6)

Type: Two-place liaison and primary trainer.
Manufacturer: Interstate Aircraft & Engineering Corp., El Segundo, California.
Total U.S. military versions: 251 (USAAF).
Unit cost: $3,000-$4,000 (est.).

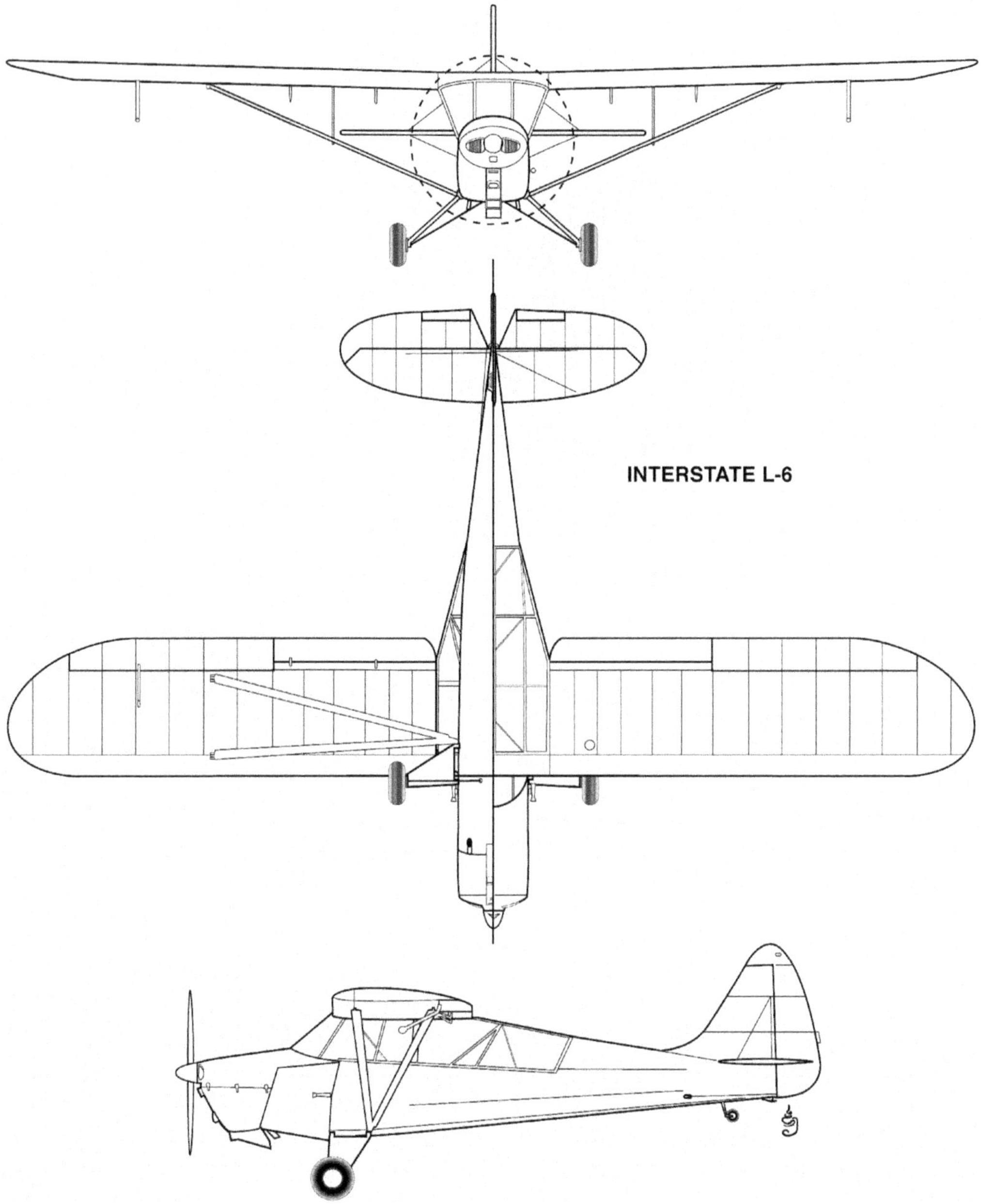

Powerplant: One 102-hp Franklin O-200–5 4-cylinder air-cooled horizontally-opposed engine driving a two-bladed fixed-pitch wooden propeller.
Performance: Max. speed 105 mph, cruise 95 mph; ceiling 16,500 ft.; range 540 mi.
Weights: 1,103 lbs. empty, 1,650 lbs. loaded.
Dimensions: Span 35 ft. 6 in., length 23 ft. 5 in., wing area 175 sq. ft.

The Interstate L-6 was the last prewar lightplane design to be considered as a potential candidate for production in the observation-liaison role. Interstate, a relatively new aircraft manufacturer, introduced its model S-1 Cadet to the civil market in 1940. While it was a conventional high-wing design with fabric-covered, steel tube and wood construction, the S-1 boasted a very

Factory new Interstate L-6 shown in 1943 making a very steep takeoff. Repeated engine cooling problems caused the type to see only brief service with the AGF Pilot School at Fort Sill, Oklahoma.

rugged airframe offering better overall performance than that of contemporaneous two-seaters like the Piper J-3, Taylorcraft D, and Aeronca T. It appears that the Franklin-powered S-1B, delivered as the XO-63 in early 1942 (XL-6 after mid 1942), was actually developed for military purposes from the outset since it arrived with a long, greenhouse canopy with canted side windows to improve downward visibility. Despite persistent engine overheating problems during acceptance trials, Interstate received a contract to deliver a further 250 S-1Bs as the L-6 on the premise the problem would be resolved. However, when repeated modifications to the cowling and addition of an oil cooler failed to work, no further production was ordered. L-6s are thought to have briefly operated with the Field Artillery School's Department of Air Training until being withdrawn sometime in 1943. None served overseas.

De Havilland PT-24–1942 (designation assigned to 200 DH-82A Tiger Moths ordered by the USAAF but never delivered)

Fairchild AT-13, AT-14 and AT-21 Gunner—1942

Technical Specifications (AT-21)

Type: Five-place gunnery and aircrew trainer.
Manufacturer: Fairchild Engine & Airplane Corp., Hagerstown, Maryland, and Burlington, North Carolina; Bellanca Aircraft Corp, Newcastle, Delaware; and McDonnell Aircraft Corp., St. Louis. Missouri.

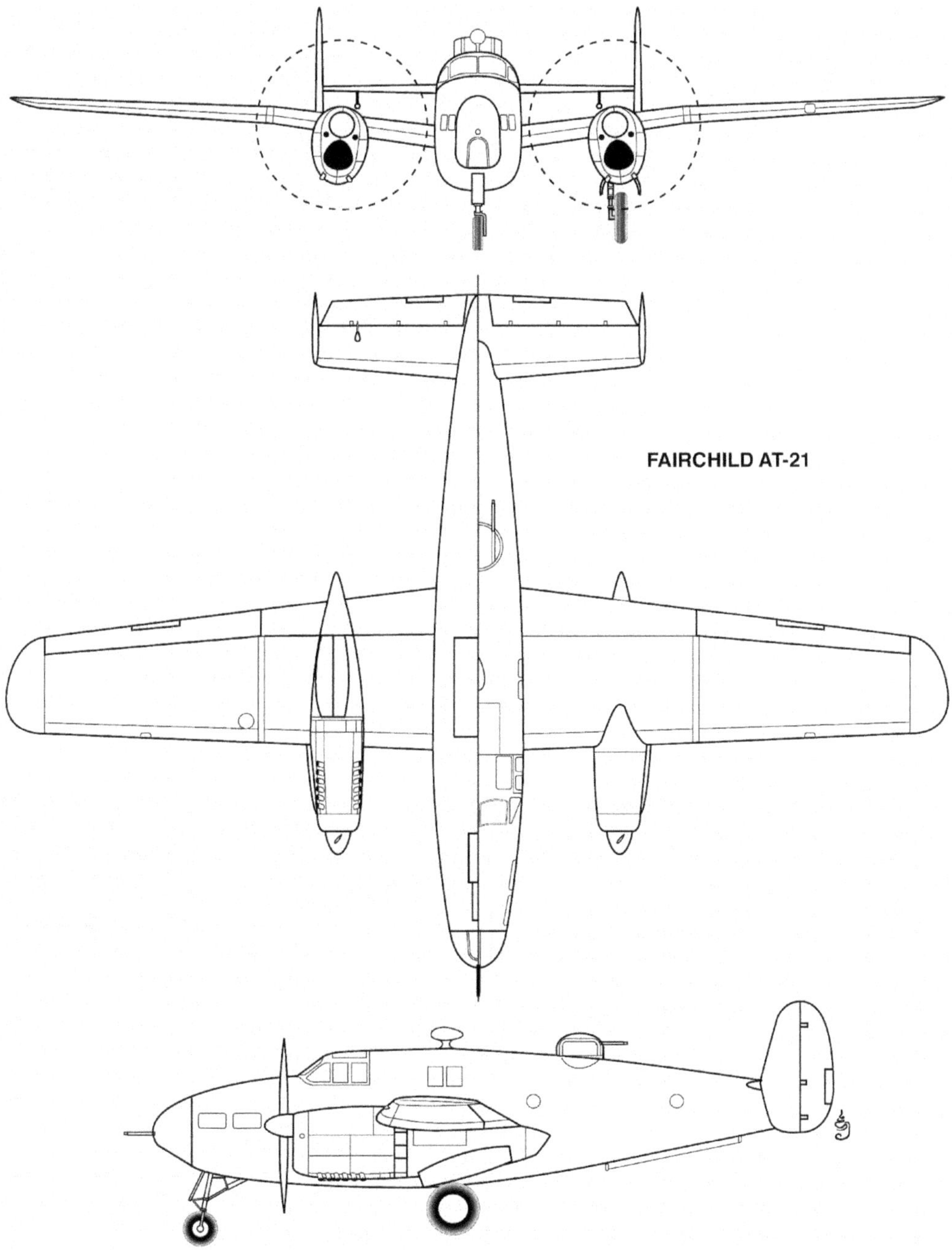

Fairchild XAT-13 shown in 1942 loading practice bombs. The aircraft appears sometime after initial testing when dihedral had been reduced in the outer wings' panels.

Total U.S. military versions: 177 (USAAF).
Unit cost: Unknown.
Powerplants: Two 520-hp Ranger V-770–11/-15 12-cylinder air-cooled inline engines driving two-bladed variable-pitch metal propellers.
Armament: One .30-calbre flexible machine gun in the nose and two .30-calibre machine guns in a power-operated dorsal turret.
Performance: Max. speed 225 mph, cruise 196 mph; ceiling 22,150 ft.; range 910 mi.
Weights: 8,654 lbs. empty, 11,288 lbs. loaded.
Dimensions: Span 52 ft. 8 in., length 38 ft., wing area 378 sq. ft.

In 1941, after soliciting proposals for a twin-engine aircrew trainer to be constructed primarily of non-strategic materials, Air Corps Material Division awarded Fairchild a contract to build two prototypes having the same Duramold plastic-bonded plywood airframe design and tricycle landing gear arrangement but with differing engine installations: the XAT-13 powered by 600-hp Pratt & Whitney R-1340-AN1 radial engines and the XAT-14 by 520-hp Ranger V-770–3/-6 inverted inline engines. Although bearing a superficial resemblance to the Lockheed model 14 (see AT-18, below), the XAT-13/-14 featured a much higher aspect-ratio wing mounted to the middle of the fuselage. Sometime in 1942, the XAT-13 prototype appeared at Wright Field equipped as a gunnery trainer with a twin dorsal turret and a single flexible mount in the nose; and the similarly equipped XA-14 arrived for testing later the same year.

Testing of the prototypes revealed marginal single-engine performance, plus inherent yaw instability due to the close proximity of the vertical tail surfaces to the wing. The yaw problem was partially resolved by reducing dihedral in the outer wing panels, giving the aircraft a slightly gulled appearance from the front. In any event, evaluating officials deemed the XA-14 to be the

The XA-14, seen in late 1942, achieved production as the AT-21. Production terminated in 1944 after only 175 had been delivered due to availability of large numbers of B-25s and B-26s for training duties.

better of the two and ordered the type into production as the AT-21. In addition to Fairchild, production was licensed to Bellanca and McDonnell; however, all AT-21 production terminated in 1944 when Fairchild had delivered only 106, Bellanca 39, and McDonnell 30. Several factors contributed to the type's demise: the stability problems and poor single-engine performance were never satisfactorily rectified; the anticipated aluminum shortage never materialized; and finally, and perhaps most importantly, significant numbers of combat aircraft such as the North American B-25 and Martin B-26 were becoming available for aircrew training. All AT-21s were withdrawn from Army Air Forces Training Command during 1944, and a number of them reportedly continued in service as target tugs. At Wright Field, the XA-14 underwent evaluations as the XA-14A after the gunnery equipment had been replaced with a bombardier's station and a navigator's astrodome, but development was discontinued after brief testing. A small number of AT-21s were sold surplus after the war ended.

Boeing-Stearman AT-15 Crewmaker—1942

Technical Specifications (AT-21)

Type: Five-place gunnery and aircrew advanced trainer.
Manufacturer: Stearman Aircraft Div. of Boeing Airplane Co., Wichita, Kansas.
Total U.S. military versions: 2 (USAAF).
Unit cost: Unknown.
Powerplants: Two 600-hp Pratt & Whitney R-1340-AN1 9-cylinder air-cooled radial engines driving two-bladed variable-pitch metal propellers.

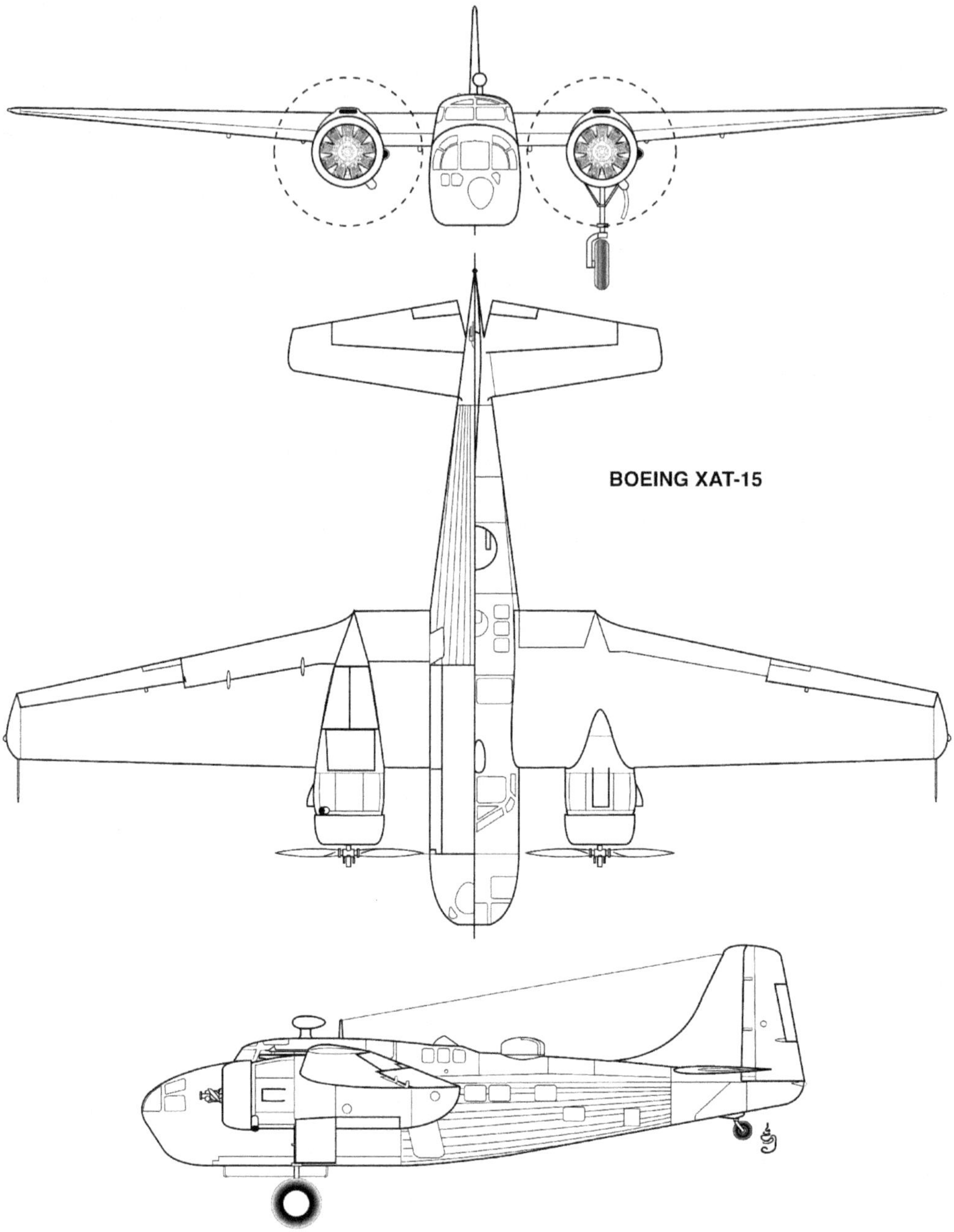

Armament: Two .30-calbre flexible machine guns in nose and waist positions, two .30-calibre machine guns in a power-operated dorsal turret, and up 100 lbs. of practice bombs.
Performance: Max. speed 202 mph, cruise 185 mph; ceiling 18,900 ft.; range 850 mi.
Weights: 10,640 lbs. empty, 12,061 lbs. loaded.
Dimensions: Span 59 ft. 8 in., length 42 ft. 4 in., wing area 457 sq. ft.

Responding to the same 1941 crew trainer requirement as the Fairchild XAT-13 and -14, Boeing assigned responsibility for design and construction to its Stearman subsidiary in Wichita after receiving a contract to build two prototypes under the designation XAT-15. Other than its

One of two Boeing XAT-15s delivered to Wright Field, Ohio, for testing in May 1942. Built almost entirely of wood by the Stearman Division, an anticipated contract to produce 1,000 or more never materialized.

wooden construction, the general design of the XAT-15 emerged with many aerodynamic similarities to Stearman's XA-21 light bomber prototype of 1939: a deep-bodied fuselage mated to shoulder-mounted wings, sharply tapered at the trailing edges, and a conventional landing gear arrangement. As per Air Corps Material Division specifications, construction consisted of a tubular steel fuselage and built-up wooden wings and tail surfaces, all skinned in plywood. The aircraft, having a built-in bomb bay, could be configured for a combination of bombardier and gunnery training. Both XAT-15 prototypes were delivered to Wright Field for testing in May 1942. Photographs indicate that they initially appeared without power-operated dorsal turrets but that turrets were retrofitted at a later date.

Available historical data fails to mention any official assessment of the XAT-15's suitability in its designed role. One source indicates that Material Division had originally envisioned giving Stearman a contract to produce 1,000 or more AT-15s but later decided against procurement when the forecast aluminum shortages failed to transpire and combat aircraft like North American B-25s and Martin B-26s were becoming available in sufficient numbers to fulfill the crew training role. Both XAT-15s were stricken from the inventory in 1944.

Ryan PT-25–1942

Technical Specifications (YPT-25)

Type: Two-place primary trainer.
Manufacturer: Aeronca Aircraft Corp., Middletown, Ohio
Total U.S. military versions: 5 (USAAF).
Unit cost: Unknown.
Powerplant: One 185-hp Lycoming O-435-1 6-cylinder air-cooled horizontally-opposed engine driving a two-bladed fixed-pitch wooden propeller.
Performance: Max. speed 149 mph, cruise 134 mph; ceiling 20,300 ft.; range 378 mi.

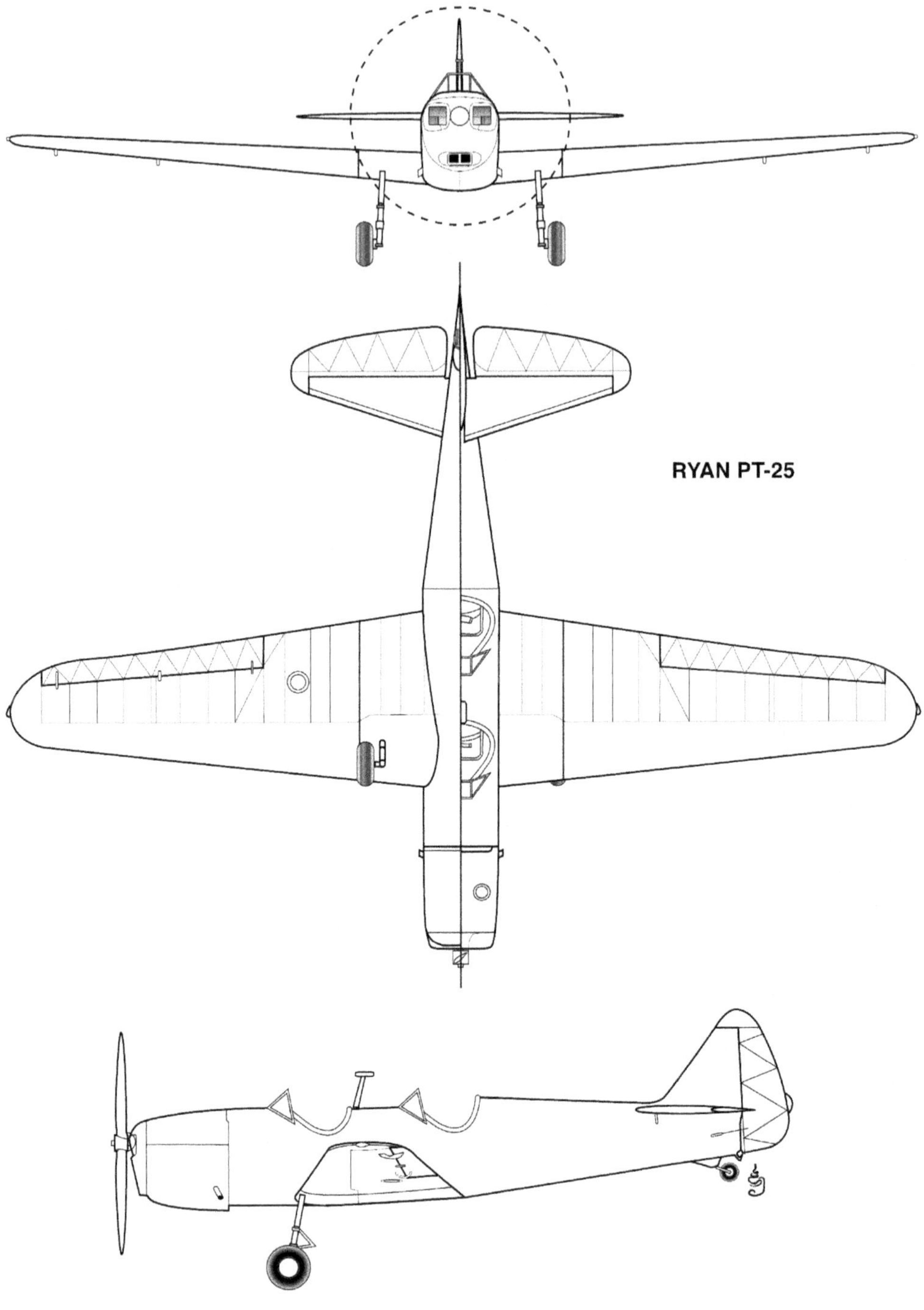

RYAN PT-25

Weights: 1,300 lbs. empty, 1,800 lbs. loaded.
Dimensions: Span 32 ft. 11 in., length 24 ft. 4 in., wing area 115 sq. ft. (est.).

Some historical references cite the PT-25 as a derivative of Ryan's PT-22 when, in fact, it represented a wholly new structural and aerodynamic design, and had it appeared a year or two earlier, might have achieved mass-production along with the PT-17 and PT-19. Resembling a

One of five Ryan YPT-25 prototypes tested in early 1943. The design was so aerodynamically clean it used a belly-mounted speed brake to slow it down. It came too late to be placed in production.

slightly scaled-down PT-19, the airframe of the Ryan model ST-4 was fabricated entirely of wood from the firewall aft. The fuselage consisted of a semi-monocoque, molded plywood shell, and its fully-cantilevered wings and empennage were made up of spruce frameworks skinned in plywood except for fabric-covered control surfaces. The design was so aerodynamically clean that it employed a belly-type speed brake to slow it down in maneuvers. Air Corps Material Division ordered five ST-4 prototypes under the designation YPT-25, but by the time they completed testing in 1943, Army Air Forces Training Command was reaching its peak in primary trainer procurement, plus Material Division had developed serious reservations about the deterioration of wooden airframe structures from moisture and heat. As a consequence, the PT-25 never achieved production, and the five prototypes were returned to Ryan for use in its flying school.

Boeing-Stearman BT-17–1942

Technical Specifications (XBT-17)

Type: Two-place basic trainer.
Manufacturer: Stearman Aircraft Div. of Boeing Airplane Co., Wichita, Kansas.
Total U.S. military versions: 1 (USAAF).
Unit cost: Unknown.
Powerplant: One 450-hp Pratt & Whitney R-985-AN1 9-cylinder air-cooled radial engine driving a two-bladed variable-pitch metal propeller.
Performance: Max. speed 190 mph, cruise 160 mph; ceiling 20,000 ft.; range (not reported).
Weights: 3,080 lbs. empty, 4,150 lbs. loaded.
Dimensions: Span 35 ft. 9 in., length 27 ft. 10 in., wing area 200 sq. ft.

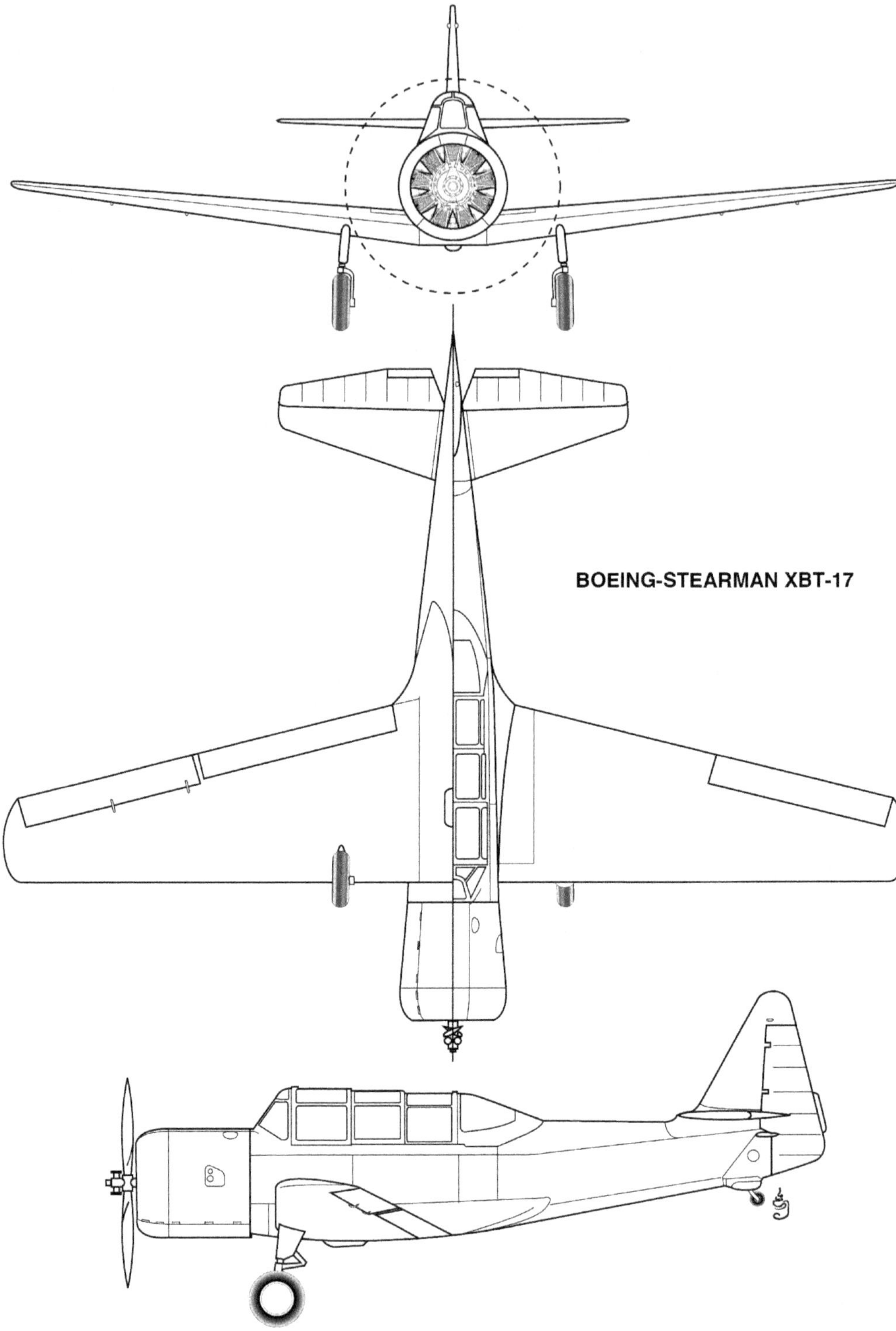

BOEING-STEARMAN XBT-17

The sole Boeing-Stearman XBT-17 prototype as seen at Wright Field, Ohio, in early 1942. The design had been extensively reworked after originally appearing in 1940 as an open cockpit primary trainer.

The Stearman model X-90 originally appeared in 1940 as an open-cockpit primary trainer powered by a 225-hp Lycoming R-680 engine. Comparable in layout to the PT-19 but dissimilar in construction, the X-90 featured an all-aluminum airframe with a semi-monocoque, stressed-skin fuselage and fabric-covered, metal-framed wings and tail surfaces. Since Air Corps Material Division was then favoring primary trainer designs fabricated mainly from non-strategic materials, it did not procure the X-90 prototype and instead, directed the company to rework the design as a basic trainer. In early 1942, Stearman redelivered a revised prototype, the Model X-91, to Wright Field for official evaluations. While keeping the X-90's general structural and aerodynamic conformation, the X-91 incorporated a 450-hp Pratt & Whitney R-985-AN1 engine, variable-pitch propeller, fully-enclosed, sliding canopies, improved instrumentation, and radios. Material Division did purchase the prototype and assign the designation XBT-17 but ultimately elected not to place the type into production.

BT-18 through BT-27 (skipped)

Lockheed AT-18–1942

Technical Specifications (AT-18)

Type: Six-place gunnery and navigation advanced trainer.
Manufacturer: Lockheed Aircraft Corp., Burbank, California.
Total U.S. military versions: 300 (USAAF).
Unit cost: Unknown.
Powerplants: Two 1,200-hp Wright R-1820–87 9-cylinder air-cooled radial engines driving three-bladed constant-speed metal propellers.
Armament: One .30- or .50-calibre flexible machine gun in the nose, two .50-calibre machine guns in a power-operated dorsal turret, and up 750 lbs. of practice bombs.

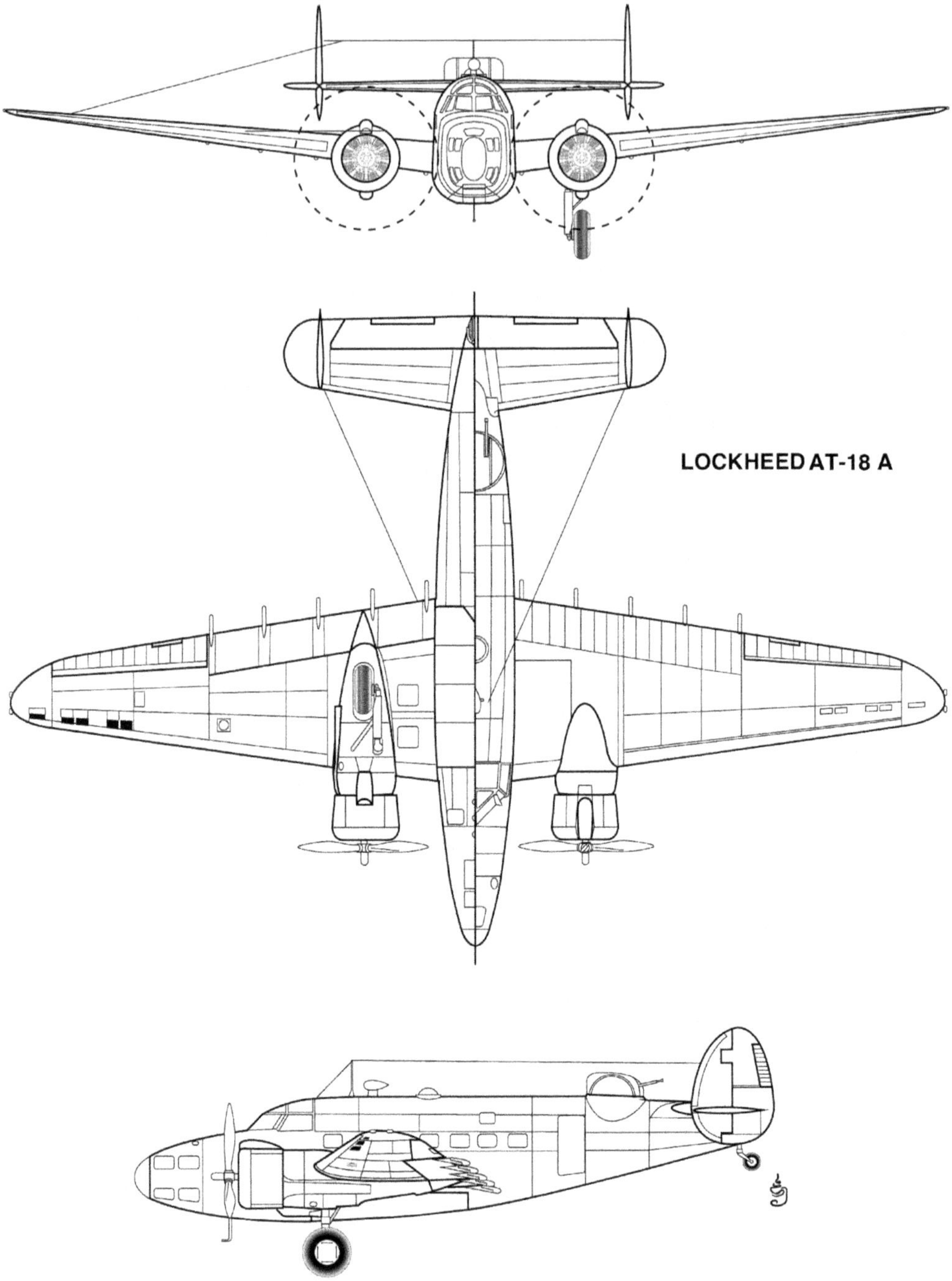

Performance: Max. speed 246 mph, cruise 206 mph; ceiling 24,500 ft.; range 1,960 mi.
Weights: 11,632 lbs. empty, 17,500 lbs. loaded.
Dimensions: Span 65 ft. 6 in., length 44 ft. 4 in., wing area 551 sq. ft.

The Lockheed model 14 "Super Electra" started life in late 1937 as a 12-passenger airliner, followed by a militarized export version that began entering service with RAF Coastal Command in 1939 as the Hudson I. For its day, the Super Electra design embodied the aeronautical state-

A factory new Lockheed AT-18 seen in early 1943. The AT-18s were equipped primarily for gunnery training while the -18As were configured to train navigators. Many used as fast target tugs in gunnery schools.

of-the-art, featuring an all-metal airframe of stressed aluminum skin and a very streamlined, elliptical-section fuselage mated to a mid-wing having compound taper at the trailing edge. To offset high wing-loading (31.5 lbs. per sq. ft.), the model 14 introduced a system of tracked Fowler-type trailing-edge flaps and fixed leading-edge slots that kept landing speeds in the 70 mph range. Its distinctive twin-fin empennage was designed to provide positive yaw control in the event of an engine failure. In late 1941, the U.S. government impounded 175 Hudson IIIAs intended for the RAF, 153 of which were redirected to the USAAF as A-29 attack bombers and 20 to the Navy as PBO-1 patrol bombers.

In May 1942, Air Corps Material Division awarded Lockheed a contract to build 300 model 14s to an advanced trainer specification under the designation AT-18. The first 217 AT-18s, outfitted for gunnery training, were delivered to Army Air Forces Training Command (AAFTC) during 1943; the next batch of 83 aircraft, designated AT-18A, lacked armament but included a dorsal astrodome and stations for an instructor and three navigator students. Because of their superior speed, many AT-18s saw use primarily as target tugs in Fixed and Flexible Gunnery Schools. According to AAFTC records, all AT-18s/-18As were withdrawn from active training duties during 1945.

Stinson AT-19 Reliant—1942

Technical Specifications (AT-19)

Type: Three to four-place instrument trainer.
Manufacturer: Stinson Aircraft Div. of Vultee-AVCO Corp., Wayne, Michigan.
Total U.S. military versions: 50 to 55 (USAAF).

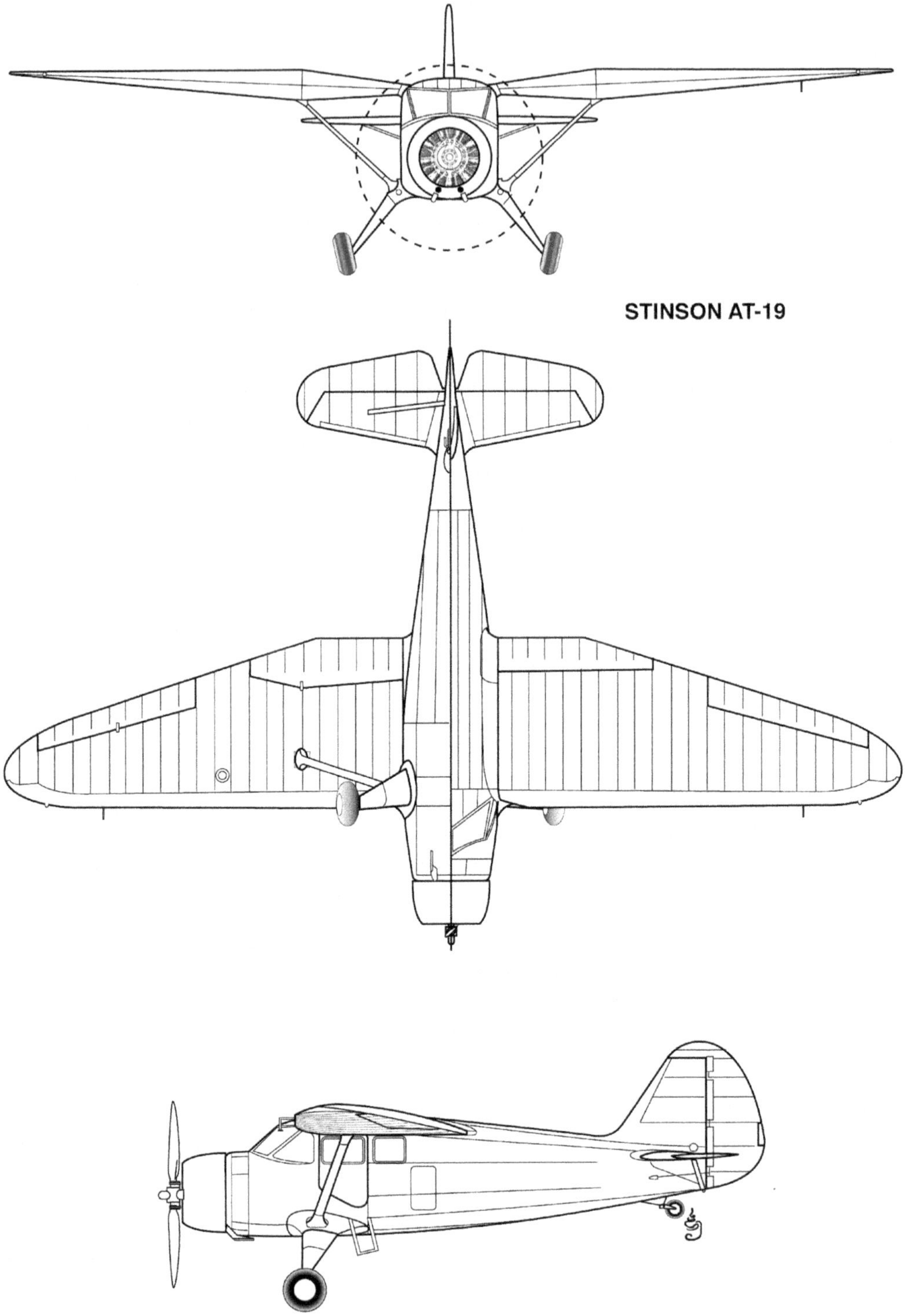

STINSON AT-19

A Stinson AT-19 seen in RAF markings before delivery. Most of the 550+ produced went to Great Britain under Lend-Lease and few, if any, actually served with Army Air Forces Flying Training Command.

Unit cost: $12,000-$13,000 (est.).
Powerplant: One 285-hp Lycoming R-680–13 9-cylinder air-cooled radial engine driving a two-bladed variable-pitch metal propeller.
Performance: Max. speed 145 mph, cruise 120 mph; ceiling 14,000 ft.; range 575 mi.
Weights: 2,800 lbs. empty, 4,000 lbs. loaded.
Dimensions: Span 41 ft. 11 in., length 28 ft. 3 in., wing area 257 sq. ft.

Of the 550+ AT-19s produced during World War II under USAAF serial numbers, the vast majority went to Great Britain under the Lend-Lease program. The Stinson model V-77 emerged in 1941 as a militarized variant of the SR-10J Reliant, a five-place, high-wing monoplane of mixed metal and wood construction. Differences from the civil model included a single door on the left side, a smooth cowling, and general airframe strengthening. One of the type's most noticeable features was wings with a sharp taper at the trailing edges and varying chord thickness from root to tip, which produced a characteristic "gull-wing" effect. Though assigned the advanced trainer designation AT-19, records of Army Air Forces Training Command are not clear as to whether any actually served in a training capacity.

Adding to the confusion, the government impressed some 20 civil Stinson Voyagers into the USAAF inventory during 1942, eight model 105s (75-hp to 80-hp Continental engines) as the AT-19A and twelve model 10As (90-hp Franklin engines) as the AT-19B, but someone apparently caught the mistake, as both were re-designated L-9A and L-9B, respectively. One source indicates that the USAAF thereafter allocated the L-9As/Bs to Civil Air Patrol for coastal patrol duties. The government also impounded a number of Lend-Lease V-77s which were subsequently modified for photo survey work and placed into USAAF service as the AT-19C. Finally, 48 civil Reliants impressed into USAAF wartime service as utility transports received the designation UC-81.

Lockheed P/RP/TP-38 and P-322 Lightning—1942

Technical Specifications (P-322/P-38F)

Type: Twin-engine fighter trainer.
Manufacturer: Lockheed Aircraft Corp., Burbank, California.
Total military versions (trainers): 200+ (USAAF).
Unit cost (new): $97,147.

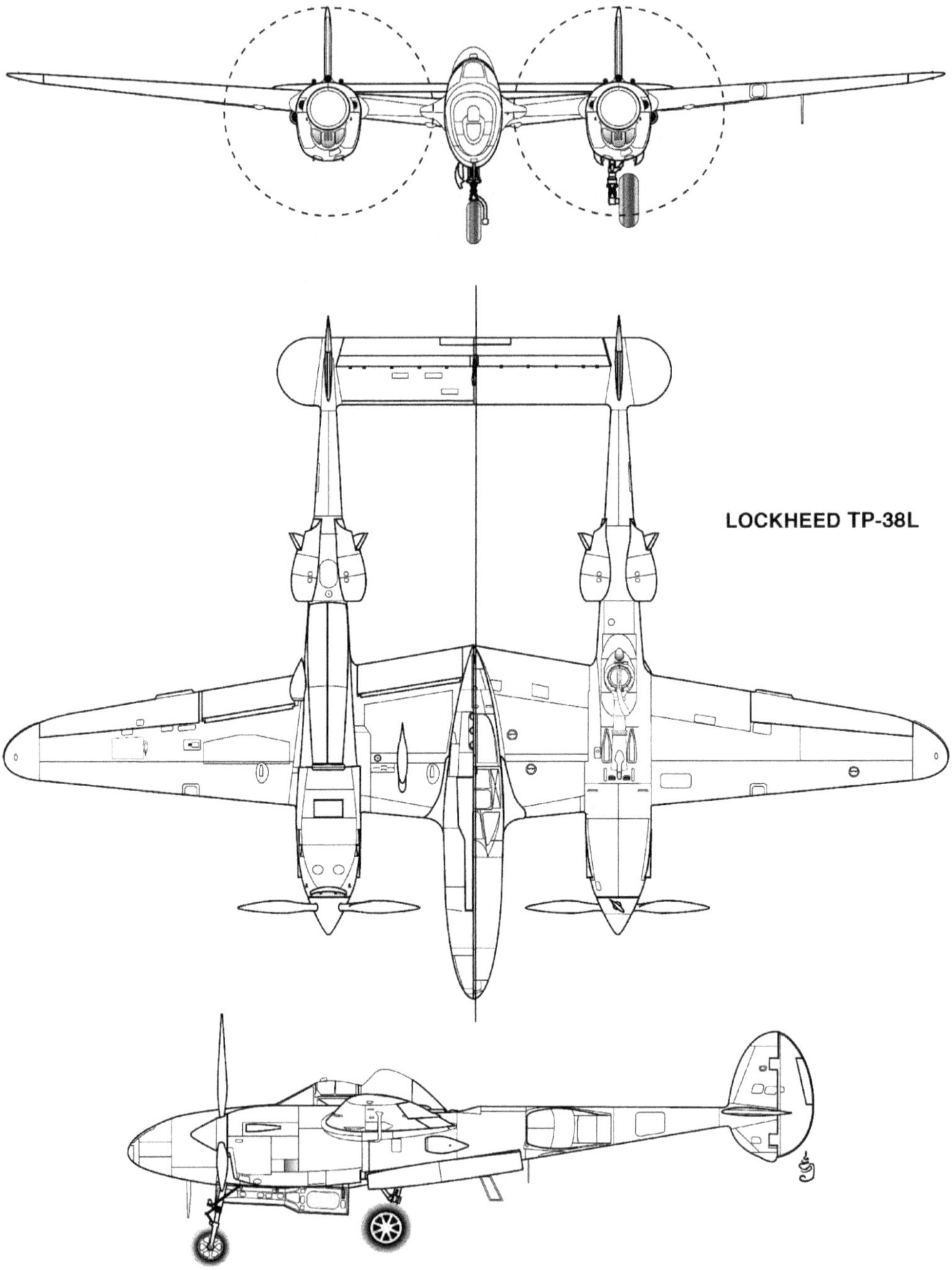

Lockheed P-322, built originally for Great Britain as the *Lightning I*, was employed as a P-38 transition trainer by Western Flying Training Command at Williams Field, Arizona.

Powerplants: Two 1,090-hp Allison V-1710–33 12-cylinder water-cooled inline engines driving three-bladed constant-speed metal propellers.
Armament (P-322): Two .30-calibre and two .50-calibre machine guns in nose.
Performance: Max. speed 390 mph, cruise 300 mph; ceiling (not reported); range 970 mi.
Weights: 10,834 lbs. empty, 14,500 lbs. loaded (est.).
Dimensions: Span 52 ft., length 37 ft. 10 in., wing area 328 sq. ft.

The Lockheed P-38 Lightning, well-known for it prowess as a USAAF fighter in every combat theatre of World War II, also saw some considerable use in the training role. Starting in 1942, 66 P-38Ds received the "R" non-combat designation prefix and were assigned to Army Air Forces Training Command as transition trainers for prospective P-38 pilots. Later the same year, when Britain cancelled its contract for 667 Lightning Is (model 322-I) and IIs (model 322-II), the USAAF took over the order. Of this total, 20 complete model 322-I airframes plus 130 still minus engines were placed on the USAAF inventory under the designation P-322, while the remaining model 322-IIs were ultimately completed to USAAF specifications and delivered to combat units as P-38Fs and P-38Gs. The first 20 model 322-Is/P-322s differed from standard P-38 variants in having no turbo-superchargers and right-hand engines on both sides; the 130 incomplete P-322s were delivered with V-1710–27/-29 engines lacking turbo-superchargers but with left and right-hand, contra-rotating engines. Historical records indicate that most of the P-322s spent their careers as P-38 transition trainers with Western Flying Training Command at Williams Field, Arizona.

Serial number records show that at least three P-38Ls became TP-38Ls, which, after being field-modified to incorporate a second seat aft of the pilot, were used to train radar operators for P-38M nightfighters. Sources also mention several similarly converted TP-38Js, however, none appear on official records.

Federal (Avro) AT-20 Anson—1943

Technical Specifications (AT-20)

Type: Three to four-place multi-engine and navigation advanced trainer.
Manufacturer: Federal Aircraft, Ltd., Montreal, Canada.

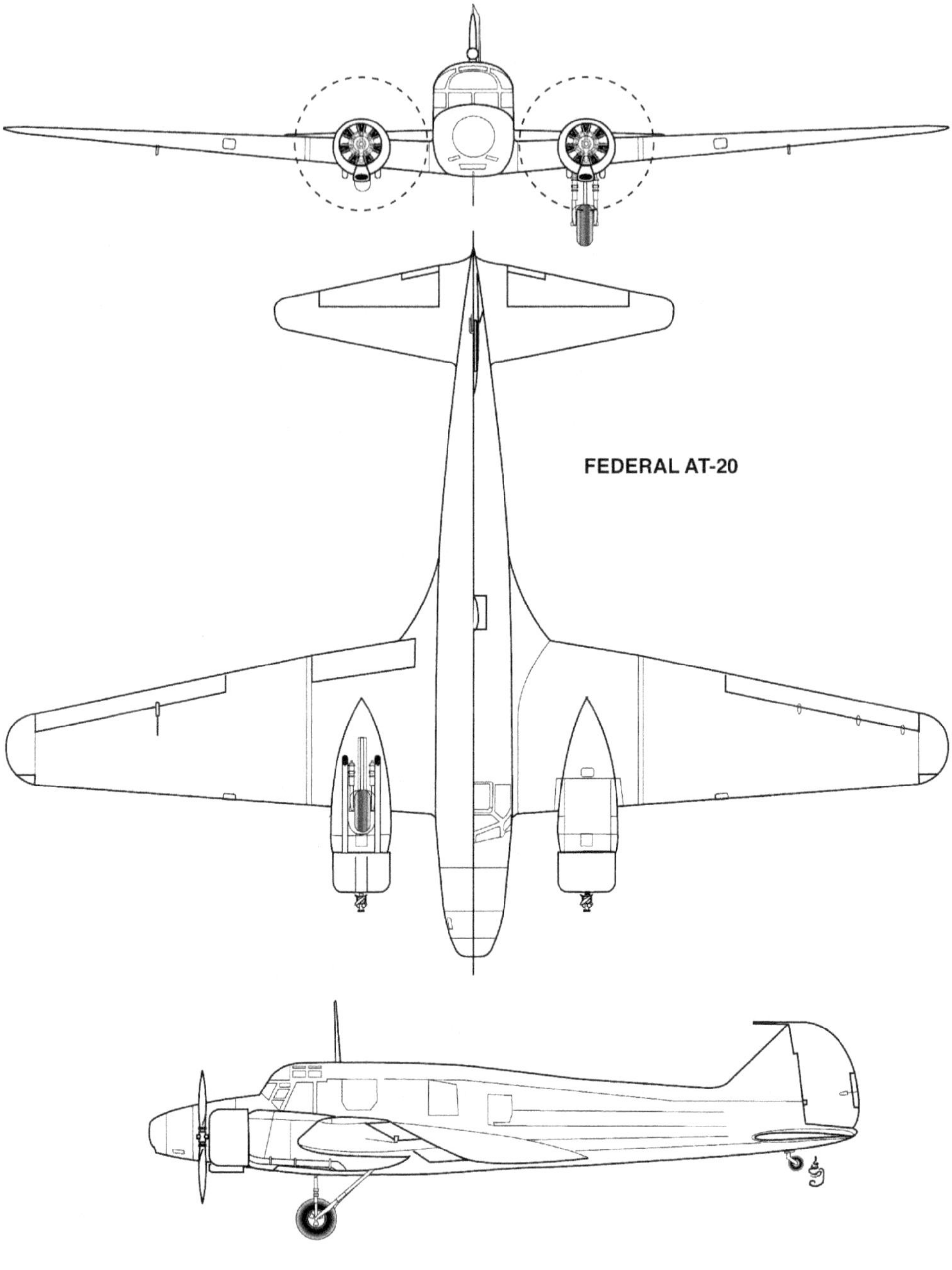

Total U.S. military versions: 50 (USAAF).
Unit cost: Unknown.
Powerplants: Two 330-hp Jacobs R-915–7 7-cylinder air-cooled radial engines driving two-bladed fixed-pitch wooden propellers.
Performance: Max. speed 188 mph, cruise 155 mph; ceiling 16,200 ft.; range 790 mi.
Weights: 5,880 lbs. empty, 7,650 lbs. loaded.
Dimensions: Span 56 ft. 6 in., length 42 ft. 3 in., wing area 463 sq. ft.

As a twin-engine aircraft of British design, the Avro Anson is second only to the Vickers Wellington in terms of total numbers built before and during the war years. It initially became

One of 50 Avro (Federal) Anson Mk. IIs delivered to Army Air Forces Flying Training Command in 1943 under the AAF designation AT-20. License-produced in Canada.

operational with the RAF in 1936 in the coastal patrol role, but, in six major variants, saw far wider application in the British Commonwealth Air Training Plan as a multi-engine, navigator, and gunnery trainer. During the course of the war, 1,822 Anson Mk. IIs were license-produced by Federal Aircraft, Ltd. in Canada, which differed from the Mk. I in having American-made Jacobs R-915–7 engines and a hydraulic gear retraction system but retained the former's fabric-covered, tubular steel fuselage frame and built-up wooden wings and tail surfaces. During 1943, 50 Mk. IIs were diverted to the USAAF under the designation AT-20 and used for multi-engine training in the Central and Western Flying Training Commands. Ultimate disposition of the type is not mentioned in any historical references.

Consolidated AT-22/TB-24 Liberator—1943

Technical Specifications (AT-22/TB-24D)

Type: Engineer, gunnery, and radar advanced trainers.
Manufacturer: Consolidated Aircraft Corp., Fort Worth, Texas.
Total military versions: 65 (USAAF).
Unit cost (new): $297,627.
Powerplants: Four 1,200-hp Pratt & Whitney R-1830–43 14-cylinder air-cooled radial engines driving three-bladed Hamilton Standard constant-speed metal propellers.
Performance: Max. speed 300 mph, cruise 215 mph; ceiling 28,000 ft.; range 2,140 mi.
Weights: 30,645 lbs. empty, 56,000 lbs. loaded.
Dimensions: Span 110 ft., length 66 ft. 4 in., wing area 1,048 sq. ft.

The Consolidated B-24 Liberator bomber and its C-87 Liberator Express counterpart hold the record of having been produced in greater numbers (18,482 built between 1939 and 1945) than any other type of American military aircraft. While the history of the bomber and transport variants has been extensively documented, the type's service as a trainer is not nearly as well known. In addition to Operational Training Units assigned to numbered Air Forces of the USAAF, unmodified B-24s served with Army Air Forces Training Command (AAFTC) as four-engine transition trainers at Fort Worth AAF, Texas, and Ardmore AAF, Oklahoma.

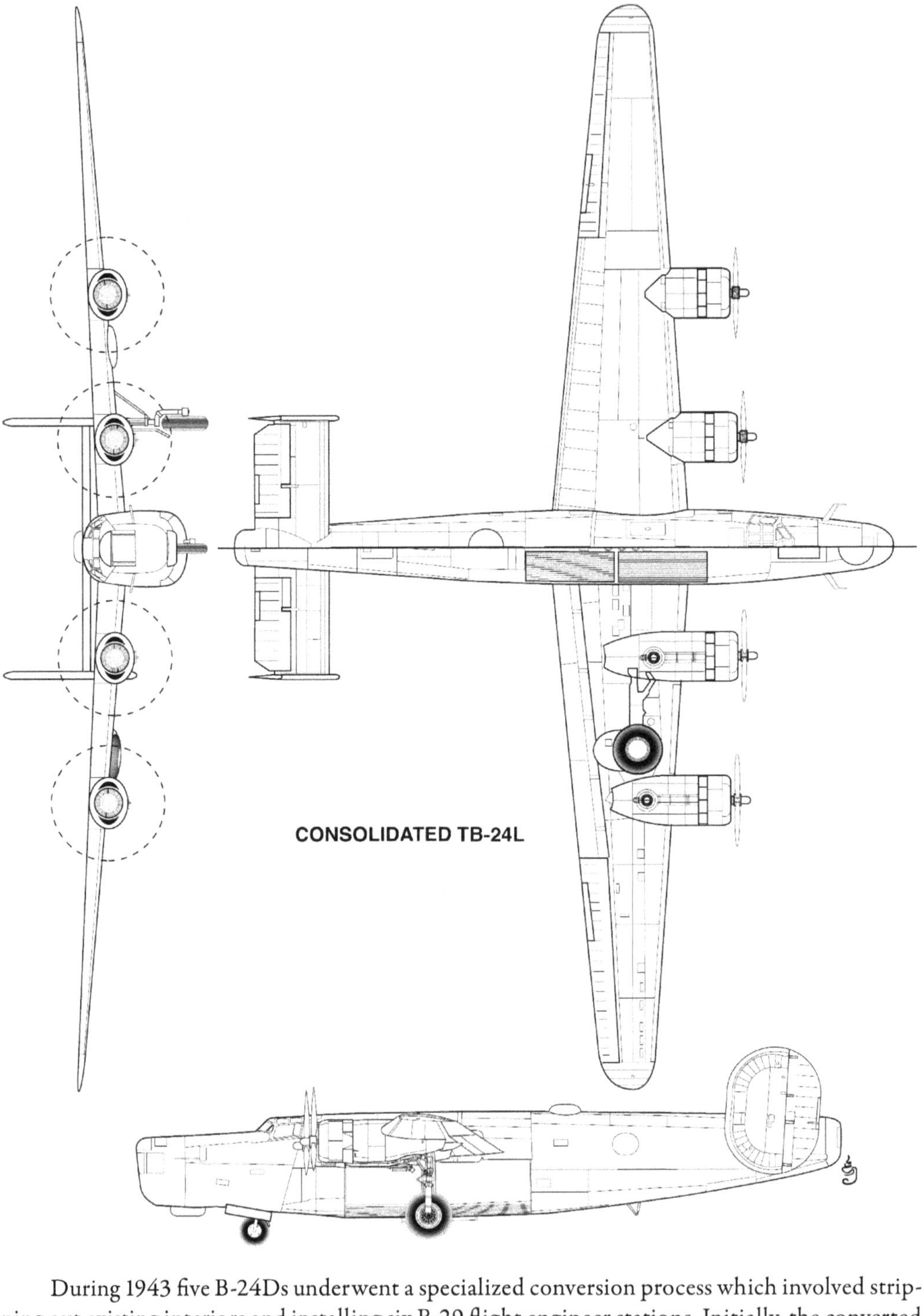

During 1943 five B-24Ds underwent a specialized conversion process which involved stripping out existing interiors and installing six B-29 flight engineer stations. Initially, the converted aircrft received the designation AT-22 but were later re-designated TB-24D. As the tempo of B-29 training increased in 1944 and 1945, serial number records indicate that at least 60 B-24Ls modified to incorporate B-29 Central Fire Control remote gunnery systems and APQ-13 radar

A Consolidated B-24D that became an AT-22 after being stripped of armament and bombing equipment. Interior housed five B-29 flight engineer stations. Later redesignated TB-24D.

bombing sets returned to service as the TB-24L. It appears that many of these TB-24Ls had previously been converted into C-109 fuel transports. Army Air Forces Training Command records indicate that B-29 flight engineer training was conducted at Amarillo AAF, Texas, and Lowry AAF, Colorado, and gunnery and specialized aircrew training at Buckingham AAF, Florida, Las Vegas AAF, Nevada, and Harlingen AAF, Texas. All TB-24s were withdrawn from training duties soon after the war ended in September 1945, with most going to one of five War Assets Administration reclamation centers to be scrapped.

Martin AT-23/TB-26 (JM) Marauder—1943

Technical Specifications (AT-23A/TB-26B)

Type: Aircrew advanced trainers and target tugs.
Manufacturer: Glenn L. Martin Co., Baltimore, Maryland, and Omaha, Nebraska.
Total military versions: 615 (343 USAAF; 272 USN/USMC).
Unit cost (new): $102,629.
Powerplants: Two 1,920-hp Pratt & Whitney R-2800–43 18-cylinder air-cooled radial engines driving four-bladed constant-speed metal propellers.
Performance: Max. speed 282 mph, cruise 214 mph; ceiling 21,700 ft.; range 1,150 mi.
Weights: 24,000 lbs. empty, 37,000 lbs. loaded.
Dimensions: Span 71 ft. 7 in., length 58 ft. 3 in., wing area 658 sq. ft.

Although its reputation was marred by a high incidence of accidents in early training, the Martin B-26 Marauder went on to fly more missions and destroy more targets, with fewer loses, than any other type of USAAF bomber participating in the European Theater of Operations

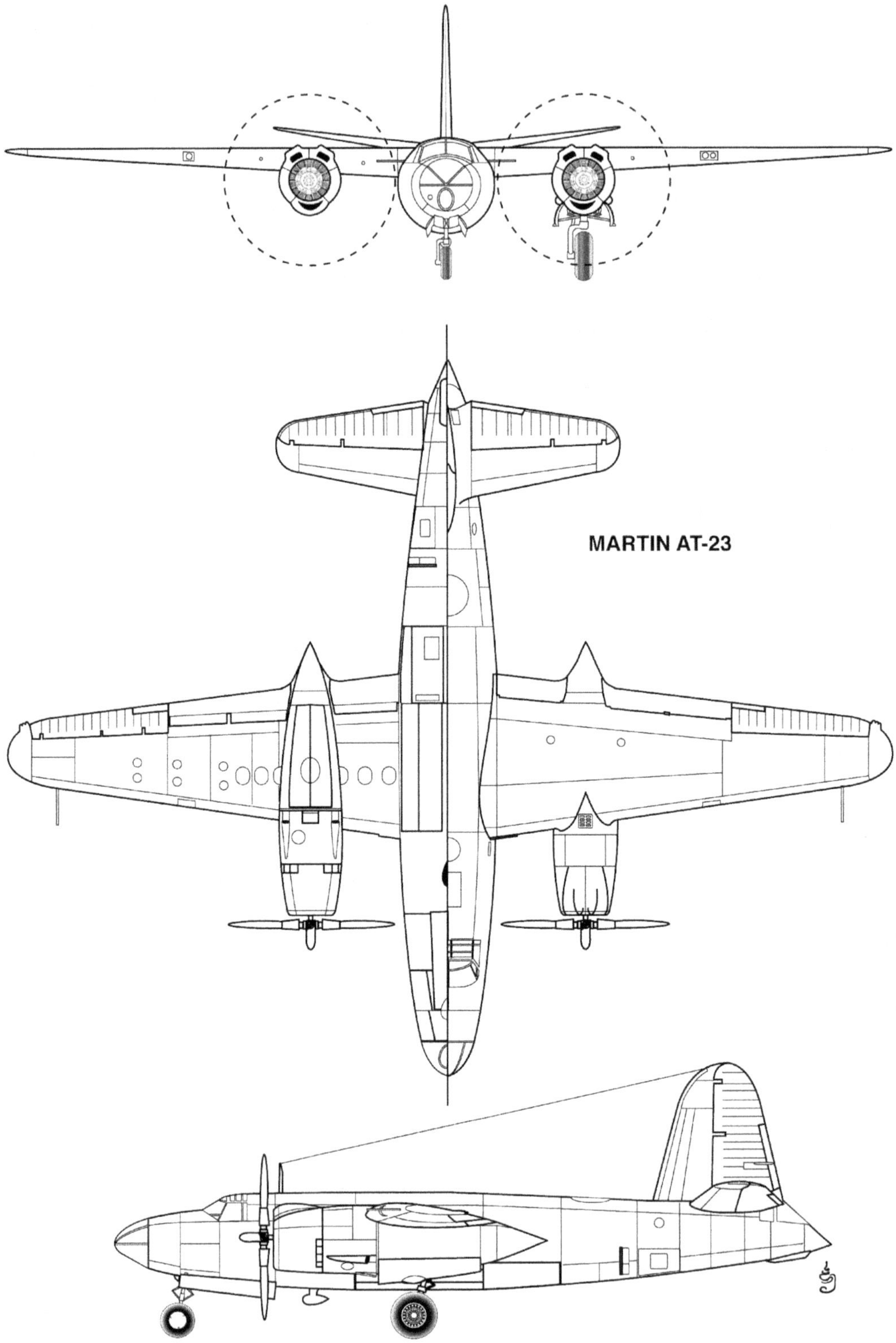

MARTIN AT-23

This Martin AT-23B is one of 208 B-26Bs stripped of armament and bombing equipment and refitted for target towing. Used for both pilot transition training and target towing at gunnery schools.

during World War II. Many unmodified Marauder variants (B-26 through B-26G) saw extensive use in Operational Training Units assigned to various numbered Air Forces of the USAAF and in multi-engine transition schools of Army Air Forces Training Command (AAFTC).

The first dedicated trainer variants, 208 B-26Bs stripped of armor, armament, and bombing equipment and refitted for target towing, reentered service during 1943 under the designation AT-23A, and 350 similarly equipped B-26Cs became the AT-23B. In mid–1944, both types were re-designated TB-26B and TB-26C, respectively. During 1944 57 B-26Gs were completed as the TB-26G and went directly to AAFTC. Historical records suggest that TB-26s were used by AAFTC for multi-engine transition schools as well as high-speed target tugs in various flexible and fixed gunnery schools. The process of phasing out TB-26s from training duties began in 1945 and all had been withdrawn before the end of the year. Most TB-26s that survived the war went to War Assets Administration sites where they were stripped of usable parts and chopped-up for scrap.

Naval Variants: Starting in 1945, 225 TB-26Cs transferred to the Navy received the naval designation JM-1, followed by 47 TB-26Gs as the JM-2. Most were used for target towing in gunnery training, though a few JM-1s became JM-1Ps after installation of camera equipment. At least one JM-1 served with the Marines as a target tug. All JMs were retired from service during 1946.

North American AT-24/TB-25 Mitchell—1943

Technical Specifications (TB-25J)

Type: Multi-engine, aircrew, and weapons system advanced trainers.
Manufacturer: North American Aviation, Inc., Inglewood, California, and Kansas City, Missouri.
Total military versions: 780 (USAAF/USAF/ANG).
Unit cost (new): $142,194.
Powerplants: Two 1,700-hp Wright R-2600–29 14-cylinder air-cooled radial engines driving three-bladed constant-speed metal propellers.
Performance: Max. speed 293 mph, cruise 242 mph; ceiling 24,500 ft.; range 1,520 mi.
Weights: 19,490 lbs. empty, 27,560 lbs. loaded.
Dimensions: Span 67 ft. 7 in., length 53 ft. 6 in., wing area 610 sq. ft.

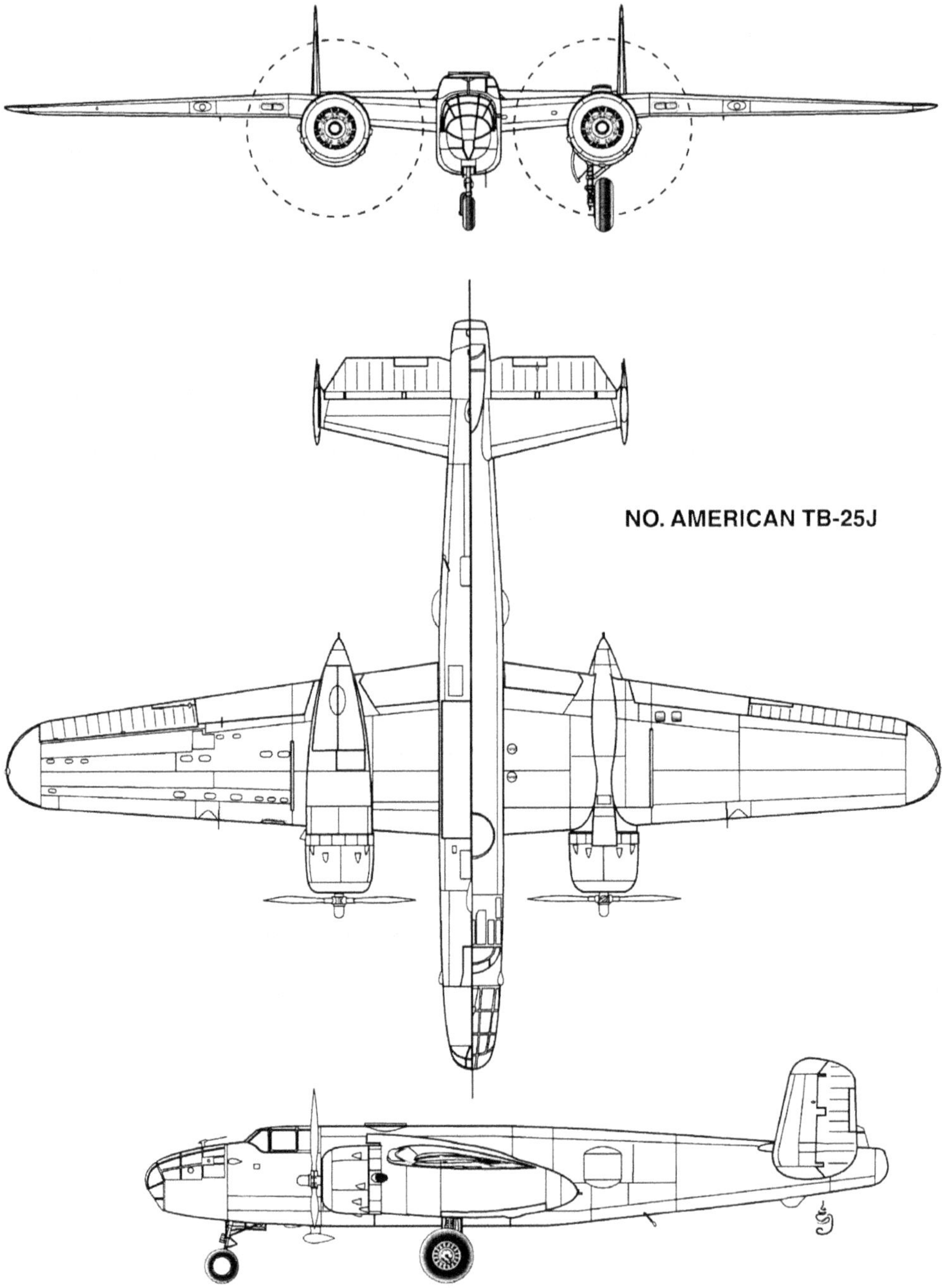

The most-produced (9,816 built) and arguably the best all-around USAAF medium bomber of World War II, the celebrated North American B-25 Mitchell actually served most of its 25-year operational career as a trainer. Besides being an excellent bombing and strafing platform, the B-25 possessed stable and predictable handling qualities that made it ideal for transitioning relatively inexperienced pilots to heavier, multi-engine aircraft. Thus, starting in 1943, older B-25 variants, removed from combat operations and stripped of armor, armament, and bombing

An early AT-24D converted from a B-25J. It was later redesignated TB-25J. In the immediate postwar era, TB-25Js became Air Training Command's principal type of multi-engine trainer.

Well-used TB-25J serving with the Missouri Air National Guard in the 1950s. The very last TB-25Ms were not withdrawn from ANG units until 1966.

equipment, were returned to duty as advanced trainers. The first batch, 60 B-25Ds, became multi-engine trainers under the new designation AT-24A, and they were followed in 1943 and 1944 by similar numbers of B-25Gs as the AT-24B, B-25Cs as the AT-24C, and B-25Js as the AT-24D. In mid–1944, the designation system changed and these aircraft became the TB-25D, TB-25G, TB-25C, and TB-25J. Early on, AT-24s/TB-25s assigned to Army Air Force Training Command (AAFTC) served primarily in multi-engine transition training but from mid 1944, they began to replace AT-9s, AT-10s, and AT-17s in advanced twin-engine schools. Though some

B-25Hs stripped of armament were re-designated TB-25H, all are thought to have served in non-training roles with the Air Weather Service.

In the immediate postwar years, as AAFTC was restructured into the much smaller Air Training Command (ATC), TB-25Js became its principal type of twin-engine advanced trainer, and from 1946 onwards, some 600 converted B-25Js returned to service as TB-25Js. Most went to ATC, but some served with other commands as instrument, proficiency, and crew trainers. After 1950, due to the Korean conflict and escalation of the Cold War, additional B-25Js were pulled from storage, stripped of armor, armament, and bombing equipment in order to be modified for training needs: 117, equipped with nose radomes and Hughes E-1 fire control systems for training bombardier-navigators in two-seat Lockheed F-94 fighter-interceptors, returned to service as the TB-25K; 90, rebuilt and modified by Hayes Aircraft Co. for specialized advanced pilot training, reemerged as the TB-25L; 40 equipped with nose radomes and Hughes E-5 fire control systems for training bombardier-navigators in two-seat Northrop F-89 fighter-interceptors, became the TB-25M; and 47, undergoing a rebuild similar to the TB-25L but upgraded with R-2600–29A engines, returned to service as the TB-25N. Some TB-25Ns were later refitted as VIP transports under the designation VB-25N.

From late 1953 to late 1954, all TB-25Js remaining in the ATC fleet underwent a refurbishment program that included an IRAN (inspect and repair as needed) and installation of autopilots, bomb bay fuel tanks, new radios, and demand oxygen systems, 60 of which were refitted with solid nose shells in place of the original transparent nose. Over the course of their careers, TB-25Js logged over 2 million flight hours and trained 30,000 new twin-engine pilots. The final examples were not retired from ATC until January 1960. Beginning in the mid–1950s, TB-25Ks and Ms were transferred to 16 different Air National Guard units to support Fighter Interceptor Squadrons equipped with Lockheed F-94s or Northrop F-89s, with the last being withdrawn sometime in 1966.

Boeing TB-17 Flying Fortress—1943

Technical Specifications (TB-17G)

Type: Multi-engine advanced trainer.
Manufacturer: Boeing Airplane Co., Seattle, Washington,; Vega Airplane Div. of Lockheed Aircraft Corp., Burbank, California; and Douglas Aircraft Co., Long Beach, California.
Total military versions: 780 (USAAF/USAF/ANG).
Unit cost (new): $204,370.
Powerplants: Four 1,200-hp Wright R-1820–97 9-cylinder air-cooled radial engines driving three-bladed constant-speed metal propellers.
Performance: Max. speed 287 mph, cruise 182 mph; ceiling 35,600 ft.; range 2,000 mi.
Weights: 36,135 lbs. empty, 65,500 lbs. loaded.
Dimensions: Span 103 ft. 9 in., length 74 ft. 4 in., wing area 1,420 sq. ft.

Many historians consider the Boeing B-17 to be the most famous heavy bomber of all time. When it appeared for the first time in mid 1935, its speed, range and payload were unmatched by any other bomber in the world. The B-17 also possessed well-mannered handling qualities that made it a good choice for transitioning new pilots to a large, four-engine aircraft. During World War II, unmodified B-17 variants (B-17B through B-17G) were utilized by Operational Training Units assigned to the various numbered Air Forces within the USAAF, but, starting in 1943, as war-weary B-17Fs became available, they were stripped of armor, armament, and bombing equipment and turned into dedicated trainers; however, the exact number re-designated as train-

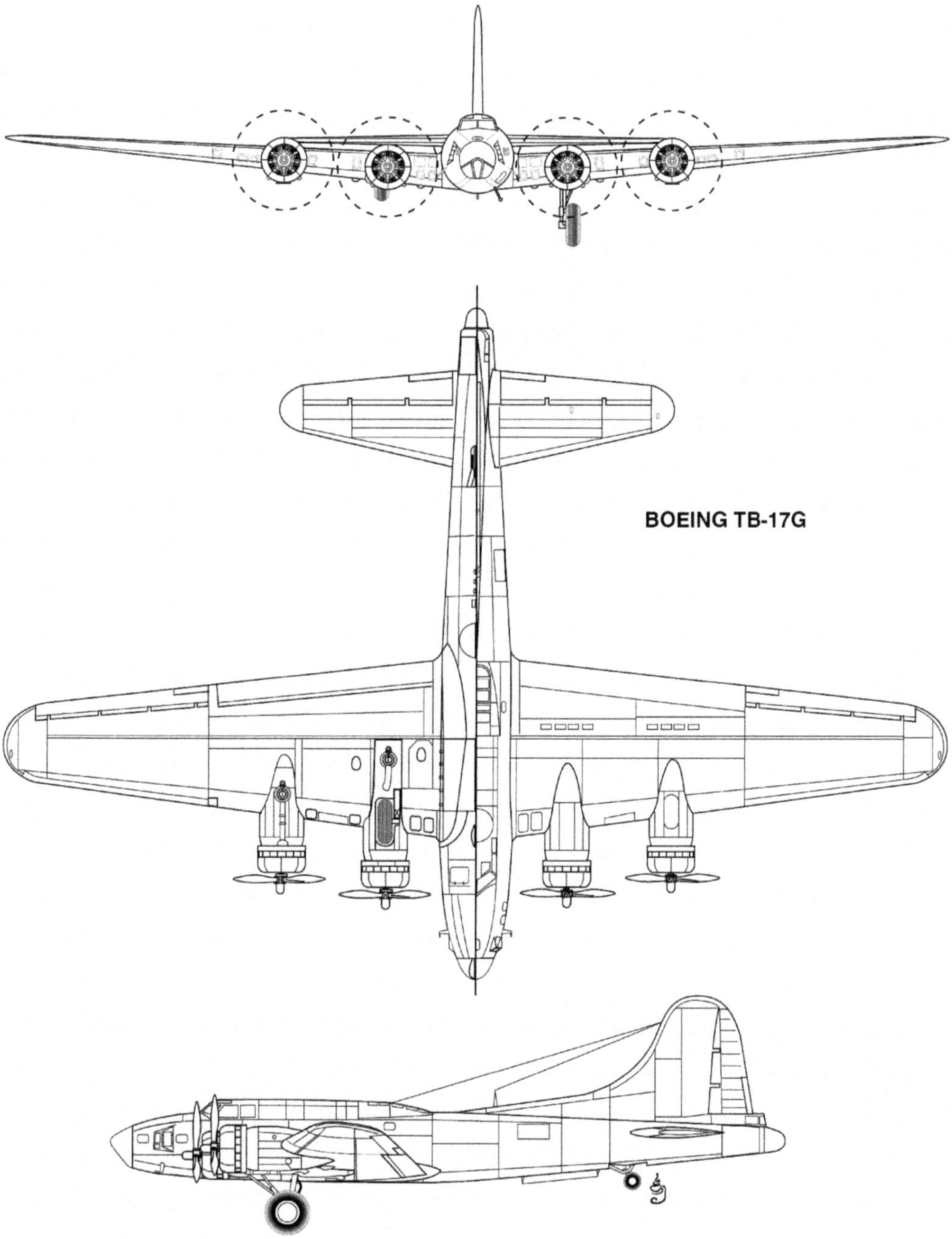

ers is not clear. While USAAF serial number records only identify five aircraft as the TB-17F and one as the RB-17F (R = non-combat status), many more are listed as not having been removed from service until mid or late 1945, well past their combat careers, which suggests that some may have been modified for the training role without an official change in designation. The records do reflect that at least 180 B-17Gs underwent a conversion process and returned to service as the TB-17G. (Note: RB-17Gs were never trainers but photo-reconnaissance F-9Cs re-designated by the USAF in 1948.)

This TB-17G saw postwar service with Air Training Command before being consigned to the Air Force Museum. Last examples were phased out of training duties in 1951.

During World War II, Army Air Forces Training Command operated TB-17Fs/Gs in seven different four-engine transition schools in Florida, Kansas, New Mexico, Ohio, Tennessee, and Texas. In the aftermath of the war, a smaller number of TB-17Gs remained active for four-engine transition training with the newly established Air Training Command (ATC). During this time, a few B-17Gs that had been modified for search and rescue as the SB-17H were converted to trainers under the designation TB-17H. ATC phased out all TB-17Gs from training duties in 1951, though some, converted into VIP transports as the VB-17G, remained active until 1960.

Bell P/RP/TP-39 Airacobra—1943

Technical Specifications (P-39Q)

Type: Single-engine fighter trainer.
Manufacturer: Bell Aircraft Corp., Buffalo, New York.
Total military versions (trainers): Unknown (USAAF).
Unit cost: $50,666.
Powerplant: One 1,325-hp Allison V-1710–85 12-cylinder water-cooled inline engine driving a three-bladed constant-speed metal propeller.
Armament: One 37-mm cannon and two .50-calibre machine guns in the nose, four pod-mounted .30-calibre machine guns in the wings, and up to 500 lbs. of ordinance carried externally.
Performance: Max. speed 376 mph, cruise 250 mph; ceiling 32,100 ft.; range 525 mi.
Weights: 5,645 lbs. empty, 8,400 lbs. loaded.
Dimensions: Span 34 ft., length 30 ft. 2 in., wing area 213 sq. ft.

Although the Bell P-39 Airacobra was one of the USAAF's principal fighters at the start of World War II, it was progressively withdrawn from combat as more capable aircraft (e.g., P-38, P-47, and P-51) arrived to take its place. An unorthodox design having its engine located behind the pilot to make space for heavy cannon armament in the nose, it suffered from lack of

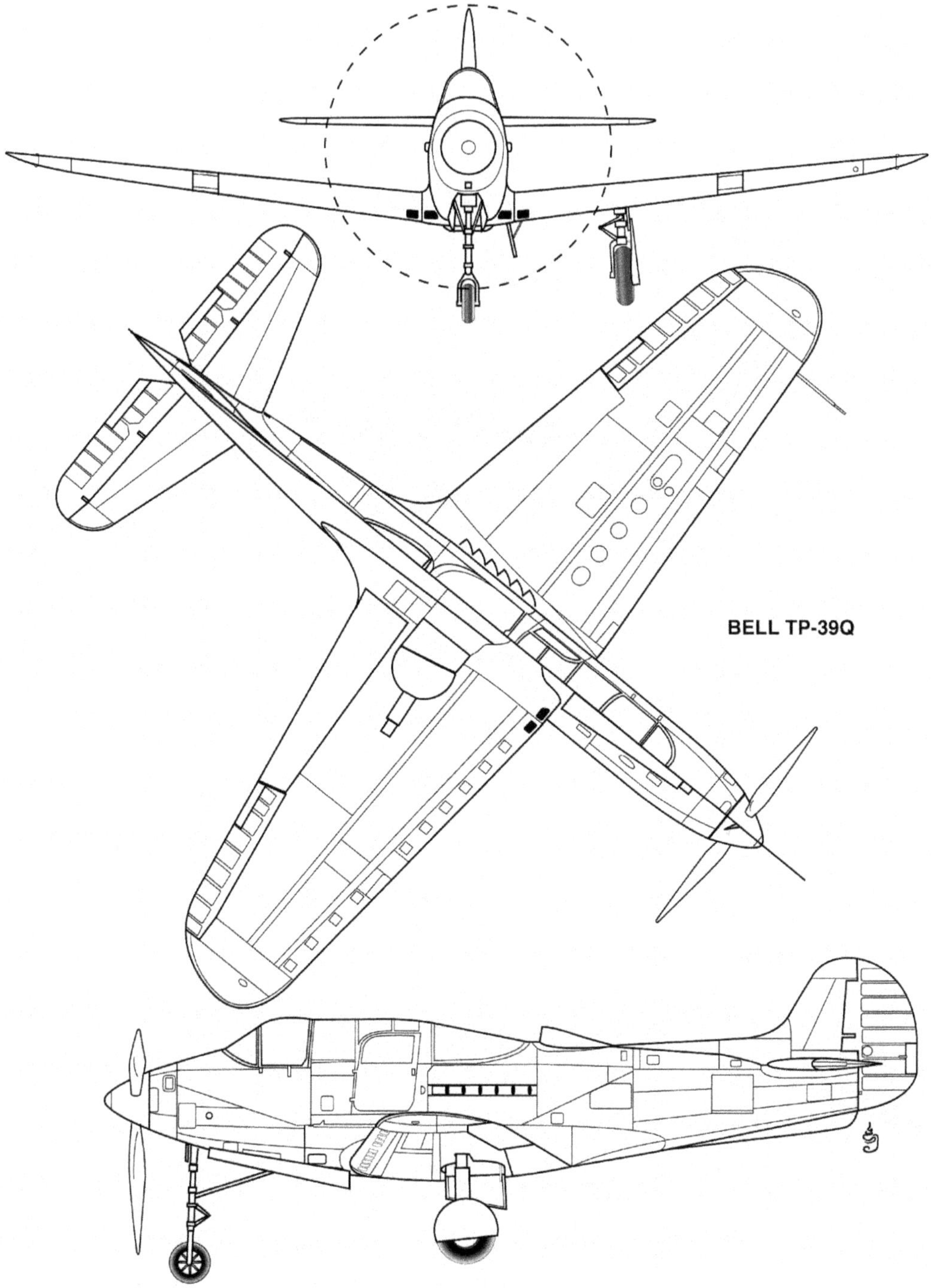

range and poor performance above 15,000 feet. As P-39s were removed from combat, they became one of the most numerous fighter types to be employed by both Operational Training Units (OTU) and by Army Air Forces Training Command (AAFTC) in its transition schools, and from late 1943, many brand new P-39Qs went directly from the Bell factory to OTUs or AAFTC. To provide a dual-control trainer in transition schools, some P-39s had their nose armament

One of an undetermined number of Bell P-39Qs that became TP-39Qs after removal of the nose cannon and installation of a second cockpit. Many brand new P-39Qs went straight from the factory to training bases.

removed, with a second cockpit position added in front of the existing cockpit. Serial number records indicate that at least 30 P-39Qs, one as the TP-39Q and 29 as the RP-39Q, were converted, though there may have been more. Two aircraft are listed as the TP-39N, but available information does not reveal whether they were actually converted into a two-seat configuration. All P/RP/TP-39s were withdrawn from training duties during 1945.

Curtiss P/RP/TP-40 Warhawk—1943

Technical Specifications (P-40N)

Type: Single-engine fighter trainer.
Manufacturer: Curtiss-Wright Corp., Buffalo, New York and St. Louis, Missouri.
Total military versions (trainers): Unknown (USAAF).
Unit cost: $44,892.
Powerplant: One 1,200-hp Allison V-1710–81 12-cylinder water-cooled inline engine driving a three-bladed constant-speed metal propeller.
Armament: Six .50-calibre machine guns in the wing and up 1,500 lbs. of ordinance carried externally.
Performance: Max. speed 343 mph, cruise 288 mph; ceiling 31,000 ft.; range 1,080 mi.
Weights: 6,200 lbs. empty, 8,350 lbs. loaded.
Dimensions: Span 37 ft. 4 in., length 33 ft. 4 in., wing area 236 sq. ft.

The Curtiss P-40 Warhawk was the USAAF's numerically most important fighter at the time the U.S. entered World War II, but like the P-39, reported above, was increasingly removed

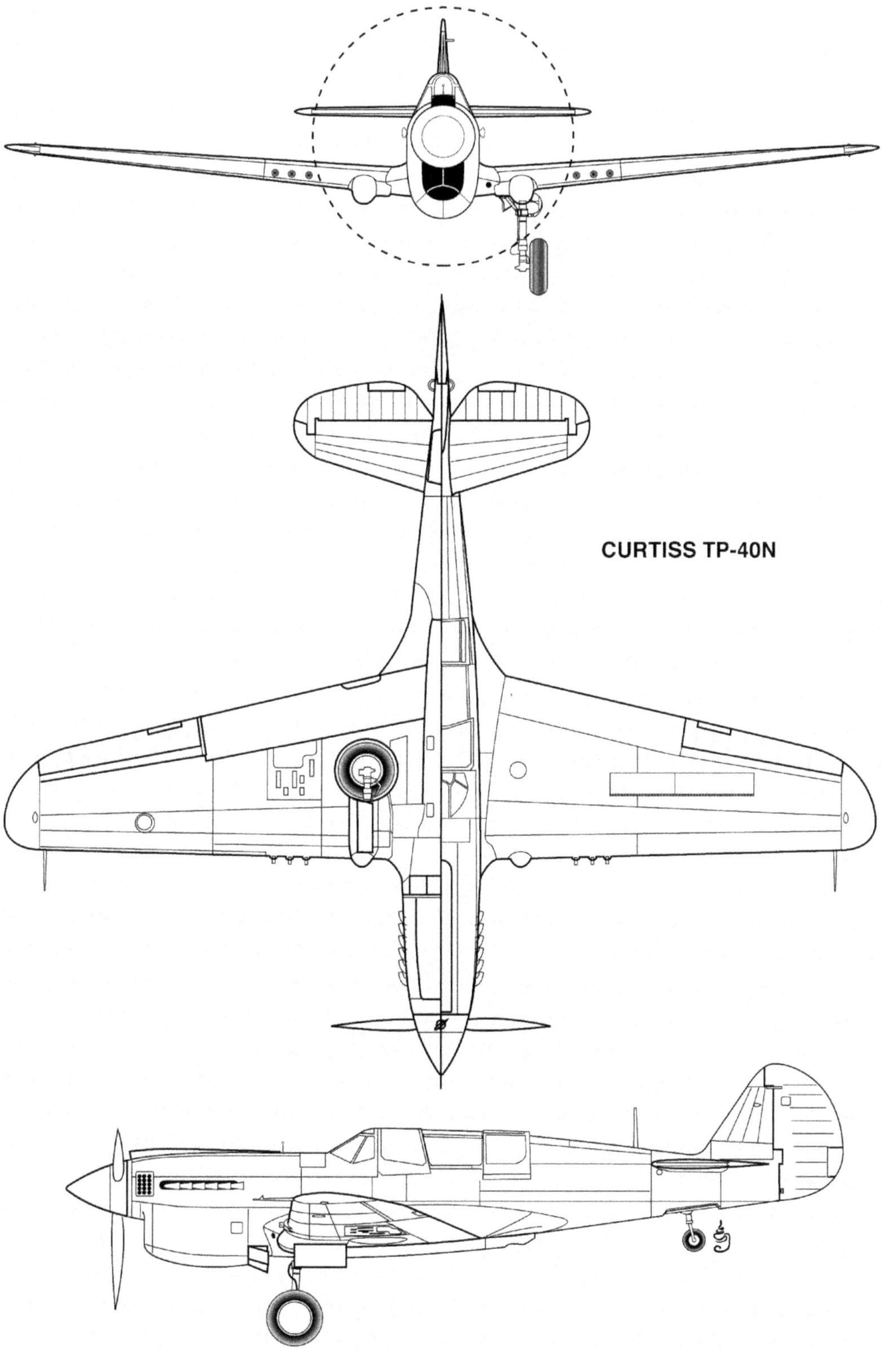

CURTISS TP-40N

At least 37, perhaps more, P-40Ns were modified at the factory for installation of a second cockpit as seen in this photograph of a TP-40N. They were most likely distributed among the five fighter transition schools.

from combat as more P-38s, P-47s, and P-51s rolled off the assembly lines. Starting in 1942, as they became available, significant numbers of unmodified P-40 variants (P-40B through P-40N) served both with Operational Training Units (OTU) and transition schools in Army Air Forces Training Command (AAFTC), and many late-production P-40Ns went directly from the factory to OTUs and AAFTC.

Some single-seat P-40Ns are reported to have been re-designated as TP-40Ns, but the limited data available does not disclose whether they were modified in any way. In any case, serial number records do indicate that at least 37 P-40Ns were factory-modified as dual-control trainers with a second cockpit installed aft of the existing cockpit, 12 as the TP-40N and 25 as the RP-40N. Specific assignment of TP/RP-40Ns is not known, but they were most likely distributed among the five AAFTC P-40 transition schools located in Arizona (Luke AAF and Williams AAF), Texas (Moore AAF), and Alabama (Tuskegee AAF). All P-40 training was discontinued in 1945.

Curtiss (Republic) TP-47 Thunderbolt—1943

Technical Specifications (TP-47G)

Type: Single-engine, two-seat fighter trainer.
Manufacturer: Curtiss-Wright Corp., St. Louis, Missouri.
Total military versions (trainers): 2 (USAAF).
Unit cost: $73,100.
Powerplant: One 2,300-hp Pratt & Whitney R-2800–21 18-cylinder air-cooled radial engine driving a four-bladed constant-speed metal propeller.

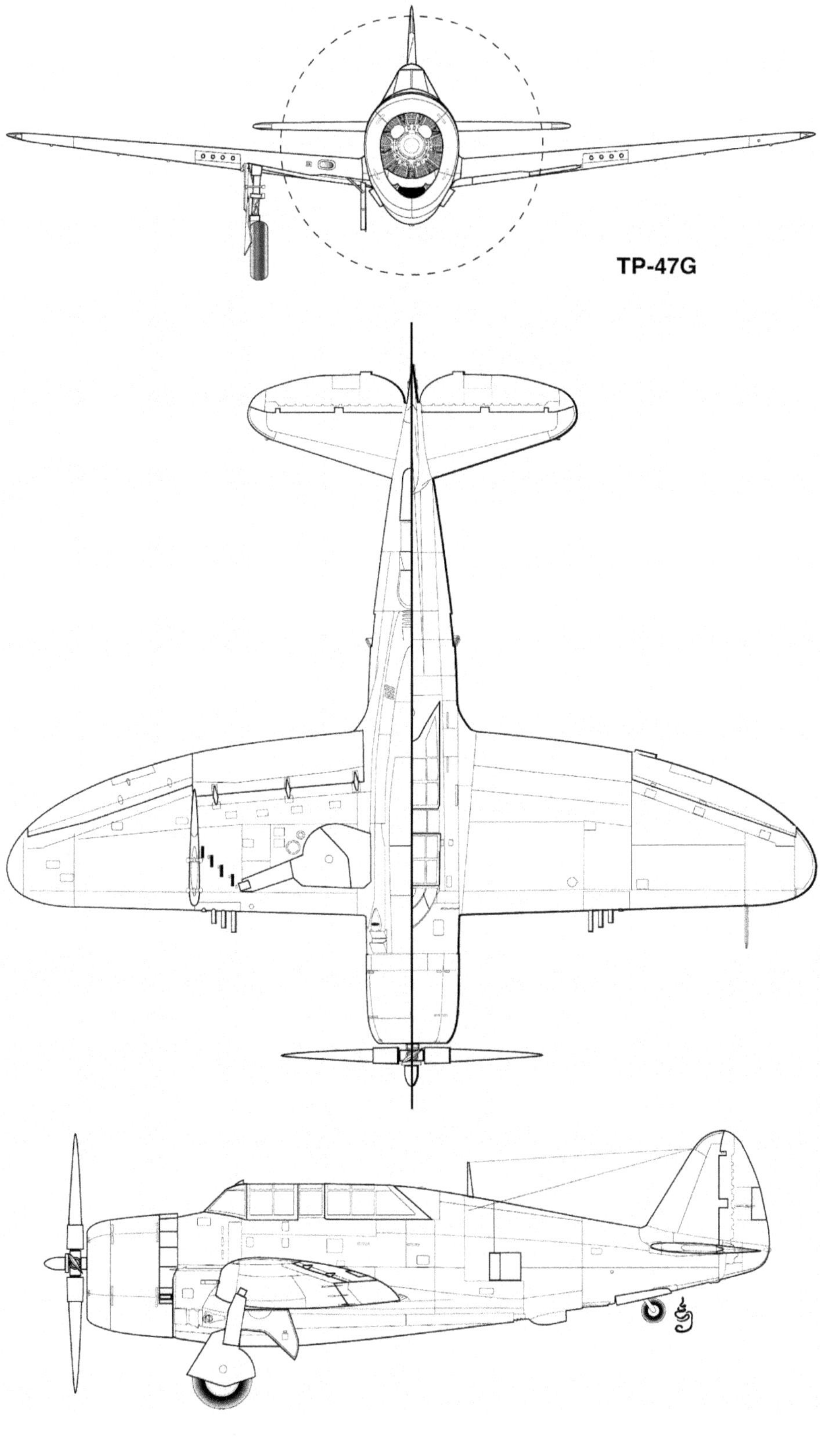

TP-47G

Unofficially known as the "Doublebolt," the photograph shows one of two Curtiss-built TP-47Gs that had been completed in a two-seat configuration on the production line.

Armament: Eight .50-calibre machine guns in the wing and up 2,000 lbs. of ordinance carried externally.
Performance: Max. speed 433 mph, cruise 231 mph; ceiling 40,000 ft.; range (not reported).
Weights: 9,000 lbs. empty, 14,000 lbs. loaded.
Dimensions: Span 40 ft. 9 in., length 36 ft. 1 in., wing area 300 sq. ft.

The P-47 Thunderbolt was not only the most-produced American fighter of World War II (15,678 built) but also flew more combat sorties (433,435) than any other type of American military aircraft. Sometime in 1942, amid plans to eventually equip Operational Training Units (OTU) and Army Air Forces Training Command (AAFTC) transition schools with P-47s, Curtiss-Wright, a license contractor, was directed to complete two P-47Gs (essentially identical to razorback P-47Ds) in a two-seat configuration with dual controls under the designation TP-47G. Changes included removal of a fuselage fuel tank and extending the canopy forward to enclose a second cockpit. The two aircraft, unofficially known as "Doublebolts," were delivered to the USAAF in 1943, but no further TP-47G production was ordered. Some sources indicate that a number of other P-47s were field-modified to add a second cockpit; however, no additional examples appear in serial number records with a T or R prefix. Unit assignment of the two TP-47Gs subsequent to delivery is not revealed on any USAAF records. The single-place razorback P-47 currently registered to the Fighter Collection in Duxford, England as G-CDVX is reported as having been sold surplus as a TP-47G.

North American TP/TF-51 Mustang—1944

Technical Specifications (TP-51D)

Type: Single-engine, two-seat fighter trainer.
Manufacturer: North American Aviation, Inc., Dallas, Texas.

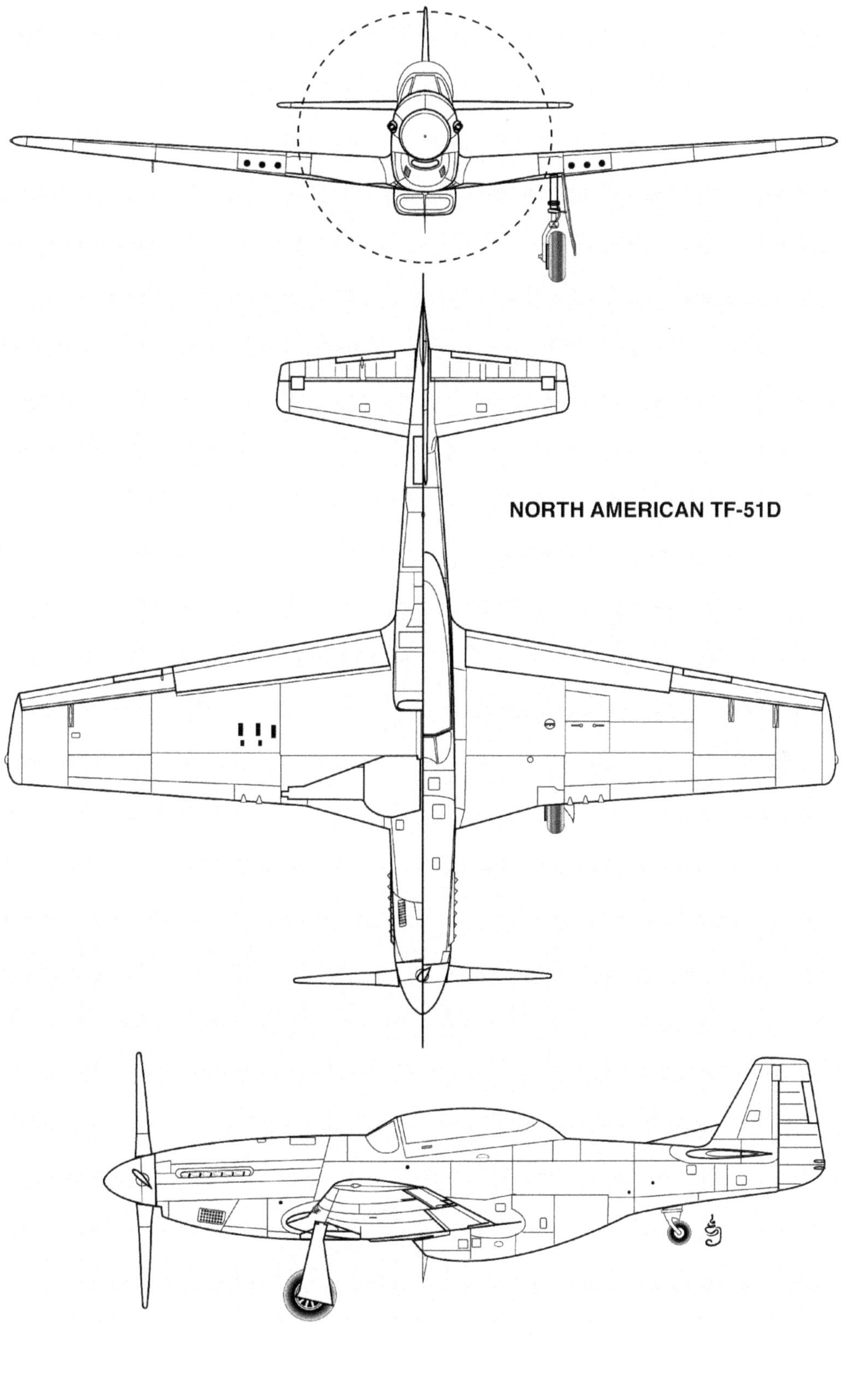
NORTH AMERICAN TF-51D

One of ten TF-51Ds modified on North American's Dallas assembly line in 1945. This example is seen in postwar USAF markings while serving with Air Training Command.

Total military versions (trainers): 25 (USAAF/USAF).
Unit cost: $50,985.
Powerplant: One 1,490-hp Packard (Rolls-Royce) V-1650–7 12-cylinder water-cooled inline engine driving a four-bladed constant-speed metal propeller.
Armament: Six .50-calibre machine guns in the wing and up 2,000 lbs. of ordinance carried externally.
Performance: Max. speed 437 mph, cruise 362 mph; ceiling 41,900 ft.; range (not reported).
Weights: 7,635 lbs. empty, 12,100 lbs. loaded.
Dimensions: Span 37 ft., length 32 ft. 3 in., wing area 235 sq. ft.

The North American P-51 is widely regarded as having been the best all-around USAAF fighter design of the World War II era. In addition to excellent performance and maneuverability, the P-51 was the first fighter in the European Theater to possess the range to escort bombers all the way to and from targets deep inside Germany. Historical records indicate that a second seat was installed behind the pilot in a few early P-51s, the most notable being a modified P-51B (AAF s/n 43–6877) used by General Eisenhower to over-fly the Normandy beachheads shortly after the D-Day invasion; however, none were fitted with dual-controls or known to have been used for training. In late 1944 and early 1945, ten P-51Ds modified on the Dallas assembly line entered USAAF service under the designation TP-51D. The straightforward conversion involved removing an 85-gallon fuel tank behind the pilot and installing in its place a second seat with a full set of controls, and a canopy bulged over the rear seat.

While the actual wartime assignment of the 10 TP-51Ds is not clear, there is a likelihood that some or all served in the Army Air Forces Training Command P-51 transition schools at Luke AAF, Arizona, Alachua AAF, Florida, and Pecos AAF, Texas. These aircraft were re-designated TF-51Ds in 1948 and are thought to have served with USAF Air Training Command during the Korean War. Some historical sources mention that several photo-recon RF-51Ds assigned to Air National Guard units were converted into a two-seat configuration as the TRF-51D during the mid–1950s, but no operational details are available. In 1956 Temco Aircraft Corp. modified 15 USAF F-51Ds, which, in addition dual controls and a second seat, featured a taller vertical fin and new bulged canopy providing more headroom in the rear seat. Although the Temco conversions received USAF serials as the TF-51D, all were allocated to Central and South American air forces under Military Assistance Programs.

Douglas TC-47 Skytrain—1944

Technical Specifications (TC-47B)

Type: Navigator and crew trainer.
Manufacturer: Douglas Aircraft Co., Santa Monica, California.

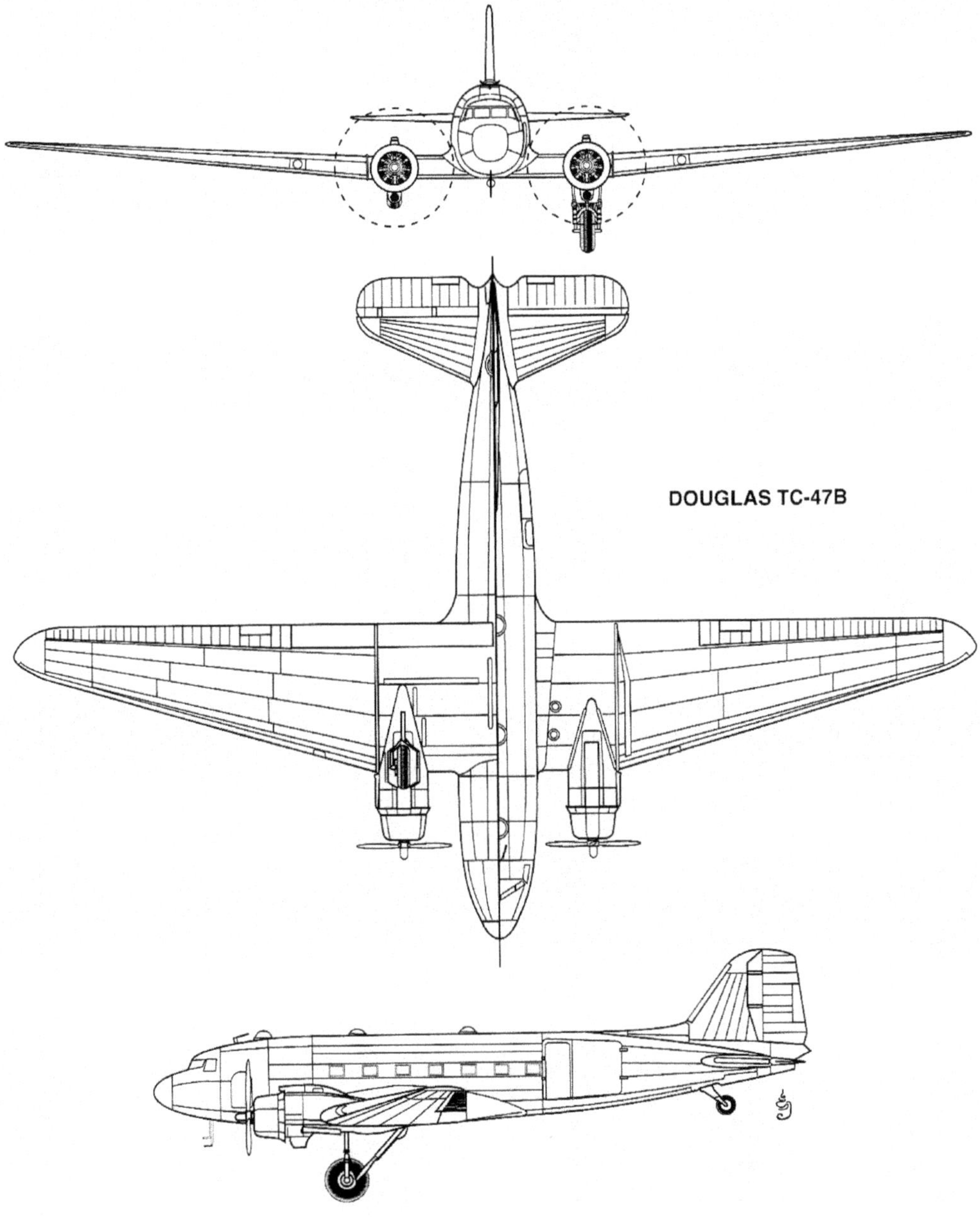

Douglas TC-47B or D seen while serving with ATC in the postwar USAF. After leaving ATC in 1954, some TC-47s continued in service as C-47Ds after being reconfigured as transports.

Total military versions: 133+ (133 USAAF/USAF; unknown USN).
Unit cost: $85,035.
Powerplants: Two 1,200-hp Pratt & Whitney R-1830–90B/C *Twin Wasp* 14-cylinder air-cooled radial engines driving three-bladed Hamilton Standard constant-speed metal propellers.
Performance: Max. speed 224 mph, cruise 160 mph; ceiling 26,400 ft.; range 2,125 mi.
Weights: 18,135 lbs. empty, 26,000 lbs. loaded.
Dimensions: Span 95 ft. 6 in., length 63 ft. 9 in., wing area 987 sq. ft.

In 1944, prior to delivery, 133 C-47Bs were set aside on Douglas's assembly line in Oklahoma City and reconfigured as navigator trainers. Designated TC-47B, the aircraft came with two additional astrodomes and three navigator stations. When delivered to Army Air Forces Training Command in 1944 and 1945, historical references indicate that the TC-47Bs were used primarily for training navigators in celestial navigation. Following World War II, an unknown number of TC-47Bs continued in service with postwar Air Training Command (ATC), and some may have become TC-47Ds when their high-altitude superchargers had been removed. C-47s were listed on ATC's aircraft roster until 1954, however, it is not certain whether any of these were actually TC-47s. Some of the TC-47Bs are known to have been converted back to a transport configuration and ended service as C-47Ds.

Naval Variants: The Navy also dedicated an unknown number of R4D-5s (AAF C-47A) and R4D-6s (AAF C-47B) to training as the R4D-5T and -6T. While historical references suggest that these aircraft were used for navigator training, there is no mention of how they may have been modified for the role. R4D-5Ts and -6Ts still on the Navy inventory in 1962 were redesignated TC-47H/J.

Consolidated TB-32–1945

Technical Specifications (TB-32)

Type: Multi-engine bomber trainer.
Manufacturer: Consolidated Aircraft Corp., Fort Worth, Texas.

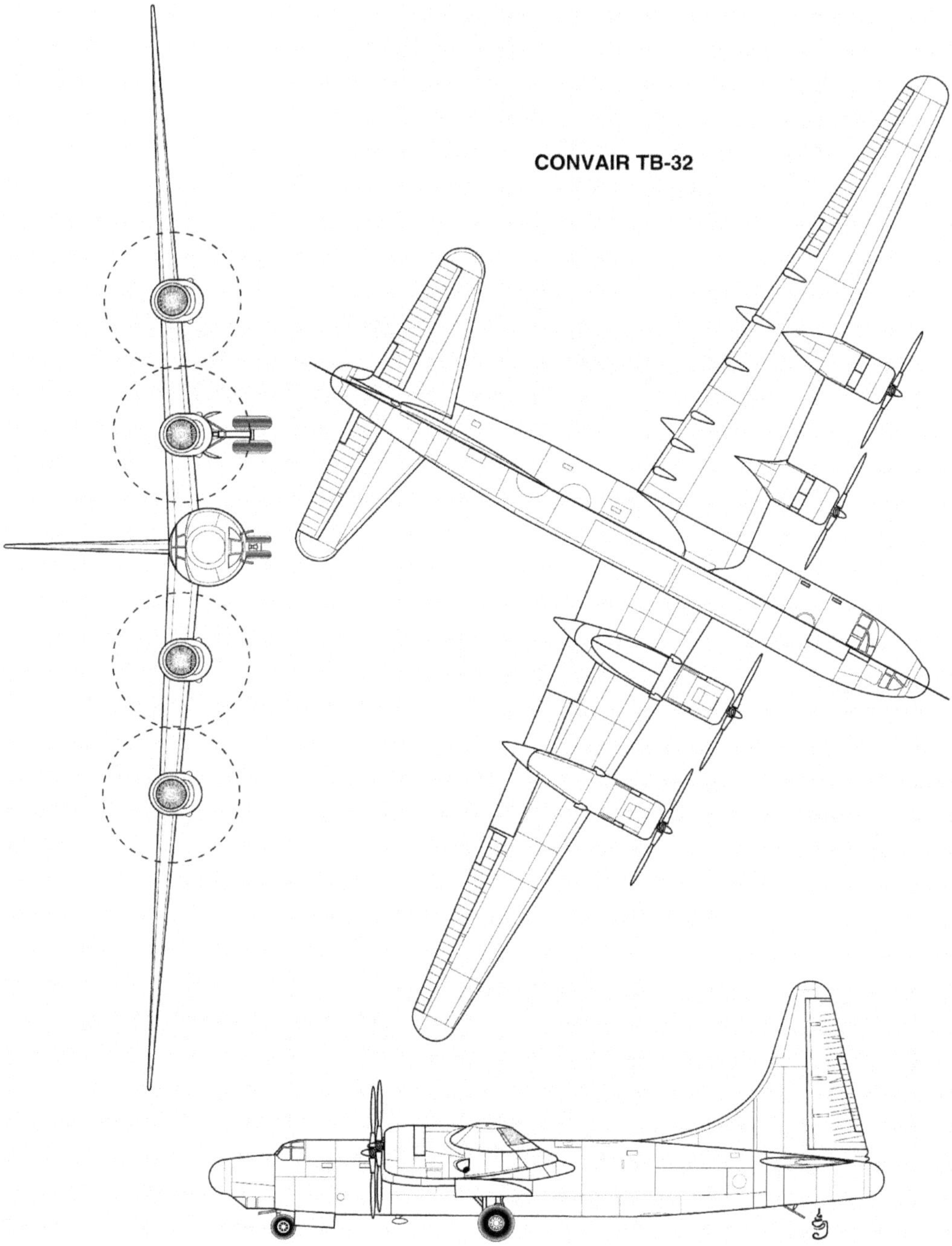

Total military versions: 29 (USAAF).
Unit cost (new): Unknown.
Powerplants: Four 2,200-hp Wright R-3350–23A 18-cylinder air-cooled radial engines driving four-bladed constant-speed metal propellers.
Performance: Max. speed 357 mph, cruise 290 mph; ceiling 30,700 ft.; range 3,800 mi.
Weights: 60,278 lbs. empty, 100,800 lbs. loaded.
Dimensions: Span 135 ft., length 82 ft. 1 in., wing area 1,422 sq. ft.

One of 29 Consolidated TB-32s completed minus armament and bombing equipment and delivered to Army Air Forces Flying Training Command at Fort Worth AAF, Texas, sometime in mid–1945.

Although ordered alongside Boeing's B-29 in 1940, the development and testing of the Consolidated B-32 became protracted to the extent that the first operational examples did not see combat until the final months of World War II. Of 1,900 B-32s ordered by the USAAF, only 118 had been delivered by V-J Day, and the balance of the contract was cancelled. The type's factory name "Denominator" was never officially adopted by the USAAF. Starting in early 1945, the Fort Worth factory completed 29 (some sources claim 40) B-32s minus armament and bombing equipment which were delivered to the USAAF under the designation TB-32. Intended for B-32 pilot and crew transition training, the TB-32s served with the 2519th Base Unit as part of the four-engine bomber school at Fort Worth AAF, Texas. Upon being withdrawn in September and October 1945, the TB-32s were flown directly to War Assets Administration sites to be scrapped.

Curtiss TC-46 Commando—1945

Technical Specifications (TC-46A)

Type: Conversion and transition transport trainer.
Manufacturer: Curtiss Airplane Div., Curtiss-Wright Corp., Buffalo, New York and St. Louis, Missouri.
Total U.S. military versions (trainers): 20 (USAAF).
Unit Cost (C-46A): $259,268.
Powerplants: Two 2,000-hp Pratt & Whitney R-2800–51 *Double Wasp* 18-cylinder air-cooled radial engines driving four-bladed electric fully-controllable metal propellers.
Performance: Max. speed 270 mph, cruise 173 mph; ceiling 24,500 ft.; range 3,150 mi.
Weights: 30,000 lbs. empty, 49,600 lbs. loaded.
Dimensions: Span 108 ft. 1 in., length 76 ft. 4 in., wing area 1,360 sq. ft.

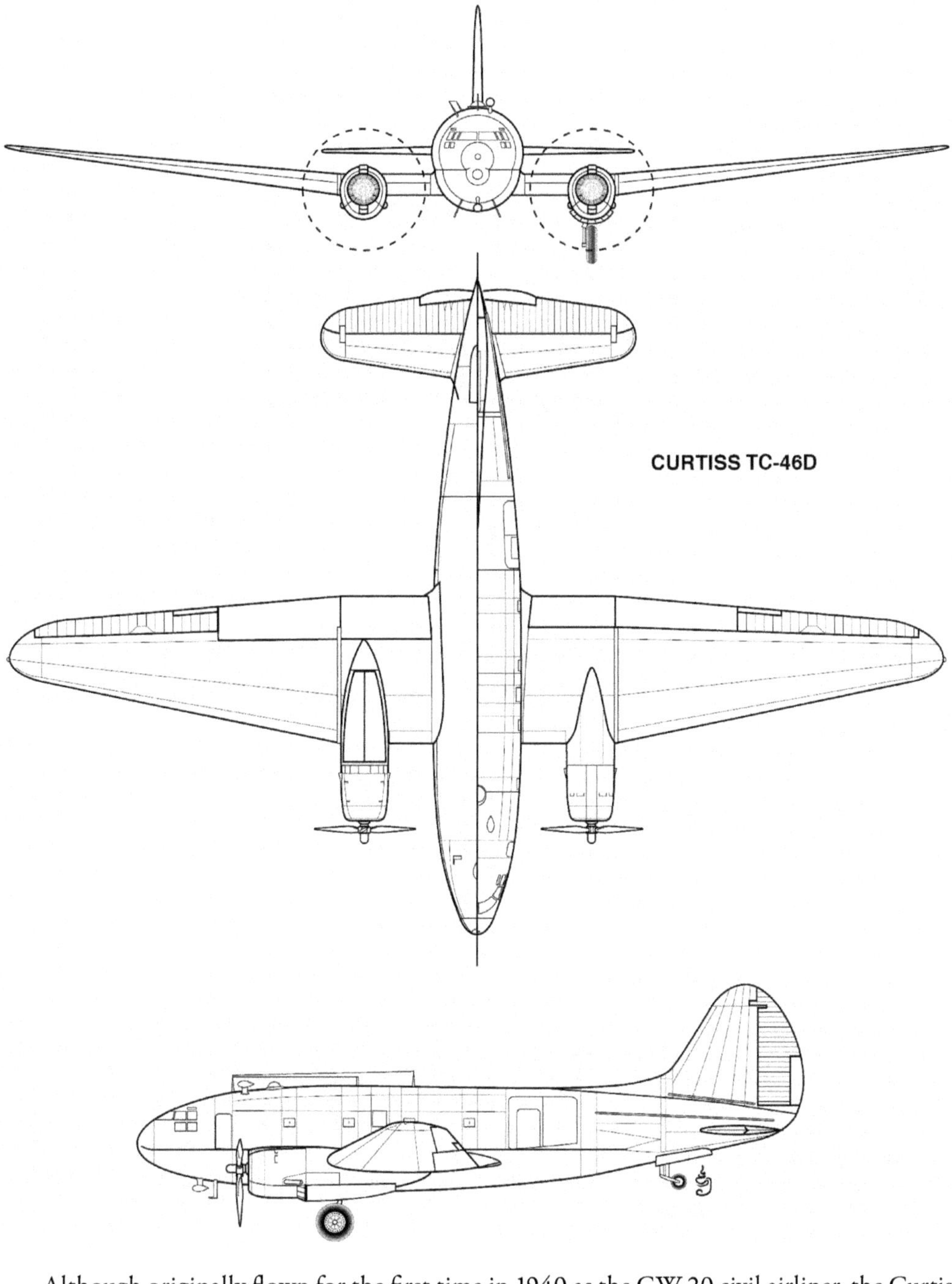

Although originally flown for the first time in 1940 as the CW-20 civil airliner, the Curtiss C-46 Commando went on to become one of the most important twin-engine transports to serve with U.S. military forces during the World War II era. It was produced in four major variants between 1942 and 1945: 25 as the C-46, 1,614 as the C-46A, 1,410 as the C-46D, and 234 as the C-46F. All C-46 production models were delivered to the USAAF except for 170 C-46As, 160 allocated to the Marines and 10 to the Coast Guard as the R5C-1. Through most of their wartime service, C-46s were needed to support combat operations, most notably the Air Trans-

Curtiss TC-46D seen in postwar USAF Air Training Command markings. Historical sources are not clear as to how long they were used or what training function they fulfilled.

port Command route over the Himalayan Mountains between India and China known as the "Hump." In 1945, a small number of C-46As and Ds, perhaps 15 to 20, were transferred to Army Air Forces Training Command for training purposes and re-designated TC-46A/D, and some continued postwar with Air Training Command until 1947 or later.

B. USN, USMC and USCG

Curtiss/Burgess N-9–1916

Technical Specifications (N-9H)

Type: Two-place floatplane trainer.
Manufacturer: Curtiss Aeroplane & Motor Co., Garden City and Buffalo, New York; the Burgess Co., Marblehead, Massachusetts.
Total U.S. military versions: 574 (560 USN, USMC; 14 USAAS).
Unit Cost: $8,000 (est.).
Powerplant: One 150-hp Wright-Hispano A (V-718) 8-cylinder water-cooled engine driving a two-bladed wooden fixed-pitch propeller.
Performance: Max. speed 80 mph at s.l.; ceiling 9,850-ft.; range 179 mi.
Weights: 2,140 lbs. empty, 2,765 lbs. loaded.
Dimensions: Span 53 ft. 4 in., length 30 ft. 10 in., wing area 496 sq. ft.

Conceived as a floatplane derivative of the Army's popular JN trainer series, the Navy Department gave Curtiss an initial contract in August 1916 for 30 aircraft as the N-9, and the

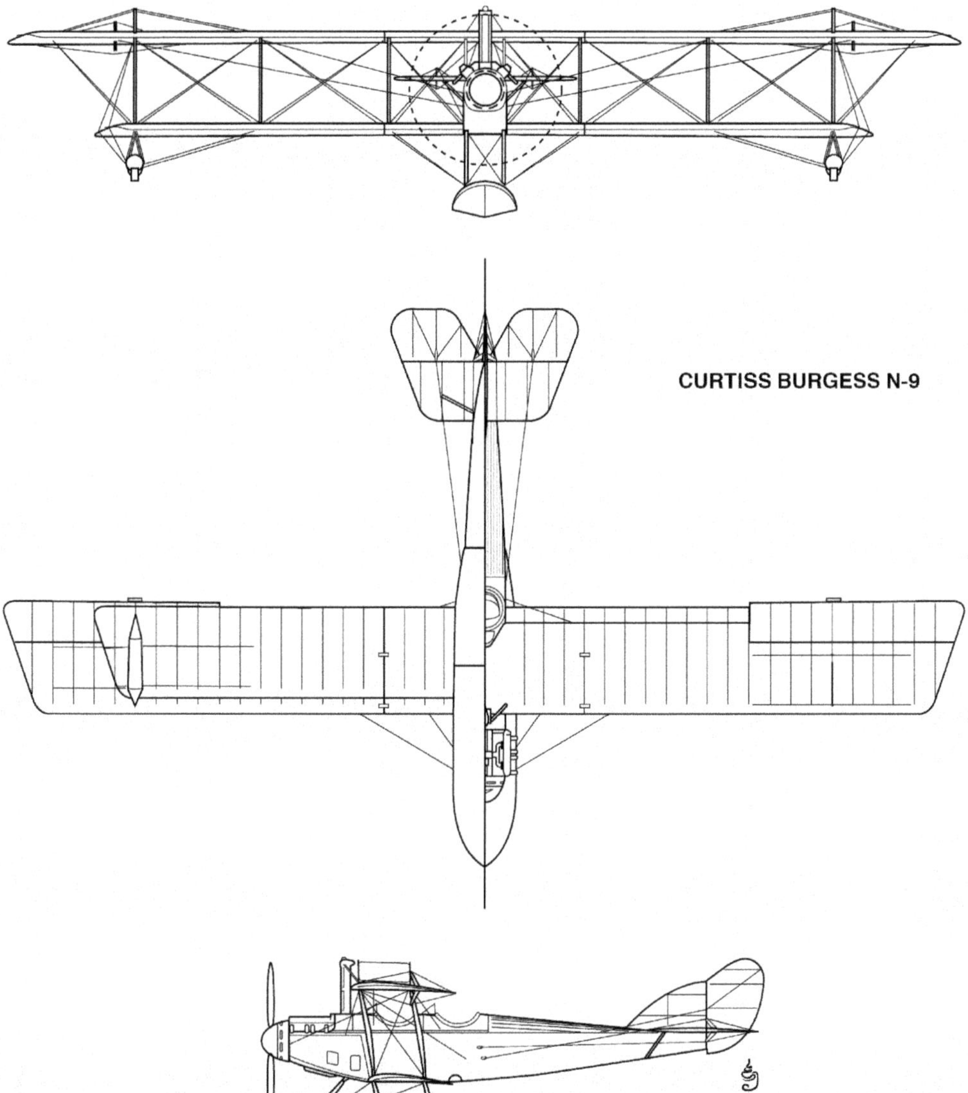

first examples appeared later the same year. It differed from the basic JN-4 series in having 10-feet of additional wingspan, larger tailplane, and a more powerful (100-hp) OXX-6 engine, which was needed to compensate for the extra weight and drag of the floats. Even so, early experience revealed that the type was still underpowered, leading to the introduction in 1917 of the N-9H with the 150-hp Wright-Hispano A (a license copy of the French Hispano-Suiza A). By the time of the armistice in November 1918, 560 of the type, primarily N-9Hs, had been delivered to the Navy, 460 of them having been completed under license by the Burgess Co. (a Curtiss subsidiary). N-9s continued to be extensively used in training and utility roles after the war, and the final examples were not withdrawn from service until 1927.

Army Variants: The Aviation Section of the U.S. Army Signal Corps took delivery of 14 OXX-6-powered N-9s in 1917 for seaplane training.

An N-9H standard World War I Navy trainer, basically a Jenny with a more powerful engine, extra bay of struts, and greater wing and rudder/fin area. Used in training until 1927.

Curtiss JN—1916
(See Curtiss JN-4, JN-6 and JNS in Series I, Part A)

Curtiss F—1917

Technical Specifications (1917 F)

Type: Two-place seaplane trainer.
Total U.S. Military (trainer) versions: 144.

A 1917 Curtiss F Boat with equal span wings. This was a standard World War I Navy trainer in the flying boat phase of pilot training, and the last examples were withdrawn during the early 1920s.

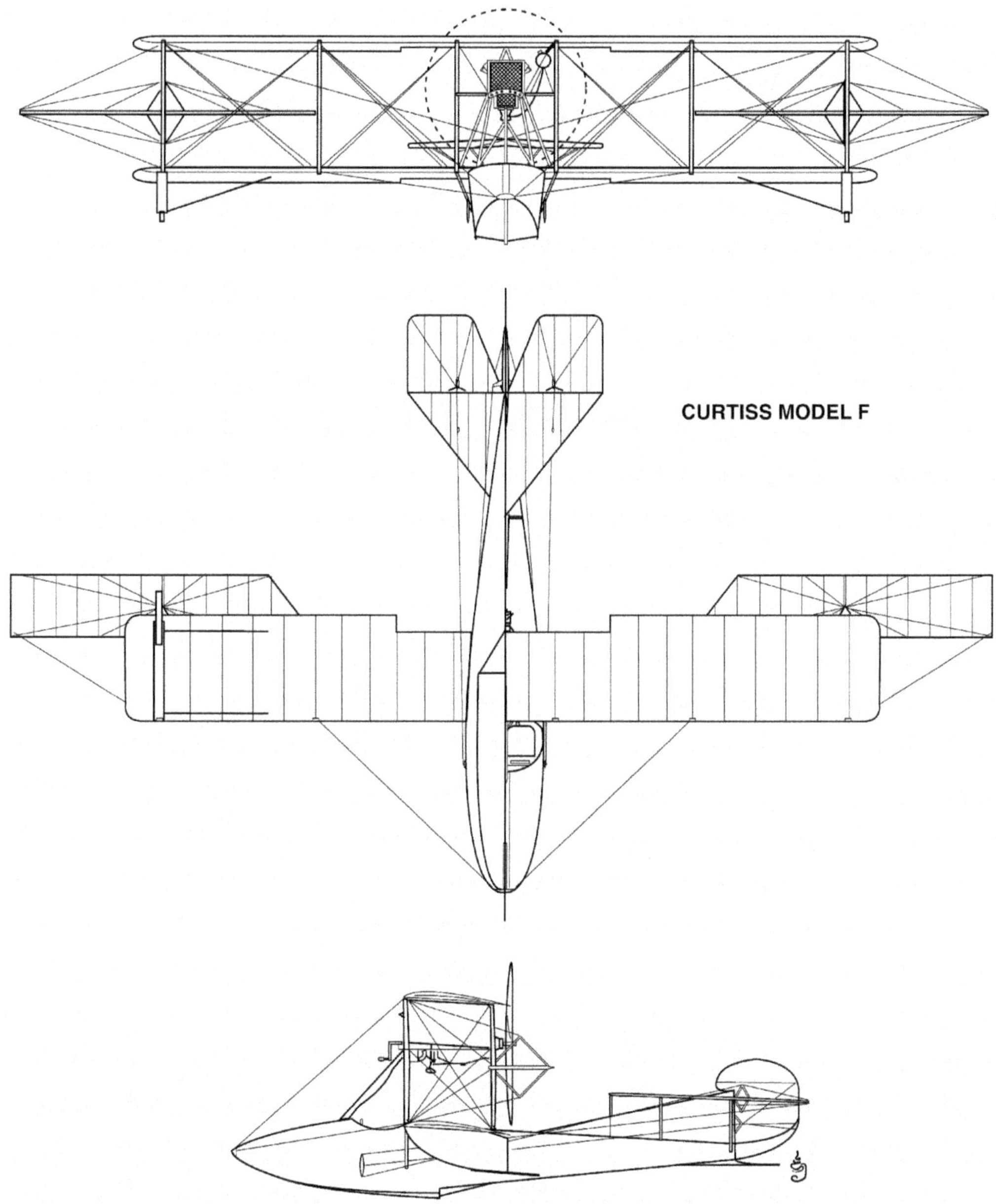

CURTISS MODEL F

Unit cost: $5,465.
Power plant: One 100-hp Curtiss OXX-3 8-cylinder water-cooled engine driving a two-bladed wooden fixed-pitch propeller.
Performance: Max. speed 69 mph; ceiling 4,500 ft.; range 390 mi.
Weights: Empty 1,860 lbs., 2,460 lbs. loaded.
Dimensions: Span 45 ft. 1 in., length 27 ft. 10 in., wing area 387 sq. ft.

Although the Navy took delivery of 13 Curtiss model F flying boats for experimental purposes between 1912 and 1916, the definitive trainer variant did not appear until 1917. The design of the model F had emerged originally as an incremental improvement over the model E of 1912, the world's first successful flying boat. Like most early flying boats, the model made use of a spruce structural framework with fabric-covered wings and tail surfaces and a wood-skinned hull. In April 1917, as a direct result of U.S. entry into World War I, the Navy Department awarded Curtiss a contract to produce 144 model Fs that would be dedicated to training purposes. By this time, the design of the F had evolved to include a 100-hp OXX-3 engine and a four-foot, nine-inch increase in the span of the upper wing, and in some late production models, trailing-edge ailerons instead of the interplane type. As the Navy's World War I pilot training program gathered pace, the model F became the most important trainer for pilots who would transition to larger flying boats like the Curtiss H-16 and HS and Naval Aircraft Factory F-5. Most model Fs were withdrawn and offered surplus once the war ended, though a few remained in service until the early 1920s.

Aeromarine 39–1917

Technical Specifications (39-B)

Type: Two-place landplane/floatplane trainer.
Manufacturer: Aeromarine Plane & Motor Co., Keyport, New Jersey.

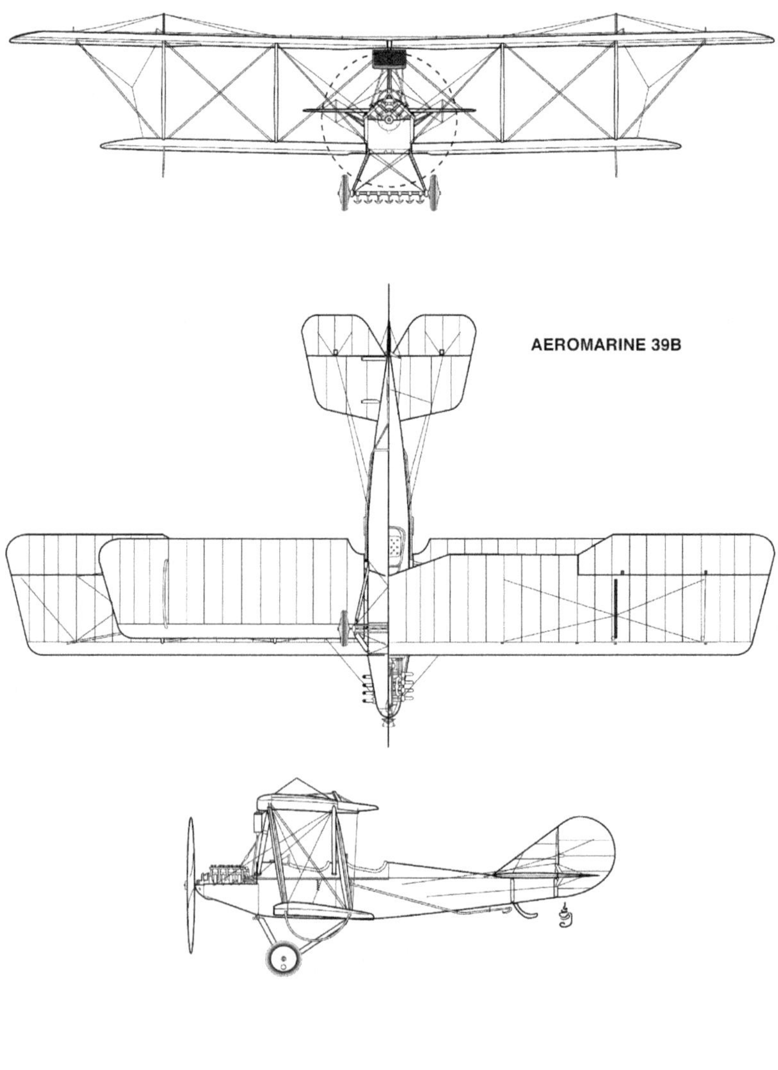

This 39-B is on final approach, making the first landing aboard the carrier *Langley* on October 22, 1922. Docile handling qualities made it a good choice for early carrier experiments.

Total U.S. military versions: 200 (USN).
Unit Cost: $9,688.
Powerplant: One 100-hp Curtiss OXX-6 8-cylinder water-cooled engine driving a two-bladed wooden fixed-pitch propeller.
Performance: Max. speed 73 mph at s.l.; ceiling 8,200-ft.; range 273 mi.
Weights: 1,467 lbs. empty, 2,017 lbs. loaded.
Dimensions: Span 43 ft. 8 in., length 27 ft. 4 in., wing area 353 sq. ft.

On October 22, 1922, an Aeromarine 39-B piloted by Lt. Cdr. Geoffrey DeChevalier became the first aircraft to land on the *Langley* (CV-1), the U.S. Navy's first aircraft carrier. In mid–1917, as a result of a successful bid on Navy Department specifications issued for a training aircraft convertible to land or sea operations, Aeromarine received a contract for 50 Model 39-As, conventional two-bay biplanes of fabric-covered wooden construction that were initially powered by 100-hp Hall-Scott 6-cylinder engines. Early evaluations with twin floats fitted led to an increase in the upper wingspan to reduce wing loading for easier water takeoffs. The 100 Model 39-Bs, delivered in 1918, introduced refinements to the basic design that included enlarged tail surfaces, a centerline float for seaplane operations, and an OXX-6 V-8 powerplant. In the period following the war, the Navy continued to operate the 39-Bs as one of its standard primary trainers.

Because of its excellent low airspeed handling qualities, the 39-B was selected for the earliest carrier experiments. The aircraft was fitted with a rear-mounted arrestor hook to engage cross-deck cables in combination with alignment hooks on the spreader bar between the main wheels to engage lengthwise wires intended to keep the plane running straight. Starting in early 1923, after successful experiments, a unit of 39-Bs began operating aboard *Langley* to test and develop deck procedures for handling larger numbers of aircraft. The last examples were withdrawn in 1926.

Curtiss/Naval Aircraft Factory MF—1918

Technical Specifications (MF)

Type: Two-place flying boat trainer.
Manufacturer: Curtiss Aeroplane & Motor Co., Garden City and Buffalo, New York; Naval Aircraft Factory, Philadelphia, Pennsylvania.
Total U.S. military versions: 96 (USN, USCG)
Unit cost: Unknown.
Powerplant: One 100-hp Curtiss OXX-3 8-cylinder water-cooled engine driving a two-bladed wooden fixed-pitch propeller.

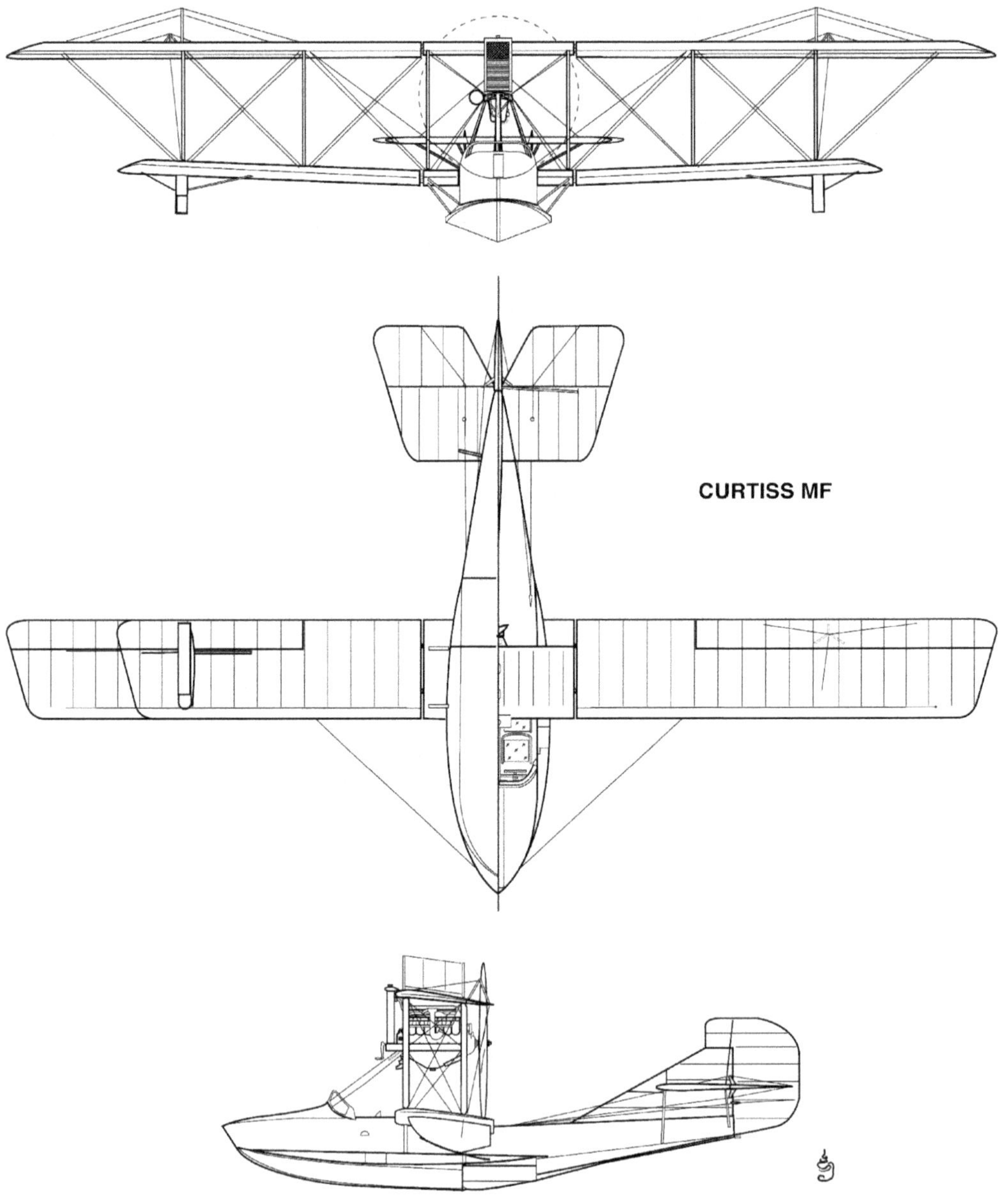

Production of the MF was foreshortened by the end of World War I. Most of the 96 built were completed by the Naval Aircraft Factory after the war. Used as a trainer until the mid-1920s.

Performance: Max. speed 72 mph; ceiling 4,100-ft.; range 345 mi.
Weights: 1,850 lbs. empty, 2,488 lbs. loaded.
Dimensions: Span 49 ft. 9 in., length 28 ft. 10 in., wing area 402 sq. ft.

The MF (or modified F) was introduced in 1918 as the successor to the Navy's F Boat trainers. Although a three-bay biplane of wooden construction like the F, it was actually an entirely new design that utilized the sponson-type hull of the H-16, new wings having trailing-edge ailerons on the upper span, along with a reshaped fin and balanced rudder. The Navy gave Curtiss an order for 47 MFs, but only 16 had been delivered when the contract was cancelled following the World War I armistice in late 1918. Eighty more examples built by the Naval Aircraft Factory were delivered to the Navy after the war, and the type remained in service as a trainer until the mid–1920s. Several MFs transferred to the Coast Guard were used in 1920 and 1921 to develop aerial search and rescue methods.

Aeromarine 40—1919

Technical Specifications (40F)

Type: Two-place flying boat trainer.
Manufacturer: Aeromarine Plane & Motor Co., Keyport, New Jersey.
Total U.S. military versions: 50 (USN).
Unit cost: $9,000.
Powerplant: One 100-hp Curtiss OXX-6 8-cylinder water-cooled engine driving a two-bladed wooden fixed-pitch propeller.
Performance: Max. speed 71 mph at s.l.; ceiling 3,500 ft.; range 250 mi. (est.).
Weights: 2,061 lbs. empty, 2,592 lbs. loaded.
Dimensions: span 48 ft. 6 in., length 28 ft. 11 in., wing area (not reported).

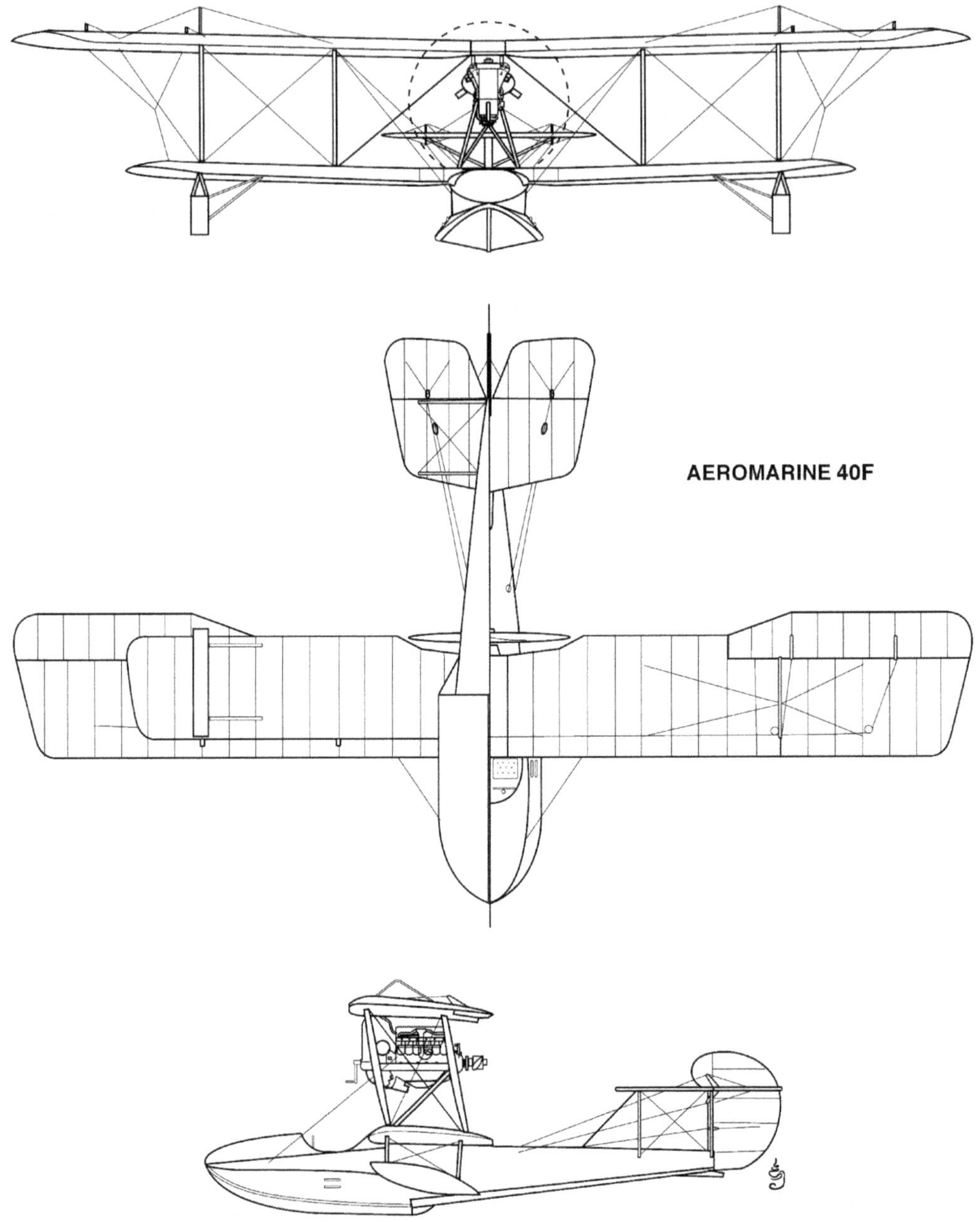

The Model 40, following the model 39 trainer, was designed and built in response to a 1918 Navy Department requirement for a two-seat seaplane trainer. Though similar in general layout to the Curtiss Model MF (see above), which it was intended to augment, it appeared as a two-bay biplane of fabric-covered wooden construction, characterized by a single-step hull having a more rounded bow and less pronounced sponsons than those seen on the Curtiss types. The Navy originally contracted for 200 as the Model 40F, but the order was reduced to only 50 after the November armistice, and the first deliveries did not occur until 1919. Some Model 40Fs were reportedly fitted with 150-hp Wright-Hispano engines to improve performance and payload. All had been withdrawn from service by the mid–1920s.

The Aeromarine 40F was intended to augment the F and MF in training, however, the contract was cut short by the end of the war. This example is on the seaplane ramp at NAS Anacostia, Maryland, in the early 1920s.

Synopsis of USN/USMC Trainer Procurement, 1922 to 1930

At the time World War I ended, the Navy boasted a force of over 1,000 aircraft dedicated to training, primarily Burgess N-9s, Aeromarine 39s, Curtiss JN-4s/-6s, and Curtiss model F flying boats. However, soon after its inception in 1921, the Naval Bureau of Aeronautics (BuAer) believed the Navy's future training needs would be better served by one type of aircraft combining the functions of landplane, seaplane, and aerial gunnery into one standard airframe. Evaluation of no less than six different trainer types took place between 1923 and 1925, with two of the entries being ordered into quantity production, 73 Boeing NB-1s and -2s and, initially, 76 Consolidated NY-1s (naval derivative of the Army PT-1 reported in Series I, Part A); and then from 1926 to 1929, a further 186 NY-2s (the largest trainer order since 1918), 25 NY-2As, and 20 NY-3s, making it the most important training aircraft in the Navy's pilot training program at NAS Pensacola from the mid–1920s to the mid–1930s.

A second round of trainer competitions initiated by BuAer in 1928 ultimately resulted in acquisition of one Boeing N2B, 52 Curtiss N2Cs for the reserves, and 19 Keystone NKs. An unrelated "lightweight trainer" evaluation in 1929 and 1930 accounted for purchase of six New Standard NTs and six Fleet N2Ys (naval derivative of Army PT-6 reported in Part A), the latter being employed as "skyhook" trainers in the lighter-than-air program.

Huff-Daland HN—1923
(See Huff-Daland TA-2 and TW-5 in Series I, Part A)

Naval Aircraft Factory N2N—1923

Technical Specifications

Type: Two-place landplane/floatplane trainer.
Manufacturer: Naval Aircraft Factory, Philadelphia, Pennsylvania.

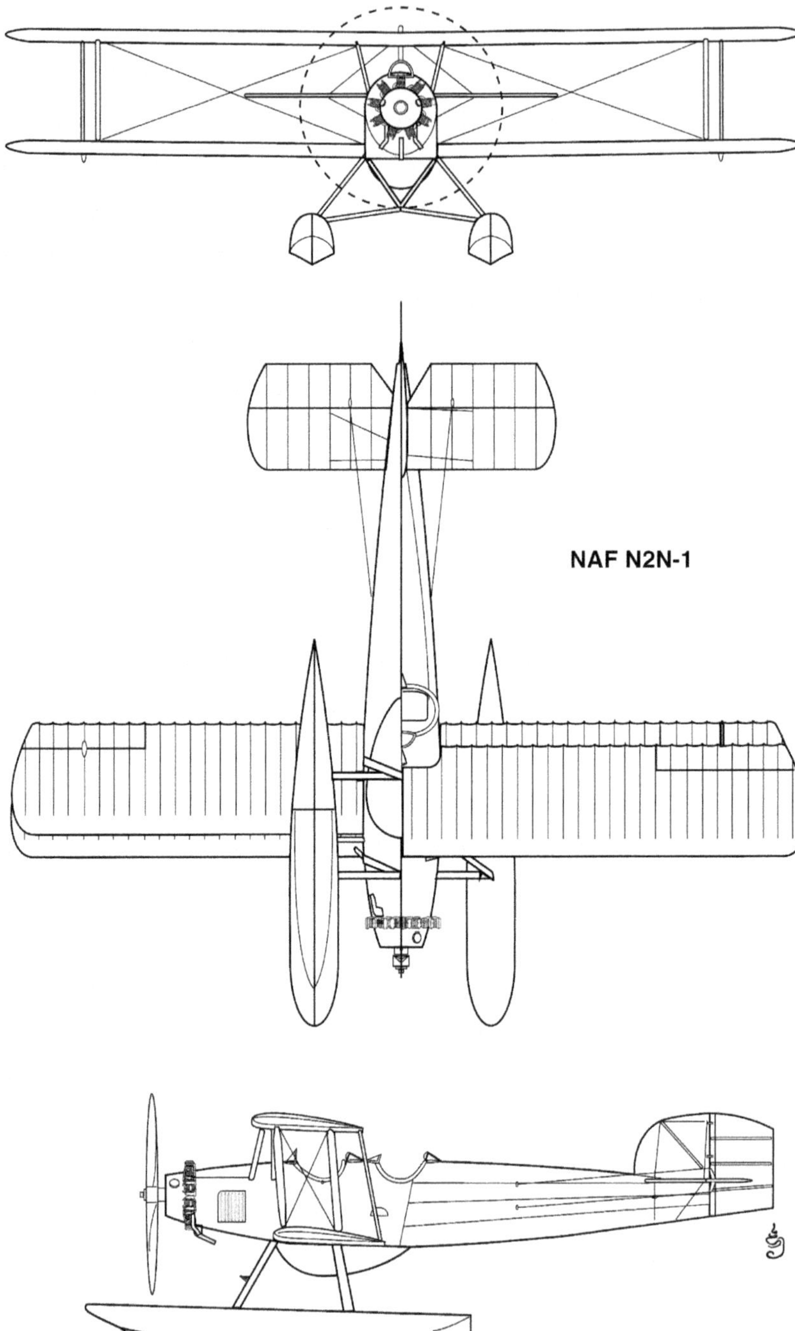

One of three N2N-1s tested in 1923, primarily to evaluate a trainer powered by a Lawrance J-1 engine. Though not ordered into production, all three subsequently served as trainers at Pensacola.

Total U.S. military versions: 3 (USN).
Unit cost: Unknown.
Powerplant: One 200-hp Lawrance J-1 R-787 9-cylinder air-cooled radial engine driving a two-bladed wooden fixed-pitch propeller.
Performance: Max. speed 106 mph at s.l.; ceiling 16,900 ft.; range (not reported).
Weights: 2,100 lbs. empty (est.), 2,640 lbs. loaded.
Dimensions: Span 33 ft. 8 in., length 26 ft. 11 in., wing area 260 sq. ft. (est.).

The N2N-1 (NN designation not used) flew for the first time in 1923 as a Naval Aircraft Factory project to develop and evaluate a float-equipped trainer powered by the new air-cooled radial engine recently introduced by the Lawrance Aero Engine Co. Its all-wood construction and general aerodynamic design were very similar to the contemporaneous TS-1 fighter, which had been designed by Rex Beisel while he was employed by BuAer. The three N2N-1s completed by NAF were briefly used for training at NAS Pensacola.

Martin N2M—1924

Technical Specifications

Type: Two-place landplane/floatplane trainer.
Manufacturer: Glenn L. Martin Co., Cleveland, Ohio.
Total U.S. military versions: 1 (USN).
Unit cost: Unknown.
Powerplant: One 200-hp Lawrance J-1 R-787 9-cylinder air-cooled radial engine driving a two-bladed wooden fixed-pitch propeller.
Performance: Max. speed 112 mph at s.l.; ceiling 19,000 ft.; range (not reported).
Weights: (Not reported).
Dimensions: Span 41 ft., length 27 ft. 10 in., wing area (not reported).

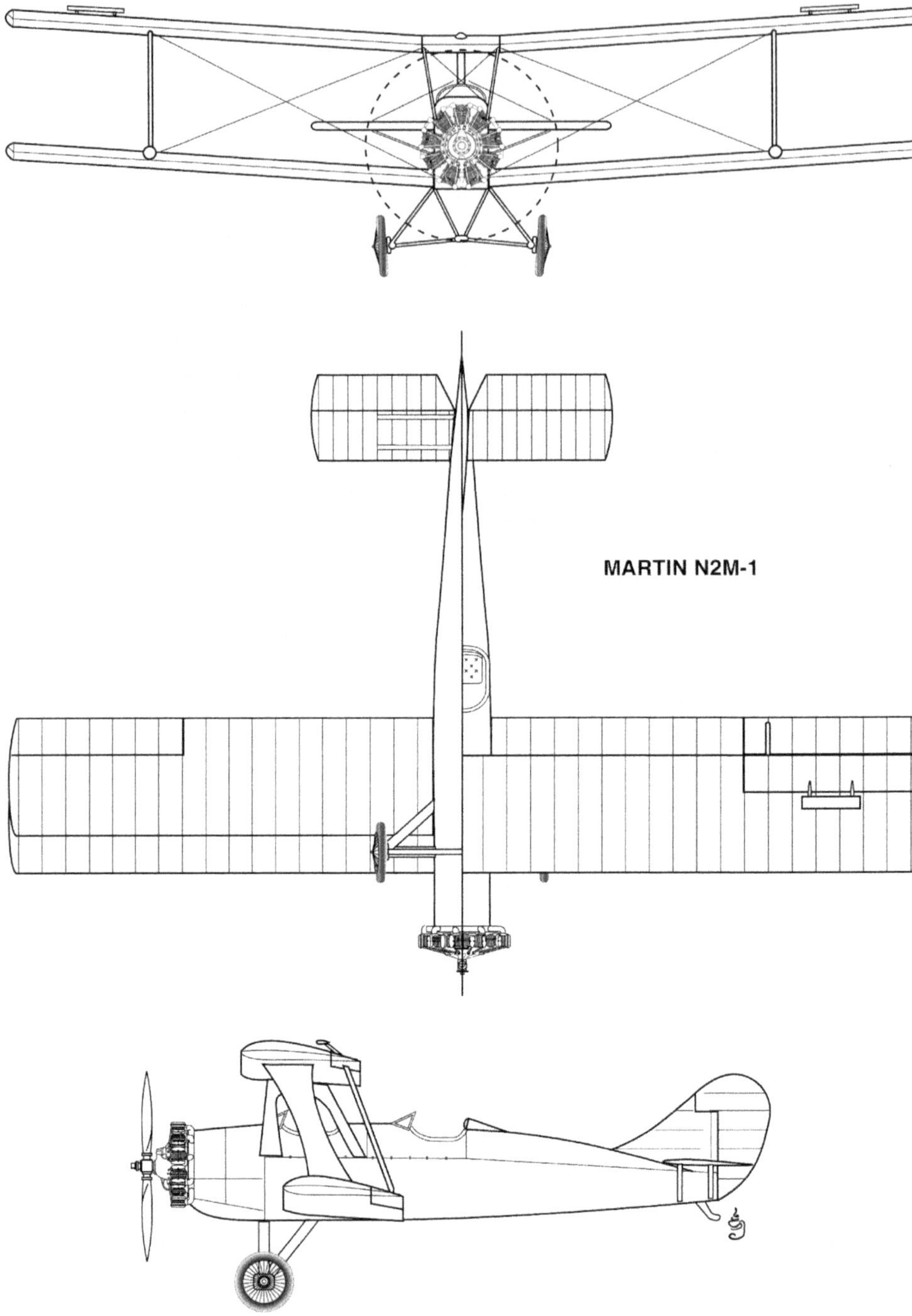

MARTIN N2M-1

In 1924 Martin offered the Navy another Lawrance-powered trainer which was based largely on the design of its Wright-Hispano-powered Model 66 Night Mail Plane of 1923. The sole example, taken into service as the N2M-1 (NM-1 had apparently been assigned to Packard for a racer), emerged as a one-bay, wood and fabric biplane characterized by low-aspect ratio wings braced by single interplane struts. After acceptance trials, the aircraft saw limited service as a trainer at NAS Pensacola until the mid–1920s.

The N2M-1 was another Lawrance-powered trainer that Martin had adapted from its Model 66 Mail Plane design. Only one was built in 1924, and it was later used as a trainer at Pensacola.

Naval Aircraft Factory TG—1924

Technical Specifications (TG-2)

Type: Two-place floatplane gunnery trainer.
Manufacturer: Naval Aircraft Factory, Philadelphia, Pennsylvania.

A liberty-powered Naval Aircraft Factory TG-4 in 1924. It was designed as a gunnery trainer. Five were built with various powerplants, but none were listed as active after 1925.

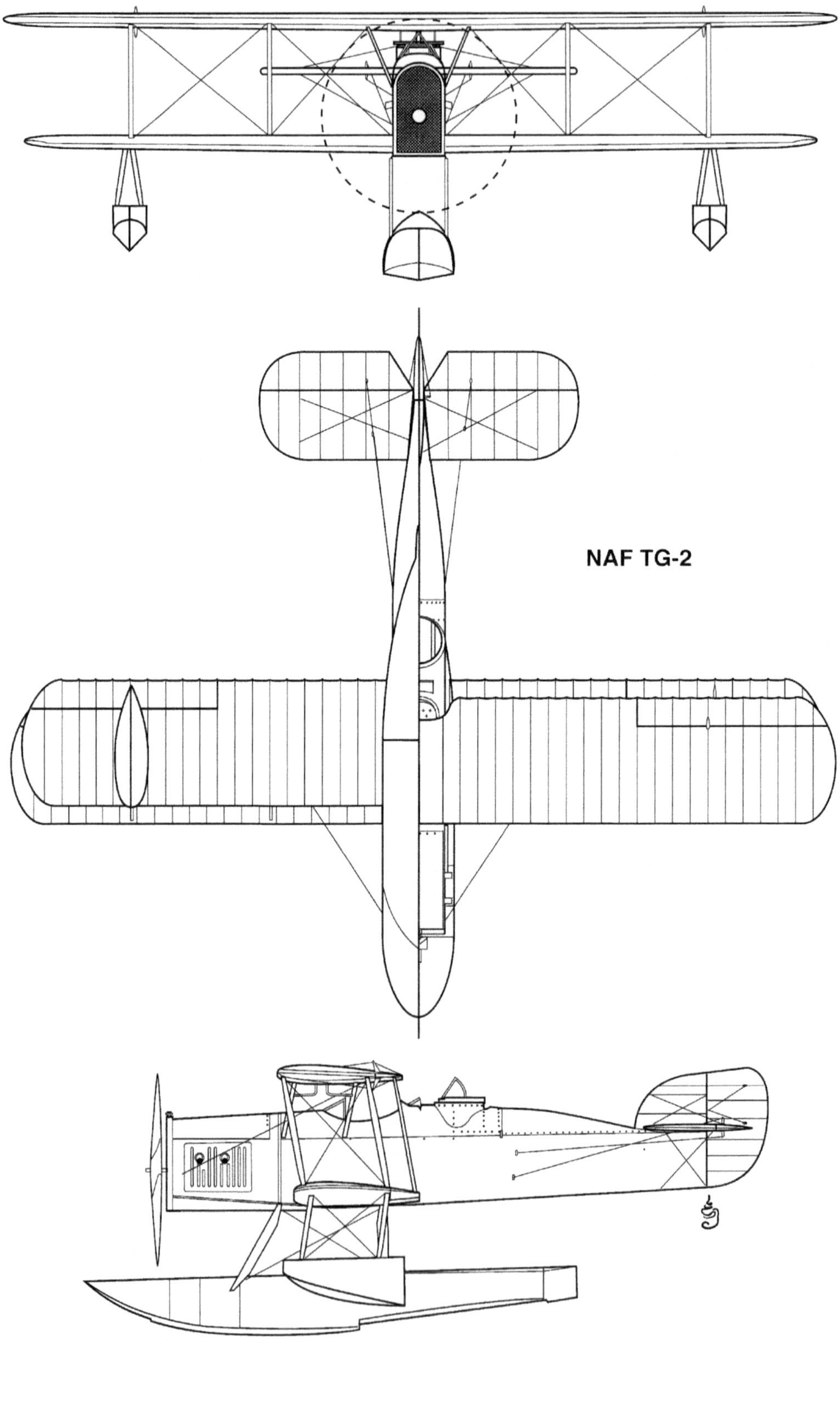

NAF TG-2

Total U.S. military versions: 5 (USN).
Unit cost: Unknown.
Powerplant: One 200-hp Liberty L 6-cylinder water-cooled engine driving a two-bladed wooden fixed-pitch propeller.
Armament: One .30-calibre machine gun on scarf ring in rear cockpit.
Performance: Max. speed 97 mph at s.l.; ceiling 13,140 ft.; range (not reported).
Weights: 2,400 lbs. empty (est.), 2,800 lbs. loaded.
Dimensions: Span 36 ft., length 30 ft. 10 in., wing area 340 sq. ft. (est.).

The Naval Aircraft Factory built five float-equipped gunnery trainers in 1924 as TG-1 through -5. They were conventional two-bay biplanes of wood and fabric construction, identical except for the powerplants installed (i.e., TG-1 and -2, 200-hp Liberty 6; TG-2 and -3, 200-hp Aeromarine T-6; and TG-5, 180-hp Wright Hispano E). The type apparently was not very successful, as none were listed as active after 1925.

Boeing NB—1924

Technical Specifications (NB-1)

Type: Two-place landplane/floatplane trainer.
Manufacturer: Boeing Airplane Co., Seattle, Washington.
Total U.S. military versions: 73 (USN, USMC).
Unit cost: Unknown.

The Boeing NB-1 was the first type of postwar Navy trainer to be ordered in quantity. Note the hand crank right behind the engine. Like most Navy trainers, it could be rigged for either floats or wheels as needed.

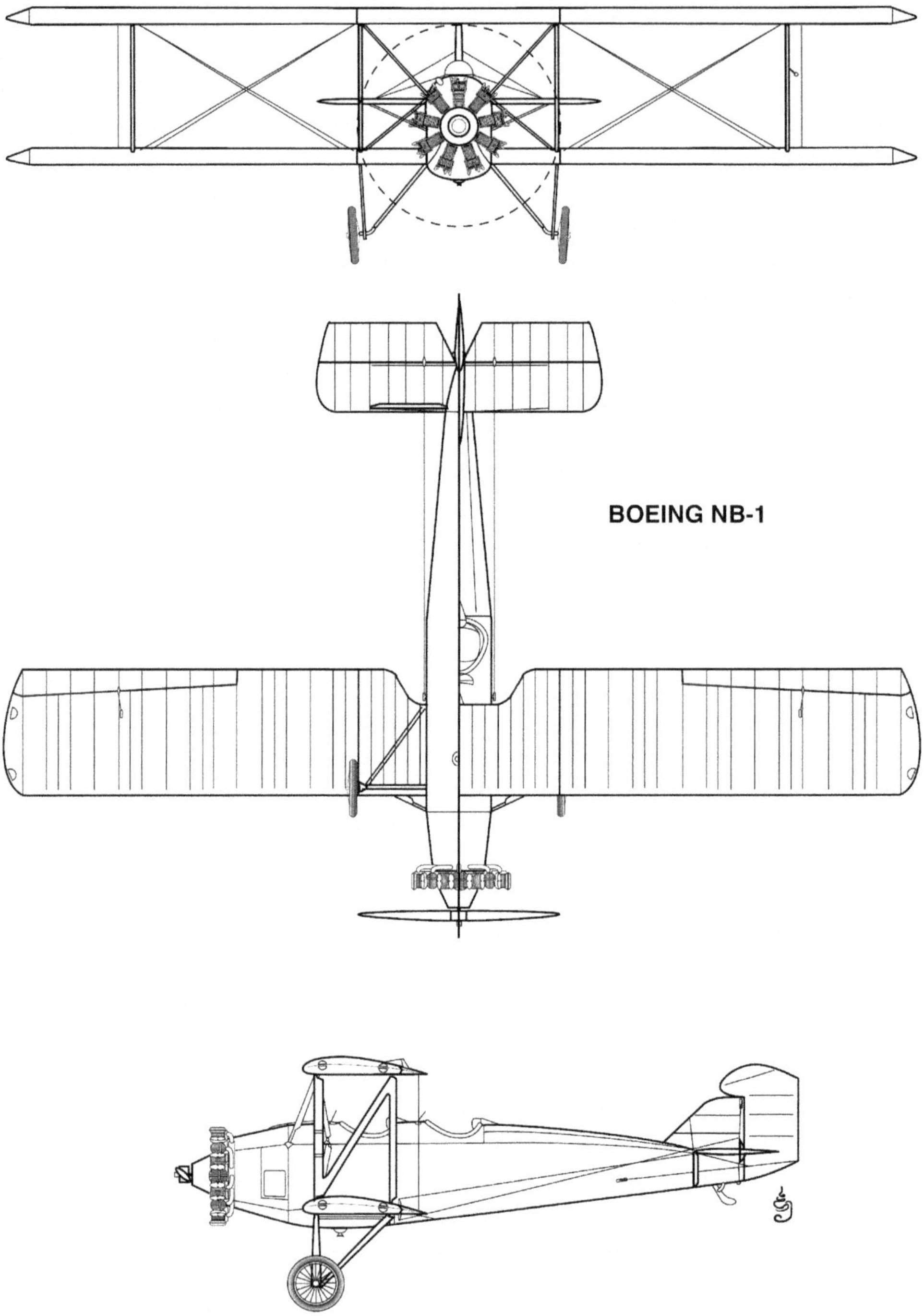

BOEING NB-1

Powerplant: One 200-hp Lawrance J-1 9-cylinder air-cooled radial engine driving a two-bladed wooden fixed-pitch propeller.
Armament (on some): One .30-caibre machine gun on scarf ring in rear cockpit.
Performance: Max. speed 100 mph at s.l.; ceiling 15,500 ft.; range 300 mi.
Weights: 2,136 lbs. empty, 2,837 lbs. loaded.
Dimensions: Span 36 ft. 10 in., length 25 ft. 5 in., wing area 344 sq. ft.

One of the first Navy trainers to be purchased in double-digit quantities following World War I, the Boeing Model 21 originated as the prototype NB-1 delivered in mid–1924. After initial testing revealed that the aircraft was actually too docile for training purposes, showing a marked reluctance to spin, it was given modifications that made it more maneuverable. After satisfactory trials, BuAer awarded Boeing a contract for 42 NB-1 production aircraft, and deliveries commenced before the end of the year. The design was characterized by such innovations as a welded, steel tube fuselage structure, single-bay wings braced by metal N-type struts, and a balanced rudder. In 1925 Boeing received a follow-on order for thirty NB-2s powered by 180-hp Wright-Hispano E-4 water-cooled engines, which, like the -1s, could be rigged with either wheels or a centerline float.

Most NB-1s and -2s were assigned to NAS Pensacola with VN-1D8 and VN-1D4, while some served with VN-6D5 at NOB Hampton Roads and VN-7D11 at NAS North Island in San Diego. A single NB-2 modified for crop dusting served with Marine squadron VO-6M during 1929 to spray the mosquito-infested swamps around Santo Domingo, Nicaragua.

Consolidated NY and N3Y—1925
(See Consolidated PT- 1 through -5 in Series I, Part A)

Boeing N2B—1928

Technical Specifications (XN2B-1)

Type: Two-place landplane/floatplane trainer.
Manufacturer: Boeing Airplane Co., Seattle, Washington.

A Boeing XN2B-1, after installation of a more conventional Wright R-540 engine, at Anacostia undergoing new trials in 1929. After no production was ordered, the Navy sold it to back to Boeing.

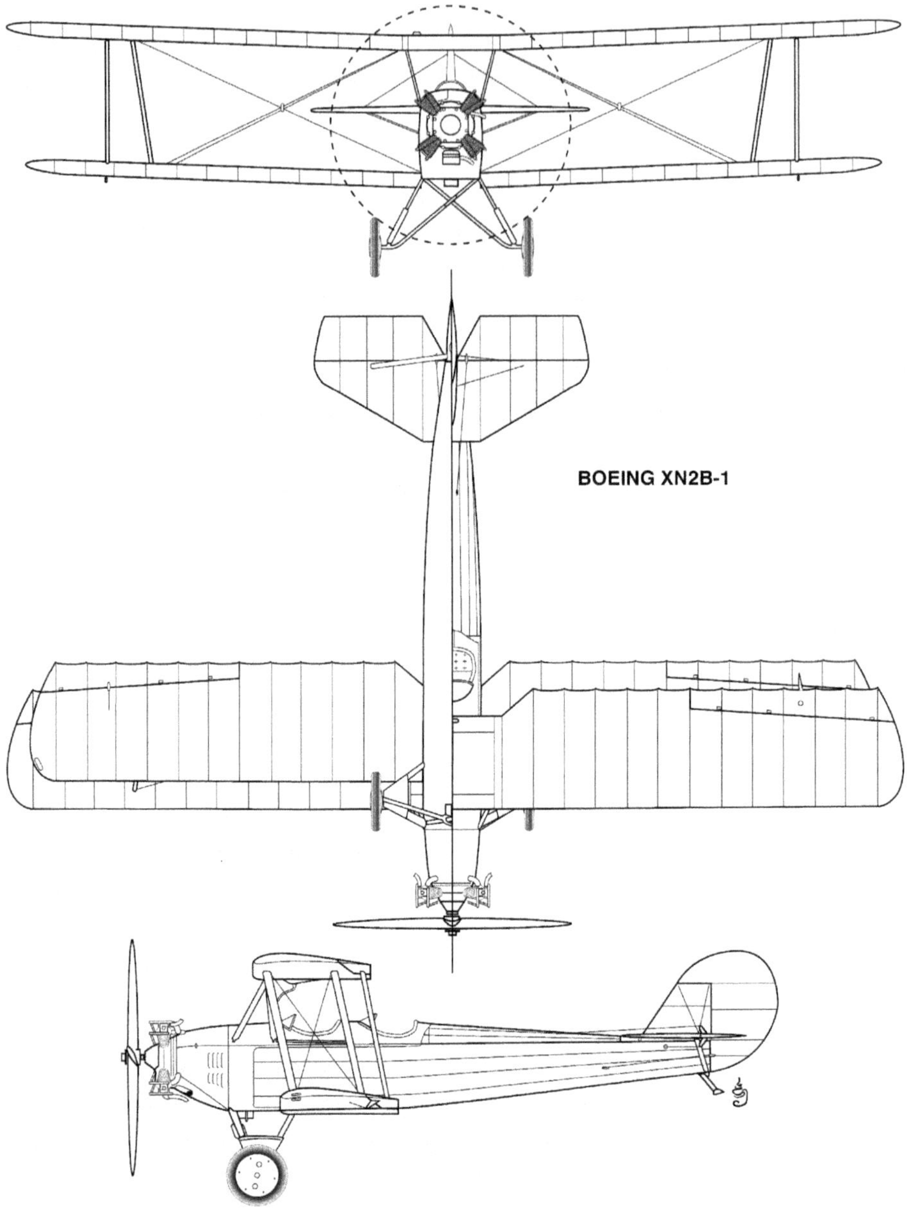

Total U.S. military versions: 1 (USN).
Unit cost: Unknown.
Powerplant: One 125-hp Fairchild-Caminez R-447 4-cylinder (crankless) air-cooled radial engine driving a four-bladed fixed-pitch wooden propeller.
Performance: Max. speed 104 mph, cruise 86 mph; ceiling 13,500 ft.; range 335 mi.
Weights: 1,652 lbs. empty, 2,178 lbs. loaded.
Dimensions: Span 35 ft. 0 in., length 25 ft. 8 in., wing area 259 sq. ft.

A trainer competition conducted by BuAer in 1928 resulted in delivery of prototypes by Boeing, Curtiss, and Keystone. Boeing's entry, designated the XN2B-1, had originated in 1926 as the Wright J-3-powered Model 64, then reappeared in 1928 as the Model 81 with a Fairchild-Caminez powerplant. This unorthodox 4-cylinder engine was a crankless air-cooled radial that developed its peak horsepower at only 1000 RPM and employed a high-pitched, four-bladed propeller to deliver thrust. The XN2B-1's fabric-covered construction was mixed, consisting of a welded, steel tube fuselage and all-wood biplane wings using an NACA M-12 airfoil section. Initial Navy trials were unsatisfactory, however, and the prototype returned to Boeing to be fitted with a more conventional 165-hp Wright J-6–5 (5-cylinder R-540) engine. Despite better performance in 1929 Navy trials, no production was ordered, and the sole XN2B-1 eventually sold back to Boeing for use in its school of aeronautics.

Curtiss N2C Fledgling—1928

Technical Specifications (N2C-2)

Type: Two-place landplane/floatplane trainer.
Manufacturer: Boeing Airplane Co., Seattle, Washington.
Total U.S. military versions: 52 (USN, USMC).
Unit cost: Unknown.
Powerplant: One 240-hp Wright R-760–94 7-cylinder air-cooled radial engine driving a two-bladed fixed-pitch metal propeller.
Performance: Max. speed 109 mph, cruise 95 mph; ceiling 17,800 ft.; range 384 mi.
Weights: 2,138 lbs. empty, 2,860 lbs. loaded.
Dimensions: Span 39 ft. 5 in., length 27 ft. 5 in., wing area 368 sq. ft.

Winner of BuAer's 1928 trainer competition, the Curtiss XN2C-1 (Model 48) began as a joint project with the Model 51 Fledgling to develop both a military prototype and a civil trainer to be used in the nationwide schools operated by Curtiss-Wright Flying Service. The new Curtiss *Challenger* twin-row, 6-cylinder radial was to power the civil model, while BuAer specified a more conventional Wright J-5 engine for its prototype. Construction consisted of a welded, steel tube fuselage and tailplane with built-up wooden wings. Its aerodynamic design was somewhat behind the times, appearing with thin-section wings that were braced by two bays of struts. Fol-

The original XN2C-1 prototype seen in the mid–1930s while serving with the reserves at NRAB New York. The type saw use in reserve pilot training right up to 1940.

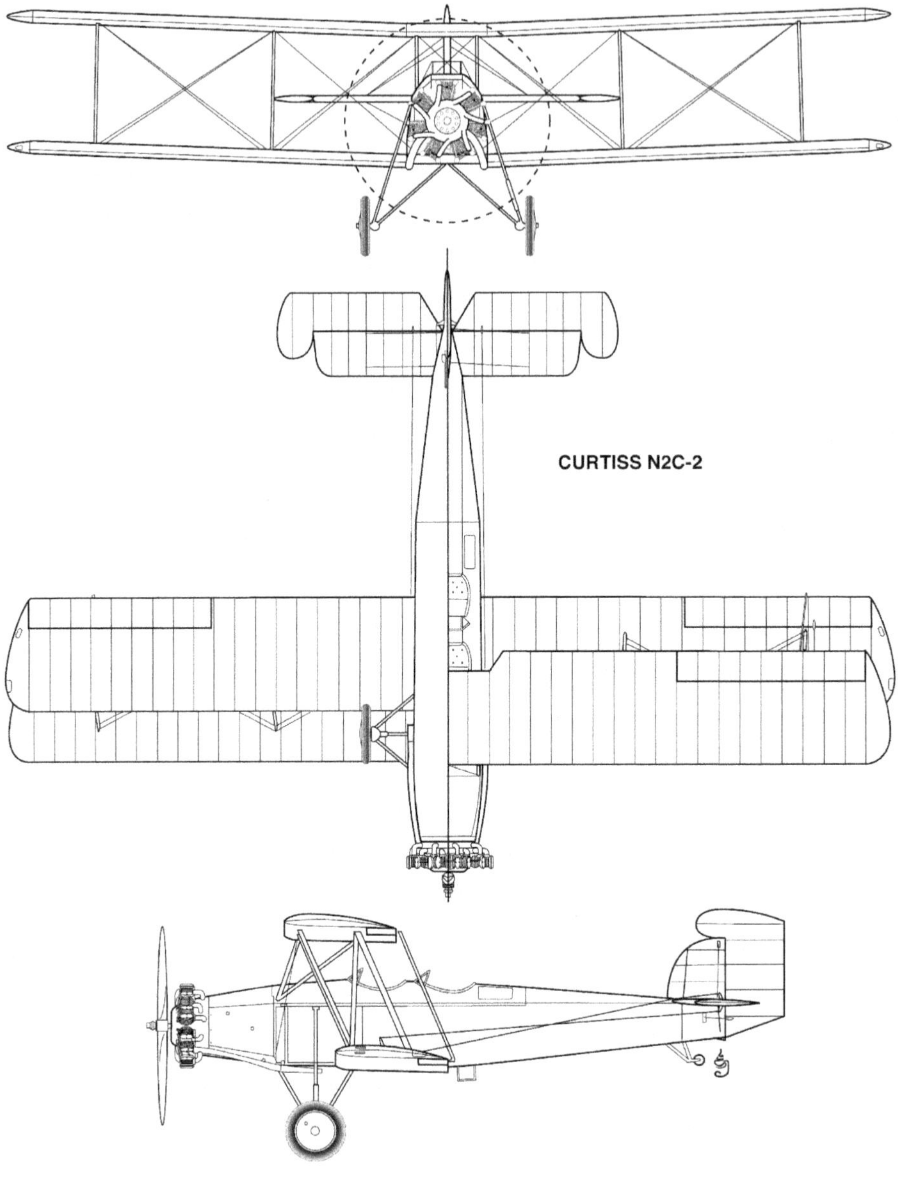

lowing evaluations in both landplane and floatplane configurations, Curtiss received a contract to build 31 production models as the N2C-1. In 1929, shortly after production of the J-5 engine had been discontinued, the Navy gave Curtiss a follow-up order for 20 aircraft to be powered by the 240-hp Wright J-6-7 (R-760-94) under the designation N2C-2.

As deliveries were completed in 1928 and 1929, most N2C-1s and -2s went directly to recently formed Navy and Marine reserve units as training aircraft. From 1931 to 1933, three N2C-2s served with Marine squadron VJ-6M in Nicaragua and Quantico and another with VJ-7M at San Diego. By mid-1937, a total of 37 N2C-1s and -2s were still listed as operating with 11 different Navy and Marine reserve units, however, all had been withdrawn from service by the end of 1939. At the very end of their careers, two N2C-2s were modified with tricycle landing gear and used as radio-controlled drones to simulate dive-bombing attacks on ships.

Keystone NK—1928

Technical Specifications (NK-1)

Type: Two-place landplane/floatplane trainer.
Manufacturer: Keystone Aircraft Corp., Bristol, Pennsylvania.
Total U.S. military versions: 19 (USN).
Unit cost: $7,900.
Powerplant: One 220-hp Wright J-5 R-790 9-cylinder air-cooled radial engine driving a two-bladed fixed-pitch metal propeller.
Performance: Max. speed 115 mph, cruise 109 mph; ceiling 13,200 ft.; range 375 mi.

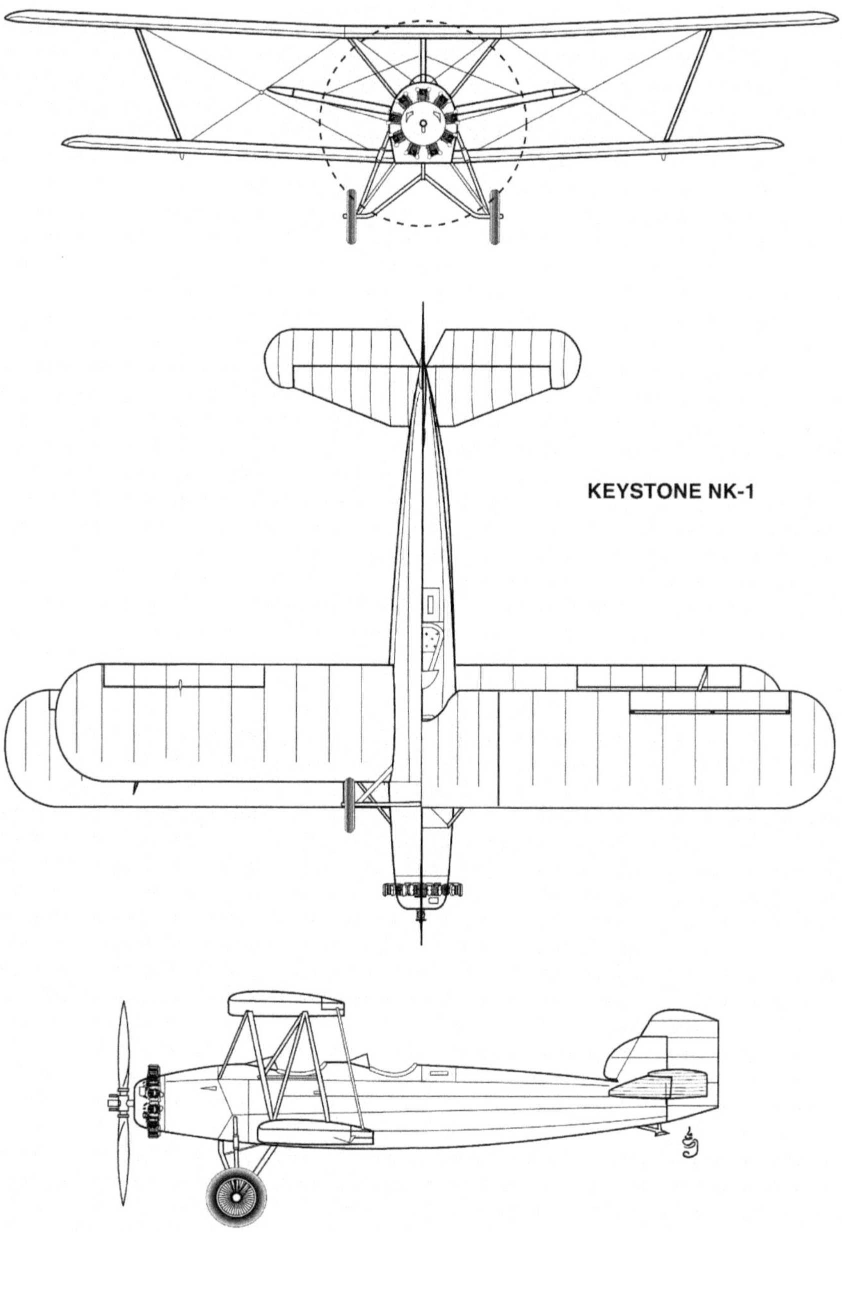

One of three XNK-1 prototypes is mounted on floats in 1928. Although the Navy accepted 16 more modified NK-1s in 1930, none were apparently ever used for training purposes.

Weights: 2,050 lbs. empty, 2,658 lbs. loaded.
Dimensions: Span 37 ft., length 26 ft. 6 in., wing area (not reported).

Developed in parallel with the civil model "Pup," Keystone delivered three essentially identical XNK-1 trainer prototypes to NAS Anacostia in 1928 to be evaluated alongside the XN2B-1 and XN2C-1 (see above). BuAer subsequently gave Keystone a contract for 16 NK-1 production models, but the expected delivery in 1929 was delayed due to modifications suggested by continued testing of the XNK-1s. When delivered in 1930, the production NK-1s differed in having an enlarged and reshaped fin and rudder, plus a small amount of dihedral added to the horizontal stabilizer. Eventual use and disposition of these trainers is not clear, but they were apparently never assigned to any regular training or reserve units.

Fleet N2Y—1929
(See Fleet PT-6 in Series I, Part A)

New Standard NT—1930

Technical Specifications (NT-1)

Type: Two-place landplane trainer.
Manufacturer: Standard Aircraft Corp., Paterson, New Jersey.
Total U.S. military versions: 6 (USN, USCG)
Powerplant: One 100-hp Kinner K-5 R-372 5-cylinder air-cooled radial engine driving a two-bladed fixed-pitch metal propeller.
Performance: Max. speed 88 mph, cruise 75 mph.; ceiling 10,000 ft.; range 300 mi.
Weights: 1,097 lbs. empty, 1,632 lbs. loaded.
Dimensions: span 30 ft., length 24 ft. 11 in., wing area 248 sq. ft.

The second "lightweight" trainer tested, six New Standard Model D-29As acquired by the Navy in 1930 were subsequently taken into the naval inventory as the NT-1. Slightly larger than the N2Y, the NT-1's fabric-covered airframe featured the typical built-up wooden wings and a

The Navy purchased six New Standard D-29As and tested them in 1930 as a "lightweight trainers" under the designation NT-1. They were afterward assigned to the Naval Aircraft Factory.

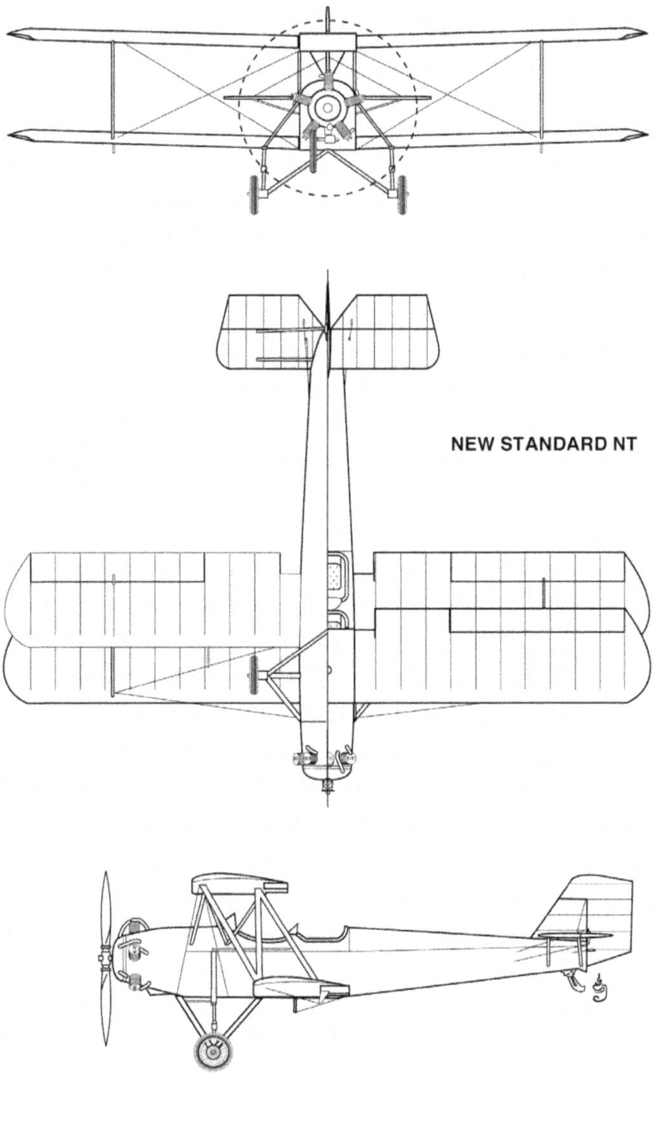

NEW STANDARD NT

very robust fuselage fabricated from riveted and bolted duraluminum girders. The six aircraft also came with the odd bathtub-type cockpits, but were later modified with more conventional separate enclosures. After testing, the NT-1s were assigned to the Naval Aircraft Factory, one on flight status with VX-3D4, and the others in storage.

Synopsis of USN/USMC Trainer Procurement 1930 to 1940

Primary Trainers: The Navy actually began a replacement program in 1932 when it acquired three Consolidated XN4Y-1s (naval derivative of the Army PT-11D reported in Series I, Part A) for evaluation purposes, but before the type could be ordered in quantity, a legislative incentive (i.e., the Vinson-Trammel Naval Act of 1934) caused BuAer to undertake a development program that would utilize the production facilities of the government-owned Naval Air Factory (NAF), which, in turn, led to testing and eventual production of 180 N3N-1s in 1935 and 1936. In order to augment N3N-1s in landplane phase of pilot training, BuAer simultaneously procured 61 Boeing-Stearman NS-1s (naval derivative of Army PT-13 reported in Series I, Part A). Passage of the Naval Expansion Acts in 1938 and 1940 produced an unprecedented surge in new primary trainer orders in 1939 and 1940: 816 N3N-3s, 250 Boeing-Stearman N2S-1s (naval derivative of Army PT-17 reported in Series I, Part A), and 201 Spartan NP-1s for Naval Reserve Primary Flying Schools.

Intermediate Trainers. Through much of this era, the Navy typically employed older operational aircraft (e.g., Vought O2U/O3U, Vought SBU, Martin T4M/TG, and Boeing F4B) to fulfill the intermediate (basic and advanced combined) training function, however, by the mid–1930s, large improvements in aircraft performance dictated a need for trainers that would provide incremental transition to new high-performance frontline aircraft, especially fighter and attack types. To this end, 40 North American NJ-1s (naval derivative of Army BT-9/-14 reported in Series I, Part A) were procured in 1936, followed by 16 North American SNJ-1s (naval derivative of Army BC-1/AT-6) in 1938 and 61 SNJ-2s in 1939. Then the Naval Expansion Acts of 1938 and 1940 generated the largest orders yet: 420 North American SNJ-3s (Army AT-6A), 150 Curtiss SNC-1s, and 1,350 Vultee SNV-1s (naval derivative of Army BT-13A).

Consolidated N4Y—1932
(See Consolidated PT-11 and -12 in Series I, Part A)

Naval Aircraft Factory N3N—1935

Technical Specifications (N3N-3)

Type: Two-place landplane/floatplane trainer.
Manufacturer: Naval Aircraft Factory, Philadelphia, Pennsylvania.
Total U.S. military versions: 998 (USN, USMC, USCG).
Unit cost: $25,000.
Powerplant: One 235-hp Wright R-760–2 (J-6–7) 7-cylinder air-cooled radial engine driving a two-bladed fixed-pitch metal propeller.
Performance: Max. speed 126 mph, cruise 95 mph; ceiling 15,200 ft.; range 470 miles.
Weights: 2,090 lbs. empty, 2,940 lbs. loaded (with floats).
Dimensions: Span 34 ft., length 25 ft. 6 in., wing area 305 sq. ft.

The N3N was a byproduct of the Vinson-Trammel Act of 1934, which, among other things, directed the Navy to generate at least 10 percent of its aircraft from the government-owned

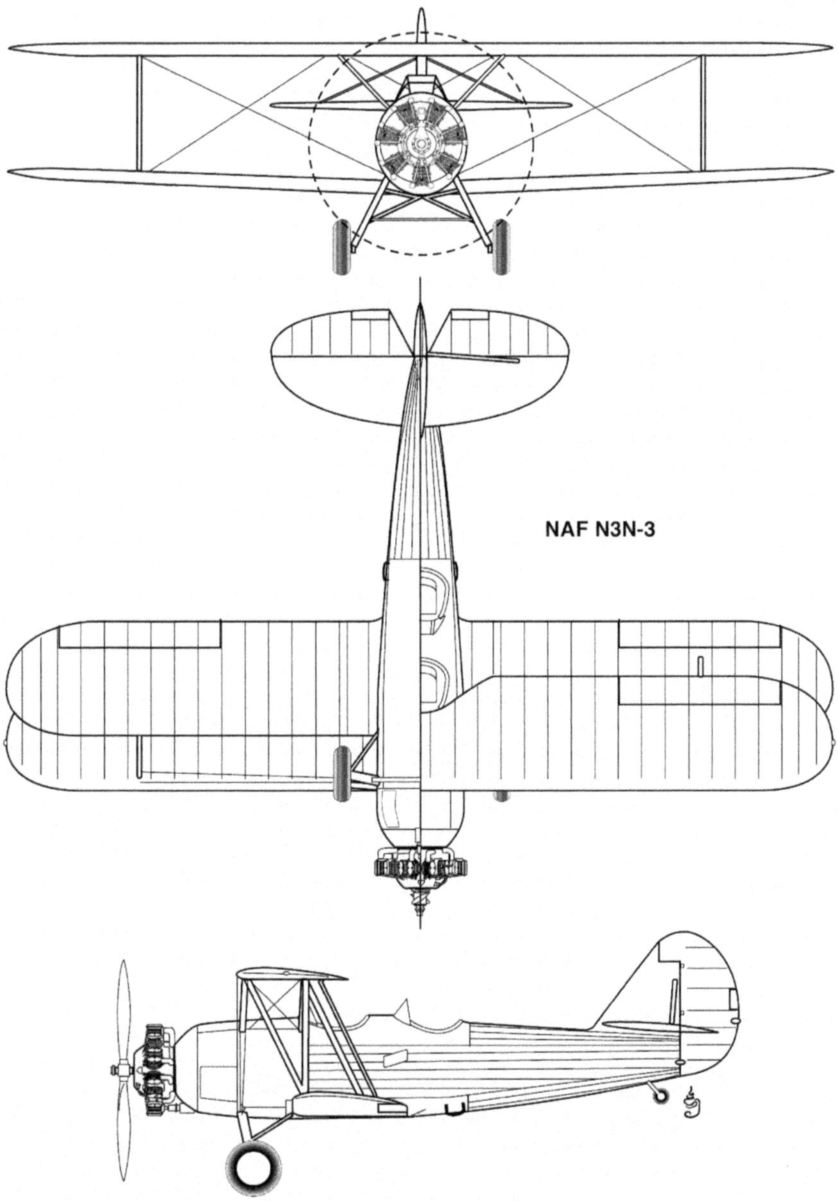

Naval Aircraft Factory. With this in mind, BuAer issued a requirement in October 1934 for NAF to design and develop a primary trainer that would replace the Navy's existing fleet of Consolidated NYs in both training centers and reserve units. The requirement, in addition to being convertible to wheel or float undercarriage, specified that the new aircraft have "maximum ruggedness and ease of maintenance," plus the "general stability for primary training purposes." Soon afterward, BuAer awarded NAF a contract to construct one aircraft as the XN3N-1, to be followed by 45 production models, with the understanding the prototype would be ready for testing by late 1935.

While the general aerodynamic design of the XN3N-1 was typical of open-cockpit, fabric-covered biplanes of the day, its structural features were more innovative: a fuselage built-up of

The 180 N3N-1s produced between 1935 and 1938 represented an across-the-board replacement of the Navy's trainer fleet. Full-chord cowling seen on this example was later removed.

riveted aluminum extrusions rather than welded, steel tubing, an all-aluminum semi-monococque tail group, and wings made up of extruded aluminum I-beams and stamped aluminum ribs. For ready access to the interior, the left side of the fuselage incorporated three removable panels. On August 23, 1935, several months ahead of schedule, the XN3N-1 was rolled out for its first flight. Early air testing, however, revealed a serious amount of tail heaviness coupled with difficulty recovering from spins. Evaluating officials also expressed concern over the lack of forward crash protection and the danger presented by the midships location of the fuel tanks. As a result, the engine mount was extended to shift the center-of-gravity forward, the stabilizer revised and rudder area increased to improve spin recovery, and the main fuel tanks relocated to the upper wings. Although the changes did not completely resolve all stability and control problems, BuAer nonetheless increased the order to 85 aircraft, and the first production N3N-1 completed acceptance trials at NAS Anacostia in mid–1936.

Through the course of 1936, as N3N-1 deliveries proceeded, NAF received two more production contracts, the first 25 of which (as with the original 85) were to be powered by 220-hp Wright R-790–8 9-cylinder engines and the remaining 80 by 235-hp Wright R-760–2 7-cylinder engines. Ten aircraft of the original order were subsequently cancelled so that 180 N3N-1s had been accepted when production ended in April 1938. All N3N-1s came with full-chord cowlings that were later removed. Continuing efforts by NAF to remedy the shortcomings of the N3N-1 resulted in the testing of the R-760-powered XN3N-2 during mid–1936. Despite further revisions to the engine mount and the movement of certain equipment forward, the XN3N-2 was judged not to offer much improvement over the -1 and never went into production.

In light of passage of the Naval Expansion Act of 1938 and the unprecedented growth it portended for naval aviation (i.e., 3,000 new aircraft), NAF officials obtained approval from BuAer to initiate a redesign program that would eliminate the N3N-1's most serious flaws, and in June 1939, received a contract to manufacture 50 new aircraft as the N3N-3. One N3N-1 returned to NAF where it was modified to incorporate a completely new tail group consisting of a larger, rounded fin and rudder and a repositioned stabilizer, a single-strut landing gear (on

A student in an N3N-3 learns how to hoist floatplanes aboard ships. N3N-3s remained in Naval Air Training throughout the war but were eventually outnumbered by Stearman N2Ss.

N3N-3s were also used for training by the Coast Guard. V-196 was the last of four N3N-3s delivered to the Coast Guard in 1941.

the landplane version), and a new nose contour which dispensed with a cowling. Other less visible improvements included changes to the controls and a strengthening of the airframe. Flight-testing of the XN3N-3, completed by January 1940 in both wheeled and float configurations, demonstrated not only excellent handling qualities but improved levels of performance that exceeded expectations. In May 1940, BuAer ordered the N3N-3 into high-rate production, initially contracting for 500 aircraft and in September, upped the order to 816. NAF thereafter manufactured N3N-3s at a rate of about 50 aircraft per month, delivering the last examples in January 1942.

As greater numbers of N3N-3s arrived for service in 1941, most N3N-1s were sent to the reserves or attached to bases for utility duties, and they began displacing N3N-1s in reserve units. In 1941, four N3N-3s were assigned to the Coast Guard, and several more to the Marines as tugs in the glider-training program. During World War II, a majority of N3N-3s were allocated between the Navy's primary flight-training centers at NAS Pensacola, Florida, and NAS Corpus Christi, Texas, but smaller numbers also served at 27 different air stations across the continental U.S. When World War II ended, most surviving N3Ns were sold surplus, however, the very last N3N-3s were not retired from operations at the Naval Academy until 1961.

Boeing-Stearman) NS and N2S—1935
(See Boeing-Stearman PT-13, -16, -18 and -27 in Series I, Part A)

North American NJ—1937
(See North American BT-9, -10 and -14 in Series I, Part A)

North American SNJ—1939
(See North American BC-1 and -2, AT/T-6 and AT-16 in Series I, Part A)

Spartan NP—1940

Technical Specifications (NP-1)

Type: Two-place primary trainer.
Manufacturer: Mid-Continent Div. of Spartan Aircraft Co., Tulsa, Oklahoma.
Total U.S. military versions: 201 (USN).
Unit cost: Unknown.
Powerplant: One 220-hp Lycoming R-680–8 7-cylinder air-cooled radial engine driving a two-bladed, fixed-pitch metal propeller.
Performance: Max. speed 108 mph at s.l.; ceiling 13,200 ft.; range 315 mi.
Weights: 2,069 lbs. empty, 2,775 lbs. loaded.
Dimensions: Span 33 ft. 9 in., length 24 ft. 3 in., wing area 301 sq. ft.

The first military aircraft ever produced by Spartan, the Navy accepted delivery of an experimental prototype in 1940 under the designation XNP-1. It emerged as a fairly conventional biplane design having an all-aluminum structure with fabric covering on the wings, tailplane, and aft fuselage. The sturdy appearing landing gear was braced by a belly pylon and side oleos.

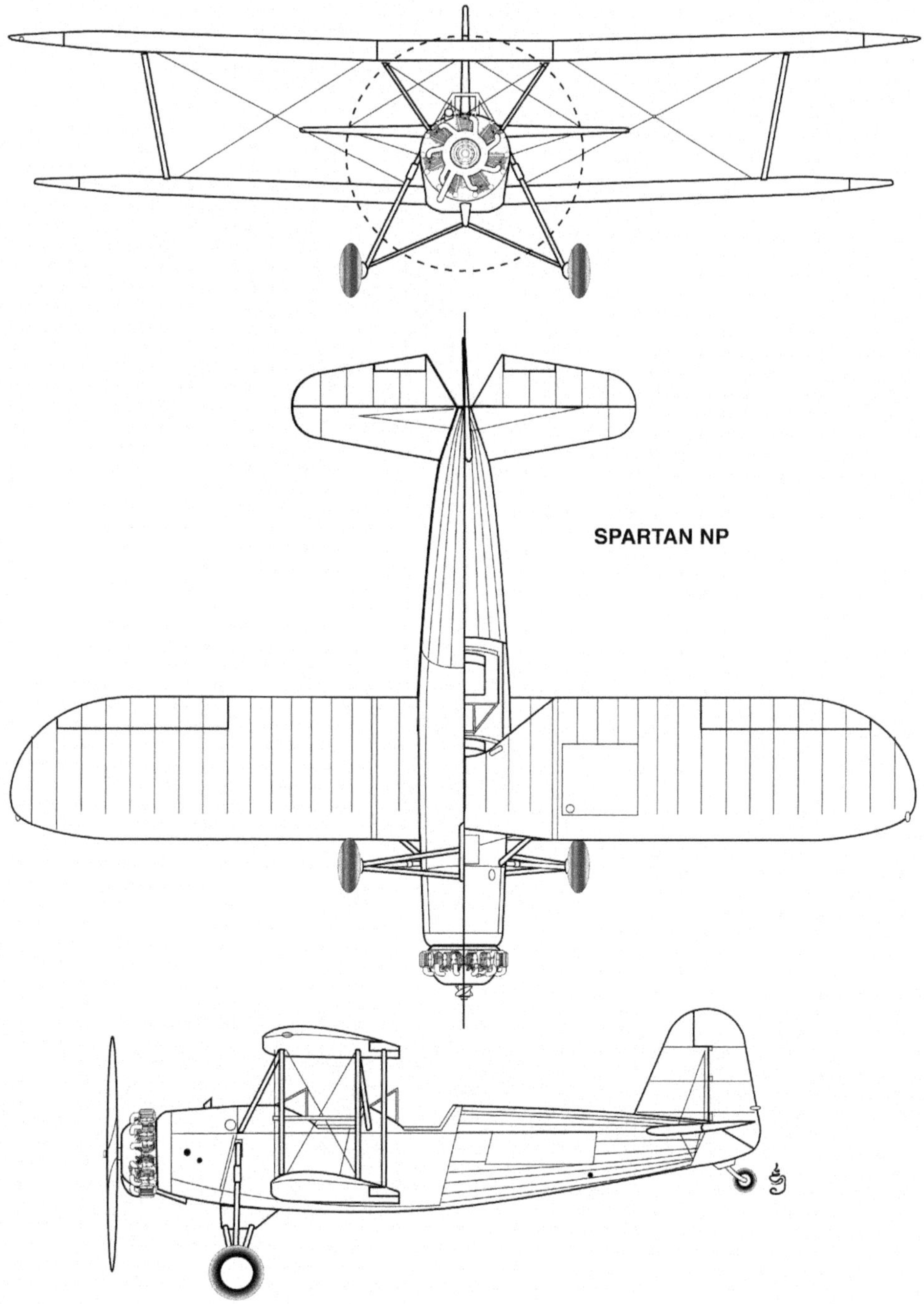

Following official trials, BuAer awarded Spartan an order for 200 more aircraft as the NP-1, with deliveries starting in 1941, 76 of which were on hand by the end of the year. All NP-1s were used to equip the new Naval Reserve Primary Flying Schools in Atlanta, Chicago, Dallas, Detroit, Kansas City, New Orleans, and St. Louis, continuing in this role throughout World War II. Notably, former President George H.W. Bush completed his first solo flight in a Spartan NP-1 sometime in 1942.

One of 201 Spartan NP-1s delivered in 1941 and 1942. All were earmarked for the new Naval Reserve Primary Flying Schools in Atlanta, Chicago, Dallas, Detroit, Kansas City, New Orleans, and St. Louis.

Synopsis of USN/USMC Trainer Procurement, 1941 to 1945

Primary Trainers: As with the USAAF, the most noteworthy characteristic of the wartime era was not development of new types of Navy primary trainers but the large numbers in which preexisting types were produced (4,900+) after 1940, namely, 500 Naval Aircraft Factory N3N-3s, 4,117 Boeing-Stearman N2S-2s, -3s, -4s, and -5s (Army PT-13 and -17 variants), 200 Spartan NP-1s, plus 100 Ryan NR-1s (Army PT-21). Although originally planned as an all-metal, monoplane successor to the N3N, the Naval Aircraft Factory XN5N-1 prototype was tested in 1941 but never placed in production. During 1942, the Navy took delivery of 240 factory-stock Piper J-3C-65 Cubs, which, after entering service as the NE-1 and -2 (minor changes), were used chiefly as "elimination" trainers at various preflight training centers. Finally, because of government concerns over potential aluminum shortages, the Navy contracted for 262 all-wood Timm N2T-1s that were delivered in 1943.

Intermediate Trainers. Although the Naval Air Training establishment did not distinguish a "basic" phase in its pilot training curriculum, it began introducing the first of 1,350 Vultee SNV-1s (Army BT-13A) as intermediate trainers in August 1941, and through the course of the war, added a further 650 SNV-2s (BT-13B) with upgraded engines and 24-volt electrical systems. Starting in 1941, an unspecified number of Vought OS2U-1s, -2s, and -3s rigged with wheels were introduced to the intermediate phase of pilot training at NAS Pensacola, and float-equipped OS2Us also served as the chief aircraft type in the operational training phase of catapult and recovery operations. Also in 1941, Curtiss received a second order for 155 Curtiss SNC-1s that were delivered in 1941 and 1942. Training of intermediate student pilots headed toward flying boats was initially carried out in obsolete types, Martin P3M-2s and Consolidated P2Y-2s and -3s, however, from 1942 onward, as they could be released from frontline duties Consolidated

PBYs were introduced to the program. From 1941 to 1945, the Navy took delivery of 125 more SNJ-3s (Army AT-6A), 2,400 SNJ-4s (AT-6C), 1,568 SNJ-5s (AT-6D), and 931 SNJ-6s (AT-6F), making it the most numerous type of intermediate trainer in naval service. Starting in 1942, the first of 320 SNB-1s (Army AT-11B) and 509 SNB-2s (Army AT-7B) began entering service with the Naval Air Training establishment as intermediate trainers for multi-engine pilots, bombardiers, navigators, and air gunners.

Operational Trainers. After being commissioned and designated Naval Aviators, pilot trainees entered approximately 100-hours of operational training in the category of aircraft they were destined to fly in combat, such as fighters, scout-bombers, torpedo-bombers, scout-floatplanes, patrol-bombers (land), patrol-bomber (sea), and transports. Training was generally accomplished in operational combat aircraft, though many were older variants that had been withdrawn from frontline service. Fighter schools progressed from Grumman F2F and F3F biplanes to Brewster F2As, Grumman F4Fs, General Motors FM-2s, and in the latter part of the war, Grumman F6Fs and Vought F4Us. Scout-bomber training began with Vought SBUs, Curtiss SBCs, and Northrop BTs, but reequipped over time with Douglas SBDs and Curtiss SB2Cs. Prospective torpedo-bomber pilots flew Douglas TBDs until Grumman TBFs and General Motors TBMs became available to replace them. Scout-floatplane pilots typically received catapult and ship-recovery training in Vought OS2Us. Operational training for land-based patrol-bomber pilots began in 1942 with Beech SNBs and Lockheed PBOs, then later progressed to Lockheed PVs; and for sea-based patrol-bombers, with Grumman JRFs, Consolidated PBYs and later, Martin PBMs. Prospective transport pilots received training in Beech SNBs and Douglas R4Ds. Pilots destined for carrier assignments also underwent carrier qualifications—a minimum of eight successful takeoff and landings—in SNJs fitted with arresting gear (re-designated SNJ-3C, -4C, and -5C) or in operational aircraft like F4Fs, FMs, SBDs, and TBFs. Although sources differ on their operational assignments, 13 Fairchild GK-1s (same as Army UC-61) entered the inventory in 1942, and are thought to have been employed for instrument training despite the light transport designation. In 1943 and 1944 BuAer procured 205 Howard NH-1s (a three-seat trainer adaptation of the GH-1 utility transport) for specialized instrument training of instructors, ferry pilots, and transport pilots.

Naval Aircraft Factory N5N—1941

Technical Specifications (XN5N-1)

Type: Two-place primary trainer.
Manufacturer: Naval Aircraft Factory, Philadelphia, Pennsylvania.
Total U.S. military versions: 1 (USN).
Powerplant: One 240-hp Wright R-760–6 7-cylinder air-cooled radial engine driving a two-bladed, fixed-pitch metal propeller.
Performance: Max. speed 135 mph at s.l.; ceiling 13,900 ft.; range (not reported).
Weights: Empty (not reported), 3,370 lbs. gross.
Dimensions: Span 42 ft., length 30 ft. 5 in., wing area 215 sq. ft. (est.).

In 1938 BuAer authorized NAF to design and build one XN5N-1 prototype with the aim of evaluating a monoplane for potential use as a primary trainer. As work progressed, the design appeared with an all-metal, aluminum-clad airframe and broad-chord, fully-cantilevered wings having sufficient area to produce a load factor similar to that of biplane trainers. Other features included split-type flaps and a full engine cowling. The intervention of other NAF projects, however, delayed the first flight of the XN5N-1 until the spring of 1941, by which time the Navy

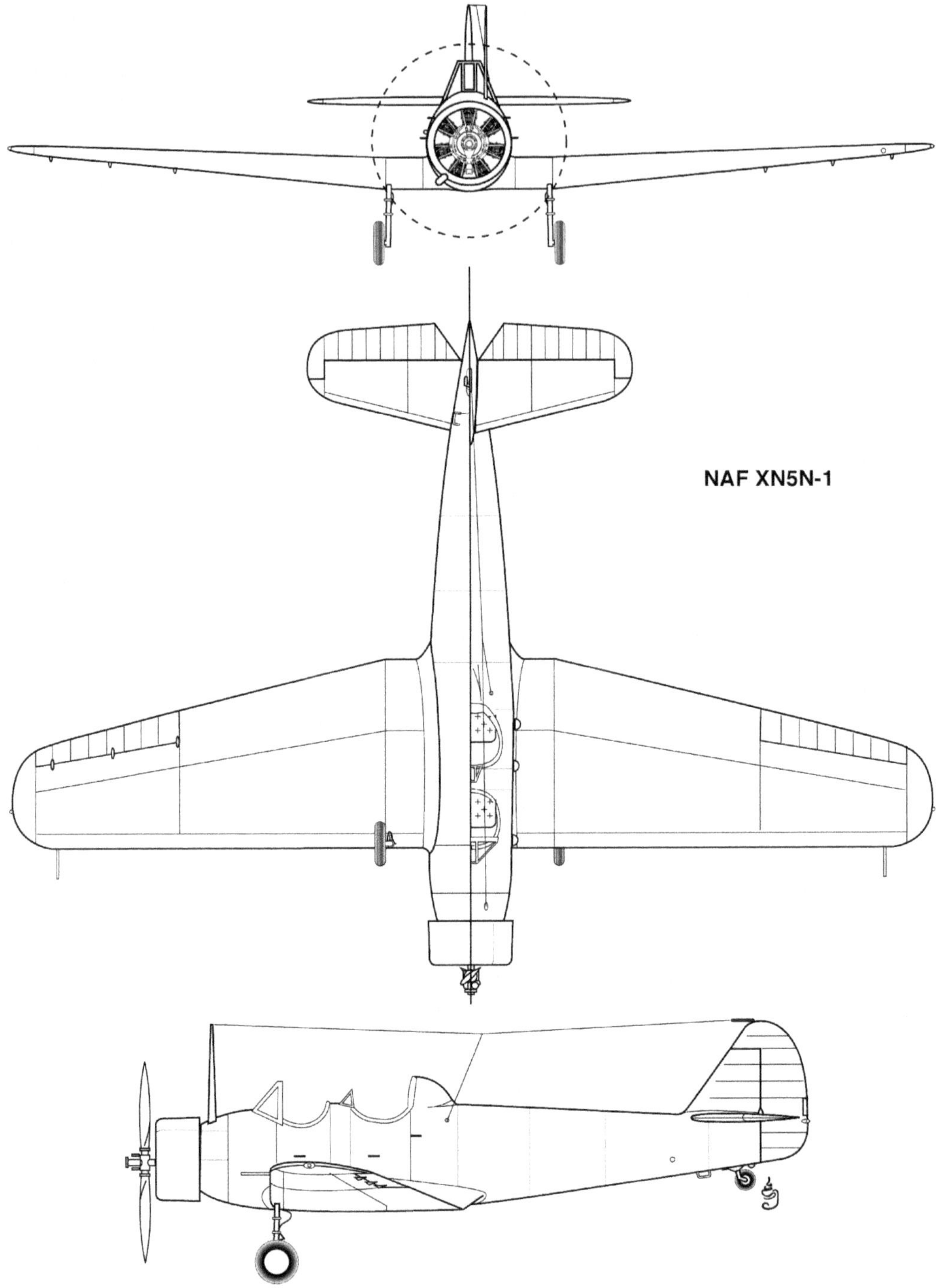

NAF XN5N-1

had placed the N3N-3 into high-rate production and contracted for large numbers of Stearman N2Ss to fulfill its long-term primary trainer needs. Even so, testing of the XN5N-1 at NAF continued after 1941, where it was used as a glider tug, and later still, fitted with a fully-enclosed cockpit.

The sole Naval Aircraft Factory XN5N-1 seen during World War II after being fitted with an enclosed canopy. A backlog of other NAF projects prevented it from reaching production.

Ryan NR—1941
(See Ryan PT-16, -20, -21 and -22 in Series I, Part A)

Piper NE—1941
(See Piper O-59/L-4 in Series I, Part A)

Vultee SNV—1941
(See Vultee BT-13, -25 and -16 in Series I, Part A)

Curtiss SNC Falcon—1941

Technical Specifications (SNC-1)

Type: Two-place intermediate trainer.
Manufacturer: Curtiss Airplane Div., Curtiss-Wright Corp., St. Louis, Missouri.
Total U.S. military versions: 305 (USN, USMC).
Unit cost: Unknown.
Powerplant: One 420-hp Wright R-975-E3 *Whirlwind* 9-cylinder air-cooled radial engine driving a two-bladed, variable-pitch metal propeller.
Performance: Max. speed 201 mph at s.l.; ceiling 21,900 ft.; range 515 mi.
Weights: 2,610 lbs. empty, 3,626 lbs. loaded.
Dimensions: Span 35 ft., length 26 ft. 6 in., wing area 174 sq. ft.

In November 1940, to augment the SNJ-3s on order, BuAer awarded Curtiss a contract to build 150 intermediate trainers under the designation SNC-1, and in 1941, added 155 more to the order. The Curtiss Model CW-22, advertised as both a light attack aircraft and a trainer, had originally been developed from the CW-21 of 1938, a *Cyclone*-powered, single-place fighter offered to the export market. Sharing the same general dimensions and aerodynamic features of

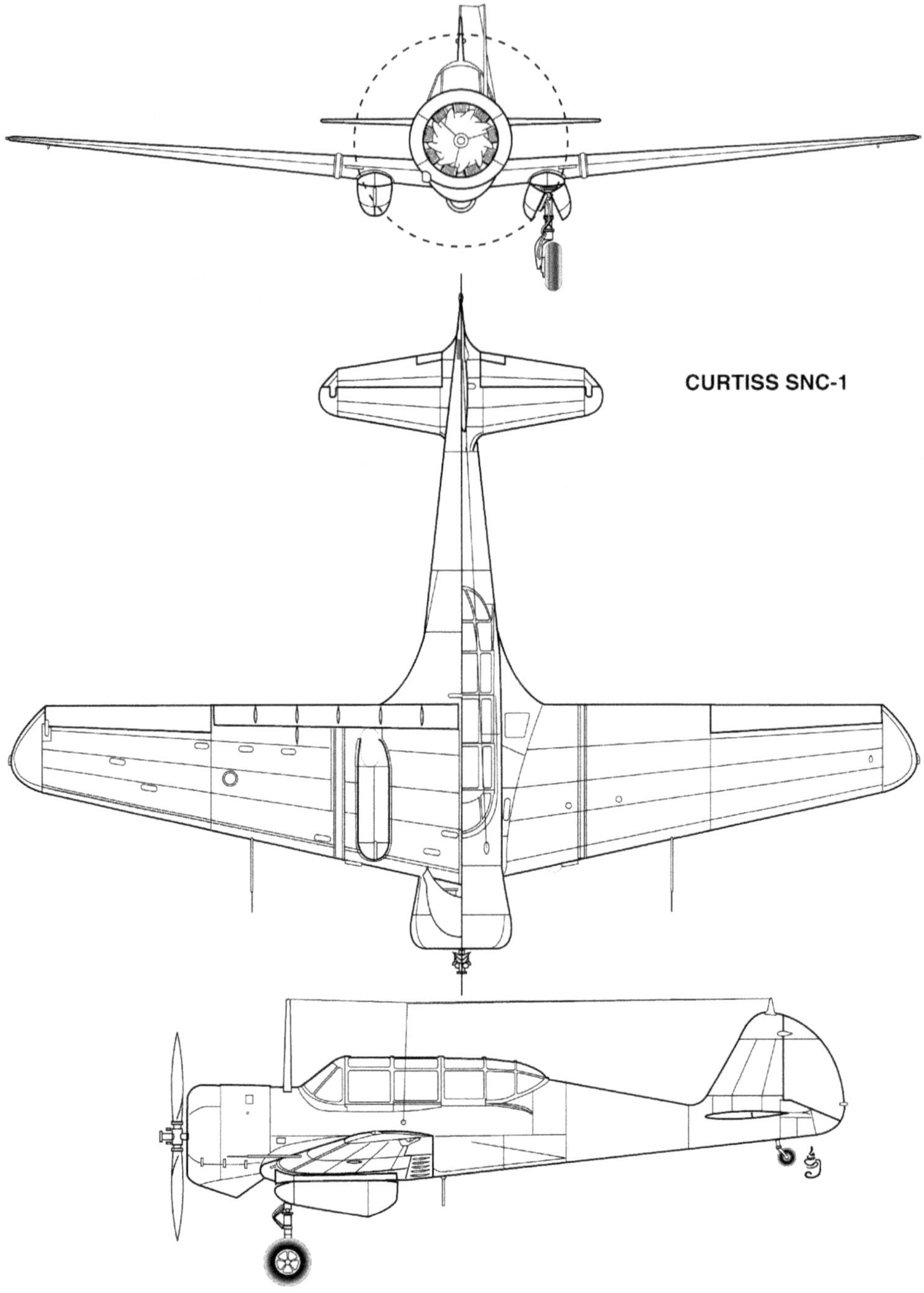

the fighter, the all-metal CW-22 was characterized by a clean, semi-monocoque fuselage structure, wings sharply tapered from the leading edges, and a rearward-folding, semi-exposed landing gear arrangement. As deliveries of SNC-1s proceeded in 1941 and 1942, they were assigned to training centers at NAS Jacksonville and NAS Corpus Christi. Several were later transferred to Marine units for liaison duties. All SNC-1s were withdrawn from service and sold as surplus soon after World War II ended.

Rapid growth of training in 1940 led the Navy to procure 305 SNC-1s to augment the SNJ-3s on order. Most were assigned to NAS Corpus Christi for the advanced phase of pilot training.

Vought OS2U Kingfisher—1941

Technical Specifications (OS2U-1)

Type: Two-place observation-scout floatplane; intermediate and advanced trainer.
Manufacturer: Chance Vought Division of United Aircraft, Inc., Stratford, Connecticut.
Total U.S. military versions: 1,518 (USN, USMC, USCG).
Unit cost: $44,000.
Powerplant: One 450-hp Pratt & Whitney R-985–48 *Wasp Junior* 9-cylinder air-cooled radial engine driving a two-bladed, variable-pitch metal propeller.
Armament: One fixed .30-calibre machine gun and one flexible .30-calibre machine gun in the rear cockpit.
Performance: Max. speed 175 mph, cruise 116 mph; ceiling 19,000 ft.; range 1,015 mi.
Weights: 3,432 lbs. empty, 5,077 lbs. loaded.
Dimensions: Span 35 ft. 11 in., length 33 ft. 7 in., wing area 262 sq. ft.

Though never designed for the role, wheeled versions of Vought's OS2U Kingfisher bridged the gap as an important intermediate trainer during the first two years of World War II. The design of the OS2U stemmed from a 1936 VOS (Heavier-Than-Air-Observation-Scout) requirement to develop a replacement for the Vought O3U biplanes then serving as gun spotters aboard Navy battleships. When the XOS2U-1 prototype emerged as the winner of the VOS fly-off in late 1938, it represented the state-of-the-art in floatplane design: with 30 percent less wing area, 25 percent less horsepower, and 15 percent less gross weight than the competing biplanes (Naval Aircraft Factory XOSN-1 and Boeing-Stearman XOSS-1), Rex Beisel's pioneering concept outperformed them in every category of flight—speed, ceiling, and range—while posting a similar landing speed (56 mph). The XOS2U-1's wide operating speeds were made possible by utilizing

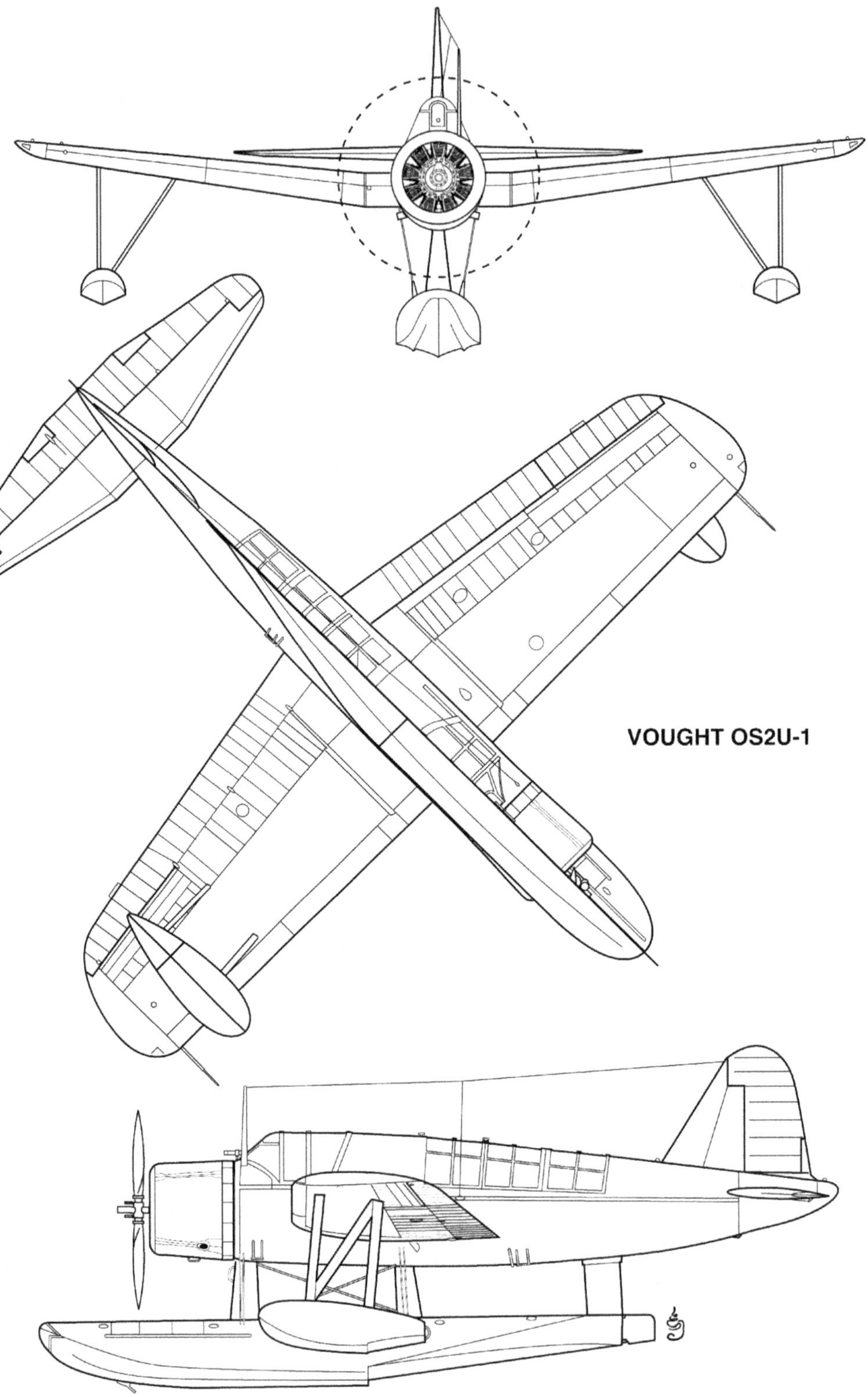

VOUGHT OS2U-1

The Vought OS2U-3, as pictured, served as an intermediate pilot trainer until enough Vultee SNV-1s arrived to take its place. Float-equipped variants also saw service in advanced training for catapult operations.

a broad-chord wing design that incorporated large, full-span flaps for slow flight and spoilers for roll control. It was also the first American military aircraft to employ spot-welding in its primary airframe structures.

Between May 1940 and May 1942 Vought delivered Kingfishers to the Navy in three production batches, 54 OS2U-1s, 158 OS2U-2s, and 1,006 OS2U-3, and between May and October 1942, a further 300 OS2U-s were license-produced by the Naval Aircraft Factory as the OS2N-1. From the outset, the OS2U had been designed with a complete set of controls in the rear cockpit in which the observer's seat could be rotated forward and a removable control stick inserted into a socket in the floorboard. By the end of 1941, 3 OS2U-1s, 99 OS2U-2s, and 157 OS2U-3s had been assigned to training duties, serving both as wheeled intermediate trainers at NAS Pensacola, Florida, and NAS Corpus Christi, Texas, and as float-equipped advanced trainers for catapult operations at NAS Jacksonville, Florida. As deliveries of Vultee SNVs and North American SNJs mounted, OS2Us were withdrawn from intermediate training and reassigned to the fleet but remained as the standard catapult trainer for the duration of the war.

Fairchild GK—1942

Technical Specifications (GK-1)

Type: Utility transport; instrument trainer.
Manufacturer: Fairchild Engine & Airplane Corp., Hagerstown, Maryland.
Total U.S. military (trainer) versions: 13 (USN).

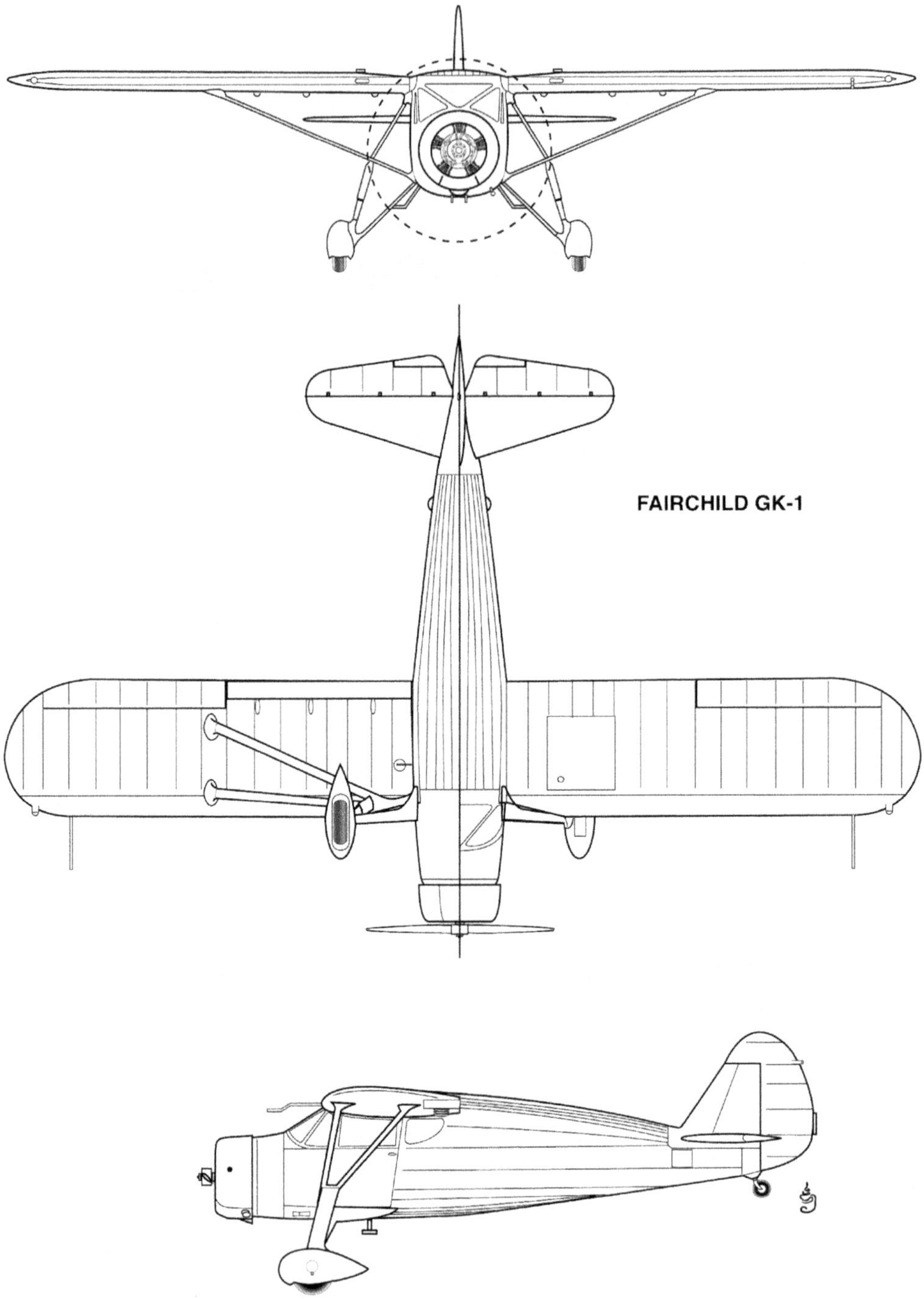

Unit cost: $6,845.

Powerplant: One 165-hp Warner R-500–1 *Super Scarab* 7-cylinder air-cooled radial engine driving a two-bladed, fixed-pitch wooden propeller.

Performance: Max. speed 130 mph, cruise 105 mph; ceiling 16,500 ft.; range 475 mi.

Weights: 1,475 lbs. empty, 2,400 lbs. gross.

Dimensions: Span 36 ft. 4 in., length 23 ft. 8 in., wing area 193 sq. ft.

A Fairchild GK-1 at NAS Pensacola, Florida, in early 1942 still in prewar markings. Red stripes on wings and fuselage denote its use as an instrument trainer.

The Fairchild Model 24C originated in 1932 as a two-seat, high-wing cabin monoplane, and as the type gained popularity, evolved into a three-seater powered by either a 145-hp Warner radial (C8E) or 145-hp Ranger inline engine (C8F). Model 24s utilized the fabric-covered construction typical for lightplanes of the day, with a welded, steel tube fuselage and built-up wooden wings, characterized by a wide-stance main landing gear that tied-in to the wing struts. The USAAF ordered Model 24s into mass production in 1941 as the UC-61 utility transports and also impressed various civilian examples as the UC-61A through -61J. First naval use of the model 24 occurred in 1936 and 1937 when the Coast Guard acquired two model 24-C8Fs as the J2K-1 and two 24Hs as the J2K-2 which were utilized mainly as small transports. In 1941 BuAer gave Fairchild a contract for 13 Model 24W-40s, which joined the service inventory in 1942 under the utility transport designation GK-1. Photographs suggest that the GK-1s received a banded trainer paint scheme and were used for instrument training, though their exact assignments are not revealed in any historical sources or serial number records.

Timm N2T—1942

Technical Specifications (N2T-1)

Type: Two-place primary trainer.
Manufacturer: Timm Aircraft Co., Van Nuys, California.
Total U.S. military versions: 262 (USN).
Unit cost: Unknown.
Powerplant: One 220-hp Continental R-670–4 7-cylinder air-cooled radial engine driving a two-bladed fixed-pitch wooden propeller.
Performance: Max. speed 144 mph, cruise 124 mph; ceiling 16,000 ft.; range 400 mi.
Weights: 1,940 lbs. empty, 2,725 lbs. loaded.
Dimensions: Span 36 ft., length 24 ft. 10 in., wing area 185 sq. ft.

The Timm N2T occupies the distinction of having been the fastest type of American-made primary trainer to achieve operational service during the World War II era. The company

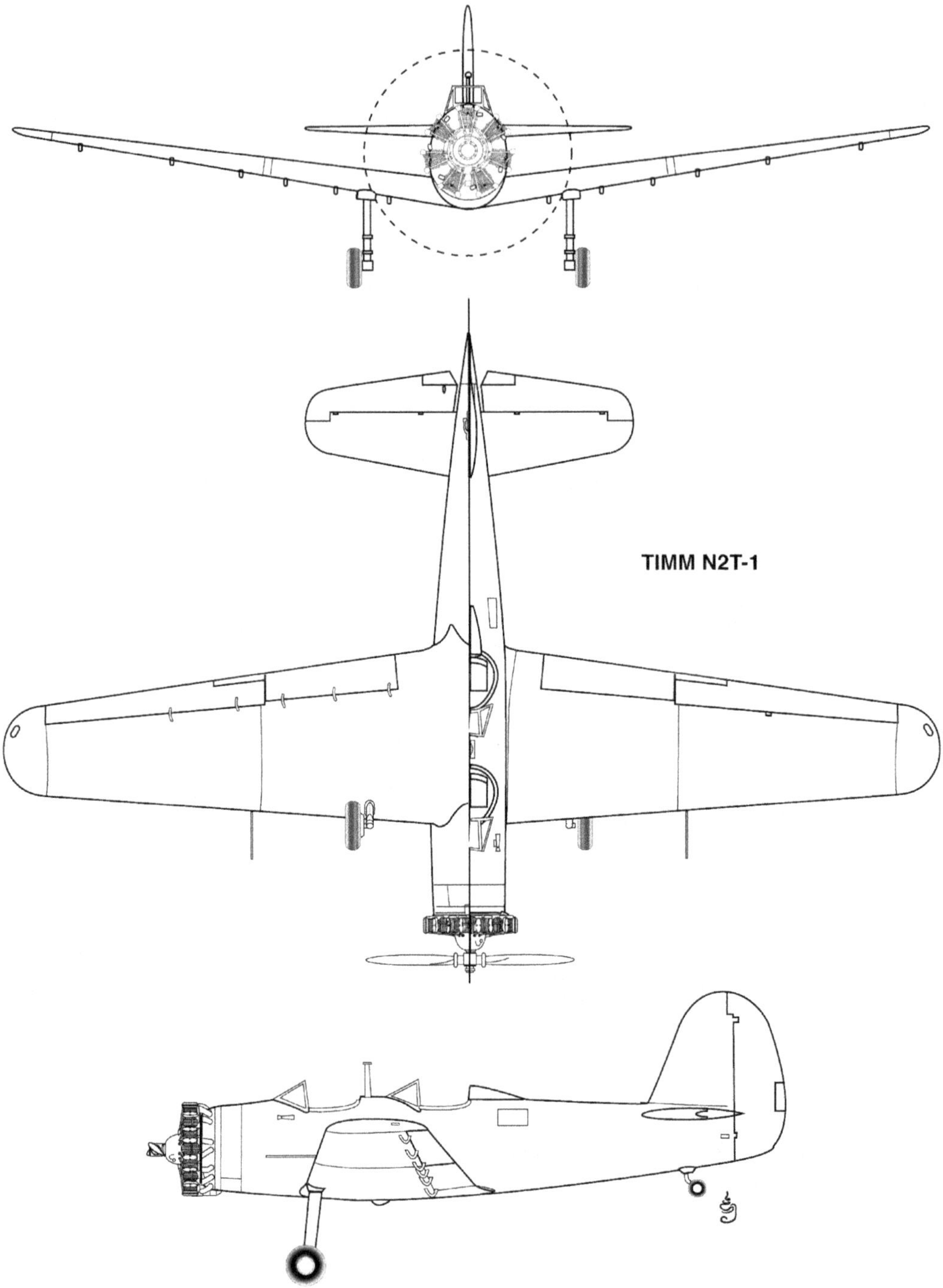

model PT-160K, powered by a 160-hp Kinner R-5 engine, first emerged in 1940 as private venture, featuring all-wood construction that consisted of a fuselage made from the patented "Aeromold" process of resin-impregnated plywood molded under pressure and built-up wooden wings and tail surfaces skinned in plywood. Because of the same concerns over aluminum shortages that induced the USAAF to procure the Ryan YPT-25 (reported in Series I, Part A), the Navy, after

A Timm N2T-1 photographed in 1943 or 1944 with red surround on national insignia. The Navy procured 262 N2T-1s in 1942 and 1943 and used them in primary pilot training until late 1944.

specifying a change to a Continental R-670 engine and a simplified landing gear arrangement, evaluated two model PT-220C prototypes in 1942, followed by a production order for 260 aircraft delivered during 1943 under the naval designation N2T-1. Plans to award Timm a follow-up contract were cancelled due to diminished demand for new primary trainers. N2T-1s initially served with Naval Air Primary Training Command but starting in late 1944, were reassigned to various bases as utility hacks. All were withdrawn and offered surplus soon after the war ended.

Beech SNB—1942 (See Beech AT/T-7 and -11 and TC-45 in Series I, Part A)

Howard NH—1943

Technical Specifications (NH-1)

Type: Three-place advanced instrument trainer.
Manufacturer: Howard Aircraft, Corp., Chicago, Illinois.
Total produced: 485 (USN, USMC, USCG)
Powerplant: One 450-hp Pratt & Whitney R-985-AN1 *Wasp Junior* 9-cylinder air-cooled radial engine driving a two-bladed, variable-pitch metal propeller.
Performance: Max. speed 201 mph at s.l.; ceiling 21,000 ft.; range 1,260 mi.
Weights: 2,700 lbs. empty, 4,350 lbs. loaded.
Dimensions: Span 38 ft., length 25 ft. 8 in., wing area 210 sq. ft.

In 1941 BuAer initially contracted for 34 Howard civil Model DGA-15s (i.e., DGA = damned good airplane) to be used as single-engine utility transports under the designation GH-1. The type's general design had been inspired by Ben O. Howard's Bendix and Thompson Trophy winning DGA-6 *Mr. Mulligan* racer of 1935. From 1942 onwards, the GH-1s were followed by

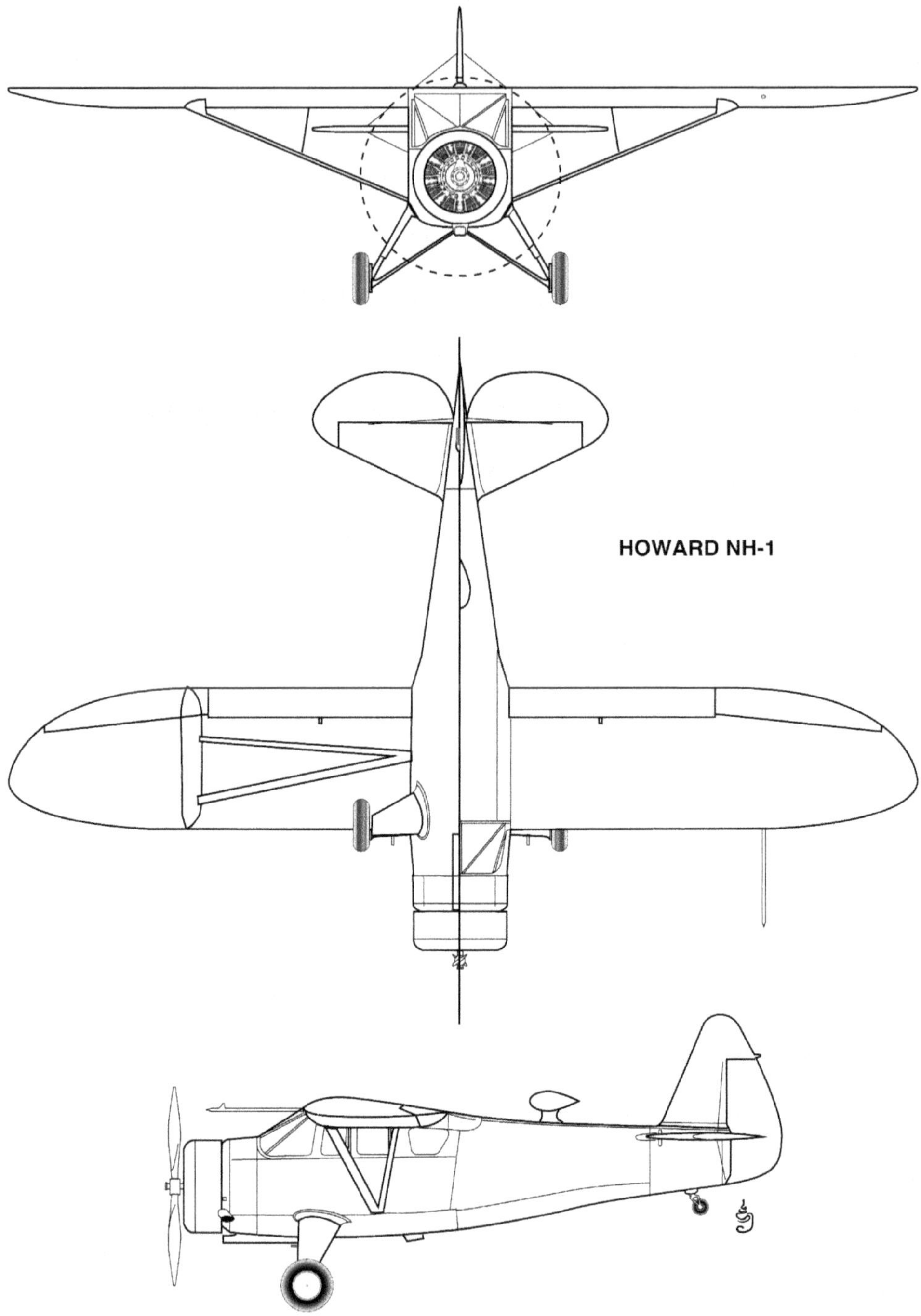

deliveries of 131 GH-2 air ambulances and 115 GH-3s with added fuel capacity. Under the new designation NH-1, BuAer awarded Howard a contract in 1942 to manufacture an unspecified number of DGA-15 airframes to be equipped with a full set of controls and instruments in the rear seat. As deliveries proceeded in 1943 and 1944, NH-1s were distributed among 22 different air stations as advanced instrument trainers for ferry pilots, transport pilots, and instrument

One of 205 Howard NH-1s procured in 1942 as instrument trainers. They were distributed among 22 bases to train ferry, transport, and instrument instructor pilots.

instructor-pilots, the largest single user being the Instrument Instructor School at NAS Atlanta, Georgia. However, BuAer cancelled the contract in 1944 when 205 NH-1s had been accepted, reportedly due to a high operational accident rate attributed to the type's landing characteristics. Most NH-1s were sold surplus soon after World War II ended, though a few examples are thought to have remained active until late 1947.

Series II

Fixed Wing Trainers, 1947 to 1962

A. USAF and U.S. Army

Synopsis of USAF and U.S. Army Trainer Procurement, 1947 to 1962

Primary and Basic Trainers. In September 1947, concurrent with the creation of the independent U.S. Air Force, a significantly downsized Air Training Command (ATC) combined primary and basic pilot training into one Basic Course, with all flight instruction being carried out in the North American AT-6D and F (re-designated T-6D/F in mid–1948). The same year, with the goal of eventually replacing the T-6 with a more modern trainer, Air Material Command (AMC) accepted competing design proposals for the North American XBT-28 and the Douglas XBT-30 (later changed to XT-28/XT-30), then in 1948, on the basis of mockups, cancelled the XT-30 and directed North American to deliver two XT-28 prototypes. But before prototypes could be evaluated, a surge in pilot training prompted by Cold War tensions and the start of the Korean War led, in 1949 and 1950, to the remanufacture of 2,068 T-6D airframes, which reentered service under new serial numbers as the T-6G. Following official trials in 1949 and 1950, AMC ordered the XT-28 into production, with 1,194 examples being delivered between 1951 and 1953 as the T-28A. The increase in pilot training, combined with the need to train pilots faster, moved ATC to reinstate the primary training phase, with the result that AMC conducted a competition between three different primary trainer prototypes between 1949 and 1951, the Fairchild YT-31, the Beech YT-34, and the Temco YT-35, ultimately awarding Beech and Canadian Car & Foundry a series of production contracts for 450 aircraft delivered between 1953 and 1956 as the T-34A.

Even while accepting deliveries of T-28As and T-34As, the USAF was actively making plans to reequip ATC with jets for the Primary and Basic Courses of pilot training, and in April 1952, with this goal in mind, AMC issued the trainer experimental (TX) requirement calling for a lightweight, jet-powered trainer equipped with two ejection seats. After considering three design proposals—the Cessna model 318, the Beech model 73 Jet Mentor, and the Temco model 51—the USAF selection board declared Cessna's twin-jet entry the winner, with the company receiving a contract to build two prototypes and one static test example under the designation XT-37. Once testing got underway in late 1954, AMC ordered 11 T-37A production models for further evaluation, followed by successive orders for a total of 444 T-37As delivered before the end of 1959. In late 1959, under new contracts, Cessna began delivering the first of 552 T-37Bs, which featured uprated J-69-T-25 engines and improved avionics, and all surviving T-37As were afterward brought up to the T-37B standard.

Advanced Trainers. Along with stock P/F-51Ds, the 10 P/TF-51Ds (see Series I, Part B) continued in postwar ATC service as advanced fighter trainers until 1952. Originally flown in 1948 as the two-place TF-80C, the Lockheed T-33A became ATC's standard advanced jet fighter trainer in 1949, with a total of 4,107 examples being procured until production ended in 1959. During the mid–1950s, AMC issued a General Operating Requirement (GOR) calling for a supersonic trainer to eventually replace the T-33, and in June 1956, after evaluating proposals from North American (TF-100C) and Northrop (N-156T), Northrop received a development contract for the YT-38A, which made its first flight in March 1959 and was subsequently ordered into mass production as the T-38A. From 1960 to 1970, Northrop delivered a total of 1,187 T-38As to the USAF, with most going into service with ATC (later renamed AETC).

The North American TB-25J (see in Series I, Part A) became ATC's standard multi-engine advanced trainer in the postwar era, with some 600 being modified from 1946 onwards, and during the early and mid–1950s, a further 294 B-25Js were reconfigured as radar-interception trainers as the TB-25K, L, M, and N. In 1951, following a design competition for a combination twin-engine advanced trainer-small transport, AMC awarded Beech a contract to build two XT-36A prototypes; however, the project was abruptly cancelled in mid–1953 before any aircraft had flown. From 1949 to 1955 the USAF procured 364 twin-engine aircraft for specialized crew training that were based upon the design of Convair's CV-240 civil airliner: two XT-29 prototypes; 45 T-29A unpressurized navigator trainers; 105 T-29B pressurized navigator/radio-operator trainers; 119 T-29C pressurized navigator/radio-operator trainers with engine upgrades; and 92 T-29D bombardier trainers. A proposed turboprop-powered T-29E version was provisionally ordered but cancelled before being built.

Operational Trainers. Although a variety of unmodified operational aircraft were employed for conversion and transition training (e.g., Douglas B-26, Boeing B-29, B-47, and B-50, Republic F-84, North American F-86 and F-100, Northrop F-89, and Lockheed F-94), a number of aircraft were re-designated, and in some cases modified, for the training role. Transport types used in training included TC-46A/Ds (see Series I, Part A), TC-47Bs (see Series I, Part A) until 1954, and Douglas TC-54Ds until about 1955. The first batch of 22 production North American B-45As delivered to Tactical Air Command (TAC) in 1948 were subsequently re-designated TB-45A transition trainers. Forty-six Boeing B-50s became the TB-50A, D, and H in 1951–1953 when stripped of armament and used for transition training in Strategic Air Command (SAC). During the mid–1950s an indefinite number of Air National Guard Douglas B-26Bs became TB-26Bs after being refitted as dual-control trainers. In 1955 and 1956, 38 dual-control Martin B-57Cs entered service as conversion trainers with tactical units operating the B-57B, followed in 1956 and 1957 by 68 B-57Es which functioned both as high-speed target tugs and dual-control trainers. As the supersonic Convair B-58 reached operational service in 1960, eight were converted or completed as dual-control TB-58A trainers and distributed among SAC operational units.

Continuing increases in jet fighter performance led to development of a number of two-seat, dual-control variants that served with tactical-fighter and fighter-interceptor units for transition and familiarization training. Two North American TF-86Fs were evaluated from 1953 to 1955, however, the USAF decided against production in favor of a two-seat F-100, with the result that 339 North American F-100Fs (fully mission-capable unlike the TF-100C) were accepted between 1957 and 1959. Likewise, other two-seat, dual-control fighter variants saw limited production during this period: 111 subsonic Convair TF-102As from 1955 to 1957; 26 Lockheed F-104Bs in 1958 and 26 F-104Ds in 1959; 79 McDonnell F-101Fs between 1959 and 1961; 64 Convair F-106Bs in 1959 and 1960; and 143 Republic F-105Fs in 1963 and 1964.

Army Aviation Fixed-Wing Trainers. Pilot and aircrew training in Army Aviation remained under overall USAF control until 1956, and primary training of its fixed-wing pilots under civilian contractors (mainly in Piper J-3 Cubs) continued until 1959. After completing fixed-wing primary, new Army pilots typically transitioned to operational liaison types like the Piper L-4 (see in Series I, Part A), Stinson L-5, Aeronca L-16 (diverted from National Guard units during the Korean War), Piper L-21 (some re-designated TL-21A), and from the late 1950s onward, Cessna L-19s. During the summer of 1959, the Army commenced its own fixed-wing primary pilot training program in pre-existing L-19As. For instrument training, the Army procured 83 Cessna LC-126s between 1949 and 1952 and 310 TL-19Ds in 1956 and 1957. Through most of this period, multi-engine conversion training was completed in operational Beech L-23A-Ds, the Army's standard twin-engine utility transport from the early 1950s to the mid–1960s; likewise, Army pilots underwent transition training to larger single-engine utility transports like the De Havilland L-20A and U-1A. In 1958, three Cessna T-37As on loan from the USAF were evaluated as potential trainers for the Army's "Long Arm" program, however, the program was cancelled within a year and the aircraft returned.

Aeronca L-16–1947

Technical Specifications (L-16B)

Type: Two-place primary trainer.
Manufacturer: Aeronca Aircraft Corp., Middletown, Ohio.

One of 509 Aeronca L-16As delivered in 1947 and 1948. It was intended to replace the Piper L-4. Some went to National Guard units while others were used for training of fixed-wing Army Aviators at Gary AFB, Texas.

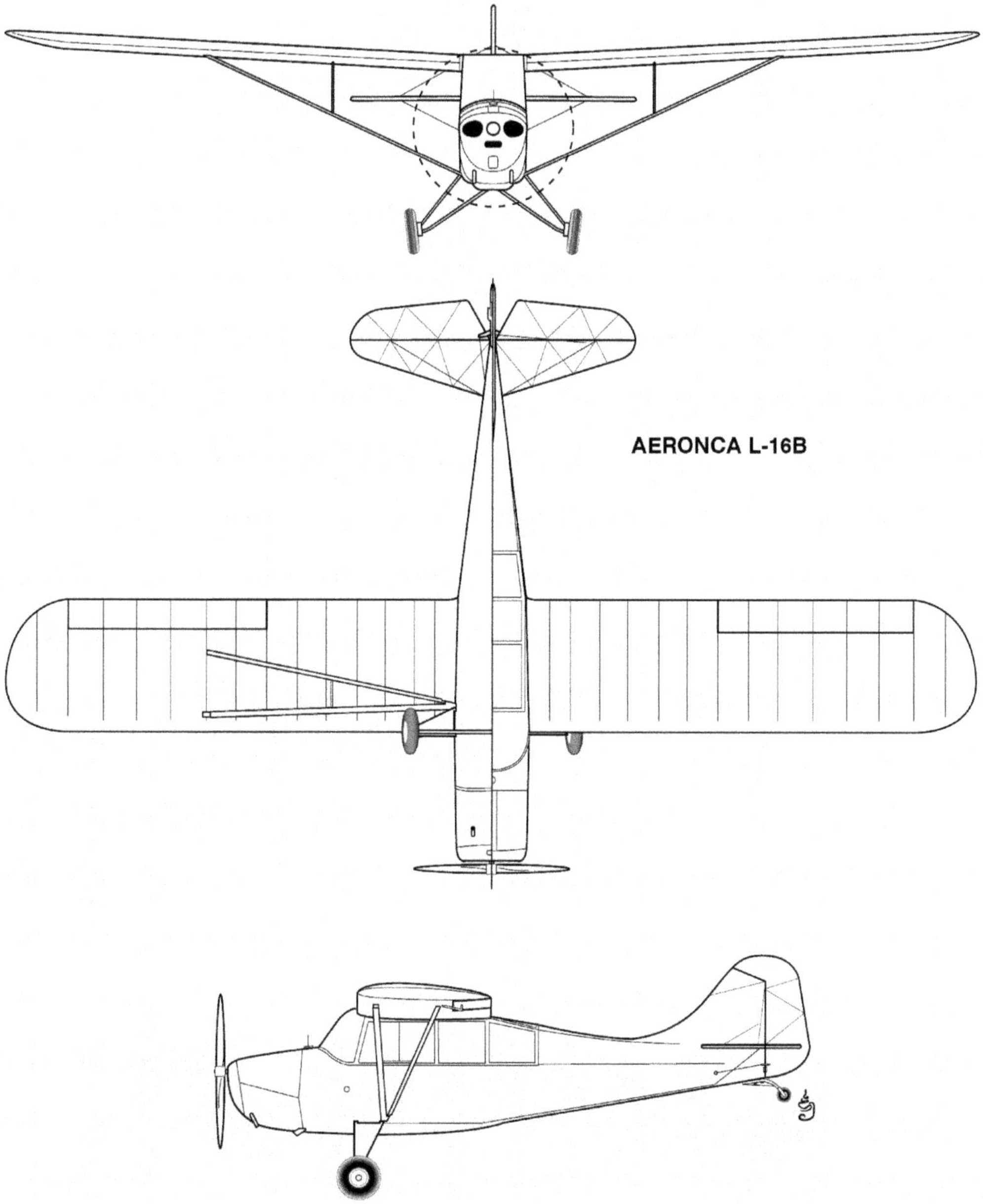

Total U.S. military versions: 609 (U.S. Army).
Unit cost (est.): $3,000.
Powerplant: One 90-hp Continental O-205–1 4-cylinder air-cooled horizontally-opposed engine driving a two-bladed fixed-pitch metal propeller.
Performance: Max. speed 108 mph, cruise 87 mph; ceiling 14,500 ft.; range 350 mi.
Weights: 890 lbs. empty, 1,450 lbs. loaded.
Dimensions: Span 35 ft., length 21 ft. 6 in., wing area 170 sq. ft.

In 1947 the Army procured 509 military derivatives of the Aeronca 7BCM "Champion" under the designation L-16A, of 376 were originally earmarked for National Guard units to replace Piper L-4s (see Series I, Part A). Powered by the same 85-hp Continental O-190–1 engines,

L-16As differed from civil models in having extended greenhouse canopy enclosures and equipment for artillery spotting. To bolster its fixed-wing training fleet, the Army added 100 improved L-16Bs to its inventory during 1948. In addition to 90-hp O-205–1 engines, L-16Bs featured a dorsal fin fillet and greater fuel capacity. L-16As and Bs were never intended for frontline observation duties, however, due to initial shortages of Army liaison aircraft when the Korean War started in mid–1950, some actually saw duty in the combat theater. As Cessna L-19s replaced them, some L-16s returned to the training role, but most were transferred to Civil Air Patrol, the last example being stricken from the Army inventory in 1956.

Lockheed TP/TF-80/T-33 (TO/TV) Shooting Star—1948

Technical Specifications (T-33A)

Type: Two-place advanced trainer.
Manufacturer: Lockheed Aircraft Corp., Burbank, California.
Total U.S. military versions: 4,806 (4,107 USAF; 699 USN/USMC).
Unit cost: $123,000.
Powerplant: One Allison J-33-A-35 turbojet engine, rated at 5,400 lbs./s.t.
Armament: Two nose-mounted .50-calibre machine guns and wing hardpoints for up to 2,000 lbs. of external ordnance.
Performance: Max. speed 600 mph, cruise 455 mph, ceiling 48,000 ft.; range 1,275 mi. (ferry).
Weights: 8,300 lbs. empty, 15,100 lbs. loaded.
Dimensions: Span 38 ft. 11 in., length 37 ft. 9 in., wing area 238 sq. ft.

Widely known as the "T-Bird," the Lockheed T-33 Shooting Star holds the distinction of having been the first type of purpose-built, jet-propelled trainer to serve in the USAF and the only such type until augmented by the Cessna T-37 in 1957. Its origins are traceable to May 1947, when Lockheed initiated an in-house design study for a two-seat variant of the P-80, followed in August by USAAF approval to modify one P-80C airframe to serve as the TP-80C prototype. To accommodate a second cockpit, the aircraft received a 26.6-inch fuselage plug forward of

An early production T-33A with under-slung tip tanks. The designation changed from TF-80C to T-33A in May 1949. The type became operational with Air Training Command as an advanced jet trainer in June 1949.

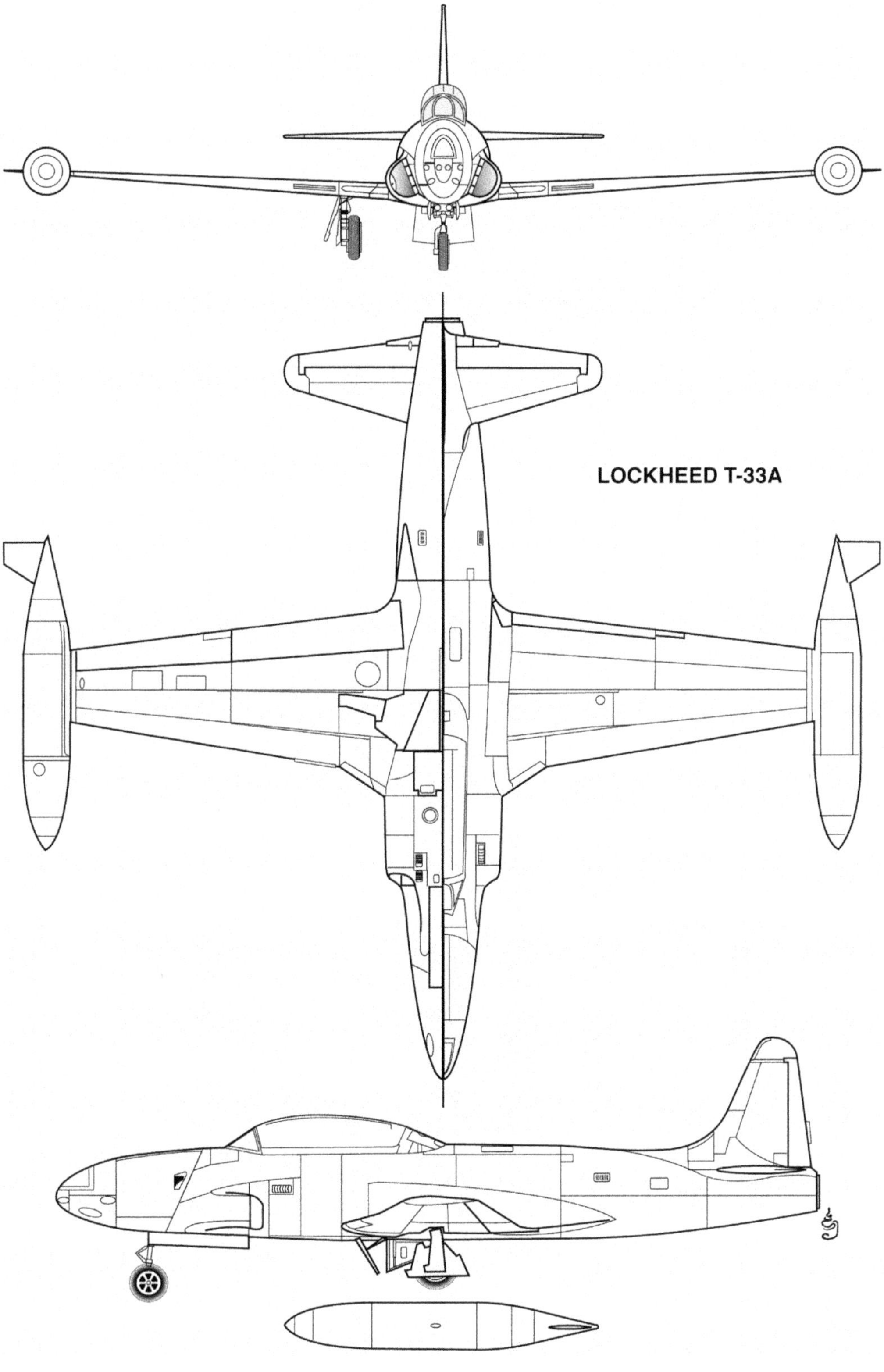
LOCKHEED T-33A

A T-33A seen late in its career while serving with the Connecticut Air National Guard. In guard and reserve units equipped with single-seat jets, T-33As were normally used for instrument proficiency.

the wing and a 12-inch plug aft, and the decreased size of the fuselage fuel tank was offset by adding wing tip tanks. The first flight of the TP-80C took place on March 22, 1948, and testing soon revealed flight characteristics virtually identical to those of the P-80C. Air Material Command (AMC) promptly gave Lockheed an initial contract to build 20 TP-80Cs for service evaluation, the designation being changed to TF-80C in June 1948, then to T-33A in May 1949. A total of 4,107 T-33As were delivered to the USAF under a succession of production contracts, with the last example being accepted in 1959. Over their production life, T-33As received a series of engine upgrades, from the 4,600-lb./s.t. J-33-A-23 to the 5,400-lb./s.t. J-33-A-35.

Production T-33As first became operational with Air Training Command (ATC) in June 1949, where they served chiefly as advanced jet trainers in Undergraduate Pilot Training (UPT), ultimately equipping eight Pilot Training Wings. In smaller numbers, the type also saw duty with various operational commands, primarily Tactical Air Command (TAC) and Air Defense Command (ADC) as instrument proficiency trainers. The phase-out of T-33As from ATC began during early 1961 as the first supersonic Northrop T-38As arrived to take their place in UPT, and the process had been completed by the end of 1964. As they left ATC, large numbers of T-33As were supplied to foreign air forces under various Military Assistance Programs (MAP), and many others went to the USAF Reserve and Air National Guard (ANG), where they served in tactical-fighter and fighter interceptor squadrons as proficiency trainers. The very last T-33As were not retired from active USAF service until March 1988.

Naval Variants: In 1948 the USAF transferred 50 single-seat Lockheed F-80Cs to the Navy that received Bureau Numbers and were placed on the inventory under the designation TO-1 (changed to TV-1 after 1950). Intended solely as land-based jet transition trainers, about 30 TO-1s were provisionally assigned to Marine squadron VMF-311 and Navy squadron VF-52 in late 1948 to prepare pilots for transition to new Grumman F9F-2s in the near future. After leaving active Navy service in the early 1950s, some TO/TV-1s continued as trainers in several reserve units a few years longer.

In 1949 the Navy initiated procurement of the first of 699 T-33As, initially designated

Lineup of TO-1s (F-80C) in 1949 with VMF-311 to transition Marine pilots to Grumman F9F-2s. The planes were later assigned to reserve units as jet trainers.

Early Lockheed TV-2 seen during the early 1950s while serving with Advanced Training Unit Three at NAS Kingsville, Texas. Though phased out of training in the early 1960s, the last TV-2s did not leave reserves until 1976.

TO-2 but changed to TV-2 with the 27th aircraft delivered. As deliveries proceeded, the Naval Aviation Training establishment integrated TV-2s into the advanced pilot training curriculum at NAS Corpus Christi, Texas, and NAS Memphis, Tennessee, for land-based jet transition training. Many other new TV-2s went directly to jet-equipped Navy and Marine fighter and attack air groups for instrument proficiency training, and starting in the mid–1950s, they were allocated to reserve units for the same purpose. All TV-2s remaining on the naval inventory in mid–1962 were re-designated T-33B. While the Phase-out of TV-2/T-33Bs from advanced training duties commenced during the early 1960s as newer jet trainers (e.g., Lockheed T2V-1/T-1A, North

American T2J-1/T-2A) entered service to replace them, the type soldiered on for many years as an instrument trainer in Navy and Marine operational units, and the very last examples did not leave the reserves until 1976.

Beech T-34 Mentor—1948

Technical Specifications (T-34A [T-34C])

Type: Two-place primary trainer.
Manufacturer: Beech Aircraft Corp., Wichita, Kansas; and Canadian Car & Foundry Co., Montreal, Canada.

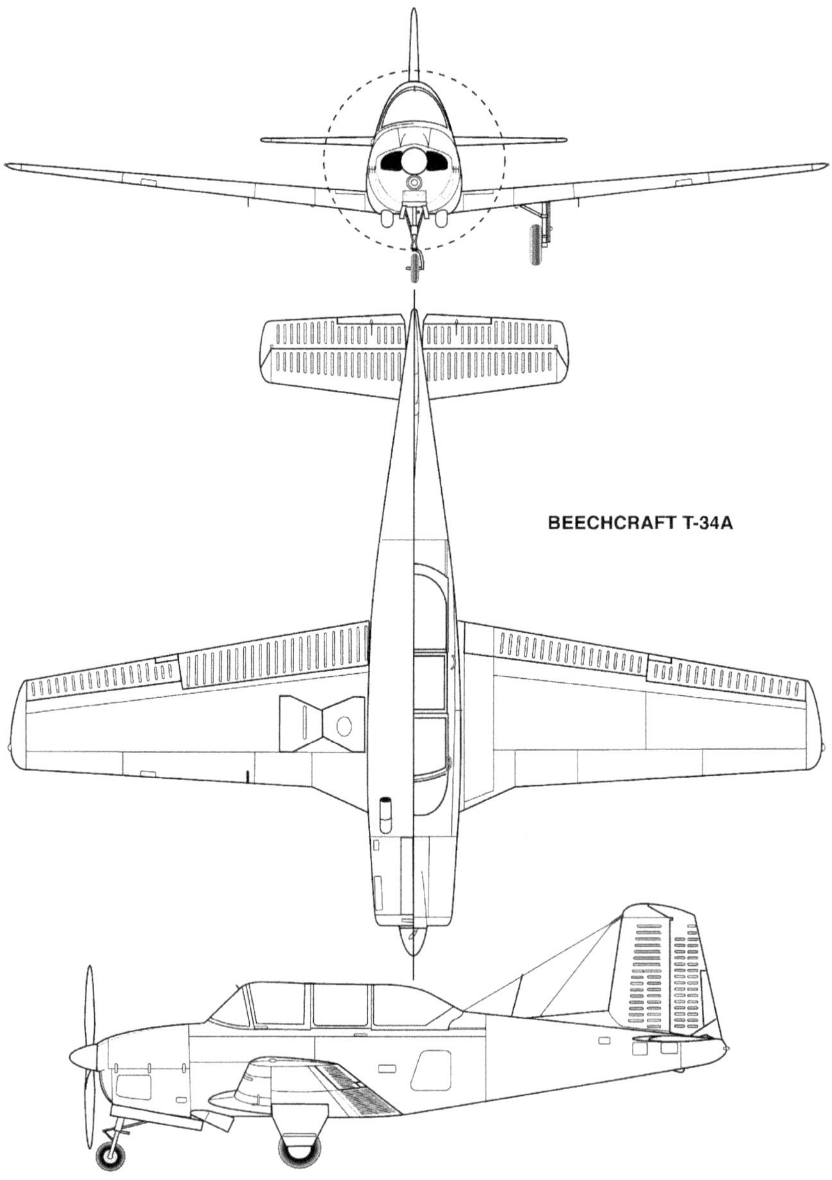

Total U.S. military versions: 1,317 (453 USAF; 864 USN/USMC/USCG).
Unit cost (T-34A): $20,000-$25,000 (est.).
Powerplant: One 225-hp Continental O-470–13 6-cylinder air-cooled horizontally-opposed [400-shp Pratt & Whitney PT-6A-25 turboprop] engine driving a two-bladed [three-bladed] constant-speed metal propeller.
Performance: Max. speed 190 mph [322 mph], cruise 173 mph [246 mph]; ceiling 20,000 ft. [30,000 ft.]; range 821 mi. [708 mi.].
Weights: 2,156 lbs. [2,960 lbs.] empty, 2,900 lbs. [4,300 lbs.] loaded.
Dimensions: Span 32 ft. [33 ft. 4 in.], length 25 ft. 11 in. [28 ft. 8.5 in.], wing area 177.6 sq. ft. [179.6 sq. ft].

It is an amazing testament that the Beech T-34 remains in active military service almost 65 years after its first flight. The model 45 arose in 1947 as a private venture instigated by company founder Walter Beech to evolve a low-cost, easy to fly military trainer based on the design of his successful model 35 Bonanza. Final design work and construction proceeded smoothly, with the model 45 prototype making its first flight from the Beech factory on December 2, 1948. Although the Bonanza's V-tail configuration had been initially considered, the model 45 finally emerged with a conventional tail group together with an airframe stressed for aerobatic maneuvers and a slimmer fuselage cross-section incorporating tandem cockpit seating. Already billed as the "Mentor" by the factory, the prototype embarked upon a demonstration tour of U.S. military bases during early 1949, then on July 4, before a crowd of thousands at the Chicago National Air Fair, was flown through a dazzling aerobatic routine by female stunt pilot Betty Skelton. At the time, the USAF already had a primary trainer design competition underway that included the Fairchild YT-31 (see in Series II, Part B as the XNQ-1) and the Temco model TE-1 (later tested as the YT-35, see below).

While the USAF selection board rated the Mentor over the YT-31 and YT-35, budget restrictions delayed procurement plans until March 1950, when Air Material Command (AMC) gave Beech an order to deliver three model 45s for competitive evaluation under the designation

A newly delivered Beech T-34A, probably in 1954. Air Training Command acquired 450 T-34As in 1954 and 1955 to reinstitute primary pilot training as a distinct phase of the syllabus.

The model 73 Jet Mentor, pictured in civil registration in 1955, was a company-funded venture intended to compete against the Cessna T-37 for the TX contract.

YT-34. The prototypes arrived in May, June, and July, the first two powered by 205-hp Continental E-185–89 engines and the last by a 225-hp E-225–8 (later changed to O-470–4). After initial military trials, the YT-34s were subjected to three years of functional testing that, among other things, included training student pilots with instructors. In March 1953 AMC awarded a production contract for 450 aircraft as the T-34A, 350 to be completed by Beech and 100 under license by Canadian Car & Foundry Co., with deliveries to Air Training Command beginning in 1954. T-34As remained the USAF's standard primary trainer until they were gradually phased out during the late 1950s in favor of jet-powered Cessna T-37As. After leaving active service, some T-34As were transferred to Air Force flying clubs and Civil Air Patrol while others went to foreign air forces.

Despite having lost the 1952–1953 TX design competition to Cessna's T-37 (see below), Beech moved forward, as a company-funded effort, with construction of its model 73 Jet Mentor, with the hope of securing future military contracts. Powered by a single Continental J-69 turbojet and sharing the major airframe components of the T-34A, the Jet Mentor was distinguished by wing-root air intakes, a single jet exhaust in the tail, and a forward-located cockpit that housed two ejection seats. After the prototype made its first flight in December 1955, it was loaned briefly to the USAF for testing, but no production was forthcoming.

Naval Variants: In September 1953, with the aim of replacing its aging inventory of SNJs, the Naval Bureau of Aeronautics (BuAer) took delivery of one of the earliest production T-34As for extensive testing. Navy requirements were more stringent, in that its pilot training syllabus called for a carrier-type, "drop-in" landing method (i.e., an abrupt stall above the ground)

Though externally similar to the T-34A, the T-34B possessed an airframe and landing gear strengthened to withstand the carrier-type, "drop-in" landing technique. They were used as primary trainers until the mid-1970s.

that imposed much more stress on the landing gear and airframe. After passing this test, BuAer awarded Beech an initial contract in June 1954 for 300 aircraft as the T-34B (no naval designation assigned). The B differed from the A in having an extra degree of wing dihedral, a swiveling (rather than steerable) nosewheel, adjustable rudder pedals instead of moveable seats, and removal of a triangular fairing beneath the rudder. The Navy order was subsequently increased to a total of 423 T-34Bs, the final examples being delivered in October 1957.

After nearly a 15-year production hiatus, the Navy returned two T-34Bs to the Beech factory in 1973 for conversion to turboprop power. Making its first flight on September 21, 1973, the YT-34C "Turbo Mentor" appeared with a 400-shp Pratt & Whitney PT-6A-25 engine having a torque limiter that restricted available power to 56 percent of maximum in order to maintain constant performance over a wide range of conditions and also extend engine life. Besides general airframe strengthening, other changes included a three-foot eight and a half-inch increase in length, a one-foot four-inch increase in wingspan, an 804-lb. increase in empty weight, and a taller fin and rudder with an enlarged fillet to offset the increase in power. Following official evaluations in 1974, the Navy ordered the type into production as the T-34C, with a total 352 being delivered under successive contracts between 1975 and 1990. Beech also developed a T-34C-1 armed attack version for export and offered a demilitarized Turbo-mentor 34C to the civil market.

T-34Bs began replacing SNJs as primary trainers at NAS Pensacola, Florida, in December 1954. The change enabled the Naval Air Training Center to reduce required flight hours in the primary syllabus by a factor of over 50 percent (from 74 to 36), and likewise reduced the average

Still in service, the career of the T-34C has spanned 38 years. Although Raytheon T-6 Texan IIs have been replacing them on a slow-rate basis since 2005, T-34Cs still equip VT-27 and VT-28.

time to solo by half, which had the combined effect of simultaneously accelerating the program, reducing costs, and lowering the accident rate.

T-34Bs enjoyed a much longer career than their USAF counterparts, remaining in service as the Navy's standard primary trainer until they were replaced by T-34C Turbo Mentors during the mid–1970s. Afterward, many T-34Bs remained operational with various recruiting commands where they were used to provide flight experience to potential pilot and flight officer candidates; and others were given to Navy flying clubs or sold surplus. Naval Air Training's fleet of T-34Cs have, at present, been in continuous service as primary trainer for 38 years. Although the Navy began the process of replacing T-34Cs with the Raytheon (Hawker Beechcraft) T-6 "Texan II" (see in Series III) in 2005, the reequipping of Training Air Wings has been slow, and at least two squadrons of T-34Cs (VT-27 and -28) remain at the time of this writing.

North American TB-45 Tornado—1948

Technical Specifications (TB-45A)

Type: Four-place tactical bomber transition trainer.
Manufacturer: North American Aviation, Inc., Inglewood, California.
Total U.S. military versions: 22 (USAF).
Unit cost: $1,100,000.
Powerplants Four Allison J-35-A-11 turbojet engines, each rated at 5,000 lbs./s.t.

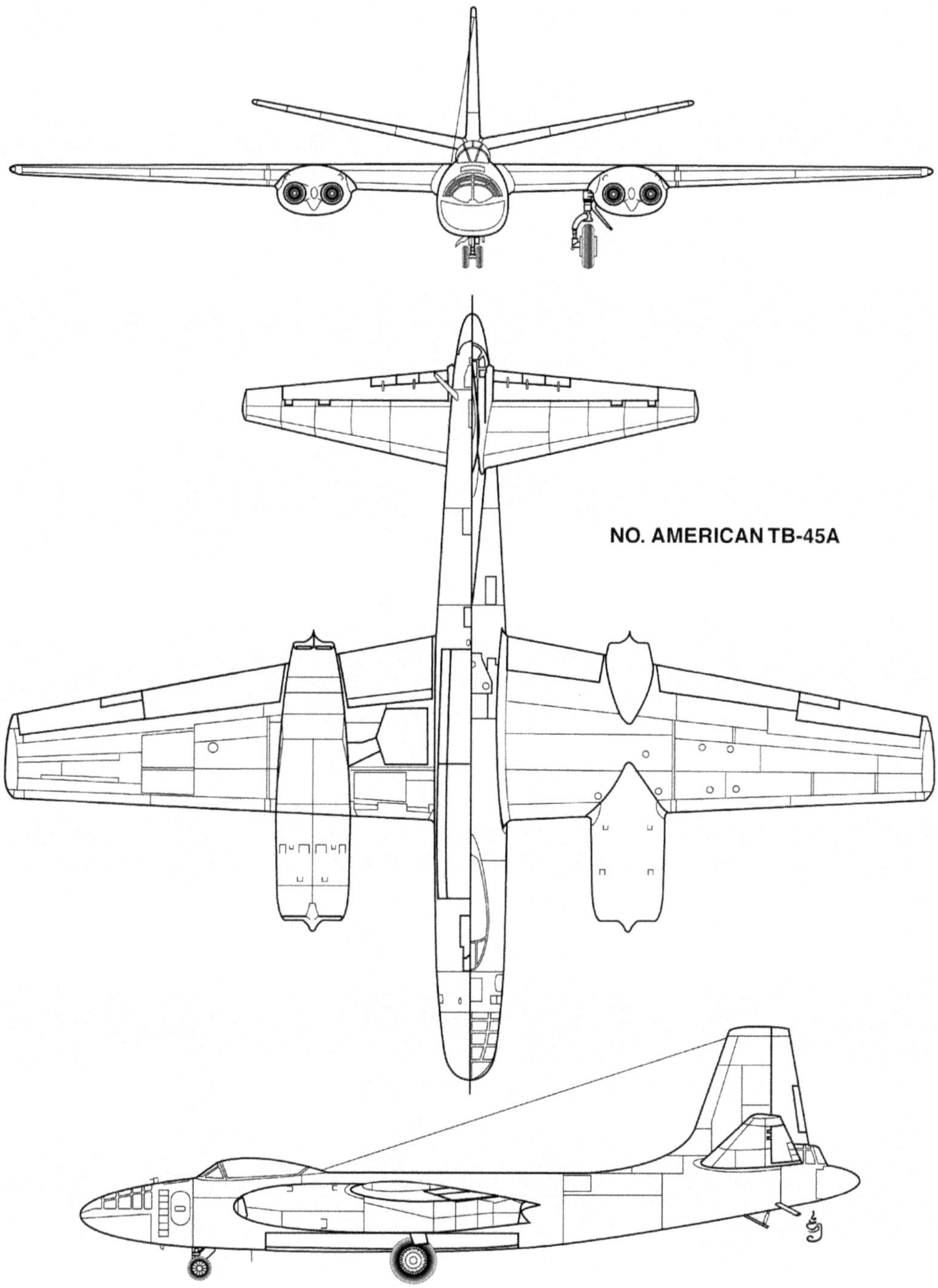

Armament: Two .50-caliber machine guns in a powered tail turret and up to 22,000 lbs. of bombs carried in an internal bomb bay.
Performance: Max. speed 543 mph, cruise 461 mph; ceiling 41,250 ft.; range 1,757 mi. loaded, 2,120 mi. max.
Weights: 47,022 lbs. empty, 86,341 lbs. max. takeoff.
Dimensions: Span 89 ft. 6 in., length 75 ft. 4 in., wing area 1,175 sq. ft.

One of 22 TB-45As in 1948. Powered by Allison J-35 engines, they initially functioned as transition trainers. Some were later retrofitted to the B-45A combat standard.

The North American B-45 was the first type of jet-propelled bomber to reach operational service with the USAF. In August 1946, after winning a design competition against Convair's B-46, it was ordered into production, with the XB-45 prototype completing its first flight in March 1947. As a consequence of delays in the availability of the more powerful General Electric J-47 engines and bombing systems specified for operational versions, the first 22 B-45A-1 production aircraft, powered by Allison J-35s were not rated for combat when deliveries commenced in April 1948. After being subjected to acceptance testing, the B-45A-1s were turned over to 47th Bombardment Wing at Barksdale AFB, Louisiana, to function as operational transition trainers under the designation TB-45A. Later, in 1951–1952, 11 of the TB-45As became combat-rated following complete engine and bombing system retrofits; the remaining 11 retained the trainer prefix after being converted into high-speed target tugs.

North American T-28 Trojan—1949

Technical Specifications (T-28A [T-28B])

Type: Two-place basic trainer.
Manufacturer: North American Aviation, Inc., Inglewood and Downey, California.
Total U.S. military versions: 1,984 (1,196 USAF; 788 USN/USMC).
Unit cost: $123,300.
Powerplant: One 800-hp Wright R-1300–1A [1,425-hp Wright R-1820–86] 7-cylinder [9-cylinder] air-cooled radial engine driving a two-bladed [three-bladed] constant-speed metal propeller.
Armament (T-28B): Two fixed .50-calibre machine guns and up to 1,800 lbs. of ordnance carried on external racks.

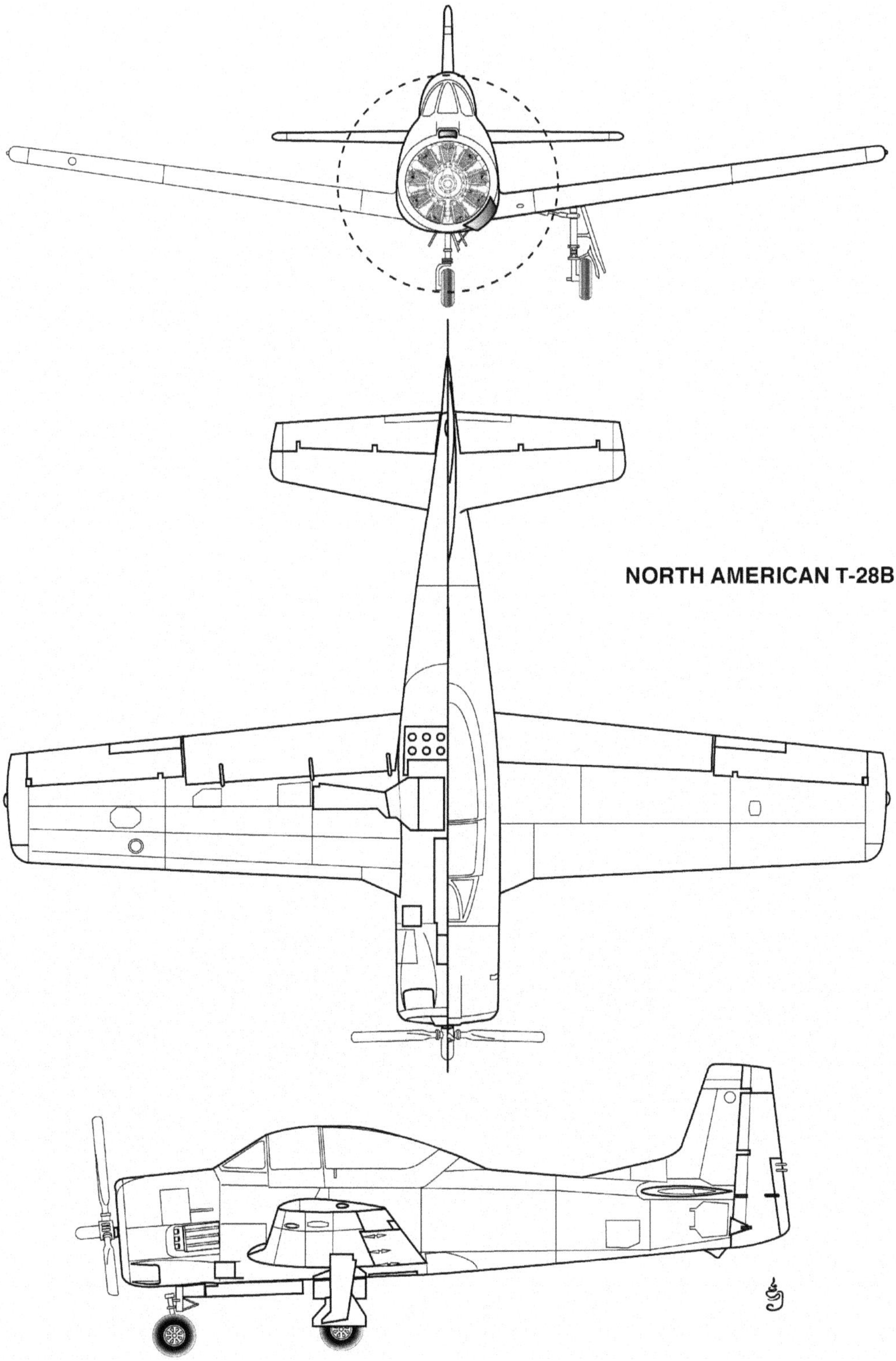

NORTH AMERICAN T-28B

Performance: Max. speed 288 mph [343 mph] , cruise 190 mph [230 mph]; ceiling 29,800 ft. [35,500 ft.]; range 1,008 mi. [1,060 mi.].
Weights: 5,111 lbs. [6,424 lbs.] empty, 6,759 lbs. [8,500 lbs.] loaded.
Dimensions: Span 40 ft. 1 in., length 32 ft. [33 ft.], wing area 268 sq. ft.

The North American T-28 Trojan, which remained in U.S. military service almost 30 years, originated in 1947 as a USAF development of the XSN2J-1 tested by the Navy in 1946. In order to better prepare pilots for the postwar jet aircraft coming into service, the USAF specification called for a tricycle gear aircraft that would be about 10 percent heavier than an AT-6. The design competition staged between North American (XBT-28) and Douglas (XBT-30) was based upon detailed drawings and mockups. In 1948, Air Material Command (AMC) declared North American's more conservative entry the winner and ordered two prototypes as the XT-28. Both XT-28 prototypes arrived for military acceptance trials in late 1949, following the conclusion of which North American received a series of contracts for 1,194 production models delivered to Air Training Command (ATC) between 1951 and 1953 as the T-28A.

As deliveries of T-28As to ATC proceeded, they ultimately replaced T-6Gs in five Flying Training Wings: the 3505th at Greenville AFB, Mississippi; the 3525th at William AFB, Arizona; the 3530th at Bryan AFB, Texas; the 3560th at Webb AFB, Texas; and the 3640th at Laredo AFB, Texas. Their active ATC career was relatively brief, however, the phase-out in favor of Cessna T-37As began in early 1957 and the final examples left during the early 1960s. As they left active service, some T-28As went to ANG units to assist in proficiency training for squadrons transitioning to tricycle-gear jets, while many others were placed in storage or transferred to foreign air forces. From 1961 to 1969, in order to develop a light attack platform for foreign air

A North American T-28A in 1951 before application of Air Training Command markings. Note higher profile canopy and smaller diameter cowling in comparison to the better-known T-28B.

One of 489 North American T-28Bs delivered to Naval Air Training between 1953 and 1955. Photograph depicts aircraft attached to Basic Training Unit One at NAS Whiting Field, Florida, probably in the late 1950s.

forces under various Military Assistance Programs, some 393 T-28As were pulled from storage and subjected to an extensive conversion from which they re-emerged as the AT-28D Nomad. The conversion entailed installation of a 1,425-hp Wright R-1820–86 engine, two .50-calber machine guns in underwing pods, and external racks for up to 1,800 lbs. of bombs or rockets. A YAT-28E powered by a 2,445-hp Lycoming T-55 turboprop engine was tested in 1963 and 1964 but never placed in production.

Naval Variants: In 1952 the Navy acquired two T-28As to evaluate whether the type would be a suitable replacement for the Naval Air Training Command's (NATC) aging fleet of SNJs. Following trials, naval officials concluded that the aircraft would need major modifications in order to meet Navy training requirements, such as: general airframe strengthening to withstand the "drop-in" landing technique; a more powerful (1,425-hp) R-1820–86 engine with a three-bladed propeller for better approach and wave-off characteristics; a hydraulically-operated speed brake in the belly; a fully-castering nosewheel (as opposed to steerable); and a lower profile canopy. Under the new specifications, BuAer awarded North American a contract in 1953 to build 489 aircraft as the T-28B (no naval designation assigned), with initial deliveries starting the same year and continuing into 1955. In order to meet a new requirement for a carrier-qualification trainer, BuAer gave North American a follow-up order in 1955 to produce 299 additional aircraft equipped with arrestor hooks, strengthened landing gear, and smaller diameter, wider-bladed propellers as the T-28C, which all entered service in 1956 and 1957.

As T-28Bs arrived for duty with NATC between 1953 and 1955, most were assigned to three Basic Training Groups (BTG) based at NAS Whiting Field near Pensacola, Florida, where

A T-28C of Basic Training Group Five seen on a ramp at NAS Saufley Field, Florida, in late 1950s. In carrier qualification training, the T-28Cs starting being displaced by North American T2J-1s in 1959.

they replaced SNJs in the advanced phase of pilot training. And when deliveries of T-28Cs commenced in 1956, they went to BTG-5 at NAS Saufley Field, Florida, for carrier-qualification training; however, their career in this role was brief, being superceded by North American T2J-1 jets (see in Series II, Part B) starting in late 1959. Afterward, the T-28Cs served alongside Bs in the BTGs at NAS Whiting and Saufley Fields. Although the phase-out of T-28Bs and Cs began in the mid–1970s as Beech T-34Cs began entering service, the final examples were not withdrawn from NATC until early 1984.

Douglas XBT/T-30—1948

Technical Specifications (XT-30)

Type: Two-place basic trainer.
Manufacturer: Douglas Aircraft Co., Santa Monica, California.
Total U.S. military versions: One mockup (USAF).
Unit cost: Unknown.
Powerplant: One 800-hp Wright R-1300-A 7-cylinder air-cooled radial engine driving a two-bladed constant-speed metal propeller.
Performance (est.): Max. speed 286 mph, cruise 190 mph; ceiling 29,600 ft.; range 1,253 mi.
Weights: Empty (not reported), 5,999 lbs. loaded.
Dimensions: Span 36 ft. 9.5 in., length 36 ft. 4 in., wing area (not reported).

The Douglas XBT-30 (re-designated XT-30 in 1948) appeared in 1947 as one of two design proposals submitted in response to a USAF requirement for a basic trainer having a tricycle gear configuration and approximately 25 percent higher wing-loading than a T-6. Its unusual design concept—an engine buried amidships driving a tractor-mounted propeller through an extension

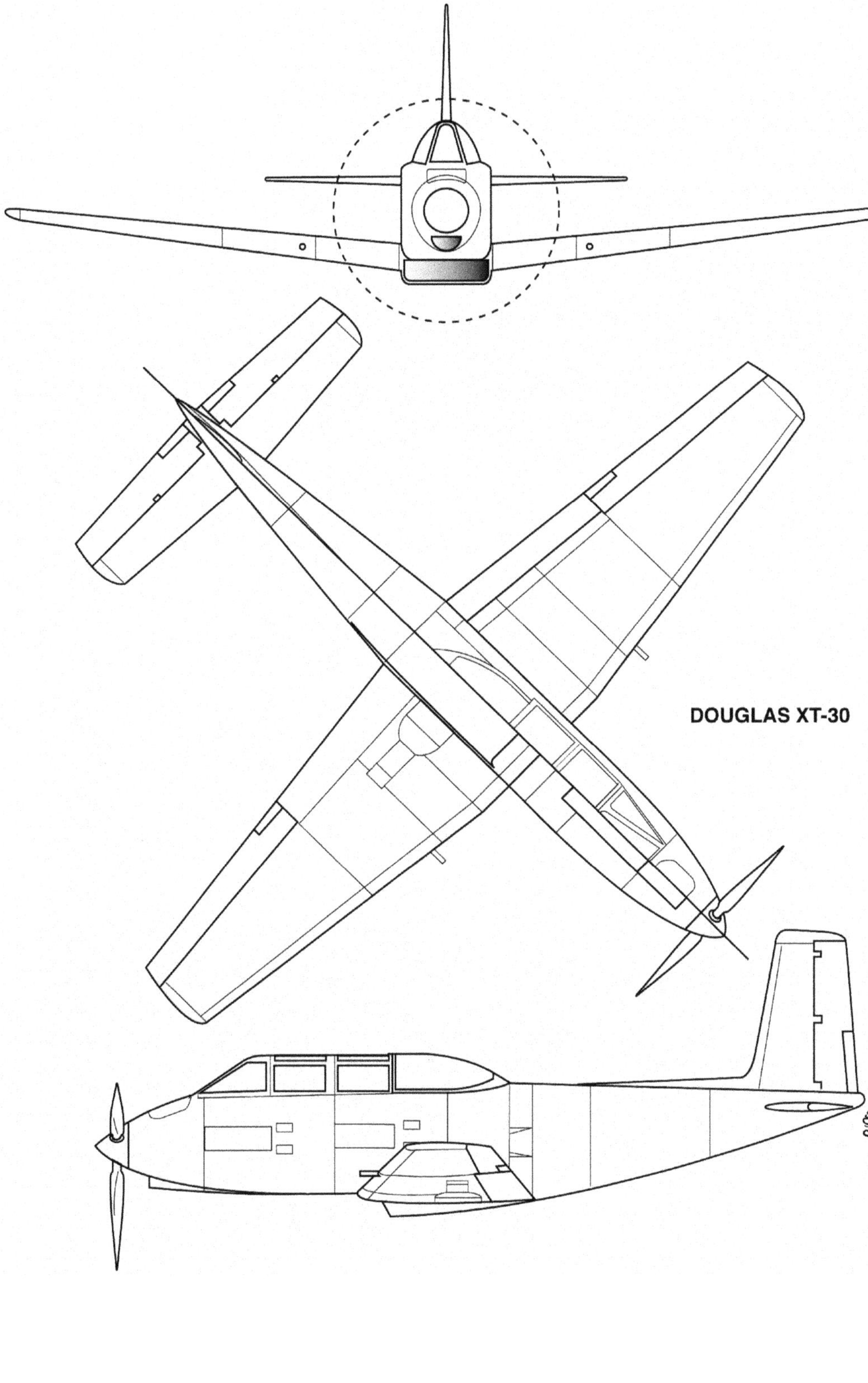

DOUGLAS XT-30

shaft—was reminiscent of the Bell P-39 (see TP-39 in Series I, Part A, above). The chief advantage of this arrangement was a forward location of the cockpit giving the pilot trainee all around visibility similar to that of the new jets. Moreover, it produced a longer wheelbase that would more closely duplicate the takeoff and landing characteristics of aircraft like the F-80, F-84, and F-86. Although Air Material Command evinced enough interest in the concept to authorize Douglas to proceed with construction of a partial mockup, the project was cancelled in favor of the more conventional North American XT-28 (see above).

Only known photograph of the XT-30 partial mockup taken at the Douglas factory in 1947 or 1948. Based upon a detailed design competition, it was cancelled in favor of the more conventional North American XT-28.

Convair T-29 and T-32 Flying Classroom—1949

Technical Specifications (T-29C)

Type: Navigator and bombardier aircrew trainer.
Manufacturer: Convair Div. of General Dynamics Corp., San Diego, California.
Total military versions: 272 (USAF).
Unit cost: $990,000.
Powerplants: Two 2,500-hp Pratt & Whitney R-2800–99W 18-cylinder air-cooled radial engines driving three-bladed, fully-reversible constant-speed metal propellers.
Performance: Max. speed 300 mph, cruise 270 mph; ceiling 30,000 ft.; range 1,800 mi.
Weights: 29,248 lbs. empty, 47,000 lbs. loaded.
Dimensions: Span 91 ft. 9 in., length 74 ft. 8 in., wing area 817 sq. ft.

Conceived originally as an intermediate range replacement for the Douglas DC-3, Convair introduced its 40-passenger CV-240 to civil airliner service in 1948, followed by the larger 52-passenger CV-340 in 1952, and finally, by the improved CV-440 in 1956, eventually selling 442 examples worldwide before production ended in the mid–1950s. In 1948, with the aim of procuring a navigator trainer to replace AT-7s and AT-11s and augment TC-47s, Air Material Command awarded Convair a contract to deliver two unpressurized CV-240s, powered by 2,400-hp R-2800–97 engines and refitted with four astrodomes and stations for 10 navigator students, under the designation XAT-29 (later re-designated XT-29). After brief testing of the two XT-29s during mid–1949, Convair received a production order for 46 aircraft with small detail changes as the T-29A, with deliveries starting in early 1950. A succession of new production contracts, brought about by the Cold War and Korea War expansion, resulted in deliveries of 317 pressurized CV-240 trainer variants between 1951 and 1954: 105 navigator trainers with three astrodomes and new stations for radio-operator students as the T-29B; 119 similar variants with an upgrade to 2,500-hp -99W engines as the T-29C; and finally, 93 lacking astrodomes and having six

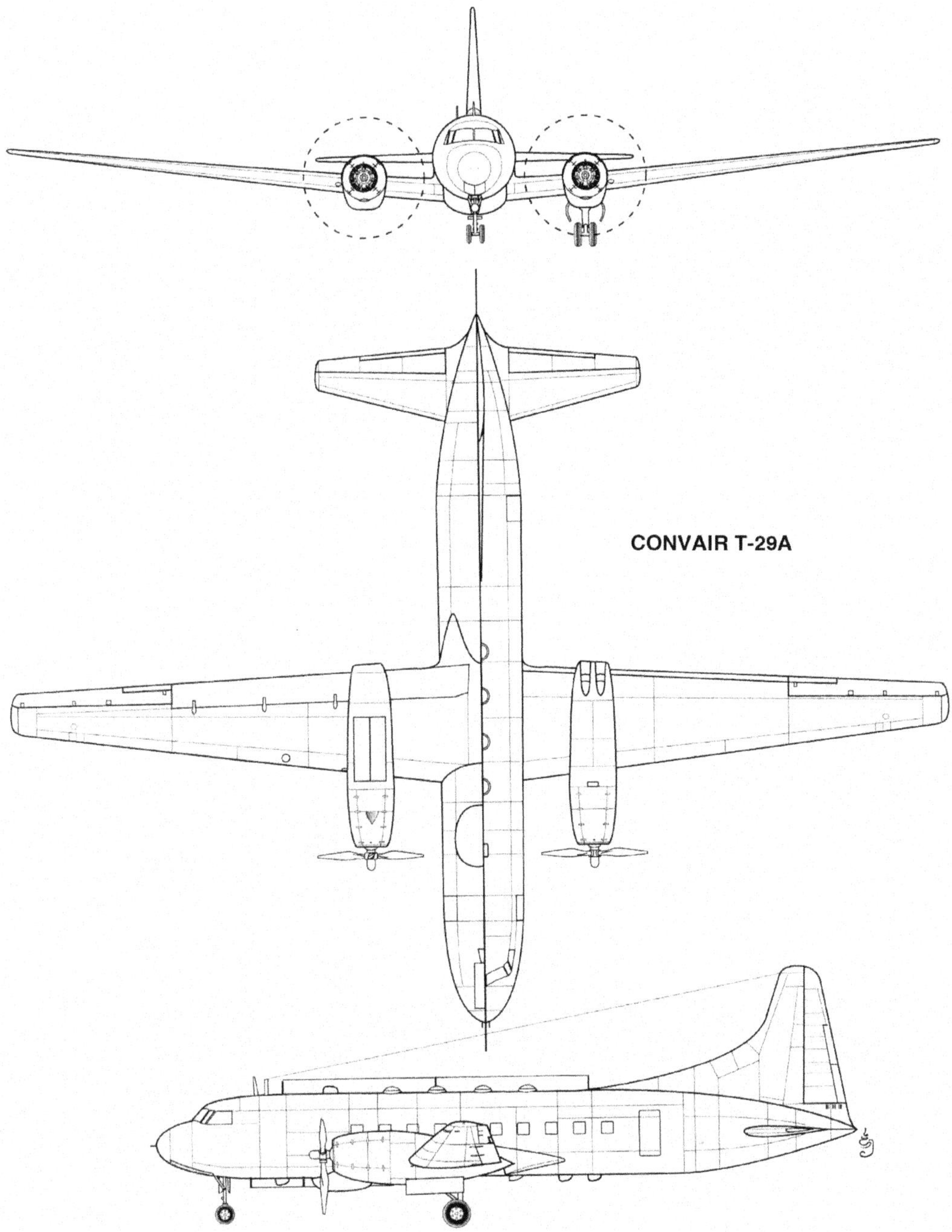

navigator-bombardier stations as the T-29D. Two other CV-240 trainer variants were proposed but never built: in 1949, a bombardier trainer with a plexiglas nose as the YT-32; and in 1952, a navigator trainer powered with 2,750-shp Allison T-38 turboprop engines as the T-29E.

As new T-29As, Bs, and Cs arrived for duty with Air Training Command (ATC) from 1950 onwards, they were assigned to three Navigator Training Wings: the 3565th at James Connally AFB, Texas; the 3605th at Ellington AFB, Texas; and the 3610th at Harlingen AFB, Texas. T-29Ds, equipped with "K" radar bombing systems, went to the 3535th Bombardier (later

One of 105 Convair T-29Bs delivered to Air Training Command in 1951 for navigator training. The B differed from the A in having a fully pressurized cabin.

One of 93 T-29D navigator-bombardier trainers delivered in the mid–1950s. Belly radome housed radar for the "K" bombing system used in B-47s. Some later converted to navigation aid checking aircraft as ET-29D.

Observer) Training Wing at Mather AFB, California. In 1959, the 3605th moved its T-29 operations to Mather AFB, and in the mid–1960s, all T-29 navigator training was consolidated at Mather. The phase-out of T-29s from ATC service began in the late 1960s, with the final examples being withdrawn in 1976. Later in their careers, a number of T-29 variants were re-designated after being modified to serve as personnel or staff transports: 26 with 25-seat interiors as the

VT-29A; 60 with 32-seat interiors as the VT-29B; two with 25-seat interiors as the VT-29C; and 46 with 25-seat interiors as the VT-29D. Some T-29C and Ds were also converted to navigation-aid checking aircraft as the ET-29C and D.

Although there were no actual naval variants of the T-29, an unspecified number of T-29Bs were loaned to the Navy during the 1960s, where they served with VT-29 at NAS Corpus Christi, Texas, to train Naval Flight Officers selected to fly land-based aircraft. The type was withdrawn from Navy service in 1975.

Fairchild T-31–1949
(See Fairchild NQ in Series II, Part B)

Temco T-35 Buckaroo—1950

Technical Specifications (YT-35)

Type: Two-place primary trainer.
Manufacturer: Texas Engineering & Mfr. Co., Inc., Dallas, Texas.
Total U.S. military versions: 3 (USAF).
Unit cost: Unknown.
Powerplant: One 165-hp Franklin 6A4-B3 (O-335) 6-cylinder air-cooled horizontally-opposed engine driving a two-bladed constant-speed metal propeller.
Performance: Max. speed 156 mph, cruise 87 mph; ceiling 17,000 ft.; range 548 mi.
Weights: 1,300 lbs. empty, 1,920 lbs. loaded.
Dimensions: Span 29 ft. 4 in., length 21 ft. 8 in., wing area 134 sq. ft.

Development of what became the YT-35 originated in early 1948 when Temco received an unsolicited inquiry from the Philippine government expressing interest in a tandem trainer derivative of the company's GC-1B Swift civil lightplane. Using the Swift as a baseline, Temco's engineering staff evolved the TE-1A trainer variant featuring a modified fuselage with tandem cockpit seating under a sliding canopy enclosure. Soon after the TE-1A flew in late 1948, Temco submitted the design to the USAF to be considered in the recently announced primary trainer competition along with the Fairchild NQ (YT-31) and the Beech model 54 (YT-34). Although the USAF selected the YT-34 as the winning design in February 1949, Temco elected to continue development of the concept based upon foreign interest and moved forward with construction of the improved TE-1B, which differed from the -1A in having a lengthened and revised fuselage cross-section, a taller fin and rudder, improved streamlining, and a more powerful 165-hp Franklin engine. When the USAF revived the trainer competition in mid–1950, Temco delivered three completed TE-1B prototypes to Randolph AFB, Texas, under the designation YT-35, factory name "Buckaroo," to undergo competitive trials against the YT-31 and YT-34.

Because of the onset of the Korean War, the test program was disrupted, with the three YT-35 being assigned to James Connally AFB, Texas. In July 1951, after 10 months of testing, the YT-35 were returned to Temco for overhaul and upon redelivery, reassigned to Goodfellow AFB near San Angelo, Texas, where they apparently languished until being sold surplus sometime later. Although the USAF had by this time placed the Beech T-34A in mass production, Temco subsequently received a government order of 10 more TE-1Bs as the T-35A, which were in turn sold and delivered to the Saudi Arabian Air Force under the Mutual Defense Aid Plan.

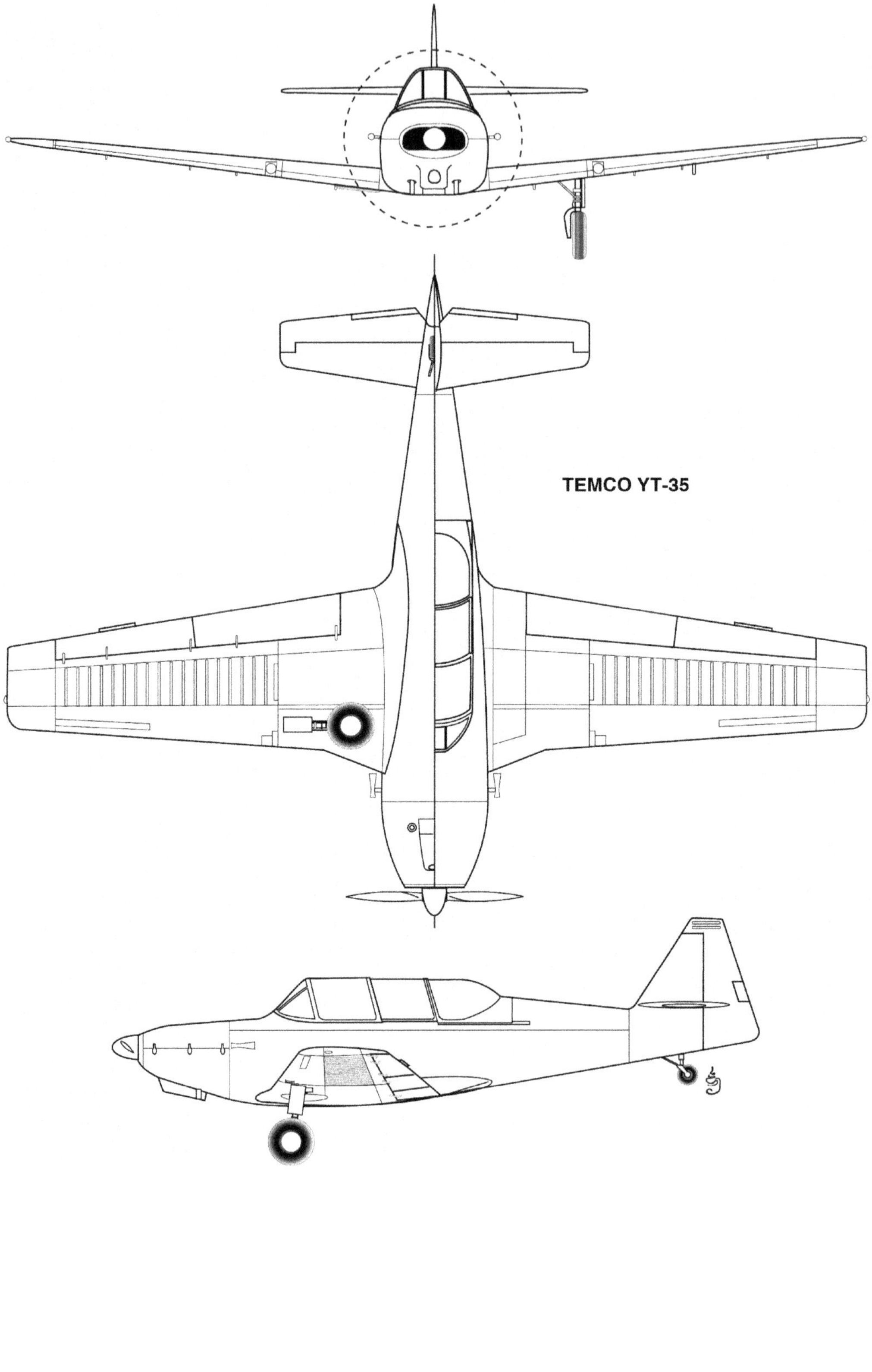
TEMCO YT-35

All three of the Temco YT-35 prototypes seen in formation, most likely in 1950 near the vicinity of Randolph AFB, Texas. The test program was interrupted by the Korean War.

Piper TL-21–1951

Technical Specifications (TL-21A)

Type: Two-place primary and transition trainer.
Manufacturer: Piper Aircraft Corp., Lockhaven, Pennsylvania.

One of 150 Piper L-21As delivered to the Army in 1951 and later redesignated as TL-21As. Most were assigned to Army Aviator fixed-wing training at Gary AFB (later Camp Gary), Texas. Their service was brief.

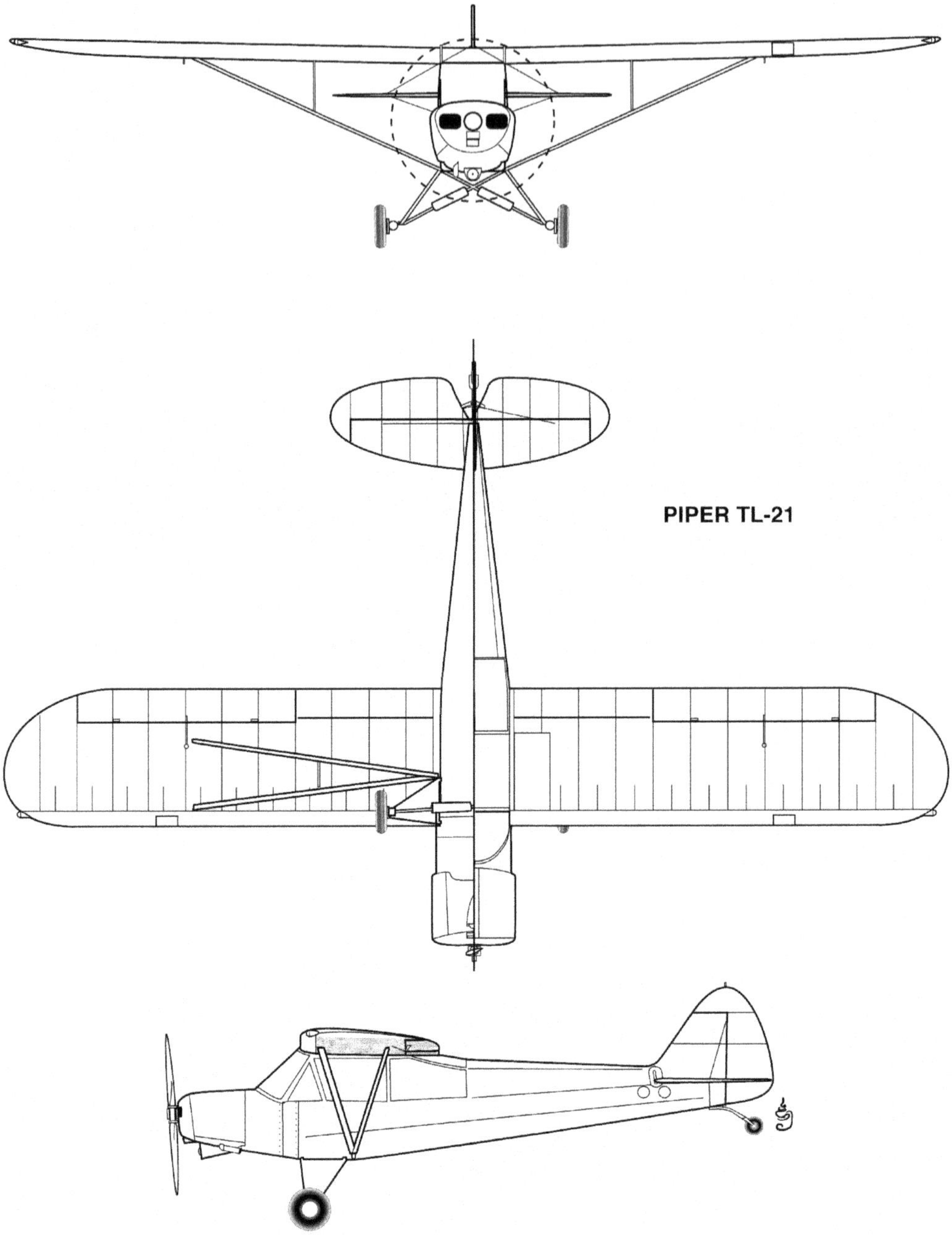

Total military versions (trainers): 150 (U.S. Army).
Powerplant: One 125-hp Lycoming O-290–11 4-cylinder air-cooled horizontally-opposed engine driving a two-bladed fixed-pitch metal propeller.
Performance: Max. speed 123 mph, cruise 108 mph; ceiling 17,100 ft.; range 400 mi.
Weights: 950 lbs. empty, 1,580 lbs. loaded.
Dimensions: Span 35 ft. 3 in., length 22 ft. 7 in., wing area 179 sq. ft.

Preceded by the ubiquitous L-4 Grasshopper (see in Series I, Part A) and L-18 (based on the PA-11 Cub 95), the YL-21 (based on the PA-18 Cub 135, later Super Cub) was tested by the

Army in 1951 as the final evolution of Piper's two-seat liaison series. Following official evaluations, the Army awarded Piper a contract for 150 aircraft under the designation L-21A, all of which were delivered by the end of 1951. After entering active Army service, most L-21As were re-designated as TL-21As and assigned to the Army's fixed-wing training program at Gary AFB (later Camp Gary), Texas. TL-21As, equipped for operational observation pilot training, served only briefly, being supplanted by Cessna L-19s during the mid–1950s.

Boeing TB-50 Superfortress—1951

Technical Specifications (TB-50H)

Type: Operational pilot, navigator-bombardier, and aircrew trainer.
Manufacturer: Boeing Airplane Co., Seattle, Washington.
Total military versions: 46 (USAF).
Unit cost: $1,480,000.
Powerplant: Four 3,500-hp Pratt & Whitney R-4360–35 28-cylinder air-cooled radial engines driving four-bladed, fully-reversible constant-speed metal propellers.
Armament: None.
Performance: Max. speed 418 mph, cruise 244 mph; ceiling 35,650 ft.; range 4,232 mi.
Weights: 60,000 lbs. empty (est.), 120,000 lbs. loaded (est.).
Dimensions: Span 141 ft. 3 in., length 99 ft., wing area 1,720 sq. ft.

The concept behind the B-50 originated in 1944 as an R-4360-powered development of the B-29. With greater power and a lighter airframe constructed of new aluminum alloys, it pos-

This Boeing TB-50D was one of 11 aircraft stripped of armament and bombing equipment and thereafter assigned to Air Training Command to train Convair B-36 navigator-bombardiers and radio operators.

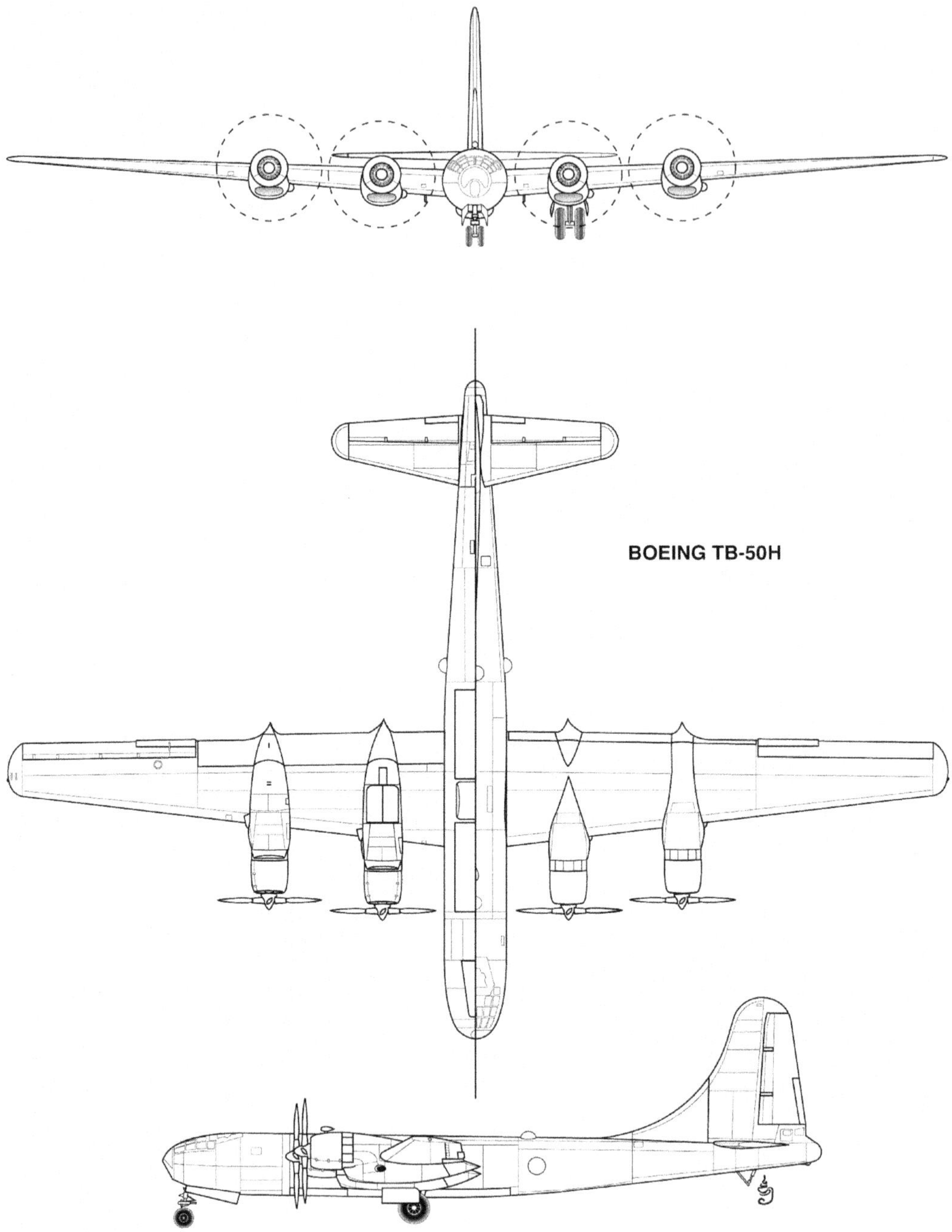

sessed the ability to carry a 30 percent heavier bomb load about 50 percent farther than a B-29. Between 1947 and 1953, the USAF accepted delivery of 370 B-50 variants to operate as temporary stopgaps in Strategic Air Command (SAC) until such time as the more-capable B-47 Stratojets became fully operational. Although most bomber variants were phased-out of frontline SAC bombardment units during 1954, tanker (KB-50J/K), reconnaissance (RB-50E/F/G), and weather variants (WB-50D) continued to serve until 1965.

As the earliest B-47s became operational in 1951 and 1952, 11 B-50As and 11 B-50Ds were withdrawn from bombardment duties, stripped of armament, and returned to service under the new designation TB-50A and TB-50D, respectively. Available references suggest that both variants were assigned to Air Training Command (ATC) after being reconfigured with the avionics and systems needed to train Convair B-36 navigator-bombardier-radar operators. They were all removed from training duties in 1955 and converted to tankers as the KB-50J. In early 1951, the USAF ordered Boeing to complete the last 24 B-50Ds on its production line to a revised specification as the TB-50H, with all examples being delivered between September 1952 and March 1953. Lacking armament, the aircraft were equipped with the "K" radar bombing and navigation systems of the B-47. In addition to the normal flight crew, TB-50Hs carried two instructors plus six B-47 system trainees. Starting in late 1952, all TB-50Hs entered operational service with ATC in the 3235th Observer Training Wing at Mather AFB, California. However, their service as trainers was brief, all 24 being withdrawn in mid–1955 to undergo a tanker conversion as the KB-50K. The TB-50H is distinguished as having been the fastest (31 mph faster than a B-50D) of all B-50 variants.

Stinson L-5 Sentinel—1952

Technical Specifications (L-5G)

Type: Light liaison and utility transport; transition trainer.
Manufacturer: Stinson Aircraft Div., Vultee-AVCO, Wayne, Michigan.
Total U.S. military versions (trainers): 100 (U.S. Army).
Unit cost: $10,000.
Powerplant: One 190-hp Lycoming O-435–11 6-cylinder air-cooled horizontally-opposed engine driving a two-bladed Aeromatic variable-pitch wooden propeller.
Performance: Max. speed 129 mph, cruise 110 mph; ceiling 15,600 ft.; range 360 mi.
Weights: 1,600 lbs. empty, 2,250 lbs. max. takeoff.
Dimensions: Span 34 ft., length 24 ft. 1 in., wing area 155 sq. ft.

This Stinson L-5 has World War II-era markings. An estimated 100 are reported to have served as operational liaison trainers for Army pilots during the early and mid–1950s.

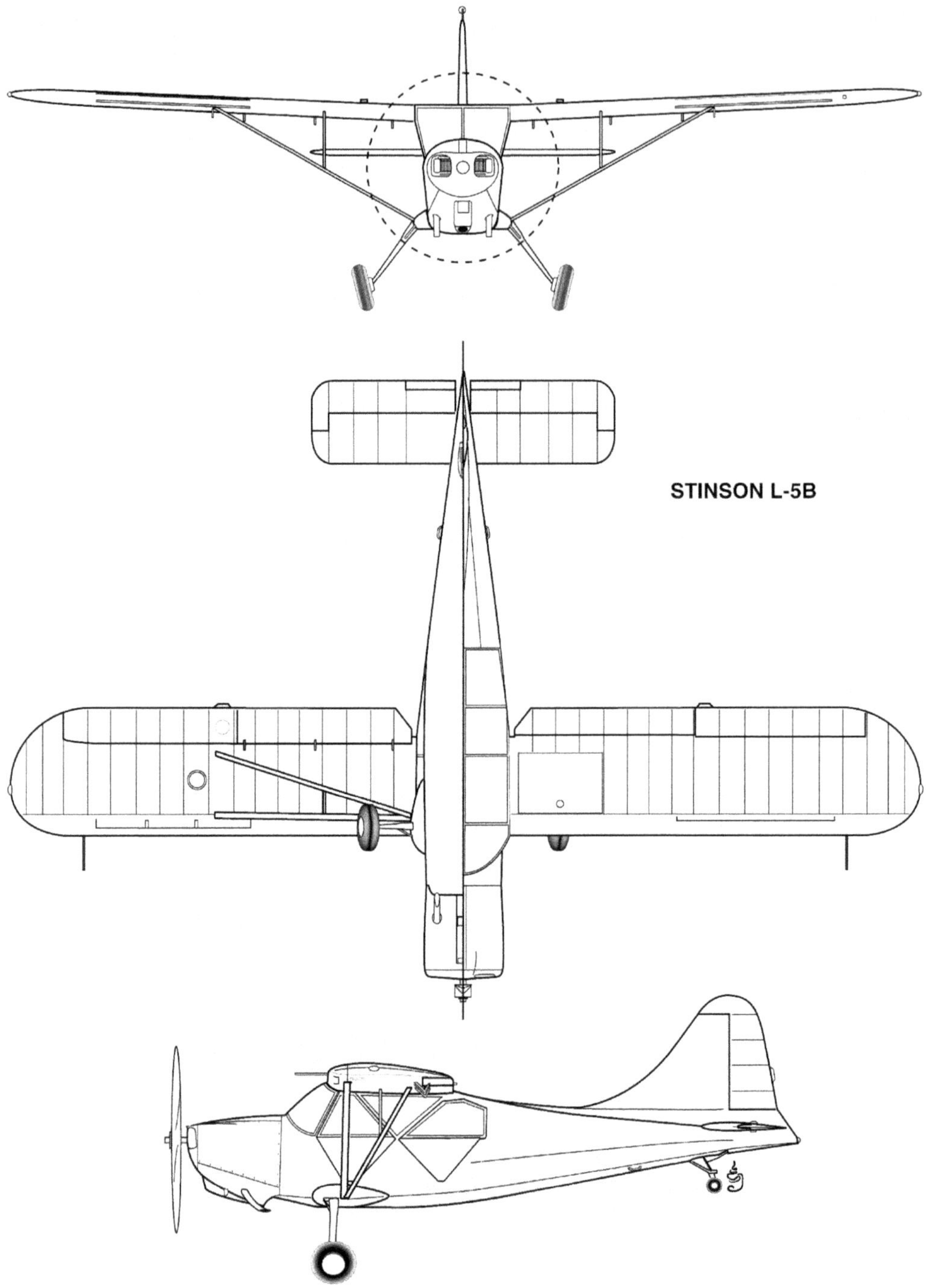

Knick-named the "Flying Jeep," the Stinson L-5 Sentinel (originally test-flown as the YO-54) holds the distinction of having been the second most-produced (Piper L-4 was first) liaison aircraft for the U.S. military services during World War II. Though sharing some design elements with Stinson's civil Model 105 Voyager, the L-5 emerged in 1942 as a purpose-built military aircraft with no civil counterpart. A wing incorporating fixed leading edge slots and three-position slotted flaps enabled the L-5 to fly as slow as 38 mph. Improved variants, which began appearing in 1943, included the L-5B with hinged panels on the side of the fuselage to allow loading of one medical stretcher case, the L-5C with provision for reconnaissance cameras, the L-5E with larger tires, stronger brakes, and drooping ailerons for improved STOL performance, and finally, the L-5G with an up-rated -11 engine and a variable-pitch propeller.

As L-5s left combat duty in the Korean war zone in 1951–1952, an estimated 100 returned to the U.S. to serve as operational liaison trainers in the Army's fixed-wing school, initially at Fort Sill, Oklahoma, then Gary AFB in Texas (later Camp Gary) until being replaced by Cessna L-19s in the mid–1950s.

Cessna LC-126/U-20–1952

Technical Specifications (LC-126C/U-20C)

Type: Four-place instrument trainer.
Manufacturer: Cessna Aircraft Co., Wichita, Kansas.
Total military versions (trainers): 63 (U.S. Army).
Unit cost: $23,500.
Powerplant: One 300-hp Jacobs R-755–11 7-cylinder air-cooled radial engine driving a two-bladed Hamilton Standard constant-speed propeller.
Performance: Max. speed 175 mph, cruise 140 mph; ceiling 19,800 ft.; range 856 mi.
Weights: 2,250 lbs. empty, 3,350 lbs. loaded.
Dimensions: Span 36 ft. 2 in., length 27 ft. 4 in., wing area 218 sq. ft.

Making its debut as Cessna's first all-metal airplane, the prototype for the model 190/195 series made its first flight in late 1944 and entered production in 1947 following type certification. The first military variants were model 195s configured as light transports, 15 procured by the USAF in 1949 as the LC-126A and five by the Army in 1951 as the LC-126B. Under a second

One of 63 Cessna LC-126Cs delivered to the Army in 1952 and 1953. They provided instrument training for Army pilots transitioning to larger aircraft like the L-20 Beaver and U-1 Otter.

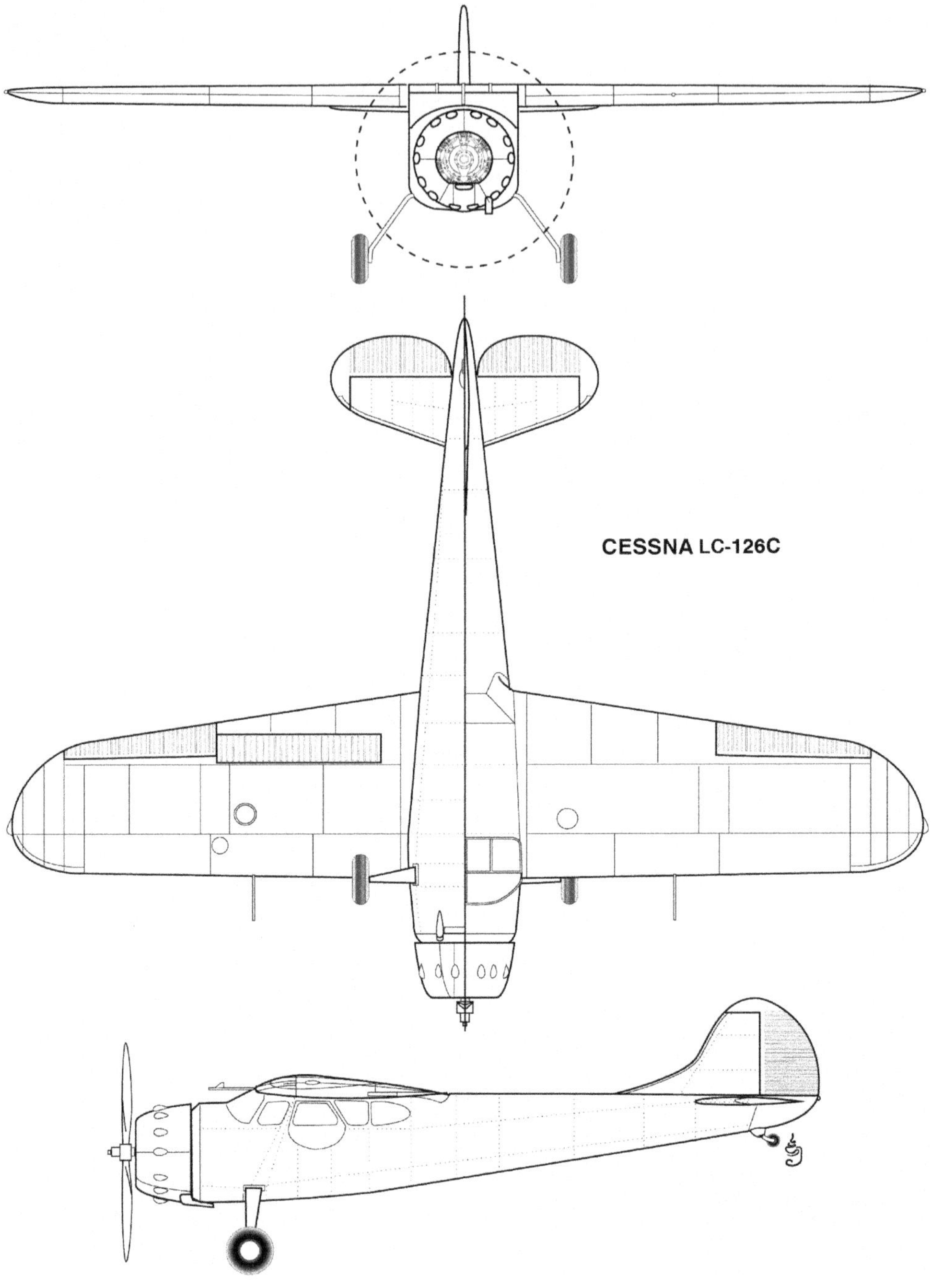

Army contract issued in 1952, Cessna delivered 63 more model 195s equipped to function as four-place instrument trainers that were placed on the inventory as the LC-126C. LC-126Cs were distributed among various Army installations to provide instrument transition training to fixed-wing pilots who were upgrading to larger aircraft like the L-20 Beaver, U-1 Otter, and L-23 Seminole. Following adoption of the Tri-Service system in 1962, LC-126Cs remaining in Army service were re-designated U-20C.

Douglas TC-54 Skymaster—1953

Technical Specifications (TC-54D)

Type: Navigator and electronic warfare officer trainer.
Manufacturer: Douglas Aircraft Co., Santa Monica, California, and Chicago, Illinois.
Total U.S. military versions (trainers): Unknown (USAF).

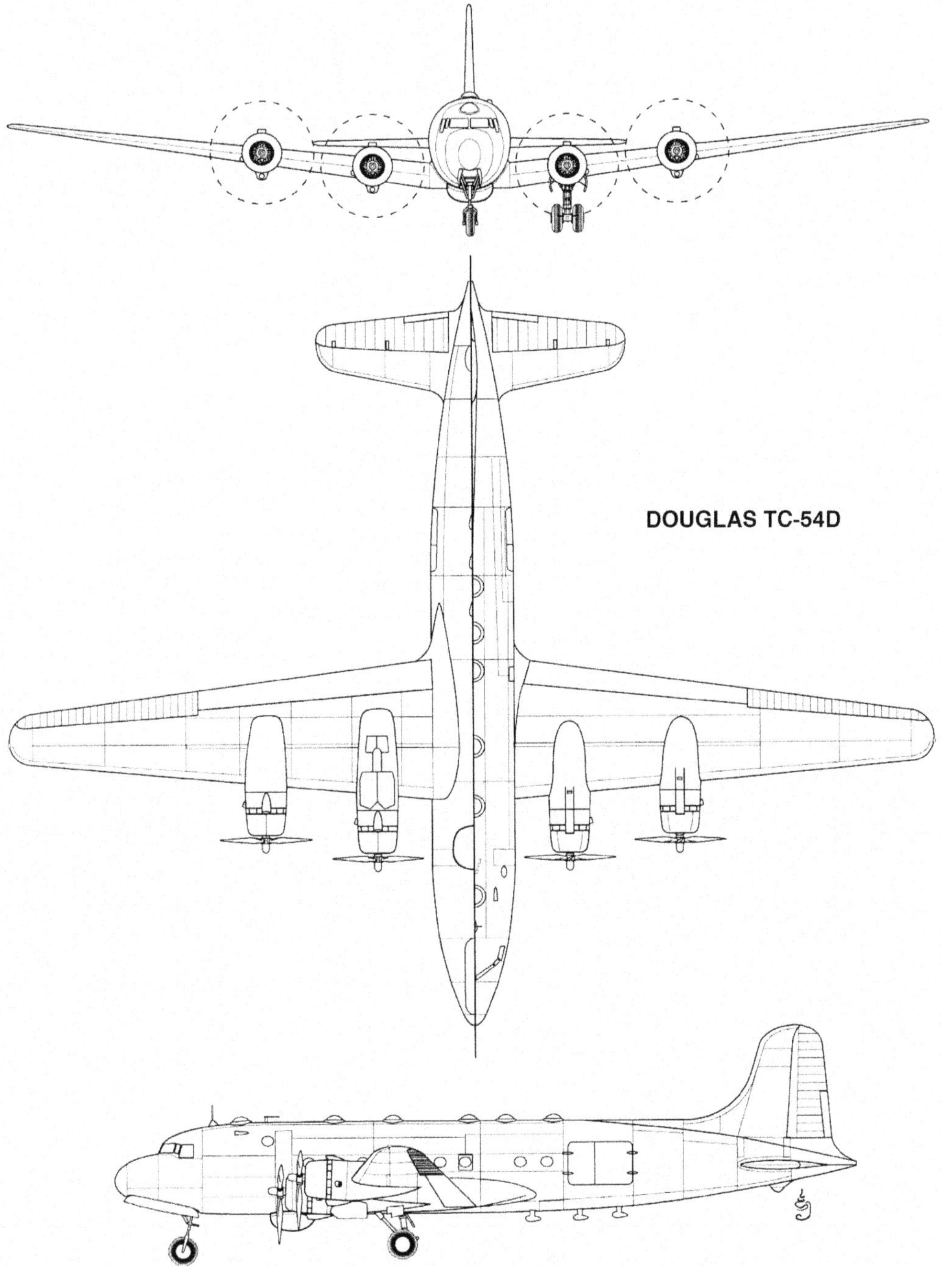

An unverified number of Douglas C-54Ds, as seen in this photograph, were modified in the 1950s as navigator trainers under the designation TC-54D. Note addition of six extra astrodomes.

Unit cost: $285,113.
Powerplants: Four 1,350-hp Pratt & Whitney R-2000–11 *Twin Wasp* 14-cylinder air-cooled radial engines driving three-bladed Hamilton Standard constant-speed metal propellers.
Performance: Max. speed 275 mph, cruise 190 mph; ceiling 22,350 ft.; range 3,100 mi.
Weights: 38,930 lbs. empty, 62,000 lbs. loaded, 73,000 lbs. max. takeoff.
Dimensions: Span 117 ft. 6 in., length 93 ft. 10 in., wing area 1,460 sq. ft.

The Douglas C-54/R5D Skymaster was arguably the most important type of four-engine American military transport produced during the World War II and remained in USAF and Navy service until the early 1970s. Official records are unclear as to the number, date, and assignment of USAF C-54Ds that were converted to training duties and returned to service with Air Training Command (ATC) under the designation TC-54D. Serial number records confirm at least eight trainer conversions, but the exact nature of the modifications is not reported. While C-54s appear as an assigned aircraft on the ATC roster from 1946 to 1972, many were most likely unmodified transports used for advanced or operational multi-engine pilot training. Unverified sources indicate that during the early to mid–1950s, TC-54Ds were modified as navigator trainers and may have been equipped with B-47-type navigation systems, and in the early 1960s, the same sources indicate that they were again modified to function as trainers for Electronic Warfare Officers.

Beech T-36–1953

Technical Specifications (XT-36A)

Type: Four-place advanced trainer and fourteen-place utility transport.
Manufacturer: Beech Aircraft Corp., Wichita, Kansas
Total military versions: 2 (USAF).
Unit cost: Unknown.
Powerplants: Two 2,300-hp Pratt & Whitney R-2800–52W *Double Wasp* 18-cylinder air-cooled radial engine driving three-bladed Hamilton Standard constant-speed propellers.

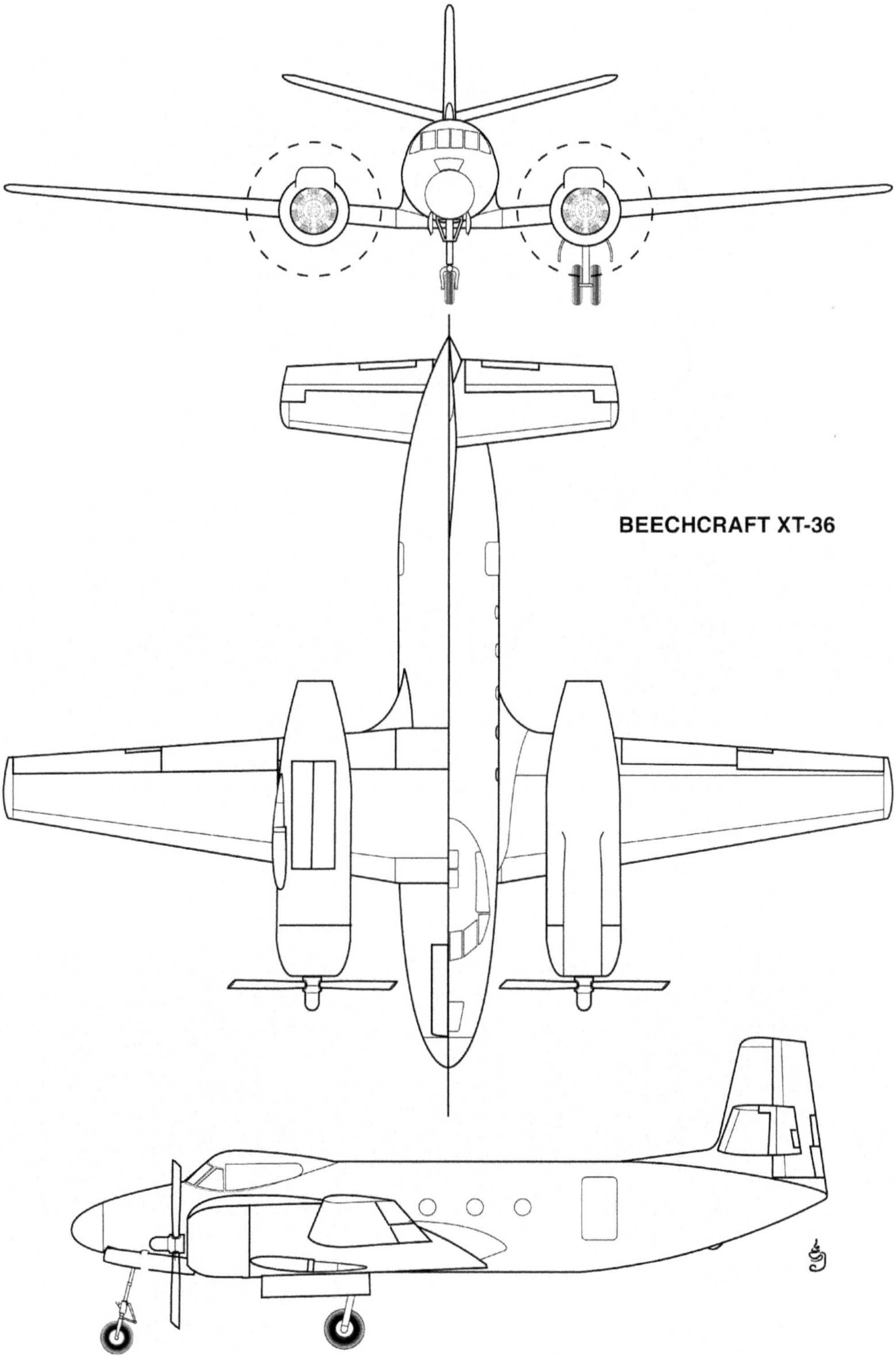

BEECHCRAFT XT-36

Artist's impression of Beech XT-36A. Although Beech completed two prototypes, the contract was cancelled in June 1953 only days before the first flight was scheduled and as a consequence, the aircraft never flew.

Performance (est.): Max. speed 300 mph; cruise, ceiling, and range (not reported).
Weights: Empty (not reported), 25,000 lbs. loaded.
Dimensions: Span 70 ft., length 52 ft. 2 in., wing area (not reported).

In 1950 the USAF circulated a design competition for a new aircraft comparable in size to a TB-25 (see Series I, Part A) which would function either as an advanced multi-engine trainer or a 12-passenger staff and utility transport. After being declared winner of the competition, Beech received a development contract in 1951 to construct one flying prototype and a static test example under the designation XT-36A, along with a provisional order for a projected production run of 195 aircraft. Unrelated to any previous Beech design, the XT-36A materialized with a circular-section fuselage, a low wing, a cruciform tail group having marked dihedral in the horizontal stabilizers, and a bulged cockpit enclosure that accommodated an instructor and three students. In June 1953, just three days prior to the scheduled first flight of the XT-36A prototype, Air Material Command terminated the contract; and the prototype was never flown. A report later issued by the Government Accounting Office cited official USAF reservations over increasing development costs due to many engineering changes, tight delivery schedules that would force Beech to subcontract much of the production work, and a general concern that the T-36 would not fulfill all of the design requirements.

Cessna L/TL-19 Bird Dog—1953

Technical Specifications (TL-19A)

Type: Two-place transition and primary trainer.
Manufacturer: Cessna Aircraft Co., Wichita, Kansas.

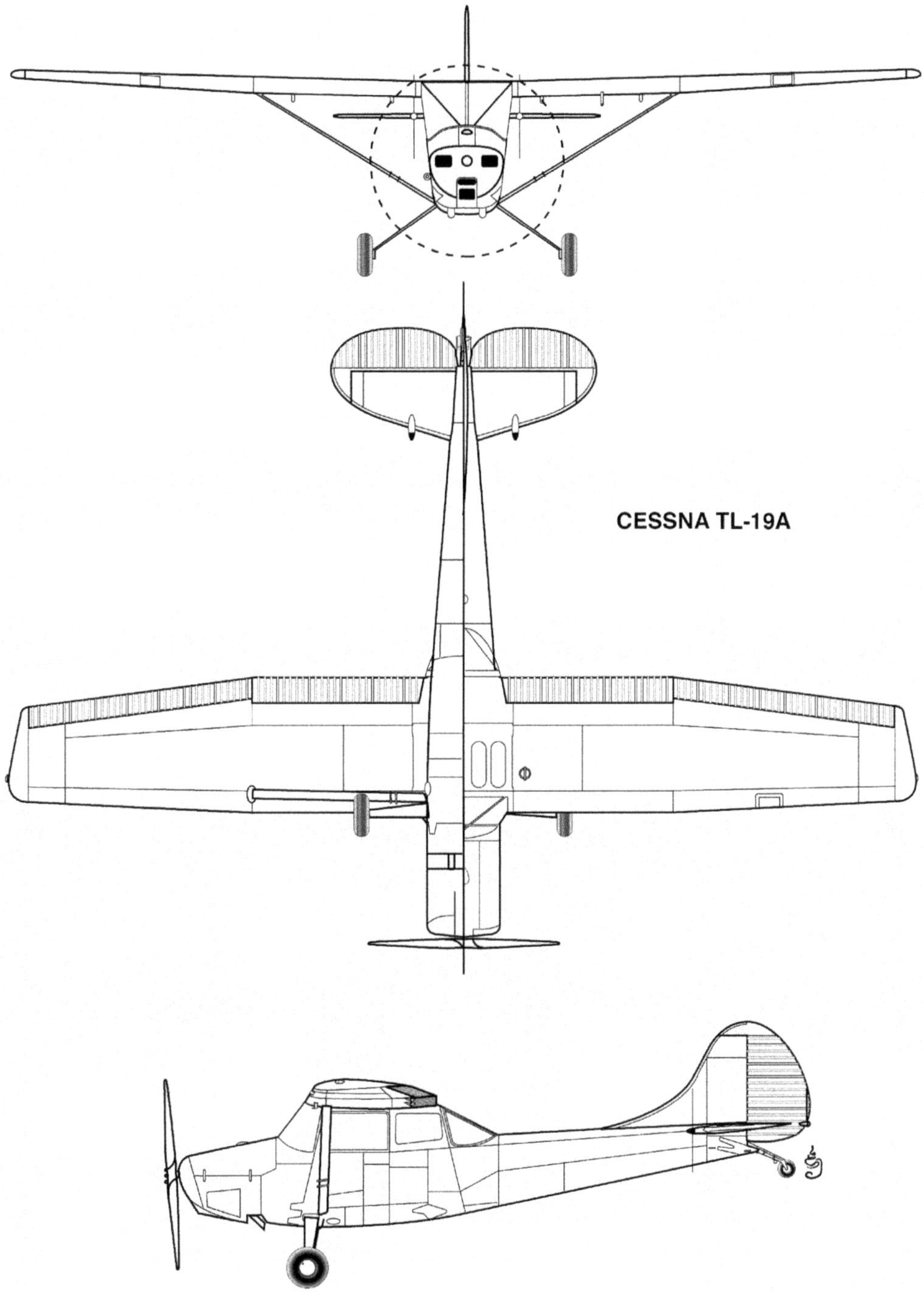

Total U.S. military versions (trainers): 340 (U.S. Army).
Unit cost: $8,400.
Powerplant: One 213-hp Continental O-470–11 6-cylinder air-cooled horizontally-opposed engine driving a two-bladed fixed-pitch metal propeller.
Performance: Max. speed 130 mph, cruise 104 mph; ceiling 20,300 ft.; range 529 mi.
Weights: 1,614 lbs. empty, 2,400 lbs. loaded.
Dimensions: Span 36 ft., length 25 ft. 9 in., wing area 174 sq. ft.

The well-known Cessna L-19/O-1 Bird Dog, originating as the model 305A in 1949, served as the Army's principal type of liaison/observation and Forward Air Control (FAC) aircraft over a period extending more than 20 years. From 1950 to 1957, Cessna produced 3,431 L-19s in three major variants for the Army, USAF, and USMC, 3,032 of which went to the Army. The first major production variant was 2,486 L-19As, followed by 310 TL-19D s and 469 L/TL-19Es. The XL-19B and C were both L-19As test-flown with experimental turboprop engines but never placed in production. When the Department of Defense implemented the Tri-Service designation system in mid–1962, the L-19A became the O-1A, the TL-19D, the TO-1D, the L-19E, the O-1E, and the TL-19E, the TO-1E. Army and USAF O-1As and O-1Es converted to fly FAC missions in Vietnam War (1962–1972) became the O-1D and -1G, respectively, and many of these ended their days in service with the Vietnam Air Force after being transferred under MAP.

L-19As began serving as transition trainers at Henry Post AAF (Fort Sill), Oklahoma, sometime in 1953. Some fitted with dual controls were re-designated TL-19A, however, the exact number of conversions is not recorded. The last 60 Army L-19As became L-19A-IT instrument trainers following modifications to install a full instrument panel and blind flying curtains in the rear cockpit, but apparently never received a "T" for trainer prefix. During 1956 the Army accepted delivery of 310 TL-19D instrument trainers that came with -15 engines, constant-speed propellers, dual controls, and a full set of instruments in both cockpit positions. Twenty L-19Es in the last

One of 310 Cessna TL-19Ds procured by the Army in 1956 as instrument trainers. Note constant-speed propeller and ADF hump on rear fuselage. These aircraft came with a full set of instruments in front and rear cockpits.

production block entered Army service as TL-19Es after being fitted with dual controls. L/TL-19 variants were the Army's principal type of transition/basic fixed-wing trainer from 1953 onwards, and undertook the primary role when the Army ceased contract training of primary pilot students in 1959. Training of fixed-wing Army Aviators during the L-19/O-1 era changed locations several times: Henry Post AAF, Oklahoma, initially; Gary AFB (later Camp Gary), Texas, from the mid–1950s to 1963; Fort Rucker AAF, Alabama, from 1963 to 1967; and Fort Stewart AAF, Georgia, from 1967 until 1971, when O-1s were withdrawn as trainers. The last O-1s left other Army service in 1974.

North American TF-86 Sabre—1953

Technical Specifications (TF-86F [Second Prototype])

Type: Two-place advanced trainer.
Manufacturer: North American Aviation, Inc., Inglewood, California.
Total U.S. military versions: 2 (USAF).
Unit cost (F-86F): $211,111.
Powerplant: One General Electric J-47-GE-27 turbojet engine, rated at 5,910 lbs./s.t.
Armament: Two fixed .50-caliber machine guns in the nose.
Performance: Max. speed 692 mph, cruise 486 mph, ceiling 48,000 ft.; range 1,293 mi.
Weights: 10,990 lbs. empty, 18,040 lbs. loaded.
Dimensions: Span 37 ft. 1 in., length 42 ft. 9 in., wing area 302 sq. ft.

In early 1953 North American tendered the USAF an unsolicited proposal for a two-seat training variant of its highly successful F-86 fighter, with the hope that it might share similar success to Lockheed's earlier conversion of the F-80 into the T-33 (see above). The following September, Air Material Command (AMC) authorized North American to proceed with conversion of one early production F-86F into the proposed two-seat configuration under the des-

The second TF-86F prototype in 1954 as evidenced by the two guns and wing pylons. An excellent airplane, it was overshadowed by the advent of supersonic trainers like the North American TF-100.

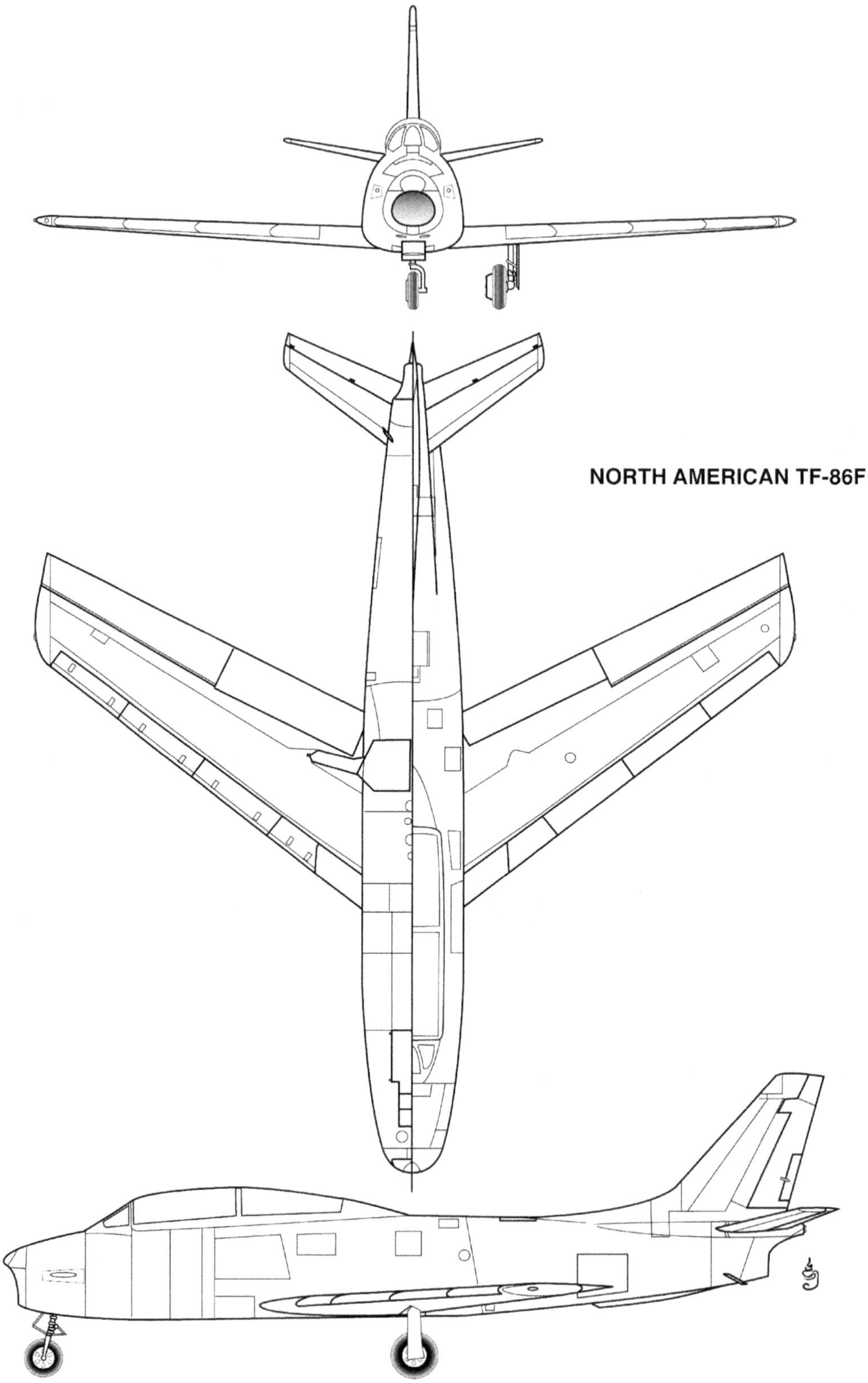
NORTH AMERICAN TF-86F

ignation TF-86F. Modifications to the basic airframe entailed stretching the forward fuselage five-feet three-inches to add a second cockpit and ejection seat in front of the original, moving the wings forward eight inches to offset the shift in center-of-gravity, and deleting all armament to conserve weight. The tandem seats were arranged under a long, clamshell-type canopy. When the TF-86F flew for the first time on December 14, 1953, it exhibited overall performance very near that of the single-seat F. However, in March 1954, before the prototype could be delivered the USAF, it was destroyed when the North American test pilot lost control during a slow roll on takeoff.

Shortly after the accident, AMC authorized North American to modify the last F-86F-35 on the assembly line into the TF-86F two-seat configuration. Differences from the first prototype included addition of two .50-caliber guns in the nose, a pair of underwing pylons for drop tanks or bombs, plus an enlarged dorsal fin fillet and ventral strake to correct stability problems that may have led to the crash of the first prototype. The first flight of the second TF-86F took place on August 5, 1954, and within a month, the aircraft began making demonstration tours of training bases as a sales promotion. Although officially accepted in early 1955, the USAF elected not to place the TF-86F in production in favor of plans to procure a two-seat version of the F-100. Thereafter, the TF-86F served as a chase plane at Edwards AFB until being scrapped some years later.

Cessna T-37 "Tweety Bird"—1954

Technical Specifications (T-37B)

Type: Two-place primary and basic trainer.
Manufacturer: Cessna Aircraft Co., Wichita, Kansas.
Total U.S. military versions: 1,269 (USAF).
Unit cost: $164,854.
Powerplants: Two Continental-Teledyne J-69-T-24 turbojet engines, each rated at 1,025 lbs./s.t.
Performance: Max. speed 426 mph, cruise 368 mph; ceiling 31,500 ft.; range 870 mi.
Weights: 4,056 lbs. empty, 6,569 lbs. loaded.
Dimensions: Span 33 ft. 9.5 in., length 29 ft. 3 in., wing area 183.9 sq. ft.

Over a career spanning 52 years, Cessna's T-37 held the record as the USAF's longest serving type of jet trainer until overtaken by the Northrop T-38 in 2013. Conducting Undergraduate Pilot Training (UPT) in an all-jet force—for student pilots slated to fly jets operationally—was an early USAF goal; thus, in early 1952, Air Material Command (AMC) solicited design competition proposals for a new Trainer Experimental (TX) requirement calling for a lightweight (i.e., 5,000 lbs. loaded), jet-powered basic trainer equipped with ejection seats. Three designs were considered: the side-by-side, twin-jet Cessna model 318; the tandem-seat, single-jet Beech model 73 Jet Mentor; and the tandem seat, single-jet Temco model 51. In early 1953, the USAF selection board declared Cessna the winner, awarding the company a contract for two flying prototypes and one static test example under the designation XT-37, followed in 1954 by a production contract to deliver 444 aircraft as the T-37A. The first prototype completed its maiden flight from the factory on October 14, 1954, and both XT-37s commenced USAF testing in early 1955. After the destruction of the first XT-37 during spin tests, strakes were added on either side of the nose and the tail surfaces enlarged and redesigned. Other changes included a speed brake fitted behind the nose wheel well door and retractable screens over the intakes to prevent ingestion of debris while on the ground.

The first batch of 11 T-37As, powered by J-69-T-9 engines, each rated at 920-lbs./s.t., began

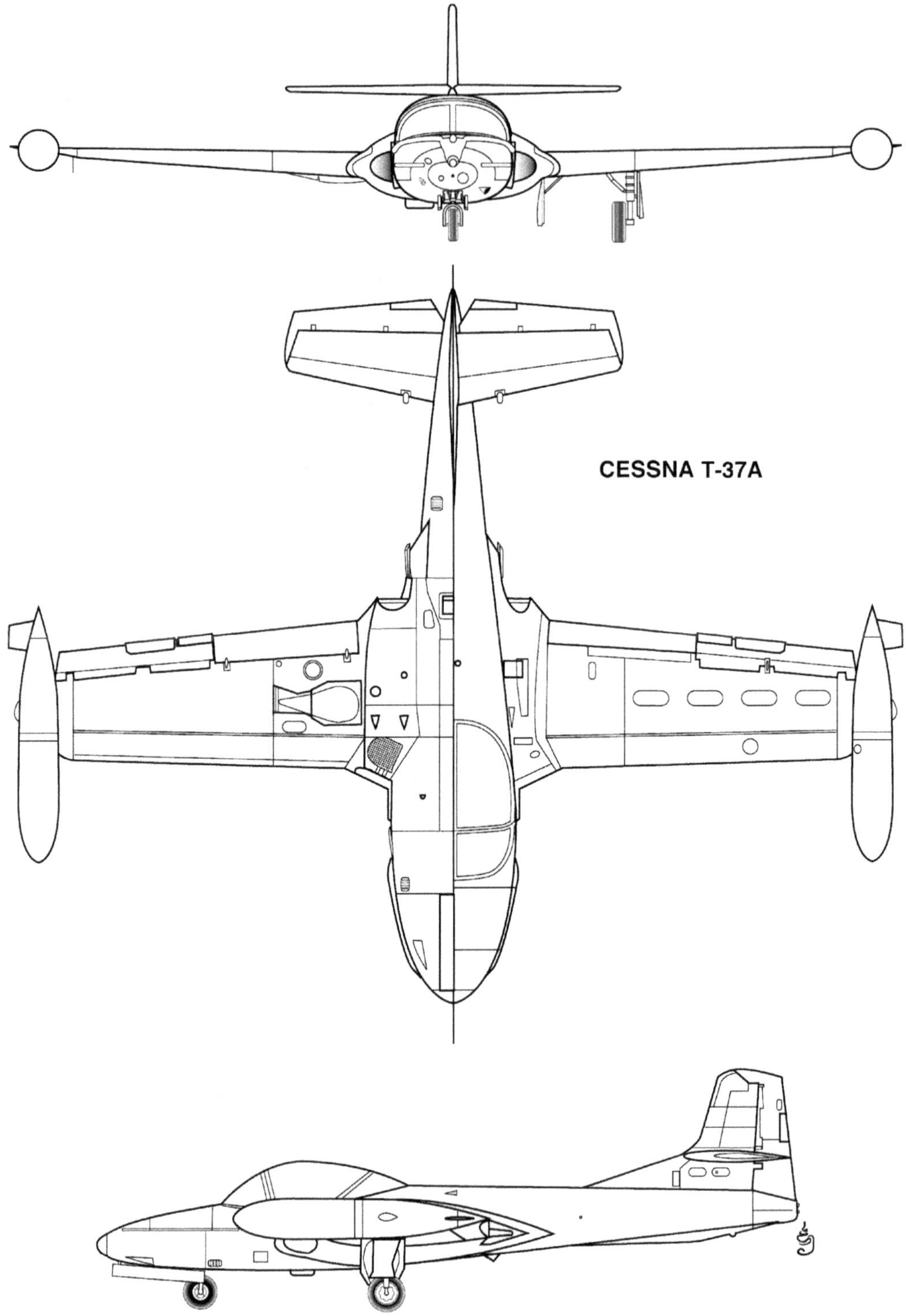

CESSNA T-37A

The first "Tweety Bird," one of two XT-37s tested by the USAF. This aircraft was destroyed during spin tests in 1955, which led to strakes being added to the nose and enlarged tail surfaces.

arriving for official service evaluations in late 1955, and after extensive trials, cockpit modifications and other small detail changes were incorporated into the production standard. Actual service introduction of T-37As into UPT took place in early 1957. The unofficial name "Tweet" or "Tweety Bird" applied to the T-37A came from the loud, high-pitched whine produced by the J-69 engines. Although the T-37A was well liked within Air Training Command (ATC) by both students and instructors, it was regarded as generally underpowered for its weight. In 1959, as T-37A deliveries were nearing completion, AMC awarded Cessna another contract to commence production of the T-37B, which differed in having uprated J-69-T-25 engines offering a 10 percent boost in power together with new avionics. Over a series of orders, Cessna delivered 552 new T-37Bs from November 1959 to January 1972, and in the interim, all T-37As were retrofitted and re-engined to the B Standard.

In 1961, as a private venture intended mainly for export, Cessna commenced development of the T-37C, an armed variant offered either as a weapons trainer or a light attack platform.

A T-37B in AETC markings of the 1960s which had been converted from the original batch of 719 T-37As. Chief difference was an engine upgrade and new avionics.

The last two T-37Bs in AETC service seen taking off from Sheppard AFB, Texas, in July 2009, marking the end of a remarkable and distinguished training career that spanned more than 52 years.

Changes from the B included a computing gunsight and gun camera, strengthened wings, 65-gallon wingtip tanks, and a weapon stores pylon under each wing. Though never procured by the USAF, Cessna produced 273 T-37Cs for foreign air forces between 1962 and 1975. However, as U.S. military involvement in Southeast Asia escalated during the early 1960s, the USAF's need for a light attack aircraft to fly counter-insurgency missions led to development of the much modified YAT-37D, which appeared in 1963 with stronger wings and landing gear, General Electric J-85–17 engines of 2,850 lbs./s.t. each, 90-gallon wingtip tanks, eight under-wing pylons for up to 6,000 lbs. of mixed ordnance, and a computing gunsight. After a three-year delay, the USAF awarded Cessna a contract to proceed with similar modifications of 39 T-37Bs that returned to USAF service in mid–1967 under the new designation A-37A Dragonfly, and then from 1968 to 1977, Cessna manufactured 577 new construction A-37Bs, of which 134 were delivered to the USAF, the balance going to nations participating in military assistance programs (MAP), mostly South Vietnam and Latin America.

As new T-37As were integrated into UPT during 1957, they began replacing T-28As in basic pilot training, and in 1958, started augmenting T-34As in the primary course. The process continued when deliveries of T-37Bs commenced in 1959, with T-37As and Bs equipping all eight of ATC's Pilot Training Wings by the end of 1961. In 1985 and 1986, when the oldest T-37s had been in service over 25 years, the USAF tested the Fairchild T-41 (see in Series III) as a planned replacement; however, the contract was cancelled before any production could begin. The USAF did not make a definite effort to replace T-37s until 1994 when it and the Navy issued requests for proposals under a program known as the Joint Primary Aircraft Training System (JPATS) which ultimately, in 2000, resulted in selection of the Beechcraft-Hawker T-6A Texan II (see in Series III). Although the phase-out of T-37Bs from Air Education and Training Command (renamed in 1993) began in 2001, the process was gradual, and the final examples did not leave service until July 2009.

Martin B/TB-57 Canberra—1954

Technical Specifications (B-57C)

Type: Two-place tactical bomber transition trainer and target tug.
Manufacturer: Glen L. Martin Co., Baltimore, Maryland.
Total U.S. military versions: 41 (USAF).

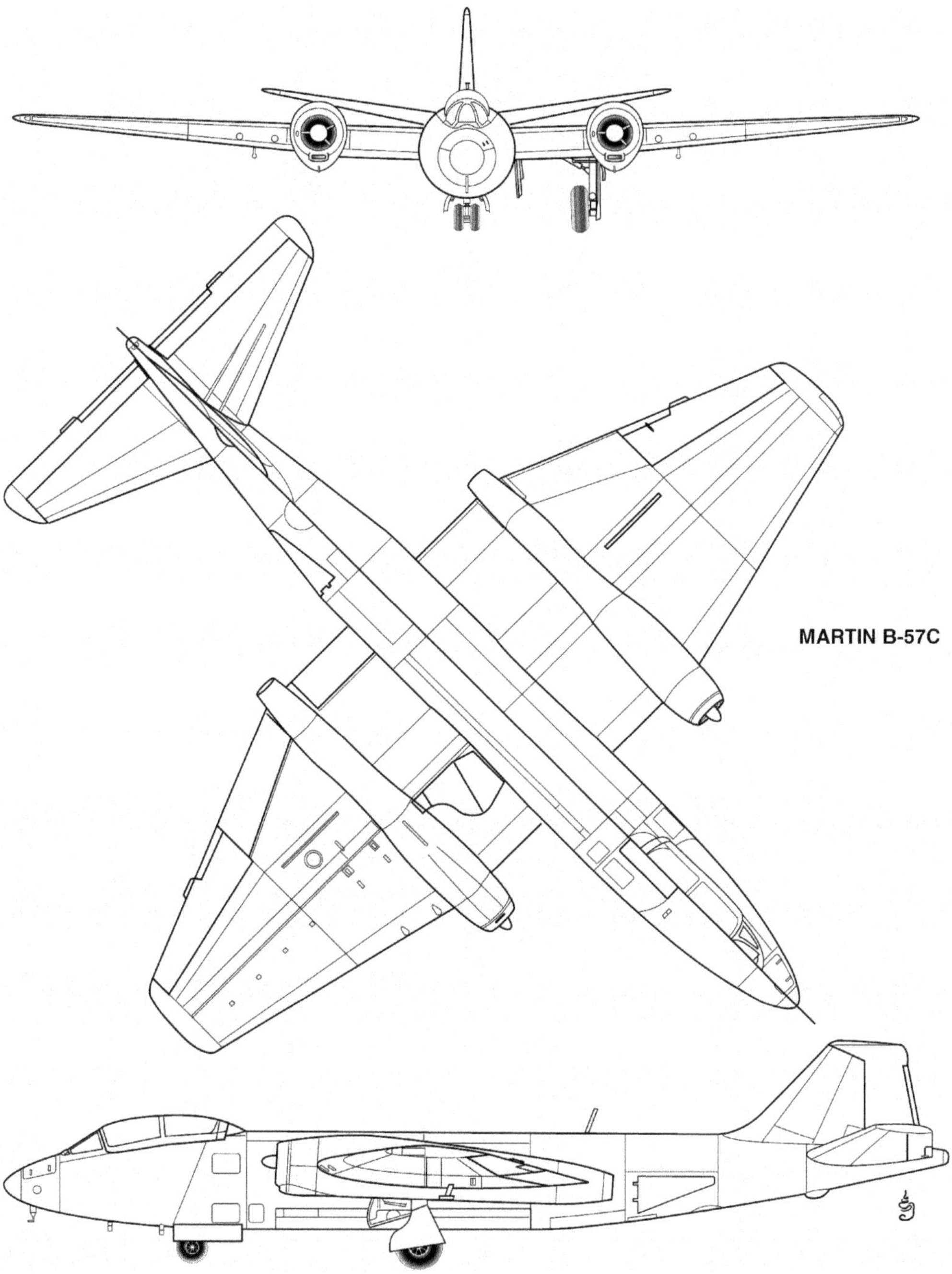

This Martin B-57C is one of 38 completed with dual controls for transition and proficiency training, at least one of which was supplied to each B-57B tactical unit.

Unit cost: $1,100,000.
Powerplants: Two Wright J-65-W-5 turbojet engines, each rated at 7,220 lbs./s.t.
Armament: Four fixed 20-millimeter cannons in wings, 16 HVAR rockets on wing stubs, and up to 6,000 lbs. of bombs carried in an internal bay.
Performance: Max. speed 598 mph, cruise 475 mph; ceiling 45,100 ft.; range 1,895 mi. loaded, 2,722 mi. max.
Weights: 28,793 lbs. empty, 56,965 lbs. max. takeoff.
Dimensions: Span 64 ft., length 65 ft. 6 in., wing area 960 sq. ft.

When the USAF selected a license-built version of the English Electric *Canberra* for production in early 1951 as the B-57, it was the first military aircraft of foreign design to be manufactured in the U.S. since the World War I-era De Havilland DH.4 (see in Series I, Part A). The USAF's departure from standard procurement policy came largely as a result of the U.S. Army's claim that the *Canberra* surpassed its American-designed rivals (e.g., Martin XB-51, North American B-45) in terms of both combat radius and the ability to maneuver within the close confines of a battlefield. Modifications bringing the basic design up to USAF combat requirements resulted in delivery of 202 B-57B production models from mid–1954 to mid–1956, which appeared with tandem crew seating, all-weather targeting systems, and a rotary bomb bay that could house tactical nuclear weapons. At the same time, the USAF also accepted delivery of 38 dual-control B-57Cs that were distributed among B-57B-equipped tactical units to serve as conversion trainers. After leaving active USAF service, a number of B-57Cs were transferred to ANG units equipped with other B-57 variants. In 1956 and 1957, Martin manufactured 68 B-57Es that were equipped to serve as high-altitude target tugs with Air Defense Command (ADC), and several are thought to have been assigned to training duties under the designation TB-57E, though historical records do not mention their assignments or whether they were modified in any way.

Douglas TB-26 Invader—1955

Technical Specifications (TB-26B)

Type: Three-place advanced multi-engine trainer.
Manufacturer: Douglas Aircraft Corp., El Segundo and Long Beach, California.

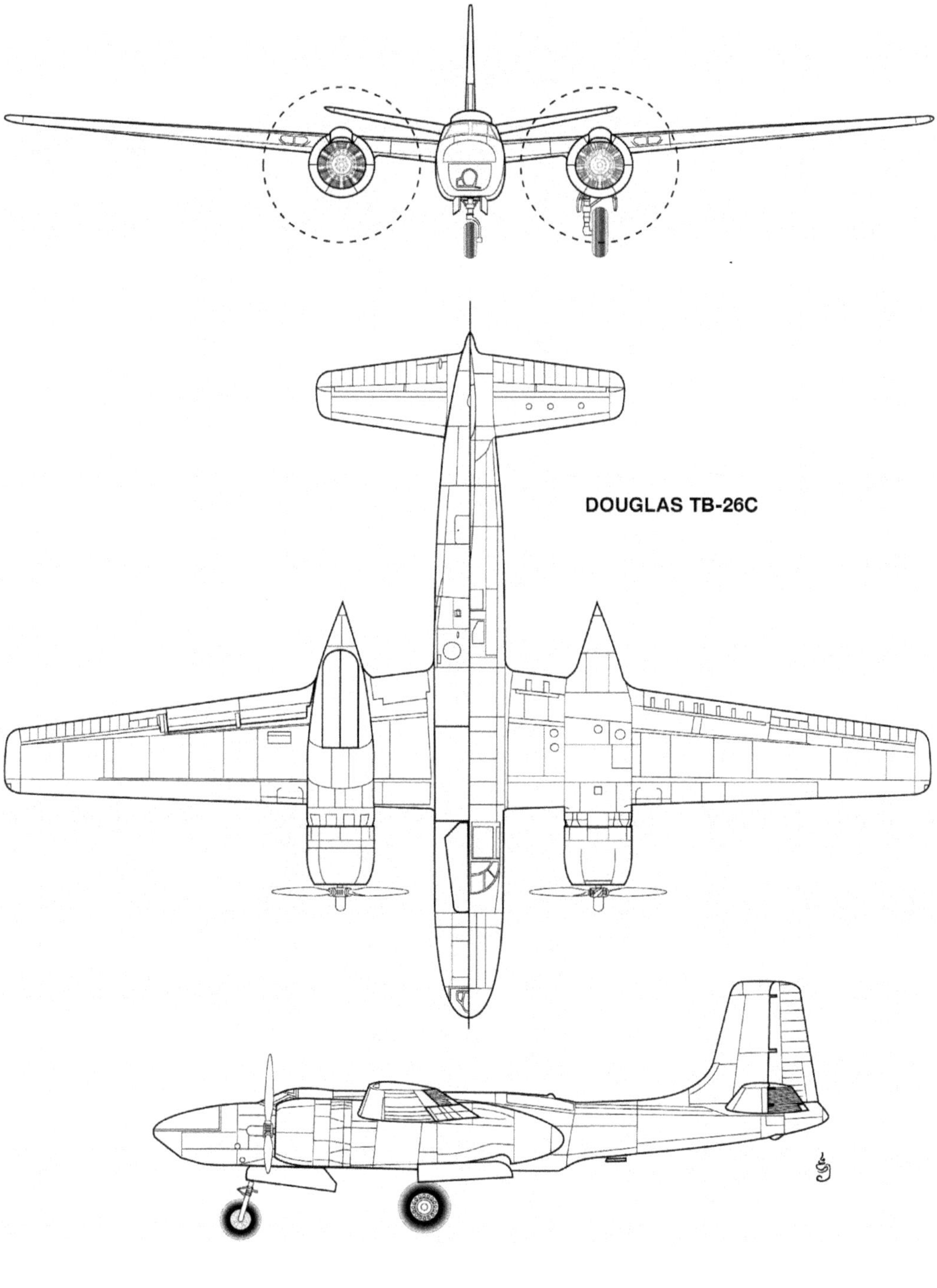

A Douglas TB-26B (solid nose) seen in the mid–1950s while serving with the 110th Bomb Squadron, Missouri Air National Guard, then based at Lambert Field in St. Louis.

Total U.S. military versions: Unknown (ANG).
Unit cost (B-26B): $192,457.
Powerplants: Two 2,350-hp Pratt & Whitney R-2800–79 *Double Wasp* 18-cylinder air-cooled radial engine driving three-bladed Hamilton Standard constant-speed propellers.
Performance: Max. speed 355 mph, cruise 285 mph; ceiling 22,100 ft.; range 3,200 mi.
Weights: 22,362 lbs. empty, 35,000 lbs. loaded.
Dimensions: Span 70 ft., length 50 ft. 8 in., wing area 540 sq. ft.

Regarded by many as the best propeller-driven light bomber ever produced, the legendary Douglas A-26 Invader (re-designated B-26 in 1948) served in three wars—World War II, Korea, and Vietnam—over a military career spanning 29 years (1944–1973). Following the end of the Korean War in 1953, many of the USAF's active service B-26Bs (solid nose) and B-26Cs (glazed nose) were transferred to the ANG, where they were redistributed among bomb squadrons in six states. Starting in 1955, an unspecified number of ANG B-26Bs and Cs were stripped of armament and equipped with dual controls for multi-engine pilot conversion training under the designation TB-26B/C (though none having this designation appear on official serial number records). When the ANG phased out all of its B-26Bs and Cs in 1957 and 1958, a few of the TB-26s are thought to have remained in service a few years longer, and the very last example, an ex–TB-26B converted to a staff transport as the VB-26B, was not retired until 1972.

Convair TF-102 Delta Dagger—1955

Technical Specifications (TF-102A)

Type: Two-place fighter transition trainer.
Manufacturer: Convair Div. of General Dynamics Corp., San Diego, California.
Total U.S. military versions: 111 (USAF).
Unit cost: $1,184,000.
Powerplant: One Pratt & Whitney J-57-P-23A turbojet engine, rated at 11,700 lbs./s.t. dry, 17,200 lbs./s.t. with afterburning.

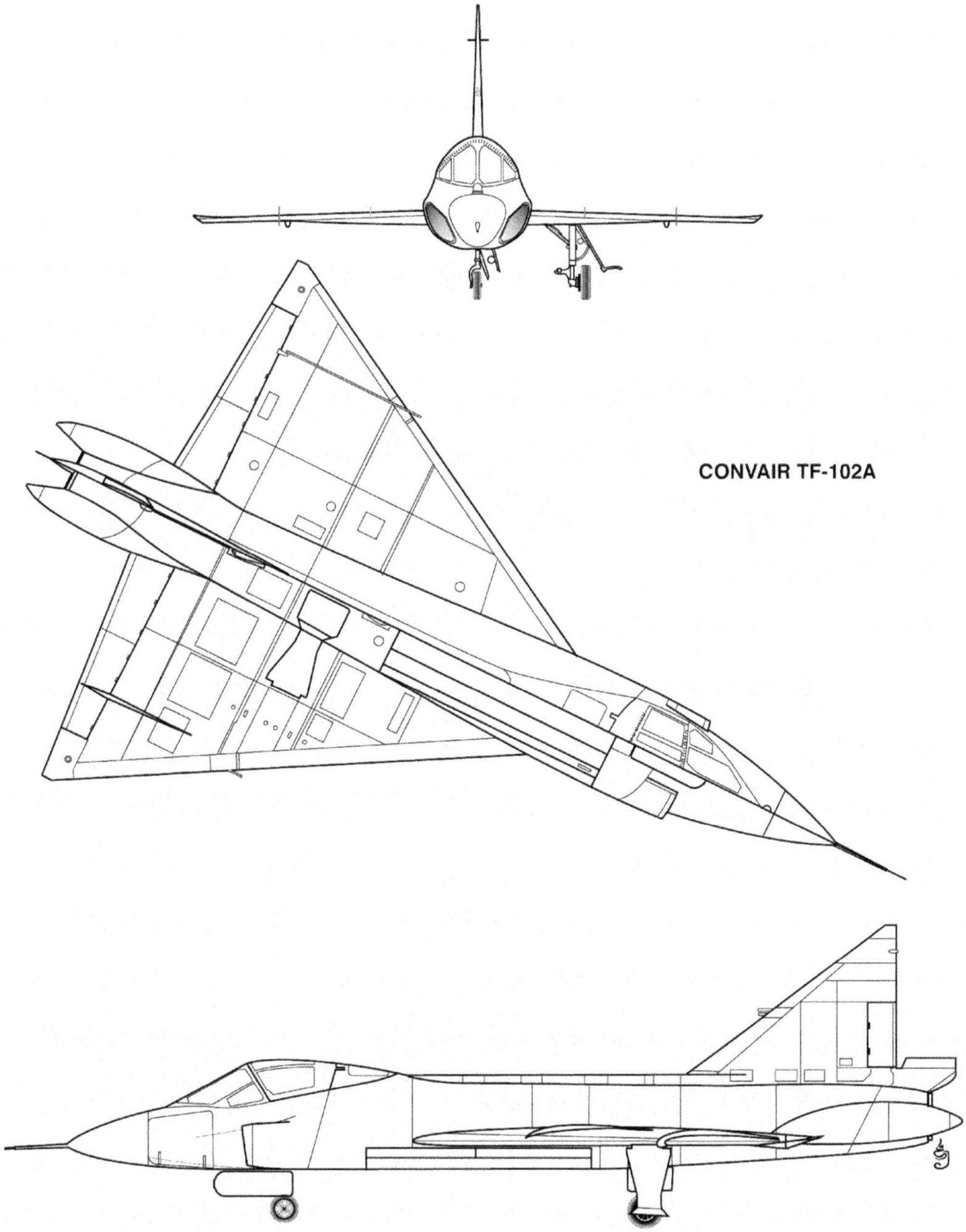

Armament: Twelve 2.75" FFAR rockets and six AIM-4/AIM-26 *Falcon* air-to-air guided missiles carried in a ventral internal bay.
Performance: Max. speed 646 mph, cruise 520 mph; ceiling 50,000 ft.; range 1,302 mi.
Weights: 23,643 lbs. empty, 27,778 lbs. loaded.
Dimensions: Span 38 ft. 2 in., length 63 ft. 4 in., wing area 661.5 sq. ft.

The Convair F-102 Delta Dagger holds the triple distinction of having been the world's first delta-winged combat aircraft to reach operational service, the first all-weather interceptor capable of achieving supersonic speed in level flight, and the first type of fighter to possess all-

The first Convair TF-102A shown in late 1955 before being retrofitted with a larger tail fin and vortex generator over the canopy framing. The original order for 150 was limited to the 111 delivered by mid–1958.

missile armament as a standard feature. In mid–1952, when the YF-102A prototype was still under construction, the USAF reached the conclusion that a trainer variant would be needed to prepare pilots for the distinctive flight characteristics of high-performance delta-winged aircraft, in particular the high angles-of-attack required for takeoff and landing and the pronounced induced drag encountered in maneuvers. Although Convair received approval to start development of a TF-102A two-seat trainer variant in September 1953, the project was subsequently delayed when wind tunnel tests demonstrated that the YF-102A would be incapable of supersonic speed in level flight. Based upon wind tunnel data, Convair revised the YF-102A's basic aerodynamic design to apply the recently discovered "area rule" principle (i.e., a coke bottle shape) to the fuselage and lengthen it seven feet, modify the camber and location of the wings, change the contour of the air intakes, replace the canopy, and move the vertical fin further aft.

In July 1954, once the USAF had approved all the revisions to the YF-102A, Convair received an initial contract for a batch of 20 two-seat TF-102As. As the trainer design progressed, it emerged with side-by-side seating out of an effort to simplify training, even though drag caused by the larger frontal aspect would render it incapable of achieving supersonic speed in level flight. The wider cockpit also required the air intakes to be positioned lower on the fuselage, but in all other respects the airframe components of the TF-102A were identical to those of the production F-102A. In September 1954, after successful flight testing of the revised YF-102A, the USAF increased the TF-102A contract to 48 aircraft, and following the maiden flight of the first TF-102A on November 8, 1955, increased it again to 150. When flight-testing of the TF-

This TF-102A is seen in 1963 while serving with the 40th Fighter Interceptor Squadron at Johnson AB, Japan. Vortex generators are discernible on the canopy rim.

102A revealed serious buffeting problems produced by airflow over the bulbous canopy at high speeds, the aircraft was initially refitted with a lower profile canopy that, while partially reducing buffeting, severely limited over-the-nose visibility. The original canopy was restored when engineers largely eliminated the problem by adding vortex generators to the cockpit framing and increasing vertical fin area. Due to delays and cutbacks in the F-102A program overall, the TF-102A contract was limited to the 111 examples delivered from late 1955 to mid-1958.

In operational service, two TF-102As were typically allocated to each of the 48 F-102A-equipped Fighter Interceptor Squadrons in Air Defense Command, Alaskan Air Command, U.S. Air Forces Europe, and Pacific Air Forces. As F-102As were phased out of active USAF service from 1960 onwards, many were transferred to ANG units, and the TF-102As went with them. At their peak in 1970, F/TF-102As served with 23 ANG squadrons in 21 different states, with the final examples being retired from the 199th Fighter Interceptor Squadron, Hawaii ANG, in January 1977.

North American F/TF-100 Super Sabre—1956

Technical Specifications (F-100F)

Type: Mission-capable two-place fighter transition trainer.
Manufacturer: North American Aviation, Inc., Inglewood, California
Total U.S. military versions: 295 (USAF).
Unit cost: $804,444.
Powerplant: One Pratt & Whitney J-57-P-21A turbojet engine, rated at 10,200 lbs./s.t. dry, 16,000 lbs./s.t. with afterburning.
Armament: Two 20-mm fixed cannons in the nose and six underwing stations for up to 5,000 lbs. of external ordnance.
Performance: Max. speed 852 mph (Mach 1.28), cruise 587 mph; ceiling 47,700 ft.; range 1,294 mi.
Weights: 21,712 lbs. empty, 39,122 lbs. loaded.
Dimensions: Span 38 ft. 9 in., length 57 ft. 1 in., wing area 400 sq. ft.

When North American F-100As began entering service with Tactical Air Command (TAC) during 1954, they became the first operational fighters in the world capable of exceeding the

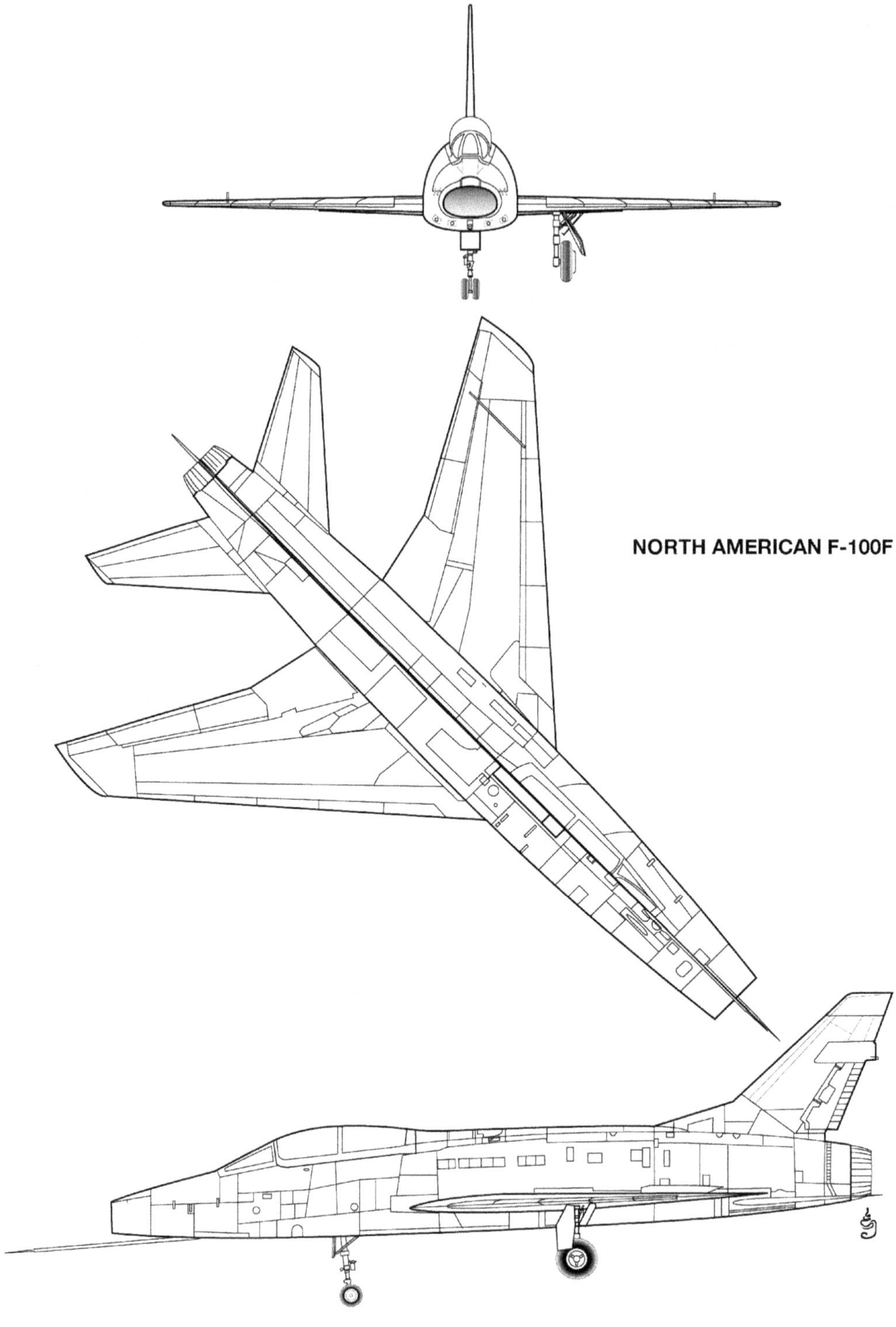

NORTH AMERICAN F-100F

A North American F-100F in Vietnam-era camouflage while serving with the 110th Fighter Squaron, Missouri Air National Guard, in the 1970s. They were retired in 1979.

speed of sound in level flight. At the same time, however, in-service accident rates pointed to the need for a two-seat trainer variant that would prepare pilots for its unaccustomed handling characteristics. To this end, in September 1955, North American agreed at no additional cost to modify one of the F-100Cs on its assembly line into a two-seat, tandem configuration with the hope that it might replace the T-33 as Air Training Command's standard advanced trainer, and the following December the company received a contract to manufacture 259 aircraft under the designation TF-100C. Flying for the first time on August 6, 1956, the prototype TF-100C appeared with a five-foot nine-inch fuselage plug to accommodate the extra cockpit but lacked the armament and tactical equipment of its single-seat counterpart.

In the interval, the USAF changed the two-seat specification to call for an armed, fully mission capable aircraft (with some reduction in payload) that incorporated the kinked wing, larger vertical tail surfaces, and other improvements of the F-100D. The production aircraft built to the revised specification flew for the first time on March 7, 1957, under the designation F-100F. Other than the extra cockpit, the only tactical difference of the F-100F from the D was two 20-mm cannons instead of four and a 2,040-lb. reduction in external weapons load. Production terminated in August 1959 with 294 F-100Fs delivered to the USAF and another 45 going directly to foreign air forces under the MDAP program. Although official records do not specify how the F-100Fs were allocated among the 30-plus active service Tactical Fighter and Fighter Bomber Wings operating the F-100C/D at various times between 1956 and 1971, the records do identify nine Tactical Fighter and Combat Crew Training Wings that in all likelihood employed two-seat Fs for both transition and operational combat training. During the Vietnam War, TAC utilized some F-100Fs in "Misty" Forward Air Control missions to identify and mark targets for incoming air strikes, and starting in 1966, specially equipped F-100Fs flying over North Vietnam pioneered the "Wild Weasel" concept of detecting and attacking enemy surface-to-air missile sites. In 1969, as the USAF commenced the process of phasing out its F-100Ds and Fs, many were transferred to ANG units, where they continued to serve in 21 squadrons located in 15 states until the final examples were retired during 1979.

Lockheed F/TF-104 Starfighter—1957

Technical Specifications (F-104B)

Type: Mission-capable two-place fighter transition trainer.
Manufacturer: Lockheed Aircraft Corp., Burbank, California.

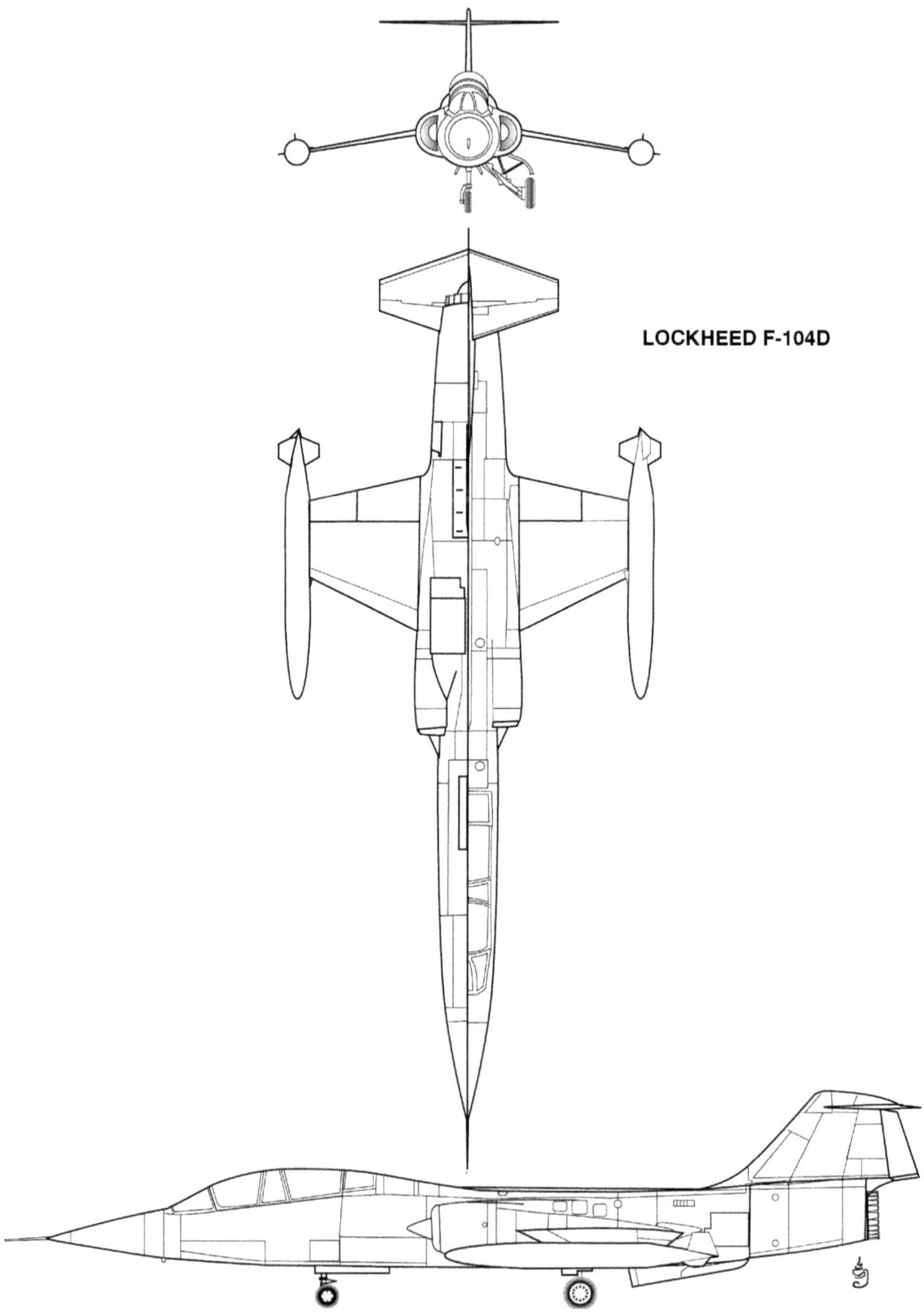

Early photograph of a Lockheed F-104B before the enlarged, power-assisted rudder had been retrofitted. A total of 26 were distributed among F-104A operational units during 1958 to be used as transition trainers.

Total U.S. military versions: 47 (USAF).
Unit cost (F-104B [F-104D]): $2,400,000 [$1,500,000].
Powerplant: One General Electric J-79-GE-3A/B turbojet engine, rated at 9,600 lbs./s.t. dry, 14,800 lbs./s.t. with afterburning.
Armament: Two AIM-9B *Sidewinder* missiles mounted on wingtips.
Performance: Max. speed 1,145 mph (Mach 1.74), cruise 592 mph; ceiling 48,600 ft.; range 1,210 mi. (unarmed, with tiptanks; 460 mi. armed).
Weights: 13,727 lbs. empty, 24,294 lbs. loaded.
Dimensions: Span 21 ft. 11 in., length 54 ft. 8 in., wing area 196.1 sq. ft.

The Lockheed F-104 is notable in having been the USAF's first operational Mach 2 fighter. Its design originated in 1952 as an unsolicited company proposal to build a lightweight fighter completely optimized for speed and climb performance. In March 1953, after reviewing the proposal, the USAF awarded Lockheed a contract to build two experimental prototypes to be powered by Wright J-65 engines as the XF-104. While the two XF-104 prototypes were still undergoing testing in late 1954, the USAF ordered 17 J-79-powered YF-104A pre-production aircraft, the first flying in February 1956. In March 1956, after brief testing of the YF-104A, the USAF ordered 153 F-104A production models with the intention that they would eventually replace F-100s in Tactical Air Command (TAC); however, by the time deliveries commenced in early 1958, they were instead allocated to Air Defense Command (ADC) as interceptors due to unplanned delays in the F-106 program (see below).

Concurrent with F-104A production, the USAF directed Lockheed to develop a two-seat TF-104 to be used purely for transition training. But before any TFs were built, the USAF changed the specification for a mission-capable, two-place aircraft, to be designated F-104B, that would equip operational F-104A squadrons (as with the F-100D and F reported above). Basic modifications to the airframe consisted of installing a second cockpit behind the existing one

One of 21 F-104Ds delivered in 1958 and 1959 to TAC units equipped with F-104Cs. Note difference in canopy configuration. This aircraft was transferred to Taiwan in 1974.

in the space normally occupied by an M-61 rotary cannon, relocating some internal electronics, reducing internal fuel capacity by 145 gallons, and reconfiguring the nosewheel to retract rearwards. Lockheed received an initial contract for six F-104Bs in April 1956, then Air Material Command added 106 more to the order sometime in 1957. As delivered, the first six F-104Bs retained the vertical tail surfaces of the F-104A but lacked the ventral strake, and soon after the first flight on January 16, 1957, testing revealed directional instability. To correct the problem, subsequent F-104B production models featured a power-assisted rudder of 25 percent increased area and a ventral strake. Other than reduced range, the F-104B exhibited overall performance very similar to that of the A. Deliveries of F-104Bs to F-104A operational units commenced in early 1958, although production was ultimately limited to the 26 examples completed by the end of the year. In connection with the 77 F-104Cs delivered in 1958 and 1959, the USAF also accepted 21 two-seat F-104Ds. The Cs, all destined for service in TAC, were dedicated fighter-bomber variants having enhanced fire-control systems, provision to carry 2,000 lbs. of bombs, rocket pods or a tactical nuclear weapon, and more powerful -7 engines. Like the B, the F-104D was fully mission-capable but lacked the rotary cannon of its single-seat counterpart. Lockheed also built 146 other two-seat variants, 30 F-104Fs and 116 TF-104Gs, which were delivered to foreign air forces under military assistance programs (MAP).

F-104Bs commenced operational service with the 83rd Fighter Interceptor Squadron (FIS) at Hamilton AFB, California, in early 1958 and went on to equip three more four ADC squadrons as deliveries continued. When ADC phased-out all of its F-104As and Bs in 1960, some were transferred to ANG FIS units in Tennessee, Arizona, and Puerto Rico while others went to Pakistan and Nationalist China under MAP. F-104Ds first became operational in late 1958 with the 476th Tactical Fighter Squadron (TFS) out of George AFB, California, and eventually equipped three more squadrons of the 479th Tactical Fighter Wing. As F-104Cs and Ds were withdrawn from TAC's active inventory in 1966 and 1967, some continued to serve with the 198th TFS, Puerto Rico ANG, until 1975.

Convair F/TF-106 Delta Dart—1958

Technical Specifications (F-106B)

Type: Mission-capable two-place fighter transition trainer.
Manufacturer: Convair Aircraft Div. of General Dynamics Corp., San Diego, California.
Total U.S. military versions: 63 (USAF).

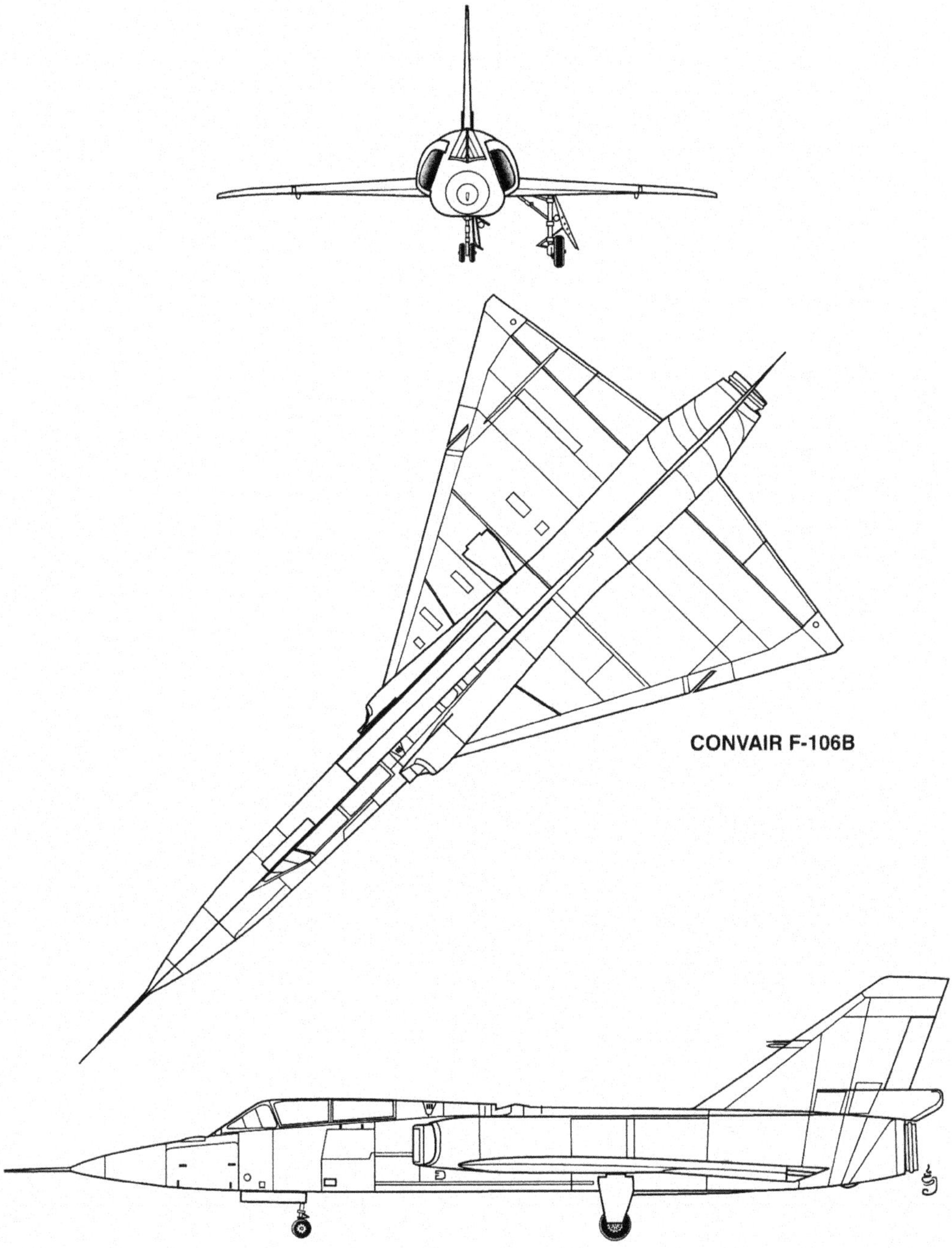

Unit cost: $4,900,000.
Powerplant: One Pratt & Whitney J-75-P-17 turbojet engine, rated at 17,200 lbs./s.t. dry, 24,500 lbs./s.t. with afterburning.
Armament: One AIR-2 *Genie* nuclear rocket and four AIM-4 *Falcon* guided missiles in an internal bay.
Performance: Max. speed 1,525 mph (Mach 2.31), cruise 650 mph; ceiling 51,400 ft.; range 1,456 mi.
Weights: 24,920 lbs. empty, 36,500 lbs. loaded.
Dimensions: Span 38 ft. 3.5 in., length 70 ft. 9 in., wing area 697.8 sq. ft.

In order to counter the threat of Soviet bombers entering North American airspace over the North Pole, efforts to develop a missile-armed, all-weather supersonic interceptor turned into one of the most complex and drawn-out programs in USAF history. It was one of the first aircraft to be designed under a "weapons system" concept in which the airframe, fire-control systems, and missile armament were all integrated into a common development scheme. A series of setbacks—aerodynamic problems with Convair's delta wing layout, an unplanned substitution of powerplants, and delays in Hughes's progress on the fire-control systems—all served to divide the program into two distinct phases: the revised Mach 1 F-102A of 1954 (see above) and the Mach 2 F-106A of 1956 (originally designated F-102B). Yet, when Convair's long-awaited F-106A Delta Dart did finally achieve operational capability in May 1959, it emerged as the fastest and most lethal interceptor the world had ever seen. Even today, the F-106 retains the world's absolute speed record (Mach 2.31) for a single-engine aircraft.

Like other Century Series fighters, the USAF made plans early in the program to procure a TF-106A (originally designated TF-102B) two-seat trainer derivative that would accompany the F-106A into production, but by the time a definitive order materialized in August 1956, the requirement had been revised to specify a fully mission-capable aircraft under the designation F-106B. To avoid the performance limitations of the TF-102A, the F-106B appeared with a tandem seating arrangement under a one-piece, clamshell-type canopy and retained the external dimensions of the A. For space reasons, one minor difference in the B was substitution of a Hughes ASQ-25 fire-control system that duplicated all of the intercept functions of the F-106A's

A Convair F-106B in markings of the 460th Fighter Interceptor Squadron, Oxnard AFB, California, circa 1970. This unit and its F-106A/Bs were deactivated in 1974 during the Aerospace Defense Command drawdown.

MA-1. The first F-106B rolled out for its maiden flight on April 9, 1958, and deliveries of production aircraft to the USAF began the following October. Due to problems revealed during acceptance testing of the F-106A, the first 12 F-106Bs were held back for further testing until they could receive the aerodynamic modifications and powerplant upgrades (-17 engine) that were being incorporated into the F-106A production standard. Initial operational capability was achieved in July 1960, and the last of 63 F-106Bs built was delivered to the USAF by the end of the year.

From 1958 to 1960, as F-106As became operational in 14 Fighter Interceptor Squadrons (FIS) within Air Defense Command (ADC), each squadron was allocated at least three two-seat F-106Bs which not only provided transition, proficiency, and combat training of pilots but flew intercept missions alongside the As. During their careers, F-106Bs received the same modifications and upgrades given to F-106As, which included retrofitting zero-zero Weber ejection seats and refinements to the fire-control and fuel systems. The process of transferring F-106As and Bs from active USAF units to the ANG began in 1972, building to a peak of six ANG squadrons by 1974. The last active USAF unit, the 49th FIS at Griffiss AFB, New York, gave up its F-106s in 1987, and the last Guard unit, the 119th FIS, New Jersey ANG, in 1988.

North American T-39 (T3J) Sabreliner—1958

Technical Specifications (T-39A)

Type: Proficiency and combat readiness trainer; utility transport.
Manufacturer: North American Aviation, Inc., Inglewood and Los Angeles, California
Total U.S. military versions: 214 (149 USAF; 65 USN).
Unit cost: $810,000.
Powerplants: Two Pratt & Whitney J-60-P-3A turbojet engines, each rated at 3,000 lbs./s.t.
Performance: Max. speed 538 mph, cruise 500 mph; ceiling 42,000 ft.; range 1,348 mi.
Weights: 9,265 lbs. empty, 18,650 lbs. loaded.
Dimensions: Span 44 ft. 6 in., length 44 ft., wing area 342.1 sq. ft.

In 1956 the USAF issued a utility-trainer-experimental (UTX) requirement calling for a twin-engine jet aircraft to operate either as a trainer for aircrews of non-tactical jets or as a fast utility and personnel transport. Two aircraft manufacturers responded with proposals: Beech with a militarized version of the French-designed Morane-Saulnier MS-760, the world's first certified private jet, and North American with a new design for the swept-wing model NA-265. By comparison to the four-place MS-760, the NA-265 offered much larger carrying capacity (seven passengers or 1,400 lbs. of cargo) and better cruising speed and range. Receiving the factory name "Sabreliner" due to its aerodynamic similarities to the F-86, the NA-265 emerged with nacelle-mounted engines located on the aft fuselage sides and wings set at 28-degrees sweepback with articulated flaps and nearly full-span automatic slats. The civil NA-265 prototype, initially powered by General Electric J-85 engines, completed its first flight on September 16, 1958, and though some sources refer to it as the XT-39, it never received a military designation. In January 1959 Air Material Command authorized USAF procurement of the NA-265 under the designation T-39A, which differed from the prototype mainly in having Pratt & Whitney J-60 engines and military avionics.

Through a series of contracts, the USAF accepted a total of 143 T-39As which commenced operational service with Air Training Command (ATC) in September 1961, and as deliveries continued, some T-39As were assigned to various USAF commands in a dual function as profi-

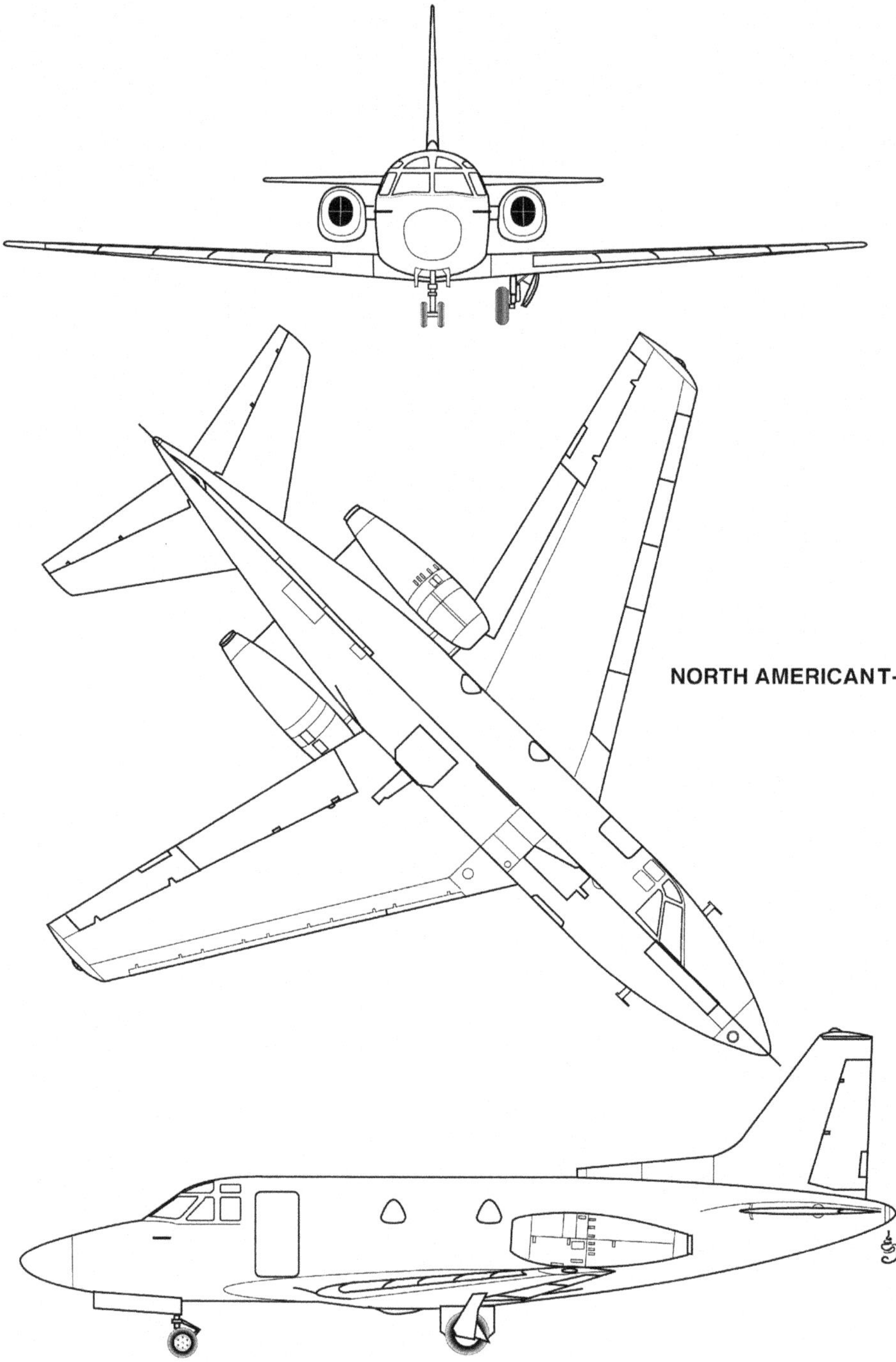

NORTH AMERICAN T-39D

ciency trainers and utility transports. The T-39As were followed by delivery of six T-39Bs outfitted with Republic F-105D radar and avionics systems for stations for three trainees. The USAF considered procuring a T-39C variant specialized for ECM and strategic bomber training, but none were ordered. During the Vietnam War three T-39As converted to electronic warfare trainers for F-105G "Wild Weasel" aircrews returned to service as the T-39F. As T-39As and Bs were

One of 143 North American T-39As that served with USAF Air Training Command from 1961 until the late 1970s. This example is seen in 1966 while visiting NAS Moffet Field, California.

The Navy T-39D, equipped to train either McDonnell F-4 radar intercept officers or Gumman A-6 bombardier-navigators, served with both VT-10 and VT-86 at NAS Pensacola, Florida, from 1963 to 1990.

released from training duties in the late 1970s, some were re-designated CT-39A/B after being refitted for transport duties. The last active CT-39s were withdrawn from USAF service during the mid–1980s.

Naval Variants: The Navy placed its first order for ten Sabreliners in mid–1961 as the T3J-1 but changed the designation to T-39D by the time the first aircraft were delivered in 1962, and through successive contracts, procured a total of 42 by the end of 1963. T-39Ds came equipped with either APQ-94 radars for training radar intercept officers in McDonnell F-4s or APQ-126 radars for bombardier-navigators in Grumman A-6s. As delivered, T-39Ds were assigned to VT-10 and VT-86 (known as "Sabrehawks"), both based at NAS Pensacola, Florida. Serial number records indicate that all T-39Ds were withdrawn from training duties between 1984 and 1990.

During the late 1960s the Navy acquired seven ex-civil Sabreliner model NA-265–40s (JT-12A-6 engines, each rated at 3,300 lbs./s.t.) that joined the inventory as utility transports under the designation CT-39E; then in 1990, the seven CT-39Es plus ten newly-acquired ex-civil NA-265–40s were remanufactured by Rockwell International and fitted with General Dynamics F-16-type radar systems to function as navigation and airborne intercept trainers under the designation T-39N. Serving with VT-86, four T-39Ns were destroyed during the course of service and the three remaining examples apparently withdrawn by the end of 2012. In yet another Sabreliner acquisition, starting in 1969, the Navy procured 12 off-the-shelf North American Rockwell NA-265–60s (three-foot two-inch fuselage stretch) that became operational as VIP transports under the designation CT-39G. In 1999, eight of the CT-39Gs were reassigned to VT-86, where, according to the U.S. Navy Fact File (see http://www.navy.mil/navydata/fact_display.asp?cid=1100&tid=1900&ct=1), they were used to train student pilots in non-tactical flight training. The current training status of the CT-39Gs is not clear, but a few may remain active.

Lockheed T-40 Jetstar—1958 (trainer variant of C-140, not ordered)

Northrop T-38 Talon—1959

Technical Specifications (T-38A)

Type: Two-place advanced trainer.
Manufacturer: Northrop Corp., Hawthorne, California
Total U.S. military versions: 1,187 (USAF).
Unit cost: $756,000.
Powerplant: Two General Electric J-85–5A turbojet engines, each rated at 2,050 lbs./s.t. dry, 3,850 lbs./s.t. with afterburning.
Performance: Max. speed 858 mph (Mach 1.3), cruise 578 mph; ceiling 50,000 ft.; range 1,140 mi.
Weights: 7,200 lbs. empty, 12,093 lbs. loaded.
Dimensions: Span 25 ft. 3 in., length 46 ft. 4.5 in., wing area 170 sq. ft.

The Northrop T-38 Talon presently holds the record as the longest-serving training aircraft in U.S. military history, having been in continuous USAF operational service since 1961. The origins of the T-38 are traceable to 1954, when Northrop began in-house development of a supersonic lightweight fighter design known as the model N-156. Though none of the U.S. military services expressed interest in the design, Northrop continued development of the project as the N-156F with the hope that it might be acquired at some future date for export under government-sponsored Military Assistance Programs (MAP). Almost coincidentally, in 1955, the USAF issued a General Operating Requirement calling for a supersonic advanced trainer to eventually replace T-33s in Air Training Command (ATC), and in response, Northrop tendered a design proposal for a stretched, two-seat trainer variant of the N-156F as the model N-156T.

Supersonic performance of the N-156T's comparatively small and lightweight airframe was

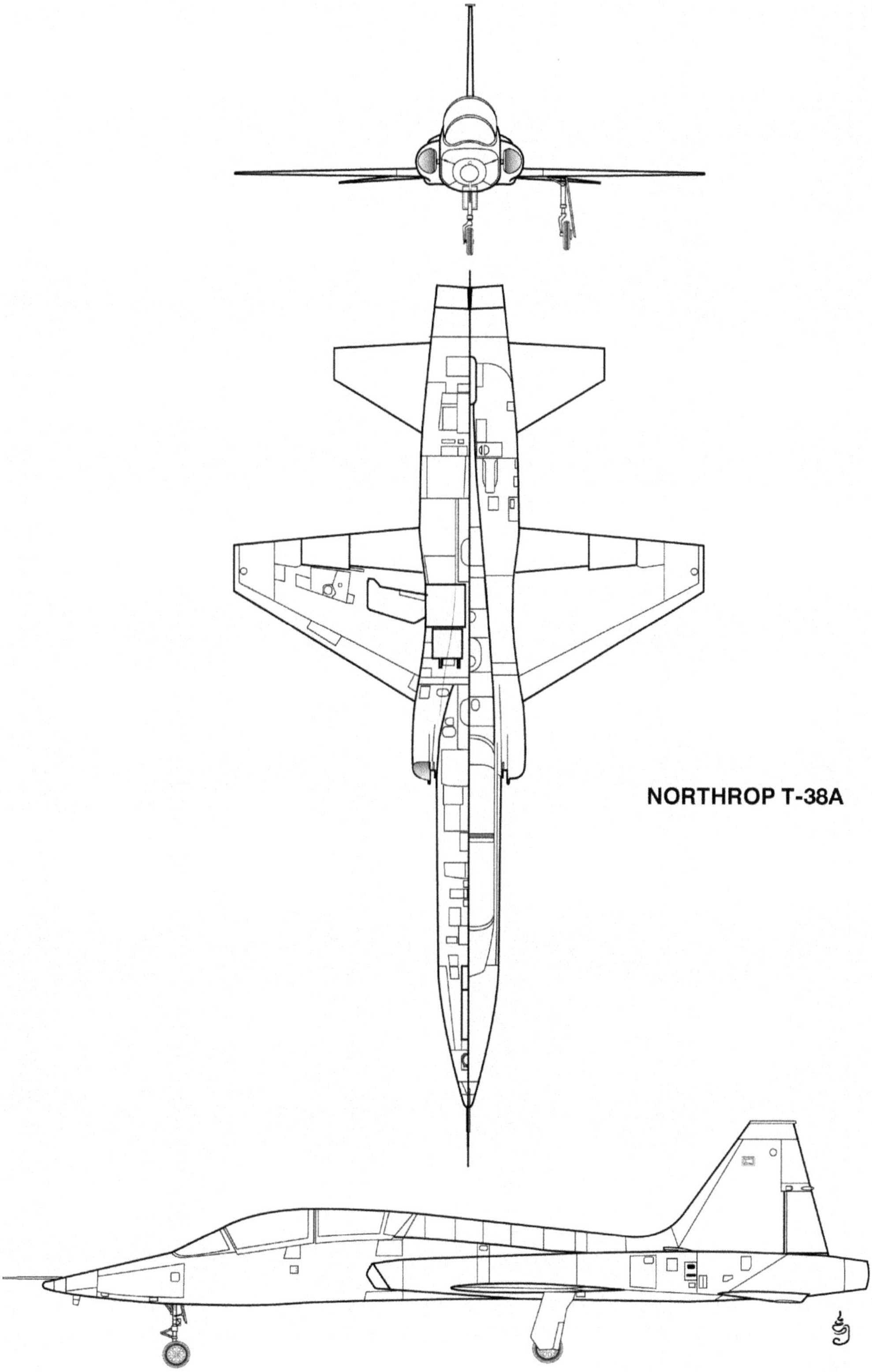

NORTHROP T-38A

One of the 1,185 Northrop T-38As delivered to the USAF between 1960 and 1972. This is the longest serving type of trainer in the U.S. armed forces to date. They served in eight Flying Training Wings at their peak.

An AT-38B with the legend "AT-38 Lead-In Fighter" painted on its side. The LIFT program was discontinued in 1993 and all AT-38Bs transferred to 12th Flying Training Wing, where they were integrated into UPT.

achieved by employing a slender fuselage cross-section in a tandem-seating configuration around the two small-diameter J-85 engines, together with low-aspect ratio, thin airfoil section wings similar in planform to the F-104. After reviewing Northrop's proposal against North American's competing TF-100C (see above), the USAF selection board concluded the N-156T not only offered lower life-cycle costs, but with its twin-engine layout and lower wing loading, potentially better safety margins in Undergraduate Pilot Training (UPT). Air Material Command (AMC) gave Northrop a development contract in December 1956 to build two flying prototypes plus one static test under the designation YT-38A. Although first YT-38A left the runway on March 10, 1959, the USAF subjected the prototypes and 32 more production T-38As to exhaustive testing and service trials before finally ordering the type into mass production in 1960. In a series of contracts awarded to Northrop over the next decade, the USAF accepted another 1,153 T-38As (1,187 altogether), the last examples being delivered in 1972.

As part of a program initiated by the USAF in 1975 under the heading of Lead-In Fighter Program (LIFT), approximately 116 T-38As became AT-38Bs when they received gun sights and a centerline belly pylon that could be rigged for bombs, rockets, or gun pods. LIFT was created to provide recent UPT graduates with a familiar aircraft for transition to tactical training; however, LIFT ended in 1993 and the tactical curriculum was incorporated into UPT. Since 1984 operational T-38As have undergone a three-phase Service Life Extension Program (SLEP) known as "Pacer Classic." The first phase involved replacement of ejection seats, fuselage longerons, landing gear and brakes, and flight controls, plus installation of a new impact resistant canopy. In Pacer Classic II, commenced in 2001, approximately 500 T-38As reemerged as T-38Cs after receiving all-new wings of high-strength alloy, replacement of fatigue-prone cockpit parts, new bird-proof polycarbonate windscreens, and modifications to improve takeoff thrust and engine durability. As of mid–2011, in Pacer Classic III, at least 125 T-38Cs have been earmarked to be

This T-38C seen taxiing in 2011 at Moody AFB, Georgia, where it is attached to the 479th Flying Training Group of AETC. The current inventory of 448 T-38Cs is slated to remain active until 2023.

retrofitted with new avionics that include digital flight displays (i.e., a "glass cockpit") and a heads-up display (HUD).

T-38As began replacing T-33As in the advanced phase of UPT in March 1961, and the process was completed in all eight of ATC's Flying Training Wings by the end of 1964. Since the early 1960s, NASA has operated a fleet of over 30 T-38As which have been used variously as astronaut pilot proficiency trainers, chase planes, and experimental testbeds. With the demise of the Shuttle program, NASA recently announced plans to reduce its Talon inventory to 16 aircraft by 2015. From 1974 to 1983, as a cost-savings measure, T-38As equipped the USAF Thunderbirds aerobatic display team. Strategic Air Command operated an unspecified number of T-38As as part of its "Accelerated Co-Pilot Enrichment Program" until its deactivation in 1991, and since that time, Air Combat Command and Air Force Global Strike Command have continued to employ Talons as proficiency trainers. During the early 1990s, in the midst of the force reductions brought on by the ending of the Cold War, many older T-38As were withdrawn from service and placed in storage, then in 1993, when ATC was merged into Air Education and Training Command (AETC), remaining T-38As were organized into three Flying Training Wings (FTW) at Columbus AFB, Mississippi (14th FTW), Vance AFB, Oklahoma (71st FTW), and Laughlin AFB, Texas (47th FTW). Today, AETC uses its Talons, primarily rebuilt T-38Cs, to prepare pilots destined to fly F-15s, F-16s, F-22s, F-35s, B-52s, B-1s, B-2s, and A-10s. As of early 2013, there were 508 Talons listed on the USAF inventory, consisting of 54 T-38As, 6 AT-38Bs, and 448 T-38Cs, and under current USAF plans, they will remain active until 2023.

McDonnell F/TF-101 Voodoo—1959

Technical Specifications (F-101F)

Type: Mission-capable two-place fighter transition and combat trainer.
Manufacturer: McDonnell Aircraft Corp., St. Louis, Missouri.
Total U.S. military versions: 79 (USAF).
Unit cost: $1,754,066.
Powerplants: Two Pratt & Whitney J-57-P-55 turbojet engines, each rated at 11,990 lbs./s.t. dry, 16,900 lbs./s.t. with afterburning.
Armament: Four AIM-4 *Falcon* guided missiles in an internal bay or two AIM-4s carried internally and two AIR-2A *Genie* rockets carried externally.
Performance: Max. speed 1,134 mph (Mach 1.7), cruise 545 mph; ceiling 58,400 ft.; range 1,930 mi.
Weights: 28,970 lbs. empty, 52,400 lbs. loaded.
Dimensions: Span 39 ft. 8 in., length 67 ft. 5 in., wing area 368 sq. ft.

The McDonnell F-101A Voodoo originated in 1952 as a one-seat, long-range fighter intended to escort Strategic Air Command bombers, but by the time it flew in 1954, its mission had evolved to that of a nuclear-armed tactical fighter or a fast photo-reconnaissance platform. Although overall performance was impressive, aerodynamic flaws—notably a dangerous pitch-up tendency at certain G-levels and angles of attack—delayed entry into operational service with Tactical Air Command (TAC) until mid–1957. Between 1955 and 1959, the USAF ultimately accepted delivery of 275 single-seat Voodoos, consisting of 77 F-101As, 47 improved F-101Cs, 35 RF-101As, and 116 RF-101Cs. Unexpected delays in the F-106 program (see above) combined with an urgent need for supersonic, all-weather interceptors in Air Defense Command (ADC) led the USAF to award McDonnell a contract in March 1955 for a two-seat interceptor version of the Voodoo under the designation F-101B, but due to the problems encountered during testing of the F-101A, production remained on hold until mid–1956. The first flight of the F-101B took

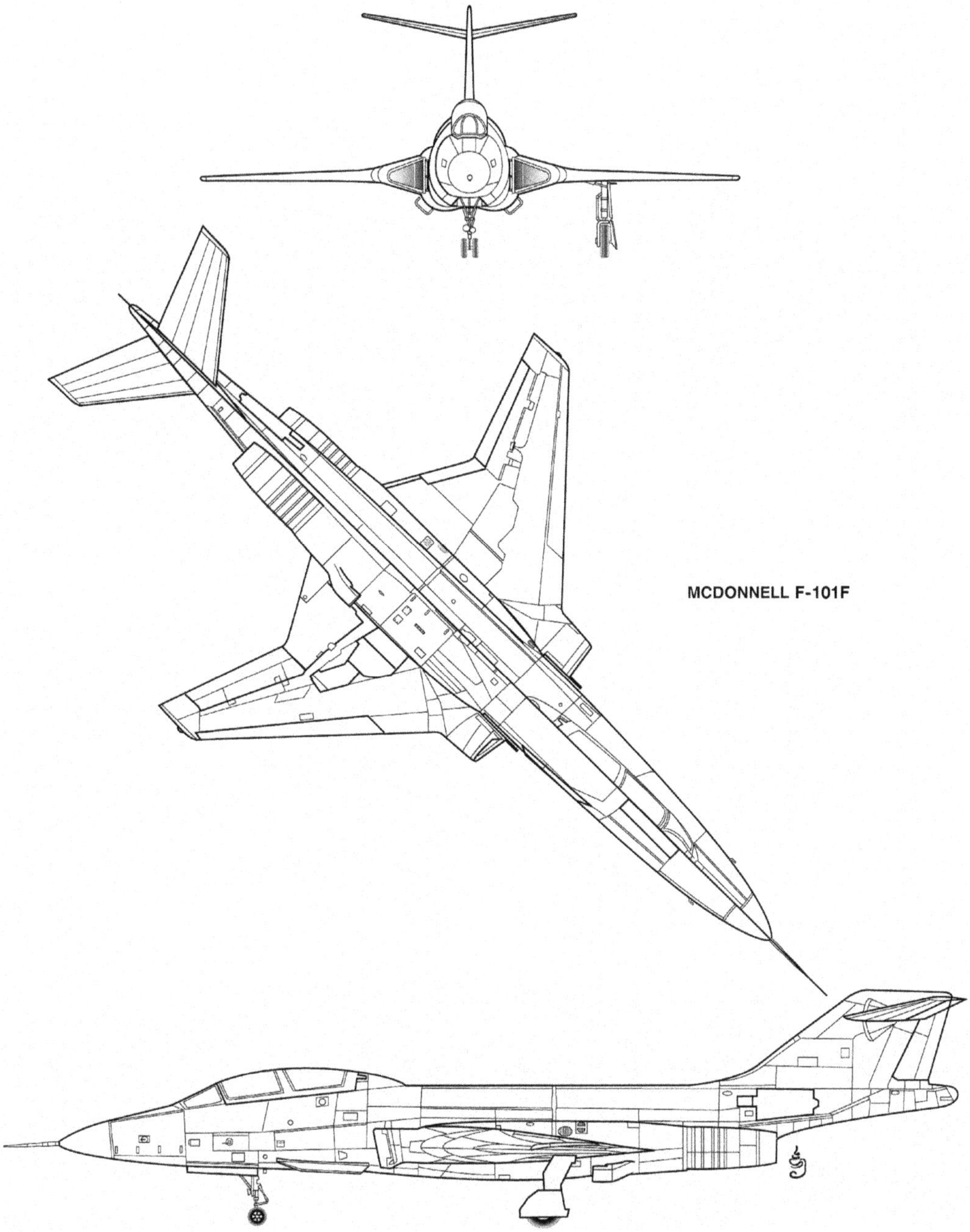

MCDONNELL F-101F

place on March 27, 1957, and the type began entering operational service with ADC in January 1959, with the last of 401 delivered in March 1961.

Originally, for transition and combat training purposes, one out of every four F-101Bs was to be equipped with dual controls as the TF-101B, but otherwise retain full combat capability, and in 1961, all 79 of the Bs produced with the dual-controls were re-designated F-101F. Although some F-101Fs were assigned directly to the 18 F-101B Fighter Interceptor Squadrons (FIS) operating within ADC, it appears that most were concentrated in Combat Crew Training Squadrons.

While the plans were originally delivered as TF-101Bs, all dual-control variants were later redesignated F-101F. Unlike other two-seat fighters allocated to operational squadrons, most F-101Fs served with Combat Crew Training Squadrons.

F-101Fs were also allocated to TAC training units for pilots converting to the RF-101A/C. Starting in 1969, as F-101Bs and Fs left active USAF units, they continued in ANG service, ultimately equipping squadrons in seven states until the last examples were withdrawn from the 111th FIS, Texas ANG, in 1982. Ten F-101Fs were supplied to the Royal Canadian Air Force along with 132 F-101Bs, where they served until 1987.

Convair TB-58 Hustler—1960

Technical Specifications (TB-58A)

Type: Three-place bomber conversion trainer.
Manufacturer: Convair Aircraft Div. of General Dynamics Corp., Fort Worth, Texas.
Total U.S. military versions: 8 (USAF).
Unit cost (B-58A): $12,440,000.
Powerplants: Four General Electric J-79-GE-5 turbojet engines, each rated at 10,500 lbs./s.t. dry, 15,600 lbs./s.t. with afterburning.
Performance: Max. speed 1,325 mph (Mach 2.03), cruise 610 mph; ceiling 64,800 ft.; range 4,400 mi.
Weights: 55,560 lbs. empty, 163,000 lbs. loaded.
Dimensions: Span 56 ft. 10 in., length 96 ft. 10 in., wing area 1,542.5 sq. ft.

Although its operational career was relatively brief (1961–1970), the Mach 2 Convair B-58 Hustler still holds the distinction of having been the first operational supersonic bomber in the world. Development of what became the B-58 began in 1949 as Weapons System 102A, and the process of bringing it from a concept on a drawing board to a flying prototype took over seven years, with the first XB-58 prototype flying on November 11, 1956. Testing comprised another five years, so that the first production B-58As were not declared fully operational until early 1961. The USAF ultimately accepted 116 Hustlers, two XB-58s, 11 YB-58A pre-production aircraft, 86 B-58As, and 17 RB-58As which differed mainly in having a ventral reconnaissance pod. Although the B-58 turned into a very expensive proposition both in terms of procurement

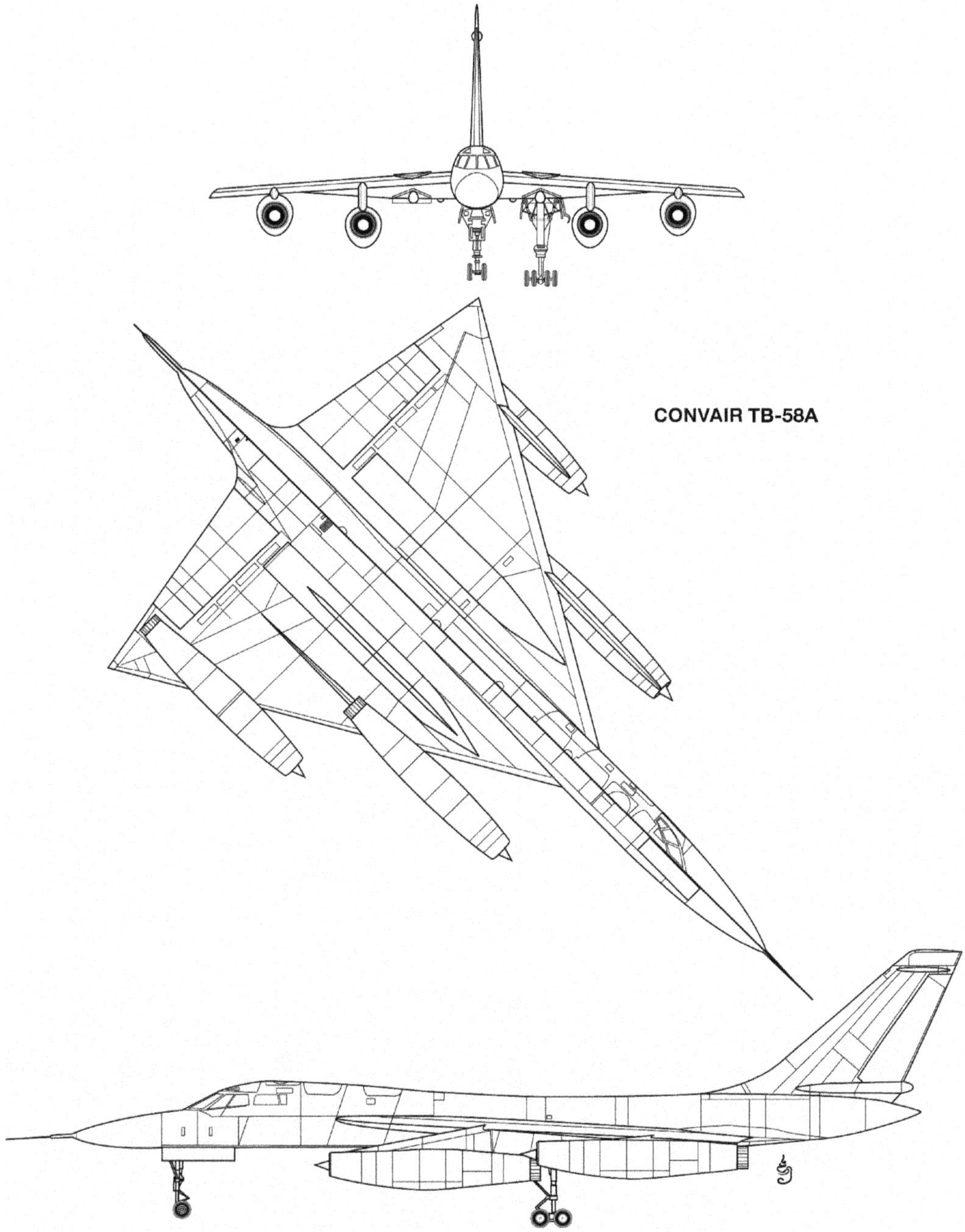

and maintenance support, the brevity of its career was attributable mainly to the advent of radar-directed surface-to-air missiles that systematically rendered the high-speed, high altitude bomber mission obsolete.

Because the B-58A's tandem cockpit arrangement allowed no access between the three crew positions, training pilots new to the airplane presented a difficult problem. In early 1959, rather than build new airframes to fulfill the training requirement, the Air Force Council directed Convair to convert nine of the YB-58As to a trainer configuration under the designation TB-58A.

The eight Convair TB-58As built in 1960 and 1961 were split between the 43rd BW at Carswell AFB, Texas, and the 305th BW at Bunker Hill AFB, Indiana. Note larger side window in second cockpit where the trainee sat.

As modified, the instructor-pilot sat in the TB-58A's forward station while the second cockpit, normally occupied by the navigator, received a second set of interconnected flight controls for a pilot-trainee, and to afford better visibility, enlarged windows were added to both sides and the seat swiveled 10-degrees to the right. The aft cockpit, normally occupied by the defensive systems operator, received a crawl space connecting it to the second cockpit, and when manned by a second pilot-trainee, it was possible (though seldom done) for pilot-trainees to change positions. All tactical equipment—navigation, bombing, and defensive countermeasures systems—was removed. One of the earmarked YB-58As was destroyed prior to conversion, but the remaining eight received the TB-58A modification, the first flying on May 8, 1960.

As deliveries proceeded, four TB-58As were assigned to combat crew training duties in the 43rd Bombarment Wing at Carswell AFB, Texas, and four to the 305th Bombardment Wing at Bunker Hill (later Grissom) AFB, Indiana. Typically, the B-58 pilot training syllabus started with familiarization flights in a TF-102A (see above), followed by a least three flights in a TB-58A. They were all withdrawn when the USAF ceased B-58 operations in January 1970.

Lockheed U/TU-2 and TR-1–1962

Technical Specifications (U-2CT [TU-2S])

Type: Two-place high-altitude reconnaissance transition and proficiency trainer.
Manufacturer: Lockheed Aircraft Corp. (Now Lockheed-Martin), Burbank, California.
Total U.S. military versions (trainers, estimated): 8 (USAF).
Unit cost: Unknown.
Powerplant: One Pratt & Whitney J-75-P-13B turbojet engine, rated at 17,000 lbs./s.t. lbs./s.t. [one General Electric F-118–101 turbofan engine, rated at 19,000 lbs./s/t/].
Performance: Max. speed 460 mph [500 mph]; ceiling 70,000 ft. [90,000 ft.]; range 2,880 mi. [4.600 mi.].

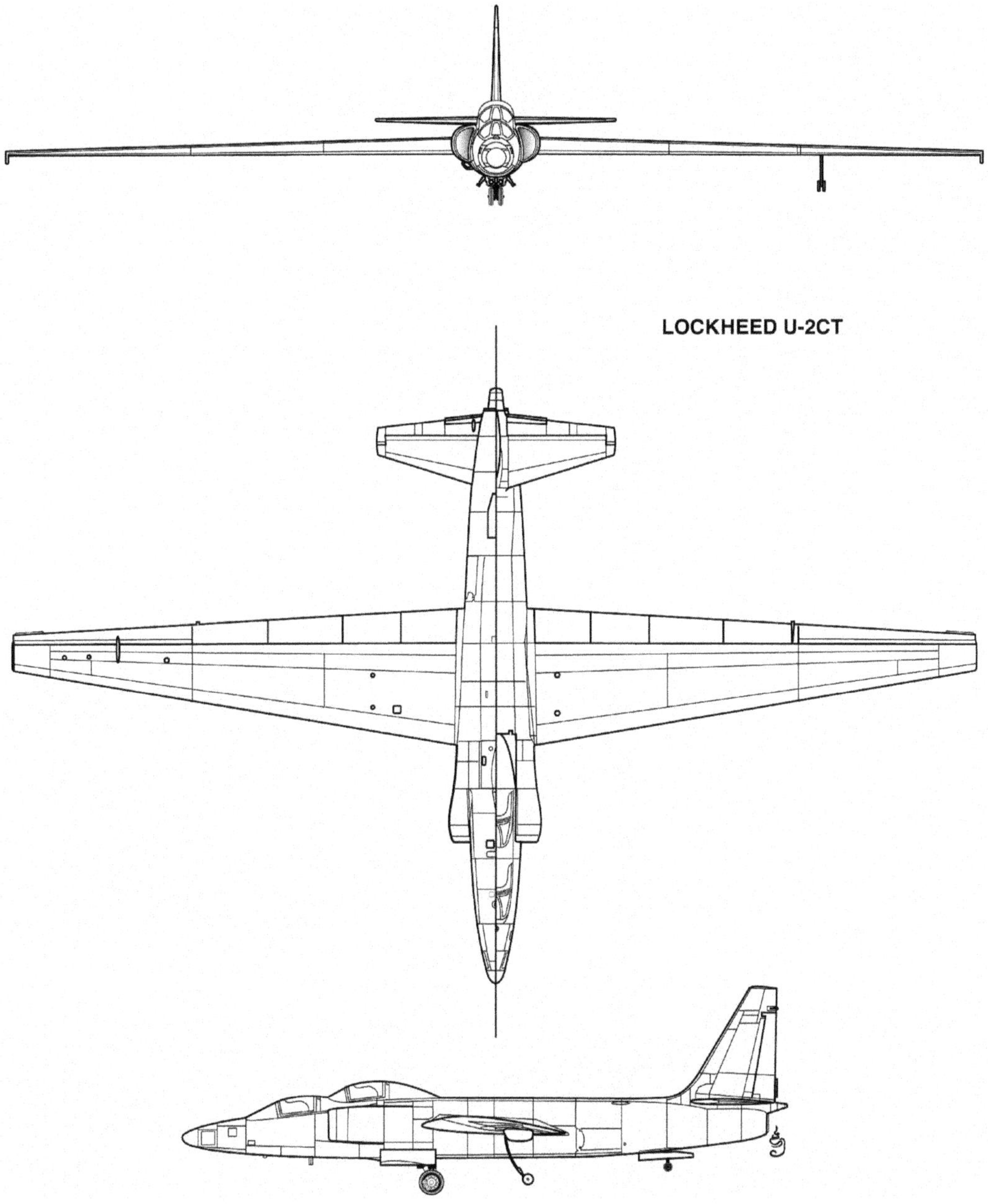

Weights: 13,071 lbs. [classified] empty, 22,542 lbs. [40,000 lbs.] max. takeoff.
Dimensions: Span 80 ft. [103 ft.], length 50 ft. [63 ft.], wing area 600 [1,000] sq. ft.

The well-known Lockheed U-2, specifically designed to fly reconnaissance missions at altitudes above enemy air defenses, completed it first flight in August 1955 and began over-flight missions in 1956. Because of factors inherent to the design—e.g., a critical envelope between stall and never-exceed speeds at operational altitude and takeoffs and landings on a single-bogie main landing gear—the U-2 was an extremely difficult aircraft to fly. Serial number records reflect that 103 U-2s have been built in two major variants: 54 U-2A/B/Ds (including the U-2 prototype) between 1955 and 1957 and 37 U-2R/ER-2/TR-1s between 1967 and 1981. Although the five

One of at least two U-2s converted into a dual-control configuration as the U-2CT. This was one of the most difficult aircraft to fly because of factors inherent to its design.

A TR-1B/TU-2S on display during the 2008 EAA Air Venture in Oshkosh, Wisconsin. The aircraft is reportedly attached to the 4029th Strategic Reconnaissance Training Wing at Beale AFB, California.

U-2Ds came with a second seat for an observer, they did not function as trainers. The actual dates are not published anywhere and authorities conflict on exact numbers built, but during the late 1950s or early 1960s, under the new designation U-2CT, at least one U-2B and one U-2D are believed to have been converted to a dual-control trainer configuration by adding a second cockpit above and behind the first. When the U-2R was introduced in 1967 with a lengthened fuselage, extended wingspan, wing pods to house new components, and many equipment and avionics refinements, one of twelve built was reportedly completed as a two-seat U-2RT trainer having a cockpit arrangement similar to that of the U-2CT. Later still, in 1980–81, Lockheed produced 35 tactical reconnaissance variants aerodynamically similar to the U-2R, comprising 33 single-seat TR-1As and three two-seat TR-1B trainers (one of which may have been derived from the single U-2RT airframe). In 1994, after undergoing a life-extension program and upgrade to F-118–101 engines, the TR-1s were re-designated U-2S and TU-2S, respectively. Today the three TU-2Ss are thought to be active with the 4029th Strategic Reconnaissance Training Squadron at Beale AFB, California.

Republic F-105 Thunderchief—1963

Technical Specifications (F-105F)

Type: Mission-capable two-place fighter transition and combat trainer.
Manufacturer: Republic Aviation Corp., Farmingdale, New York.
Total U.S. military versions: 143 (USAF).
Unit cost: $2,200,000.
Powerplant: One Pratt & Whitney J-75-P-19W turbojet engine, rated at 17,200 lbs./s.t. dry, 26,500 lbs./s.t. with afterburning.
Armament: One 20-mm M61A1 rotary cannon in the nose, up to 8,000 lbs. of bombs in an internal bay, and up to 6,000 lbs. of ordinance carried on five external stations.
Performance: Max. speed 1,386 mph (Mach 2.1), cruise 596 mph; ceiling 52,000 ft.; range 2,070 mi.
Weights: 28,655 lbs. empty, 54,580 lbs. loaded.
Dimensions: Span 34 ft. 11 in., length 69 ft. 4 in., wing area 385 sq. ft.

Unidentified Republic F-105F in Vietnam-era markings, but according to serial number records, may be the aircraft now on display at Paris–Le Bourget Airport in France.

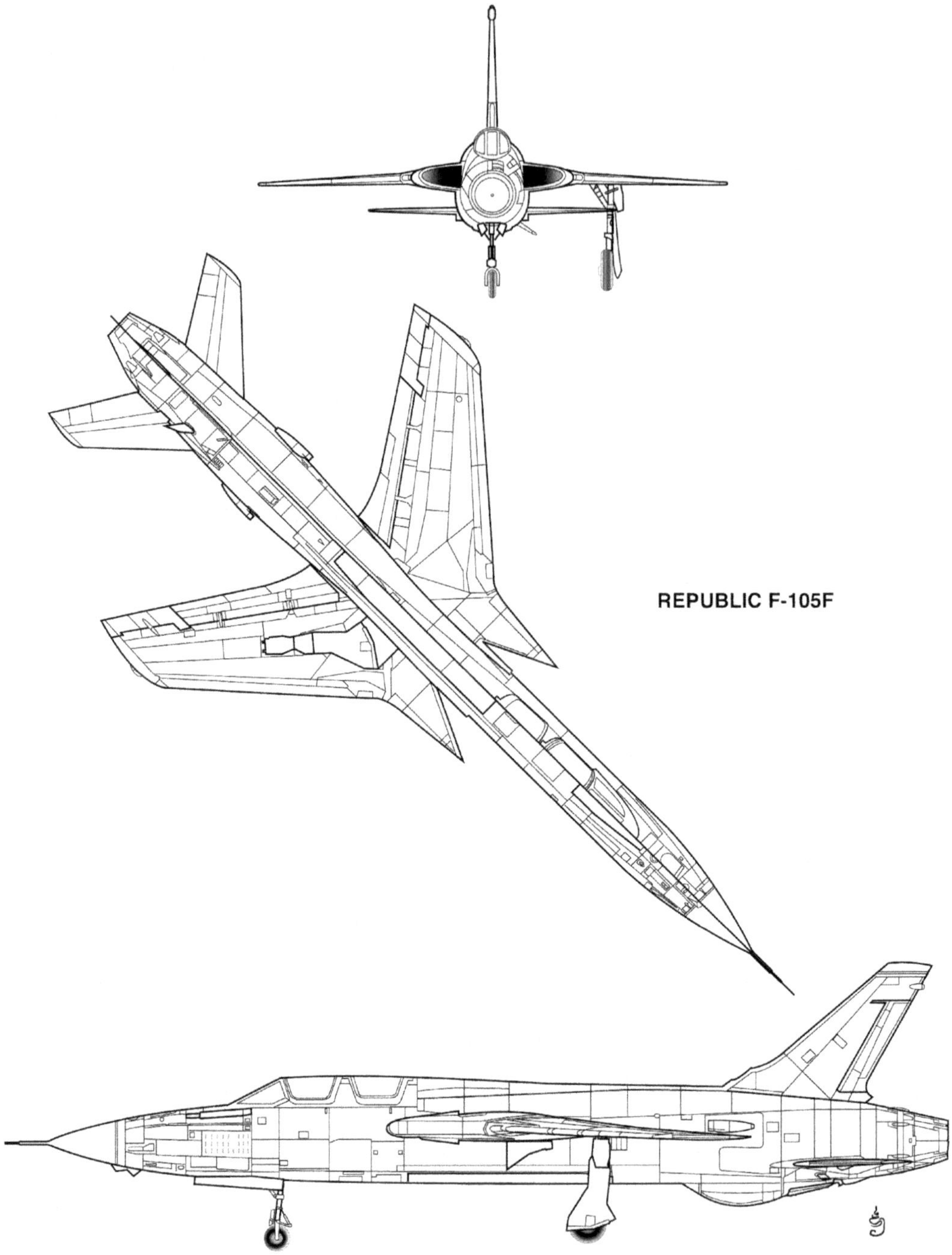

REPUBLIC F-105F

Remembered by its aircrews as the "Thud," Republic's F-105 Thunderchief was the first supersonic USAF aircraft designed from the ground up as a dedicated fighter-bomber, and when it became operational in 1958, the fastest military aircraft in the world at sea level. From its beginnings in 1952 as the model AP-63, the design was optimized for high-speed, low-level penetration in the tactical nuclear strike role. Although the two-seat F-105F did not fly until 1963, it appears here because of its organic connection with the Century Series fighters of the 1947–

1962 era. Two earlier two-seat F-105 variants had been studied, the F-105C in 1957 and the F-105E in 1959, but both were deemed impractical for transition and combat training on the basis of overall program cost; however, as all-weather F-105Ds began entering service in 1961, operational experience demonstrated the need for advanced bombing and navigational training of pilots, and as an interim measure, six T-39Bs equipped with F-105D radar and avionics (see above) were delivered to Tactical Air Command (TAC) during 1962. As a long-term solution, the Defense Department awarded Republic a supplemental contract in May 1962 to convert the last 143 F-105Ds into a two-seat configuration under the designation F-105F. Modifications to the basic F-105D airframe entailed a five-foot one-inch fuselage stretch to house a second, identical cockpit behind the first and adding six inches to the height of the tail fin. Other than being 1,800 lbs. heavier and having slightly reduced range, the F-105F was identical to the D in terms of performance and combat capability.

The maiden flight of the first F-105F took place on June 11, 1963, deliveries of the first production aircraft to the USAF began the following December, and the type was declared operational later the same month with the 4th Tactical Fighter Wing (TFW) at Seymour-Johnson AFB in North Carolina. As deliveries continued through January 1965, F-105Fs were distributed between the six TFWs operating the F-105D, plus the 4520th Combat Crew Training Wing at Nellis AFB, Nevada. In January 1966, after F-100Fs (see above) had proven the viability of the "Wild Weasel" concept over North Vietnam (i.e., detecting and attacking enemy surface-to-air missile sites), 85 F-105Fs were selected for a modification program that included installation of radar homing and warning receivers (RHAW) plus the plumbing needed to carry AGM-45A *Shrike* antiradar missiles. The first modified F-105Fs began flying Wild Weasel missions in the combat theater during the summer of 1966, and all 85 had been modified by early 1968. In Wild Weasel III, commenced sometime in 1967, another 45 to 50 F-105Fs received upgraded RHAW systems, internally-mounted self-protection jammers housed in bulged fairings along the lower fuselage, and the hardware needed to carry AGM-78A/B *Standard ARM* missiles. As they returned to service, these aircraft were initially carried as an EF-105F but later re-designated F-105G. Although all F-105Ds were withdrawn from the combat theater during 1970, the F-105Fs and Gs remained there until hostilities ended. Since some of the modified F-105Fs are known to have received the G upgrade, the exact number of F-105Fs modified for the Wild Weasel mission is not clear. All F-105Fs left active USAF service in 1972 and the last F-105Gs in 1980. F-105Fs and Gs continued in service with two USAF reserve squadrons until 1982 and with two ANG squadrons until 1983.

B. USN and USMC

Synopsis of USN and USMC Trainer Procurement, 1946 to 1962

Primary and Basic Trainers. Like the USAAF, the Navy and Marine Corps underwent drastic force reductions following the end of World War II. In the immediate post-war period, the much-reduced Naval Air Training establishment retained Boeing-Stearman N2Ss (AAF PT-13/-17) for primary pilot training, however, all Vultee SNVs (AAF BT-13) were withdrawn and North American SNJs (AAF AT-6) used in a combined basic and advanced syllabus, and starting in the late 1940s, the N2Ss were likewise withdrawn and SNJs used for the entire syllabus. Although the Navy tested new types of primary and basic trainers in 1946 and 1947, specifically the Fairchild XNQ-1 and North American XSN2J-1, restraints on naval spending at the time prevented either type from being placed in production. No new types of primary or basic trainers were procured until 1952 and 1953, when the Navy evaluated USAF versions of the North American T-28A and Beech T-34A. Both types were subsequently ordered into production under specifications revised to meet naval requirements: as basic trainers, 489 T-28Bs delivered between 1953 and 1955; and as primary trainers, 423 T-34Bs delivered between 1954 and 1957. For carrier qualification in advanced units, BuAer also procured 299 T-28Cs with arresting gear delivered in 1956 and 1957, however, as jet aircraft (e.g., Lockheed T2V-1 and North American T2J-1) began taking their place, T-28Cs were reassigned to basic training units alongside T-28Bs. Both T-28B/Cs and T-34Bs remained active in their respective training roles until the mid–1970s.

Advanced Trainers. In smaller numbers, Beech SNB-1s and -2s (AAF AT-11 and -7) continued postwar for multi-engine, navigator, bombardier, and aircrew training. In 1947, BuAer awarded Beech a contract to rebuild and modernize 117 SNB-1s that returned to service as SNB-4 multi-engine and instrument proficiency trainers. Likewise, the Naval Air Training establishment retained reduced numbers of Douglas R4D-5Ts, -6Ts, and -7Ts (AAF C-47A/B and TC-47B) for training purposes. In 1948, to prepare a new generation of Navy and Marine jet pilots, BuAer acquired 50 ex–USAF single-seat Lockheed F-80Cs that were placed on the inventory as the TO-1. The largest advanced trainer contract since the war came in 1949 when BuAer procured the first of 699 TO/TV-2s (USAF T-33A) which remained active as advanced trainers until the early 1960s and as instrument proficiency trainers in reserve units until 1976. In 1951 an unspecified number of Beech JRB-3/4s and SNB-2s reemerged as SNB-5s after being completely rebuilt and modernized, and some, specialized for photographic training, became SNB-5Ps. Advanced Training Units (ATU), established in the early 1950s, utilized many types of operational aircraft released from frontline duties, namely: Grumman F6Fs, Grumman F8Fs, Vought F4Us, Douglas ADs, Grumman TBMs, Grumman F9Fs (both straight-wing Panthers and swept-wing Cougars), Douglas F3Ds, Douglas R4Ds, Douglas R5Ds, Douglas A3Ds, Grumman UF-1s, Consolidated PBYs, Martin PBMs, Martin P5Ms, Lockheed P2Vs, Grumman S2Fs, and some, after being modified, received trainer suffixes (e.g., UF-1T, R4D-8T, F3D-2T, P2V-6T, A3D-2T, S2F-1T, and P5M-1T and -2T). In 1954, to follow TV-2s in advanced jet training with a carrier-capable aircraft, BuAer procured 150 Lockheed T2V-1s, though they were replaced after brief service by the 217 North American T2J-1s that entered service between 1959 and 1961. To provide dual transition and combat training alongside single-seat aircraft in ATUs, the Navy contracted for 400 Grumman F9F-8Ts that arrived for duty from 1956 to 1960. The Navy tested a two-seat Crusader, the F8U-1T, from 1962 to 1964 but elected not to place it into production.

Fairchild NQ (T-31)—1946

Technical Specifications (XNQ-1)

Type: Two-place primary and basic trainer.
Manufacturer: Fairchild Aircraft & Engine Corp., Hagerstown, Maryland.
Total U.S. military versions: 2 (2 USN and 1 ex–USN to USAF).
Unit cost: Unknown.
Powerplant: One 320-hp Lycoming R-680–13 9-cylinder air-cooled radial or 350-hp Lycoming GSO-480 8-cylinder horizontally-opposed engine driving a two-bladed constant-speed metal propeller.
Performance: Max. speed 175 mph, cruise 166 mph; ceiling 16,000 ft.; range 955 mi.

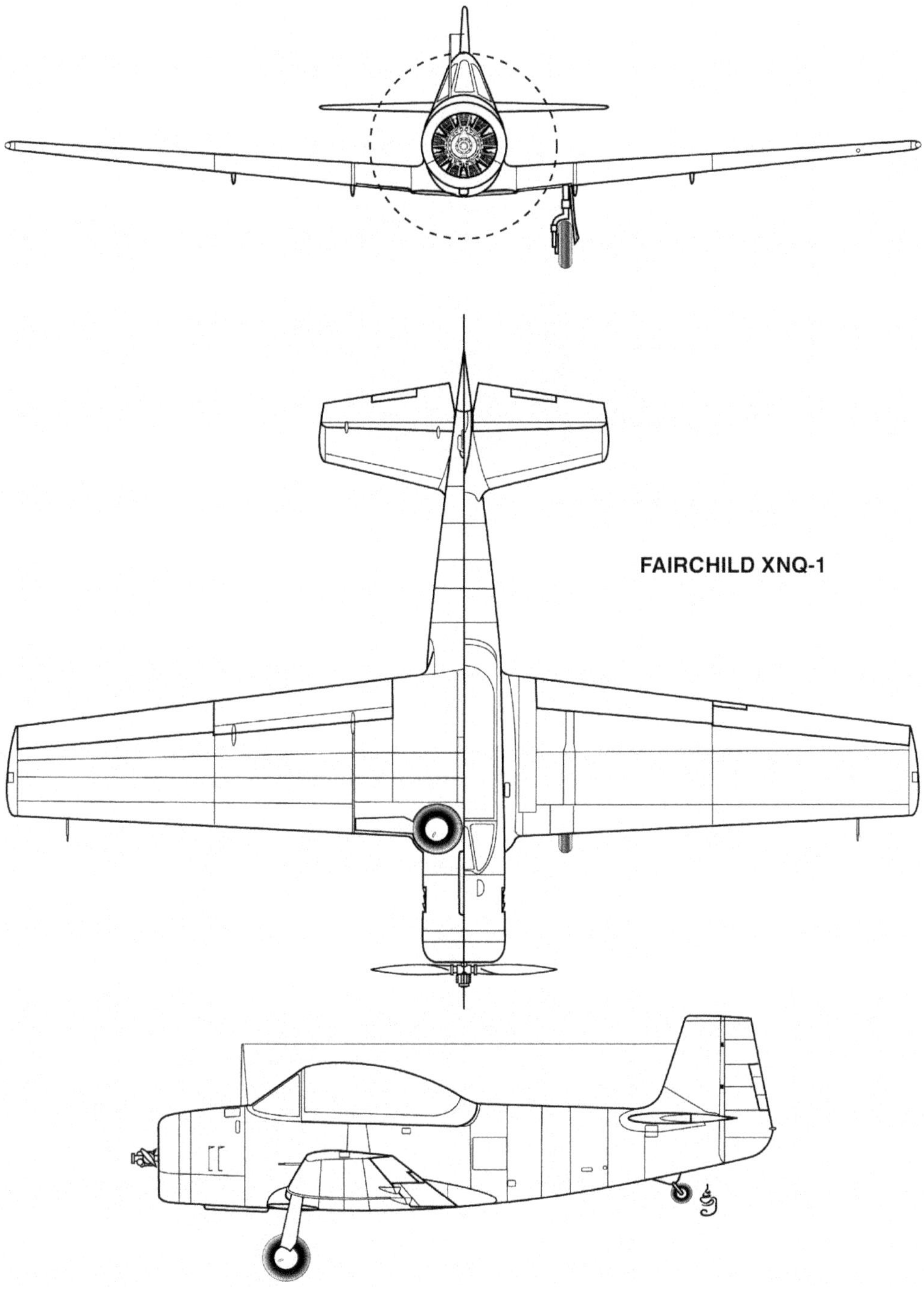

Weights: 2,974 lbs. empty, 3,898 lbs. loaded.
Dimensions: Span 41 ft. 6 in., length 27 ft. 3 in., wing area 236 sq. ft.

In April 1945 the Naval Bureau of Aeronautics (BuAer) issued requirements for a new type of trainer that would replace existing aircraft in both the primary (e.g., N3N and N2S) and basic phases (e.g., SNV and SNJ) of pilot training. Shortly thereafter, Fairchild received a contract to

One of the two Fairchild XNQ-1 prototypes delivered to NAS Anacostia, Maryland, in late 1946. Although the aircraft performed well, naval officials decided against production. Also tested by the USAF in 1949.

deliver two flying prototypes of its proposed model M-92 under the assigned naval designation XNQ-1. Conceived by Fairchild's chief engineer, Armand Thiebold, the XNQ-1 emerged with an exceptionally clean, all-metal airframe having a fully retractable main landing gear in a conventional (tailwheel) arrangement. Student and instructor, seated in tandem, were afforded excellent visibility under a frameless bubble canopy. The first XNQ-1 flew from the factory on October 7, 1946, and both prototypes were delivered to the Navy at NAS Anacostia, Maryland, for official trials during 1947. Although the XNQ-1 demonstrated good overall performance for its power and weight and excellent maneuverability, naval officials declined to place the type into production. The first prototype, after being repowered with a 350-hp Lycoming GSO-580 engine, was test-flown as the XNQ-1A but destroyed in a crash sometime in 1950. Ultimately, in 1953, BuAer followed the USAF's lead and ended up procuring the Beech T-34 in quanity.

USAF Variants: The second XNQ-1 prototype, still powered by an R-680 but refitted with a larger horizontal stabilizer, was transferred to the USAF in early 1949 to be evaluated as a primary trainer under the designation YT-31 along with the Beech YT-34 and the Temco YT-35. On March 1949, following brief trials, the USAF board selected the YT-31 for production and gave Fairchild a provisional order for 100 aircraft as the T-31A; then in late 1949, before Fairchild had tooled-up for production, the selection board cancelled the contract in favor of the Beech YT-34. The YT-31 was ultimately sold surplus and placed on the civil registry.

North American SN2J—1947

Technical Specifications (XSN2J-1)

Type: Two-place primary and basic trainer.
Manufacturer: North American Aviation, Inc., Inglewood, California.
Total U.S. military versions: 2 (USN).

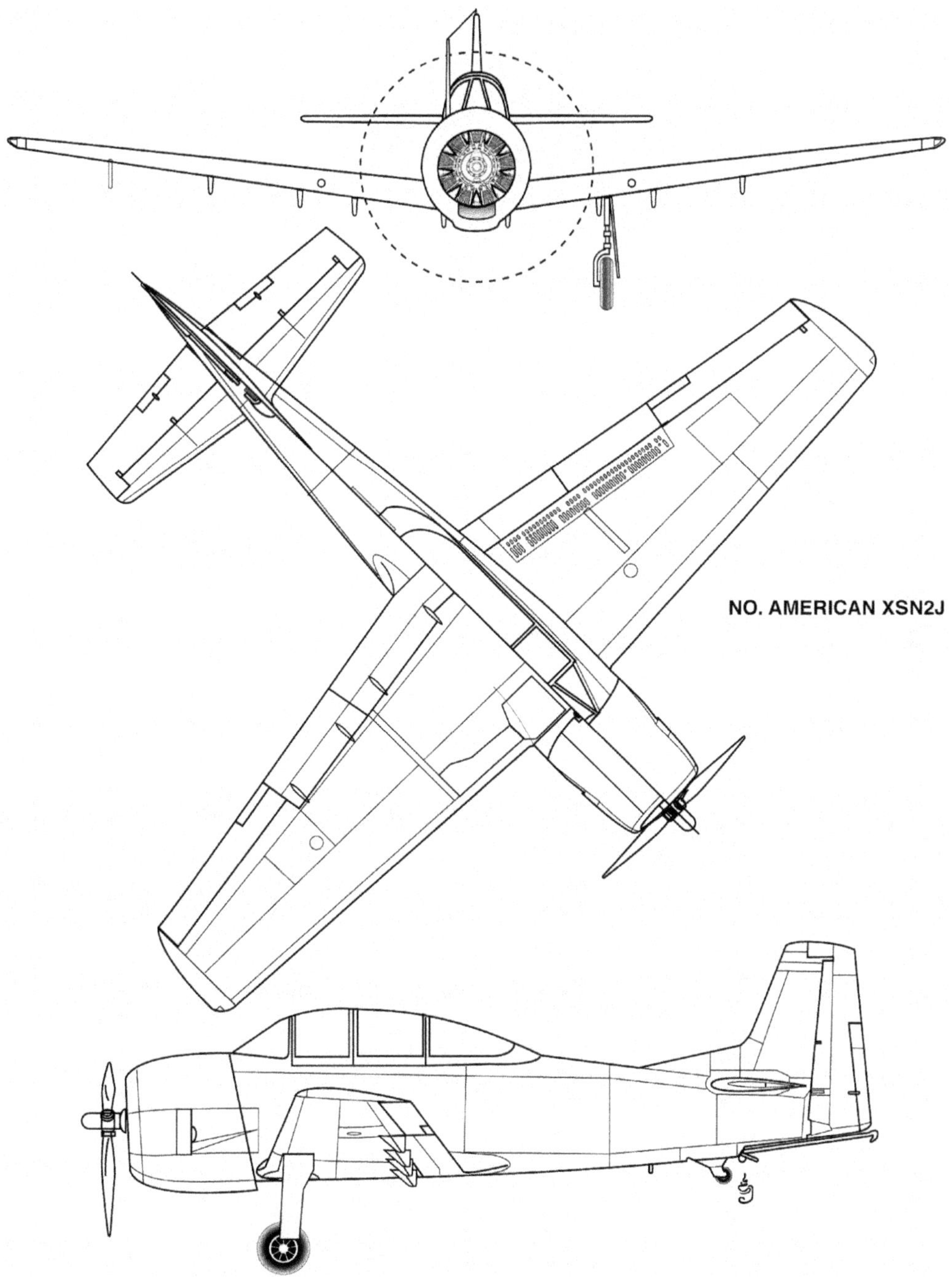

One of two North American XSN2J-1 prototypes as seen on the ramp at North American's plant in early 1947. Although budgetary restraints prevented procurement, the design soon reemerged as the T-28.

Unit cost: Unknown.
Powerplant: One 1,100-hp Wright R-1820–78 9-cylinder air-cooled radial engine driving a three-bladed constant-speed metal propeller.
Performance: Max. speed 270 mph, cruise 190 mph; ceiling 30,000 ft.; range 1,600 mi.
Weights (est.): 5,500 lbs. empty, 7,500 lbs. loaded.
Dimensions (est.): Span 41 ft., length 32 ft., wing area 236 sq. ft.

The North American model NA-142, coming as the second proposal in response to the Naval Bureau of Aeronautics (BuAer) 1945 requirement for a new primary and basic trainer, differed from Fairchild's competing XNQ-1 in offering a much heavier aircraft, more comparable to the SNJ in handling, but promising much better performance as a result of more advanced aerodynamic design. In 1946 BuAer awarded North American a contract for two aircraft to be delivered under the naval designation XSN2J-1, and the first prototype completed its maiden flight on February 10, 1947. In general design the XSN2J-1 appeared with conventional (tailwheel) landing gear in which the main struts retracted into the wing while the tailwheel remained fixed. Student and instructor sat in tandem under a two part bubble canopy, and to permit carrier approach speeds similar to the SNJ, the wing design combined slotted trailing edge flaps with cheese grater-type dive brakes just forward of the flaps. Although the XSN2J-1s met or exceeded requirements, BuAer was at the time constrained from placing the type in production due to restrictions on naval appropriations, and all development was discontinued in 1948. While the XSN2J-1s were still undergoing testing in 1947, North American was simultaneously working on the model NA-159 for the USAAF under the designation XBT-28 (XT-28 in 1948; see in Series II, Part A). Though similar in appearance to the XSN2J-1, the XT-28 was in fact a completely different design having a tricycle landing gear configuration. In 1953, four years after the USAF ordered the XT-28 into production as the T-28A, BuAer ordered 489 navalized variants as the T-28B, followed in 1955 by 299 T-28Cs equipped with arrestor hooks.

Lockheed TO/TV Shooting Star—1949 (See Lockheed T-33 in Series II, Part A)

North American T-28 Trojan—1953 (See North American T-28 in Series II, Part A)

Beech T-34 Mentor—1953 (See Beech T-34 in Series II, Part A)

Lockheed T2V/T-1 Seastar—1953

Technical Specifications (T2V-1/T-1A)

Type: Carrier-capable basic and advanced trainer.
Manufacturer: Lockheed Aircraft Corp., Inglewood, California.
Total U.S. military versions: 150 (USN).
Unit cost: Unknown.
Powerplant: One Allison J-33-A-24 turbojet engine, rated at 6,100 lbs./s.t.
Performance: Max. speed 580 mph, cruise 370 mph`; ceiling 40,000 ft.; range 973 mi.
Weights: 11,965 lbs. empty, 15,800 lbs. loaded.
Dimensions: Span 42 ft. 10 in., length 38 ft. 6.5 in., wing area 240 sq. ft.

In 1952, as a purely private venture, Lockheed initiated work on the model L-245 to develop its highly successful T-33A/TV-2 design (see in Series II, Part A, above) into a trainer suitable for carrier takeoffs and landings. Besides strengthening the airframe and landing gear and adding catapult fittings and an arrestor hook, alterations to the basic design included modifying the fuselage mid-section to elevate the rear cockpit for improved visibility, reshaping the side-intakes for greater efficiency at lower speeds and higher angles-of-attack, and most notably, to enhance low-airspeed performance at high angles-of-attack, a wing of three-feet eleven-inches greater span that incorporated leading-edge flaps and "blown" trailing-edge flaps (the first Navy aircraft to possess this innovation). With this flap arrangement, bleed-air produced by the engine compressor was piped across the flaps to delay boundary layer separation and thus increase the lift coefficient at lower airspeeds.

Lockheed's L-245 demonstrator (civil registry no. N125D) flew for the first time on December 15, 1953, and was subsequently loaned to the Navy for testing and evaluations. In May 1954 the Naval Bureau of Aeronautics (BuAer) was sufficiently impressed to award Lockheed an initial contract to deliver eight aircraft under the designation T2V-1. Meanwhile, further trials with the L-245 at NTC China Lake, California, while indicating the aircraft was generally suitable for carrier operations, also demonstrated it would need further airframe strengthening and a larger fin fillet to improve directional stability, and Lockheed subsequently incorporated the improvements into the T2V-1 production standard. Following completion of successful carrier trials of the first T2V-1 in 1956 aboard the USS *Antietam*, BuAer awarded Lockheed a series of production contracts that resulted in delivery of a further 142 T2V-1s in 1956 and 1957. However, Navy plans to acquire another 240 T2V-1s were cancelled in favor of a decision to procure the North American T2J-1 in quantity (see below).

When the type became fully operational with the Naval Air Training establishment in May 1957, all T2V-1s were detailed to serve with Basic Training Group Nine (BTG-9) based at NAS Pensacola, Florida. Although the Seastars proved to be docile in terms of carrier deck handling,

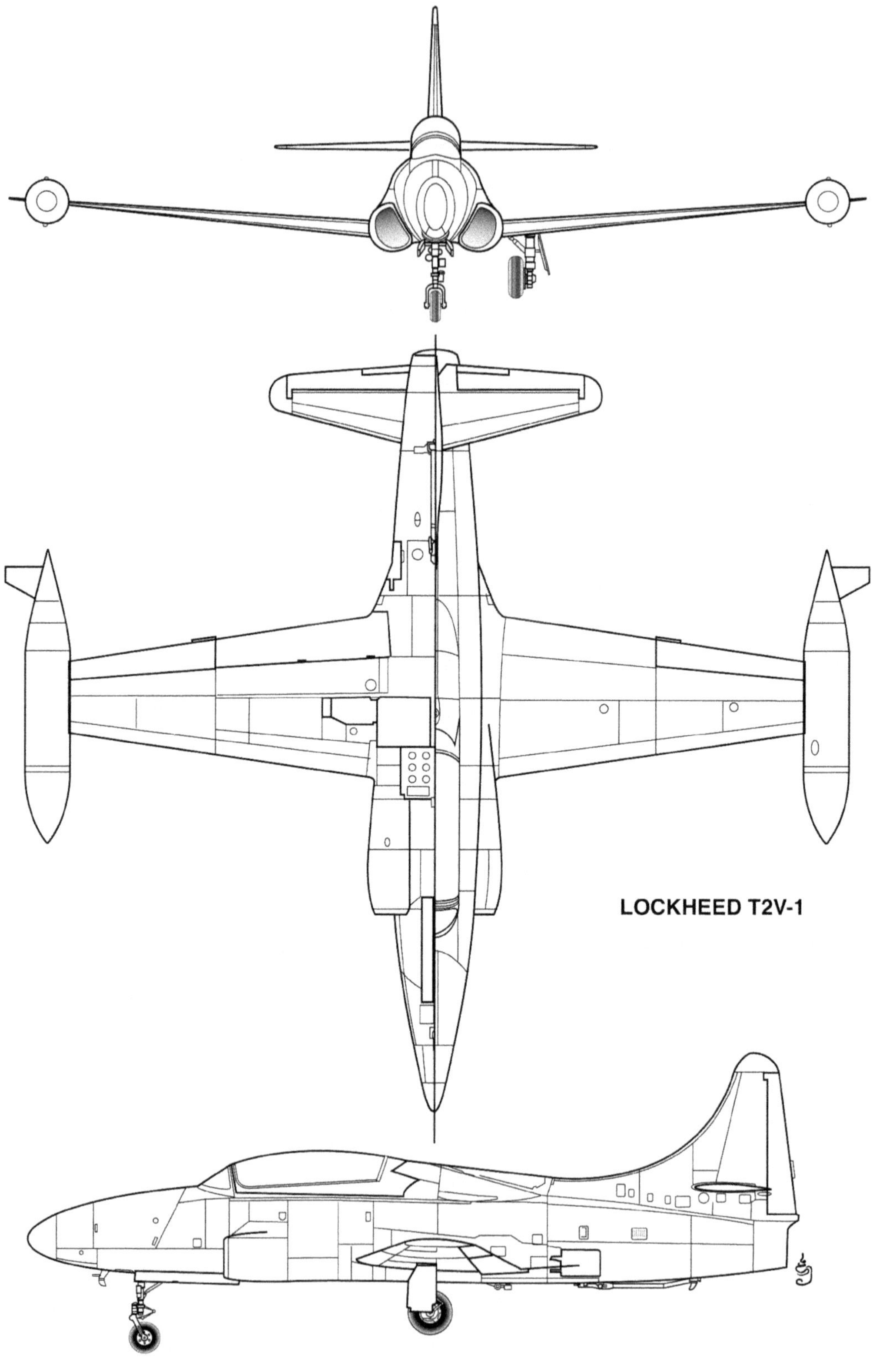

LOCKHEED T2V-1

Factory new Lockheed T2V-1, probably in 1956, seen in trainer colors (orange and white) before application of unit markings. Persistent maintenance problems with blown flap system led to early retirement.

their blown flap systems were a source of chronic maintenance problems to the extent that the Navy started replacing them with T2J-1s as early as 1959. All T2V-1s on the naval inventory in September 1962 were re-designated T-1A in the new sequential series. After leaving BTG-9, a number of T-1As were assigned to land-based training of Naval Flight Officers (F-4s and A-6s) with VT-10, also at NAS Pensacola, and as jet proficiency and instrument trainers, others went to various Navy Base Flights and Marine Headquarters units. The last active examples were struck from the inventory in 1970 and 1971.

Grumman UF-1T/TU-16C Albatross—1953

Technical Specifications (UF-1T)

Type: Utility amphibian transition and crew trainer.
Total U.S. military versions (trainers): 5 (USN).
Unit cost: $510,000.
Powerplants: Two 1,225-hp Wright R-1820–76 Cyclone 9-cylinder radial engines driving three-bladed, constant-speed metal propellers.
Performance: Max. speed 236 mph, cruise 150 mph, ceiling 21,500 ft.; range 2,415 mi.
Weights: 20,800 lbs. empty, 29,500 lbs. max takeoff (water).
Dimensions: Span 80 ft. 0 in., length 60 ft. 7 in., wing area 883 sq. ft.

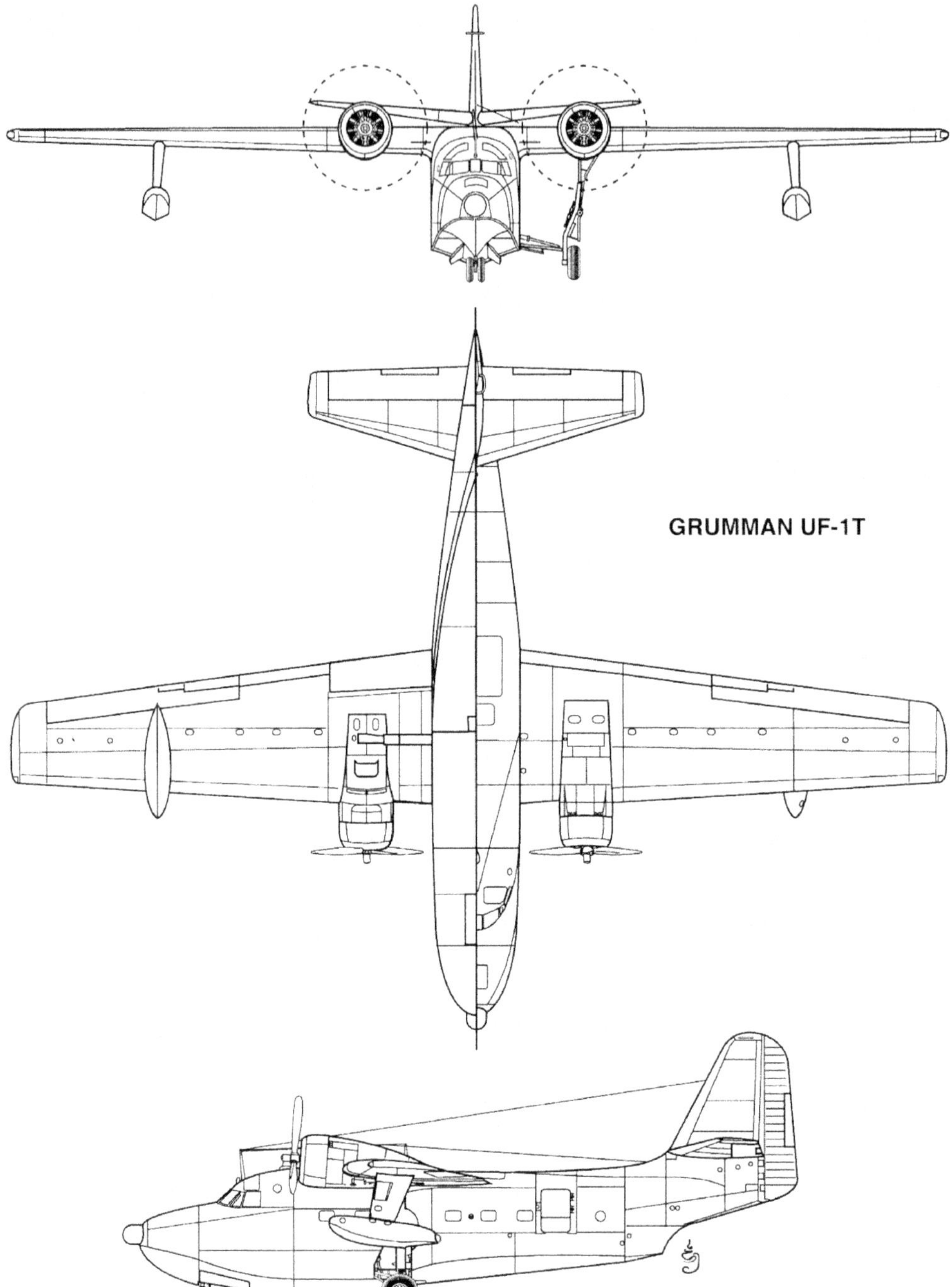

The last type of large, multi-engine flying boat to be mass-produced by any manufacturer in the United States, the design of Grumman's model G-64 arose from a 1944 Navy requirement calling for a twin-engine utility amphibian with a capacity for four crewmembers and ten passengers or 5,000 lbs. of cargo. In late 1944 the Naval Bureau of Aeronautics (BuAer) authorized Grumman to built two prototypes as the XJR2F-1, however, by the time construction was actually underway (apparently delayed by Grumman's other wartime production priorities), the war had

One of five Grumman UF-1Ts assigned to the U.S. Naval Academy in 1953 for midshipman training. The last examples were stricken from the naval inventory in 1966.

ended. In any case, development and construction continued, with the maiden flight of the first XJR2F-1 taking place on October 24, 1947. Favorable evaluations of both prototypes led to Navy contracts for 138 examples delivered between 1948 and 1954 that entered service as the UF-1, plus five more equipped with dual controls as the UF-1T. In a similar timeframe the Coast also procured eleven G-64s specially equipped for search and rescue as the UF-1G and the USAF 305 very similar G-64s as the SA-16A.

Historical records suggest that when delivered to the Navy in 1953, the five UF-1Ts were assigned to the U.S. Naval Academy at Annapolis, Maryland, to be used for midshipman training. In 1962 all UF-1Ts remaining on the inventory were re-designated TU-16C under the tri-service system, and serial number records reflect that all had been stricken from the inventory by 1966.

Douglas R4D-8T/TC-117D—1955

Technical Specifications (R4D-8T/TC-117D)

Type: Transport and navigation trainer.
Manufacturer: Douglas Aircraft Co., Santa Monica, California.
Total U.S. military versions (trainers): 6 (USN).
Unit cost: $250,000 (for conversion).
Powerplants: Two 1,475-hp Wright R-1820–80 *Cyclone* 9-cylinder air-cooled radial engines driving three-bladed Hamilton Standard constant-speed metal propellers.
Performance: Max. speed 270 mph, cruise 238 mph; ceiling 22,000 ft.; range 2,500 mi.
Weights: 19,537 lbs. empty, 31,000 lbs. loaded.
Dimensions: Span 90 ft., length 67 ft. 9 in., wing area 969 sq. ft.

Douglas introduced the "Super" DC-3 in 1949 as a potential conversion upgrade to airline companies operating DC-3s. Modifications entailed a fuselage lengthened four feet, new outer wing panels having squared tips and four degrees of sweepback, enlarged tail surfaces with squared edges, more powerful engines, fully enclosed wheels, and other streamlining enhancements, all

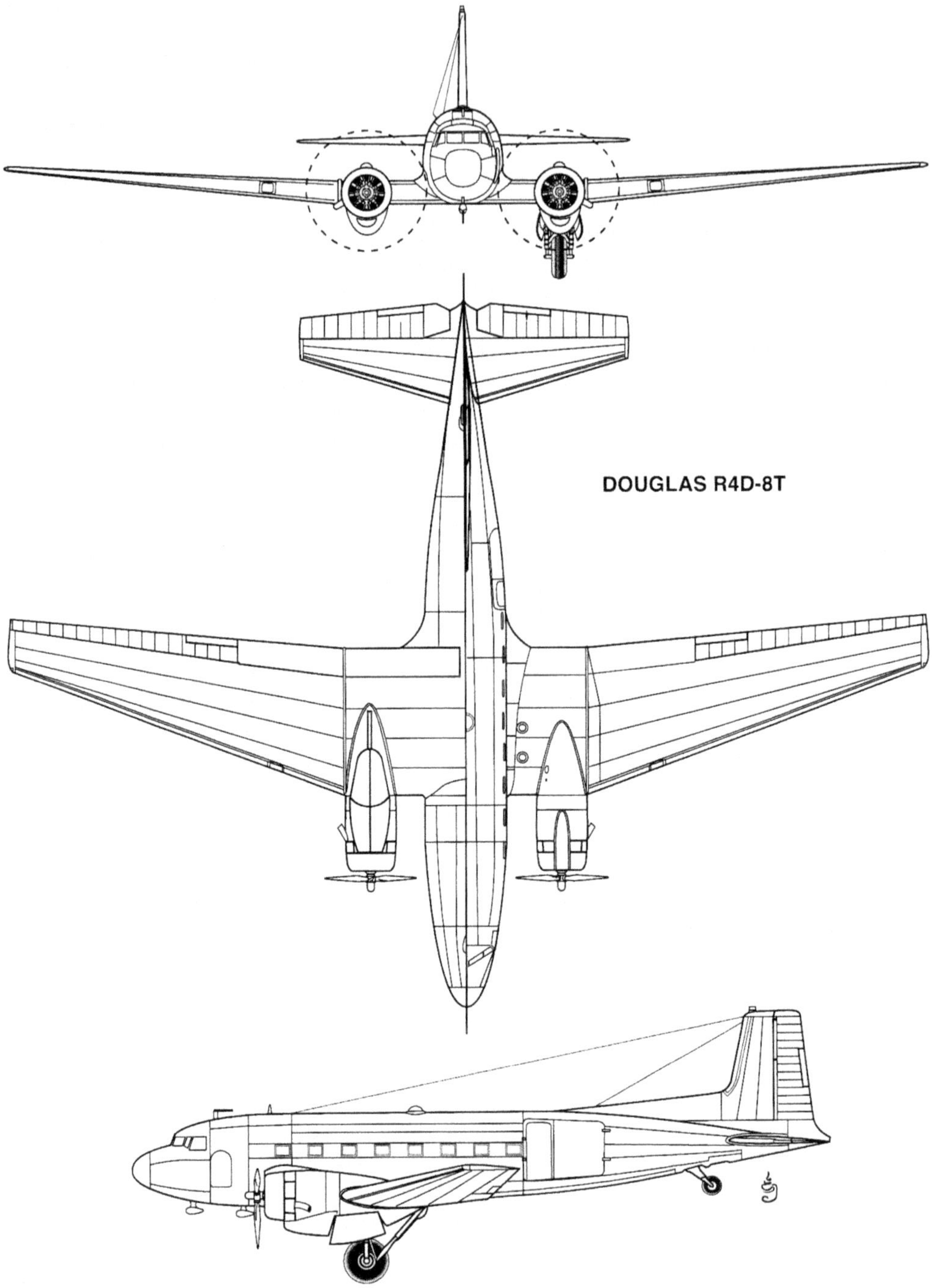

of which would enable it to carry eight more passengers at a 20 percent improvement in speed and range over a stock DC-3. In 1951, after the airlines and the USAF had shown little interest, the Naval Bureau of Aeronautics (BuAer) purchased the Super DC-3 prototype and subsequently contracted with Douglas to convert 100 R4D-5s, -6s, and -7s to be redelivered under the new designation R4D-8.

A Douglas TC-117D at NAS Midway Island during the 1960s. Most trainer variants operated with VT-27 at NAS Corpus Christi, Texas.

As they returned to Navy service, most R4D-8s served as transports with various Fleet Logistical Support Squadrons (VR), and a smaller number were employed in a similar capacity with several Marine Headquarters Squadrons (HEDRON). In the mid- to late 1950s, at least six R4D-8s, after undergoing a navigation trainer conversion that involved installation of two additional astrodomes and cabin stations for eight students, reemerged as R4D-8Ts, and in 1962, under the new tri-service designation scheme, became TC-117Ds. The six R4D-8Ts/TC-117Ds entered service with VT-29 out of NAS Corpus Christi, Texas, which also operated TC-47Ks and T-29Bs for aircrew and navigator training. All TC-117Ds were retired from training duties in the mid–1970s, though the last C-117D transport variant did not leave Marine service until 1982.

Douglas F3D-2T/TF-10B Skynight—1955

Technical Specifications (F3D-2T2/TF-10B)

Type: Night fighter and radar-intercept trainer.
Manufacturer: Douglas Aircraft Co., El Segundo Div., El Segundo, California.
Total U.S. military versions (trainers): 60 (USN/USMC).
Unit cost: Unknown.
Powerplant: Two Westinghouse J-34-WE-36 turbojet engines, each rated at 3,400 lbs./s.t.
Performance: Max. speed 580 mph, cruising 404 mph; ceiling 44,000 ft.; range 1,375 mi.
Weights: 18,160 lbs. empty, 23,575 lbs. loaded.
Dimensions: Span 50 ft., length 45 ft. 5 in., wing area 400 sq. ft.

Flown for the first time on March 23, 1948, the Douglas F3D Skynight was the first type of Navy jet aircraft to be designed from the ground up as a night fighter. Its large-diameter fuselage was dictated by the size the APQ-35 radar system housed in the bulbous nose radome. In order to free fuselage space for electronics, the two engines were externally mounted in nacelles

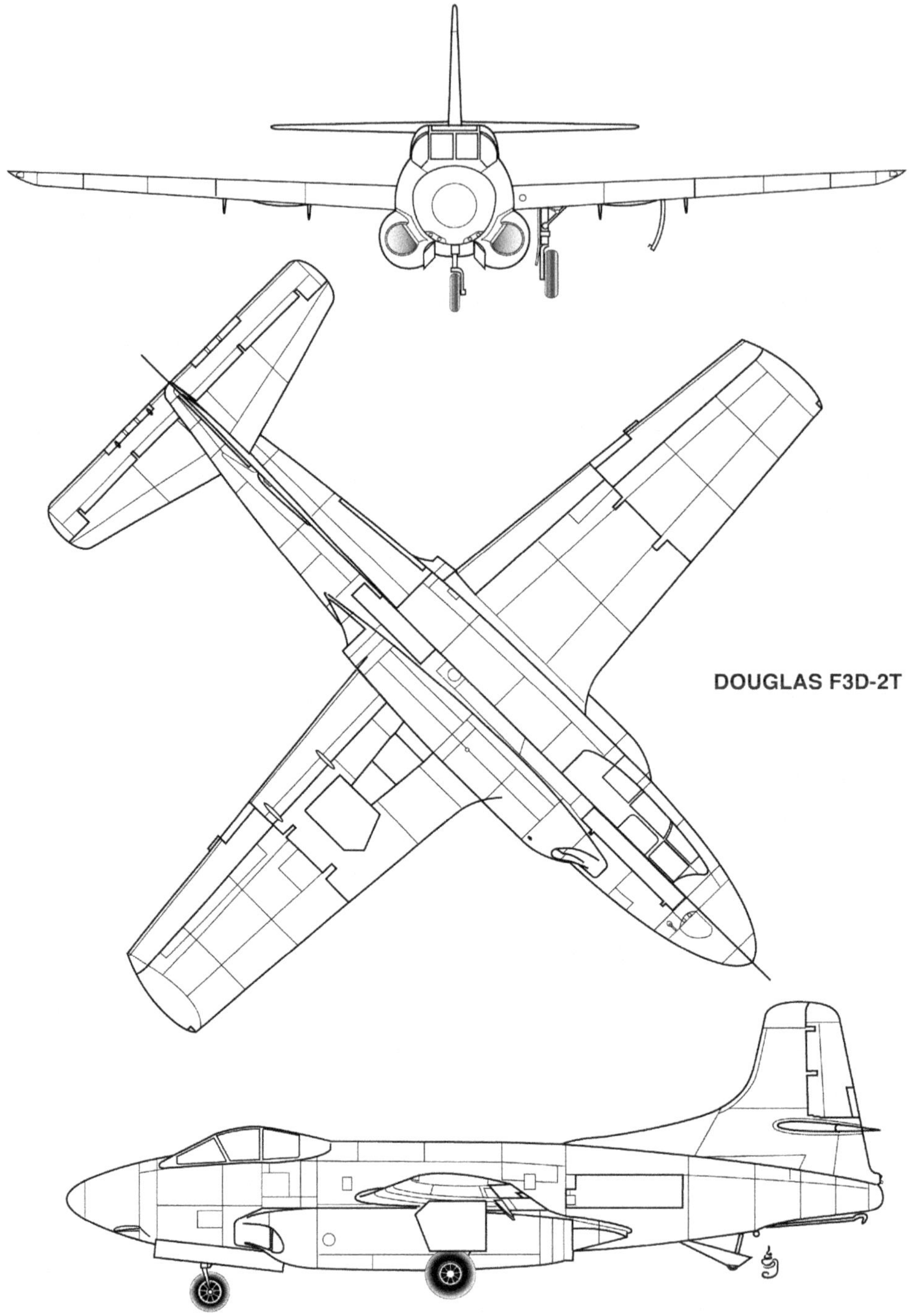

DOUGLAS F3D-2T

on the lower fuselage sides. In 1952 F3D-2s became operational with Navy Composite Squadrons (VC) and Marine night fighter squadrons (VMF [N]) in the Korean combat theater. (Note: F3D-2s were intended to be equipped with more powerful J-46 engines, but development delays led to retention of J-34s.)

In the mid–1950s five Marine F3D-2s received trainer modifications and reentered service

One of 55 F3D-2T2s modified to train pilots of McDonnell F3H-2s and Douglas F4D-1s in radar interception techniques. They were split between VF-101 at NAS Key West, Florida, and VF-121 at NAS Miramar, California.

as F3D-2T night fighter trainers with VMFT(N)-20 at MCAS Cherry Point, North Carolina. During the late 1950s, 55 F3D-2s became F3D-2T2s after being modified to train pilots of single-seat McDonnell F3H-2 Demons and Douglas F4D-1 Skyrays in radar interception techniques. Most were assigned to VF-101 at NAS Key West, Florida, and VF-121 at NAS Miramar, California, though a few were employed as radar testbeds with VX-4 and VX-5. In September 1962, under the tri-service system, all F3D-2T2s received the designation TF-10B. When retired from VF-101 and -121, some TF-10Bs continued as Naval Flight Officer (NFO) trainers with VT-86 at NAS Glynco, Georgia, until withdrawn in the late 1960s.

Lockheed P2V-6T/TP-2F Neptune—1955

Technical Specifications (P2V-6T/TP-2F)

Type: ASW patrol Crew trainer.
Manufacturer: Lockheed Aircraft Corp., Inglewood, California.
Total U.S. military versions (trainers): 4–5 (USN).
Unit cost: Unknown.
Powerplants: Two 3,250-hp Wright R-3350-WA *Double Cyclone* 18-cylinder air-cooled radial engines driving four-bladed Curtiss Electric constant-speed metal propellers.
Performance (P2V-5): Max. speed 353 mph, cruise 207 mph; ceiling 26,000 ft.; range 3,885 mi.
Weights: 39,900 lbs. empty, 77850 lbs. loaded.
Dimensions: Span 103 ft. 10 in., length 78 ft. 3 in., wing area 1,000 sq. ft.

At the time of its first flight in March 1945, the Lockheed P2V/P-2 Neptune had been envisaged as a land-based patrol-bomber, but in later variants (i.e., P2V-5/SP-2E and P2V-7/SP-2H), evolved into the Navy's standard land-based antisubmarine (ASW) platform from its introduction during the early 1950s until the Lockheed P-3 Orion began taking its place in the early 1960s, though the very last ASW SP-2Hs did not leave the reserves until 1978. The 35 P2V-6s delivered to the Navy in 1952 and 1953 were originally developed as a multi-purpose variant equipped to perform either minelaying or photoreconnaissance missions. During the mid- or

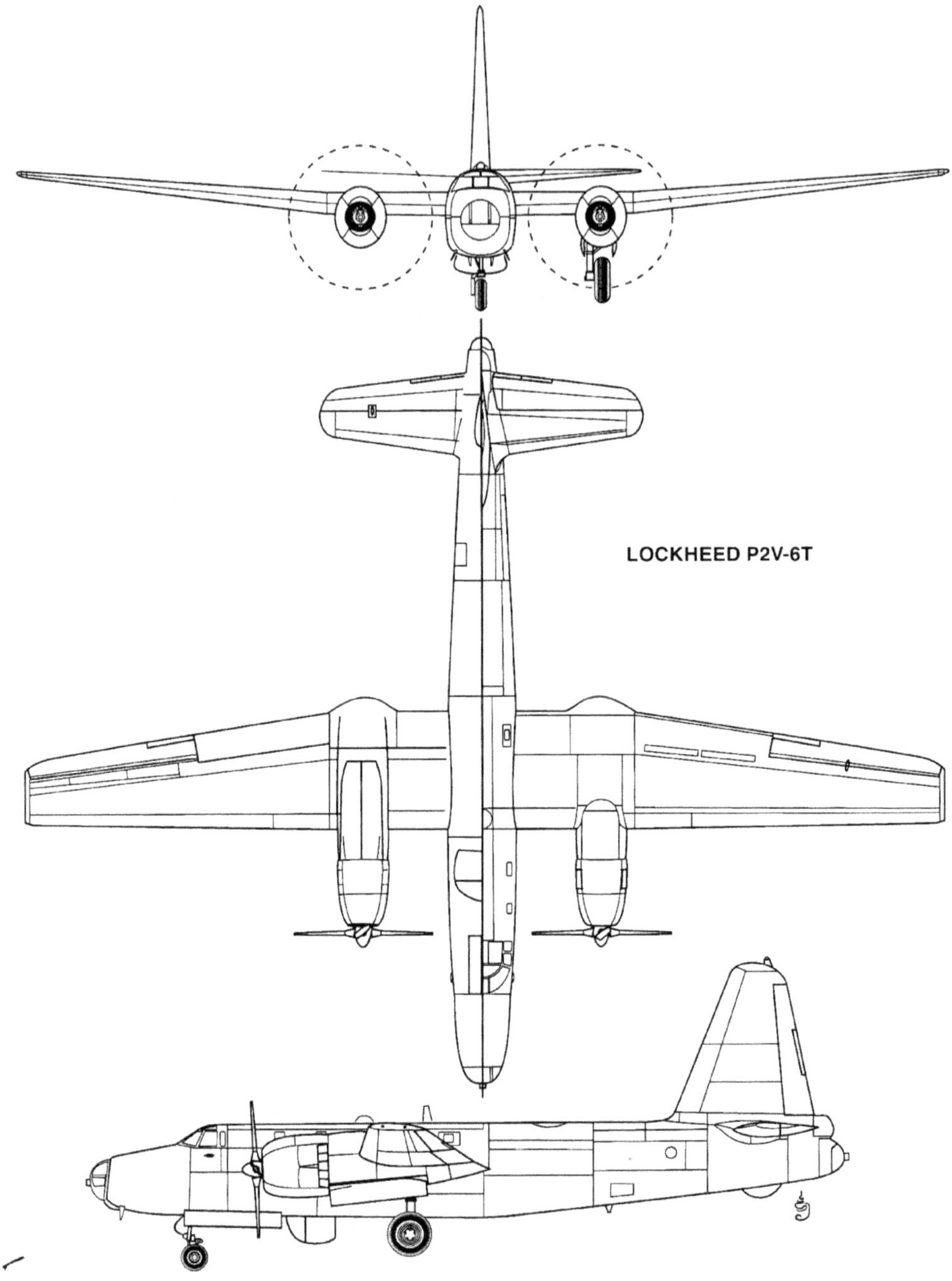

late 1950s, according to serial number records, at least three P2V-6s were stripped of all armament and wingtip tanks and reassigned to training duties under the designation P2V-6T. All appear to have served as ASW crew trainers with ATU-501 (later VT-27) at NAAS New Iberia, Louisiana, and ATU-601 (later VT-31) as NAS Corpus Christi, Texas. In September 1962, under the tri-service system, P2V-6Ts were re-designated TP-2F.

One of three converted Lockheed P2V-6T crew trainers as seen in the late 1950s at NAAS New Iberia, Louisiana, while serving with Advanced Training Unit 501. Redesignated TP-2F in mid–1962.

Grumman TF Trader—1955 (Anomalous designation applied to the carrier-onboard-delivery variant of the S2F Tracker; changed to C-1A in 1962)

Temco TT Pinto—1956

Technical Specifications (TT-1)

Type: Two-place primary trainer.
Manufacturer: Temco Aircraft Corp., Dallas, Texas.
Total U.S. military versions: 15 (USN).
Unit cost: Unknown.
Powerplant: One Continental J-69-T-9 turbojet engine, rated at 1,025 lbs./s.t.
Performance: Max. speed 345 mph; ceiling 32,200 ft.; range 450 mi.
Weights: 3,064 lbs. empty, 4,325 lbs. loaded.
Dimensions: Span 29 ft. 9 in., length 30 ft. 7 in., wing area 150 sq. ft.

Mentioned earlier with reference to the Cessna T-37 (see in Series II, Part A), the pure-jet Temco model 51 originated in 1953 as a design proposal in the USAF's TX competition. After losing to the T-37, Temco elected to continue development with company-funded construction of a model 51 flying prototype in hopes of interesting the Navy. Named "Pinto" by the factory, the prototype completed it first flight on March 25, 1956, under civil registration (N78856) and was thereafter consigned to the Navy for testing and evaluation as a primary trainer. Later the same year, in order to conduct a more extensive functional evaluation of the feasibility of using pure-jets as primary trainers, the Naval Bureau of Aeronautics (BuAer) gave Temco a contract to deliver 14 more Pintos under the naval designation TT-1.

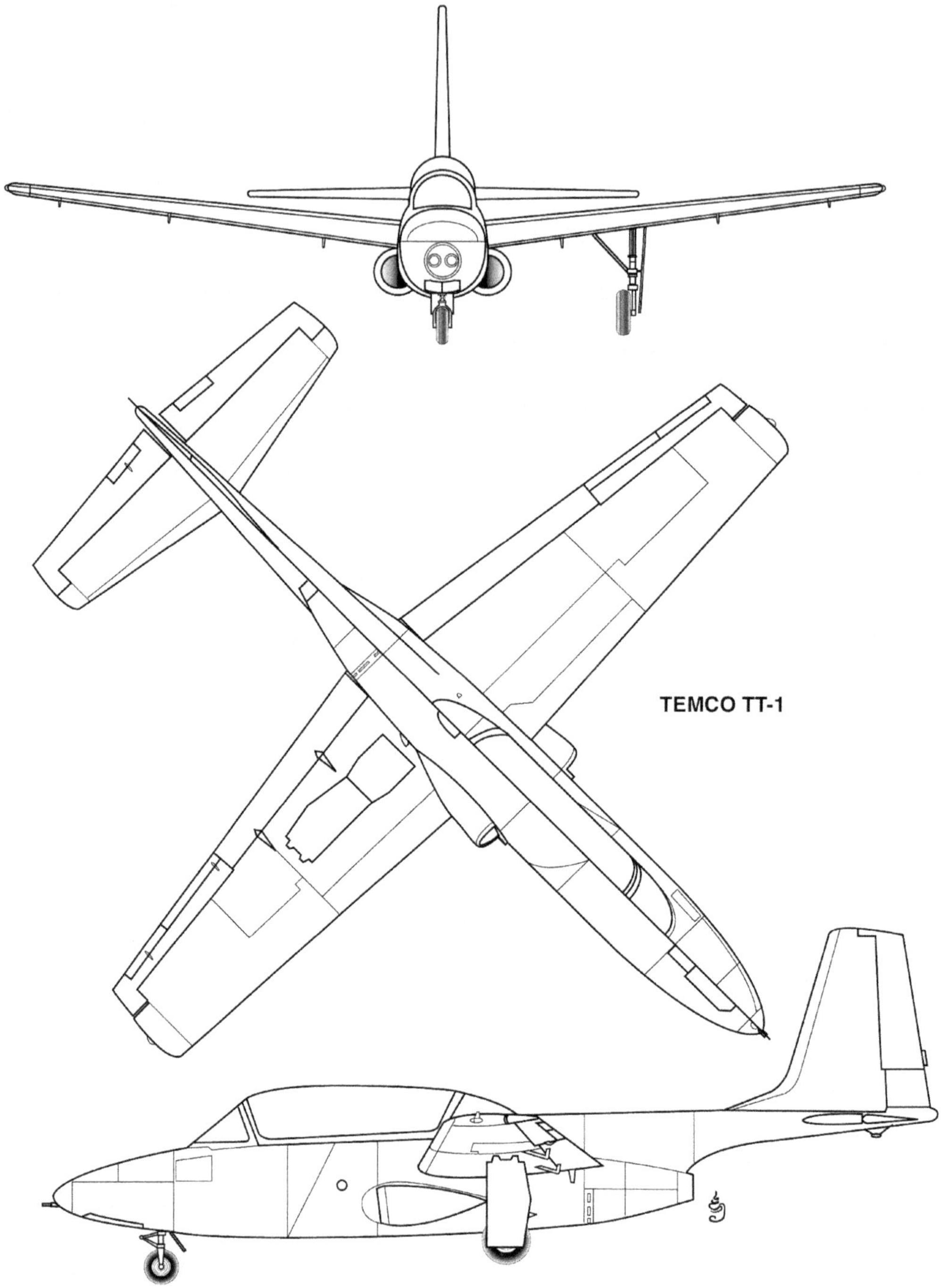

Following delivery in 1957, the 14 TT-1s, along with the prototype, underwent a series of official acceptance trials with the Naval Air Test Center (NATC) at NAS Patuxent River, Maryland, to confirm general suitability for the Navy's primary pilot training syllabus in terms of structural integrity, handling characteristics, and serviceability. In early 1959, once testing was complete, the TT-1s entered service with Air Training Wing Five (ATW-5) at NAS Pensacola,

One of the first 14 Temco TT-1s that underwent trials at NAS Patuxent River, Maryland, during 1957. The planes afterward served in Air Training Wing Five from early 1959 until withdrawn in late 1960.

Florida, for functional training of primary pilot candidates, and in March, the first student pilot, Aviation Cadet E. R. Clark, completed his first solo in a TT-1 without any previous experience in propeller-driven aircraft. TT-1s continued with ATW-5 until the end of 1960, when they were all withdrawn and removed from the naval inventory. Although the type proved to be docile and forgiving in the hands of student pilots, naval officials ultimately concluded that the TT-1's limited range (average flight durations of about one hour with reserve) rendered it impractical as a trainer.

Grumman F9F-8T/TF-9J Cougar—1956

Technical Specifications (F9F-8T/TF-9J)

Type: Two-place carrier-capable advanced and combat trainer.
Manufacturer: Grumman Aircraft Engineering Corp., Bethpage, New York.
Total U.S. military versions (trainers): 400 (USN/USMC).
Unit cost: Unknown.
Powerplant: One Pratt & Whitney J-48-P-8A turbojet engine, rated at 7,250 lbs./s.t.
Armament: Two fixed 20-mm cannons in the nose, two AIM-9 *Sidewinder* missiles, or up to 1,000 lbs. of bombs or rockets on wing pylons.
Performance: Max. speed 647 mph, cruise 516 mph; ceiling 42,000 ft.; range 1,050 mi.
Weights: 11,866 lbs. empty, 24,763 lbs. loaded.
Dimensions: Span 34 ft. 6 in., length 44 ft. 4 in., wing area 337 sq. ft.

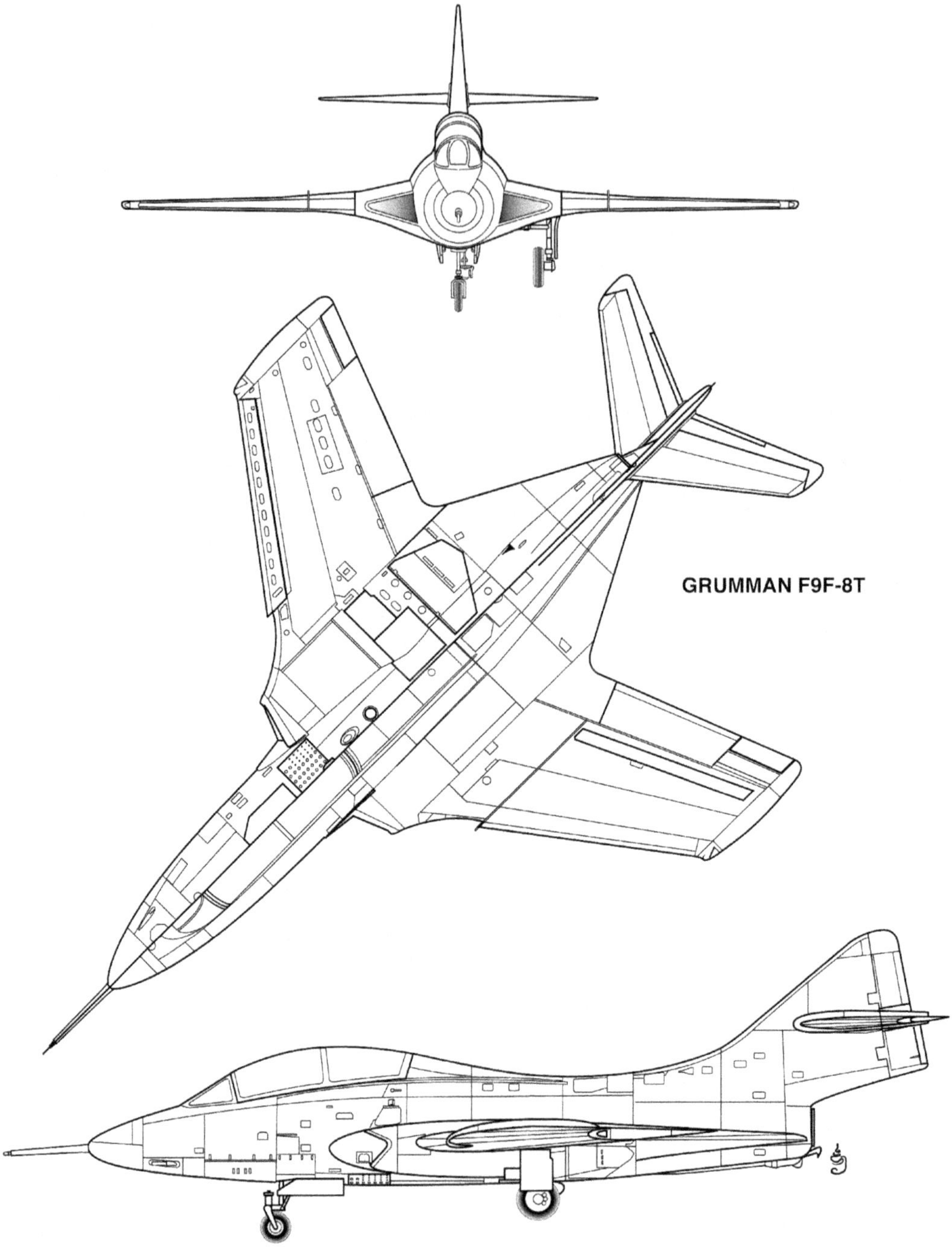

GRUMMAN F9F-8T

It is a tribute to Grumman's engineering staff that so many of their fighter designs—F4F Wildcat, F6F Hellcat, F8F Bearcat, and F9F Panther/Cougar—possessed such stable and honest handling characteristics that they saw wide service in Naval Air Training as advanced trainers. As a company-initiated effort, Grumman commenced work on a two-seat trainer variant of the F9F Cougar in 1953 on two premises: the Navy in all likelihood would procure it as a swept-wing transition and combat trainer for fleet squadrons operating single-seat Cougars and possibly,

A factory new Grumman F9F-8T. The Navy ultimately took delivery of 399 F9F-8Ts (TF-9J after 1962), the majority of which served as advanced trainers from mid–1956 until early 1974.

in Naval Air Training Command as a carrier qualification and in-flight refueling trainer. But at the time, because the Naval Bureau of Aeronautics (BuAer) was then focused on development of the Lockheed T2V-1 (see above), it envisaged no carrier qualification role for a two-seat Cougar. Nonetheless, in 1955, after F9F-8s had begun equipping both Navy and Marine fighter squadrons, BuAer authorized Grumman to proceed with the modifications needed to convert one F9F-8 airframe into a two-seat, dual control configuration. Changes to the basic airframe consisted of lengthening the fuselage two feet ten inches to accommodate the second cockpit, reducing fuselage fuel capacity by 296 gallons, and deleting two of the four nose cannons to save weight. Other differences included strengthened landing gear, an increase in wheel size, and a long, one-piece sliding canopy with an elevated rear seat and an auxiliary windshield between the cockpits to permit the canopy to be opened in flight. The first flight of the converted aircraft, designated YF9F-8T, took place on February 29, 1956.

In the interim, while the YF9F-8T was being tested, recurring problems with the T2V-1's blown flap system, together with its inability to serve as a weapons trainer, caused BuAer to reconsider an expanded training role for the two-seat Cougar. As a result, 399 F9F-8Ts produced under a series of contracts were delivered to the Navy between July 1956 and February 1960. Production F9F-8Ts featured additional structural strengthening and in later blocks, nose refueling probes and the wiring to carry *Sidewinder* missiles. In 1961 Grumman proposed an upgraded F9F-8T with a Pratt & Whitney J-52 engine and other improvements, however, the Navy elected not to pursue development and instead ultimately procured the Douglas TA-4F/J (see in Series III). Following adoption of the tri-service system in September 1962, all F9F-8Ts were redesignated TF-9J.

Over the course of their 17-year operational career, F9F-8T/TF-9Js served with a variety of Navy and Marine units: in Naval Air Training Command with VT-4 and VT-10 at NAS Pensacola, Florida, VT-21, VT-22, and VT-23 at NAAS Kingsville, Texas, and VT-24, VT-25, and VT-26 at NAS Chase Field (Beeville), Texas; and in Marine training units with VMT-1 at MCAS Cherry Point, North Carolina, VMT-2 and VMIT-10 at MCAS El Toro, California, and VMT-103 at MCAS Yuma, Arizona. In their role as proficiency and combat trainers, F9F-8Ts/TF-9Js served at times with 23 other Navy and Marine frontline squadrons. Marine TF-9Js were the only Cougar variant to see combat when four examples attached to H&MS-13, acting as a forward air controllers, directed air strikes against enemy positions over South Vietnam during 1966 and 1967. An F9F-8T/TF-9J also belonged to the Blue Angels flight demonstration team from 1957 to 1968. The phase-out of TF-9Js from training duties began in 1969, with the final examples being retired from VT-4 in February 1974.

North American (Rockwell) T2J/T-2 Buckeye—1958

Technical Specifications (T-2C)

Type: Two-place carrier-capable basic trainer.
Manufacturer: North American Aviation, Inc. (Rockwell International), Columbus, Ohio.
Total U.S. military versions: 542 (USN).
Unit cost: Unknown.
Powerplants: Two General Electric J-85-GE-4 turbojet engines, each rated at 2,950 lbs./s.t.
Performance: Max. speed 522 mph, cruise 400 mph; ceiling 40,400 ft.; range 1,047 mi.
Weights: 8,115 lbs. empty, 13,179 lbs. loaded.
Dimensions: Span 38 ft. 1.5 in., length 38 ft. 3.5 in., wing area 255 sq. ft.

As a military trainer whose operational career spanned 45 years, the North American T-2 Buckeye is exceeded in longevity only by the Cessna T-37 and Northrop T-38 (see both in Series II, Part A). It came into being as the winning entry of a design competition staged by the Naval Bureau of Aeronautics (BuAer) in 1956 to find a carrier-capable trainer to augment and eventually replace the Lockheed T2V (see above). Requirements for the design were stringent, placing emphasis on safety, durability, maintainability, and airframe life as well as the ability to take pilot students from basic through the advanced training syllabus, which included carrier qualifications and combat tactics. The design of the model NA-280 utilized the proven straight-wing layout of North American's earlier FJ-1 Fury, mounted at mid-fuselage to allow waist-level access to the aircraft's major systems and components. Student and instructor sat in tandem under a single clamshell canopy, with the rear seat raised for better visibility, and it was the very first type of trainer to be equipped with rocket-powered LS-1 ejection seats that permitted zero-zero (air-speed-altitude) ejections. Power was derived from a single Westinghouse J-34-WE-48 engine fed by bifurcated side intakes and rated at 3,400 lbs./s.t. Other advanced features included large, hydraulically actuated trailing-edge flaps and speed brakes to insure good low-airspeed handling qualities.

In the initial contract award, BuAer opted to go straight to the pre-production phase, ordering six aircraft under the designation YT2J-1. The factory name "Buckeye" signified the nickname of the state where the Columbus plant was located. After the first YT2J-1 completed it maiden flight on January 31, 1958, the Naval Air Test Center (NATC) at NAS Patuxent River, Maryland, commenced official acceptance testing, which included deck trials aboard the USS *Antietam*. In the interim, North American received a contract to manufacture 211 T2J-1 production models, with all deliveries being completed between July 1959 and April 1961. At the end of the T2J-1

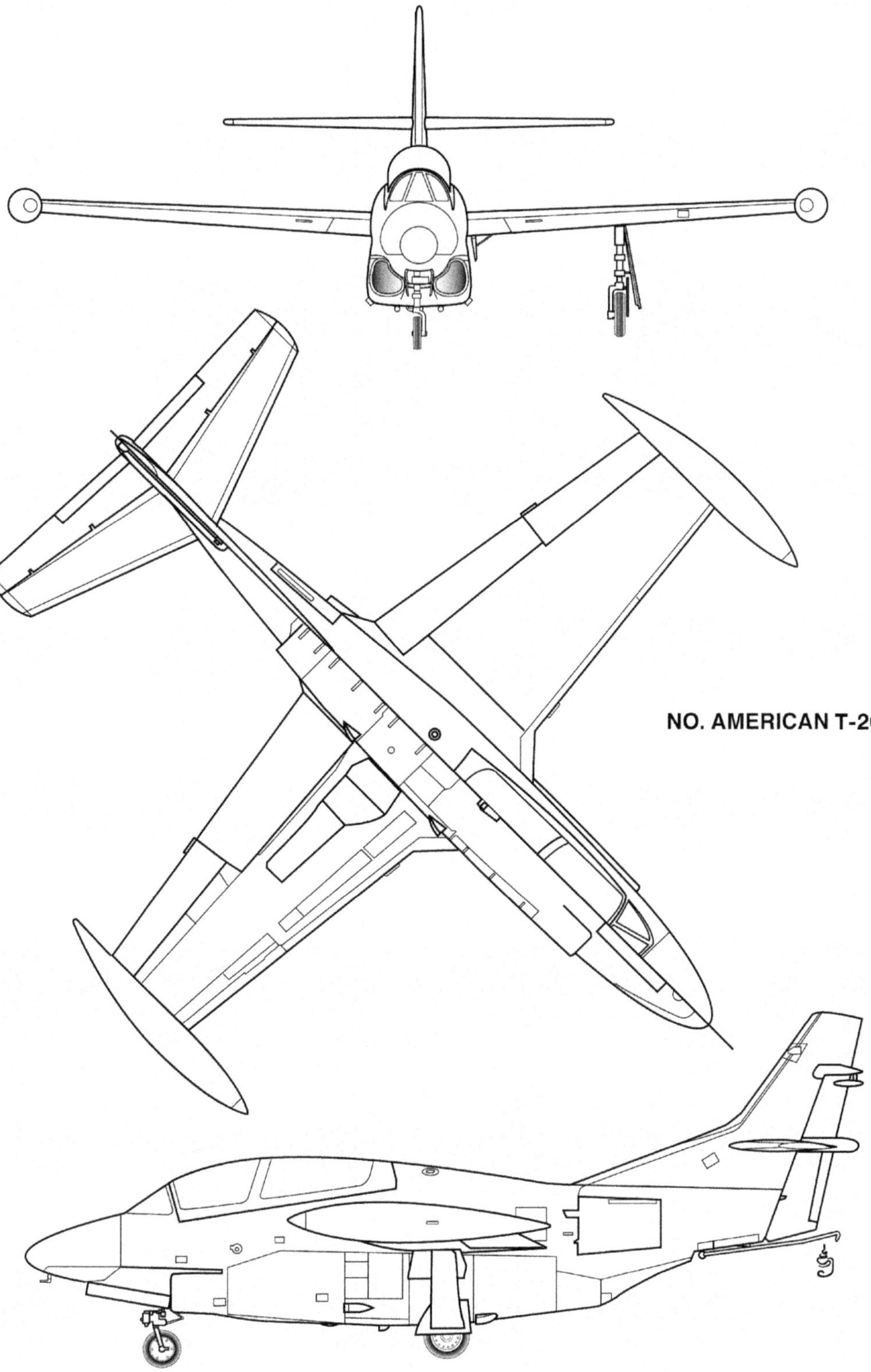

NO. AMERICAN T-2C

One of six YT2J-1s delivered to the Navy for testing in 1958. After official acceptance testing, North American received a contract to produce 211 T2J-1s, which began entering operational service in mid–1959.

production run, BuAer authorized North American to modify two aircraft, one YT2J-1 and one T2J-1, for installation of two Pratt & Whitney J-60-P-6 engines, each producing 3,000 lbs./s.t., thus increasing the power-to-weight ratio 43 percent and adding the safety factor of a second engine. Moreover, the increase in speed (+46 mph) and climb (+1,560-fpm) would better prepare students for higher-performance aircraft. The first modified example, designated YT2J-2, flew for the first time in August 1962 and was re-designated YT-2B in September, and both aircraft began evaluations with Naval Air Basic Training Command (NABTC) in April 1964. Following these trials, North American received a succession of contracts for T-2B production aircraft, with total of 97 being delivered between May 1965 and April 1968. In 1967, as a cost savings measure, the Navy directed North American, now part of Rockwell International, to convert one T-2B to accept two General Electric J-85-GE-4 engines, each rated at 2,950 lbs./s.t. After the converted aircraft, designated YT-2C, completed its first flight in April 1968, testing revealed performance nearly identical to that of the T-2B. In yet another series of production contracts extending into the early 1970s, the Navy accepted a total of 231 T-2Cs.

The first T2J-1s arrived for duty with NABTC during July 1959, and actual training of students began the following November with Basic Training Group Nine (became VT-4) at NAS Pensacola, Florida. As deliveries continued, T2J-1s (T-2A after mid–1962) went on to equip four more NABTC squadrons, VT-10 at NAS Pensacola and VT-7, VT-9, and VT-19 out of NAS Meridian, Mississippi. Newly built T-2Bs began augmenting T-2As in the pilot training syllabus in late 1965, serving initially with VT-4 and VT-7. As new T-2Cs began reaching operational service in early 1970, they initiated an across-the-board phase-out of both T-2As and Bs in

This T-2C is taking off from NAS Patuxent River, Maryland, in August 2005. T-2Cs were withdrawn from training in 2005 but continued with the Naval Test Pilot School until 2008.

NABTC units, with the process being completed by mid–1973. In addition to the units mentioned above, T-2Cs also equipped VT-23 at NAS Kingsville, VT-26 at NAS Chase Field, and VT-86 at NAS Pensacola. In 1981 and 1982, due to shortages in the T-2C fleet, a number of T-2Bs, after being withdrawn from AMARC storage, were reconditioned and returned to active duty with VT-10 where they served until retired a second time in 1991. The process of gradually replacing T-2Cs with McDonnell Douglas (Boeing) T-45 Goshawks (see in Series III) began during the early 1990s, and the very last carrier qualification course flown in T-2Cs was completed by VT-9 in July 2004. Small numbers of T-2Cs remained active in VT-86 for Naval Flight Officer training until 2005 and with the Naval Test Pilot School until 2008.

Douglas A3D-2T/TA-3B Skywarrior—1959

Technical Specifications (A3D-2T/TA-3B)

Type: Mission-capable two-place fighter transition and combat trainer.
Manufacturer: Douglas Aircraft Co., El Segundo, California.
Total U.S. military versions (trainers): 12 (USN).
Unit cost: Unknown.
Powerplants: Two Pratt & Whitney J-57-P-10 turbojet engines, each rated at 10,500 lbs./s.t.
Performance: Max. speed 610 mph, cruise 520 mph; ceiling 41,000 ft.; range 2,100 mi.
Weights: 39, 409 lbs. empty, 81,829 lbs. loaded.
Dimensions: Span 72 ft. 6 in., length 76 ft. 4 in., wing area 812 sq. ft.

Flown for the first time in late 1952, the active career of the Douglas A3D/A-3 Skywarrior spanned over 35 years (1956–1991) in a variety of tactical and support roles. It was originally conceived as a carrier-based strategic bomber with the range to carry a nuclear payload deep into Soviet territory. In 1957, as A3D-2s were becoming operational with eight heavy attack squadrons (VAH), the Naval Bureau of Aeronautics (BuAer) directed Douglas to complete 12 aircraft as bombardier-navigator trainers under the designation A3D-2T. Modifications entailed installing a pressurized cabin in place of the bomb bay with positions for one instructor and four

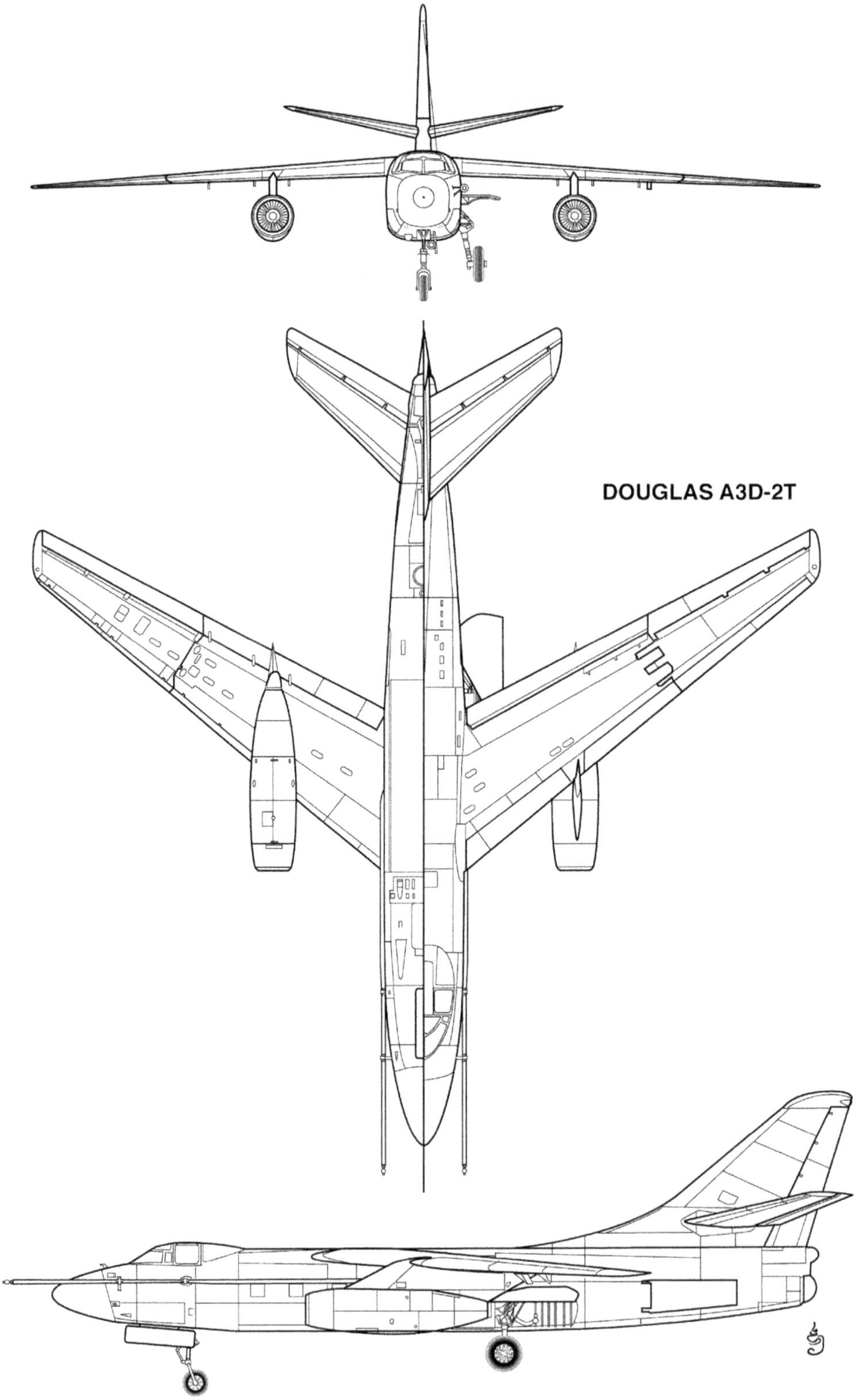

DOUGLAS A3D-2T

Douglas TA-3B photographed in the 1970s at NAS Norfolk, Virginia, while attached to VAQ-33. The aircraft probably had been converted to a VIP passenger configuration at the time the photograph was taken.

bombardier-navigator trainees. Bomb damage assessment cameras and recorders were fitted in a tail cone assembly normally occupied by two radar-directed 20-mm cannons.

The first A3D-2T flew on August 29, 1959, and as deliveries proceeded, the -2Ts were divided between VAH-3 and VAH-123, which, together with the A3D-1s, functioned as replacement training units for the A3D-2-equipped fleet squadrons. A3D-2Ts became TA-3Bs when the tri-service designation system was adopted in September 1962. During the mid–1960s as A-3Bs were withdrawn from their strategic bomber role and converted to tanker (KA-3B), reconnaissance (RA-3B), and tanker/electronic warfare (EKA-3B) platforms, five of the TA-3Bs were refitted as VIP transports, receiving airliner-type accommodations for up to nine passengers. Two received the designation UA-3B, but the other three were unchanged. Other TA-3Bs continued to function as trainers for RVAH squadrons equipped with North American RA-5C Vigilantes until the early 1970s. According to serial number records, the last transport variants were withdrawn and placed in AMARC storage in mid–1990.

Grumman S2F-1T/TS-2A Tracker—1960

Technical Specifications (S2F-1T/TS-2A)

Type: Advanced multi-engine and ASW crew trainer.
Manufacturer: Grumman Aircraft Engineering Corp. (later Grumman Aerospace), Bethpage, New York.
Total U.S. military versions (trainers): 200+ (USN).
Unit cost: Unknown.
Powerplants: Two 1,525-hp Wright R-1820–82WA *Cyclone* 9-cylinder air-cooled radial engines driving three-bladed Hamilton Standard constant-speed metal propellers.

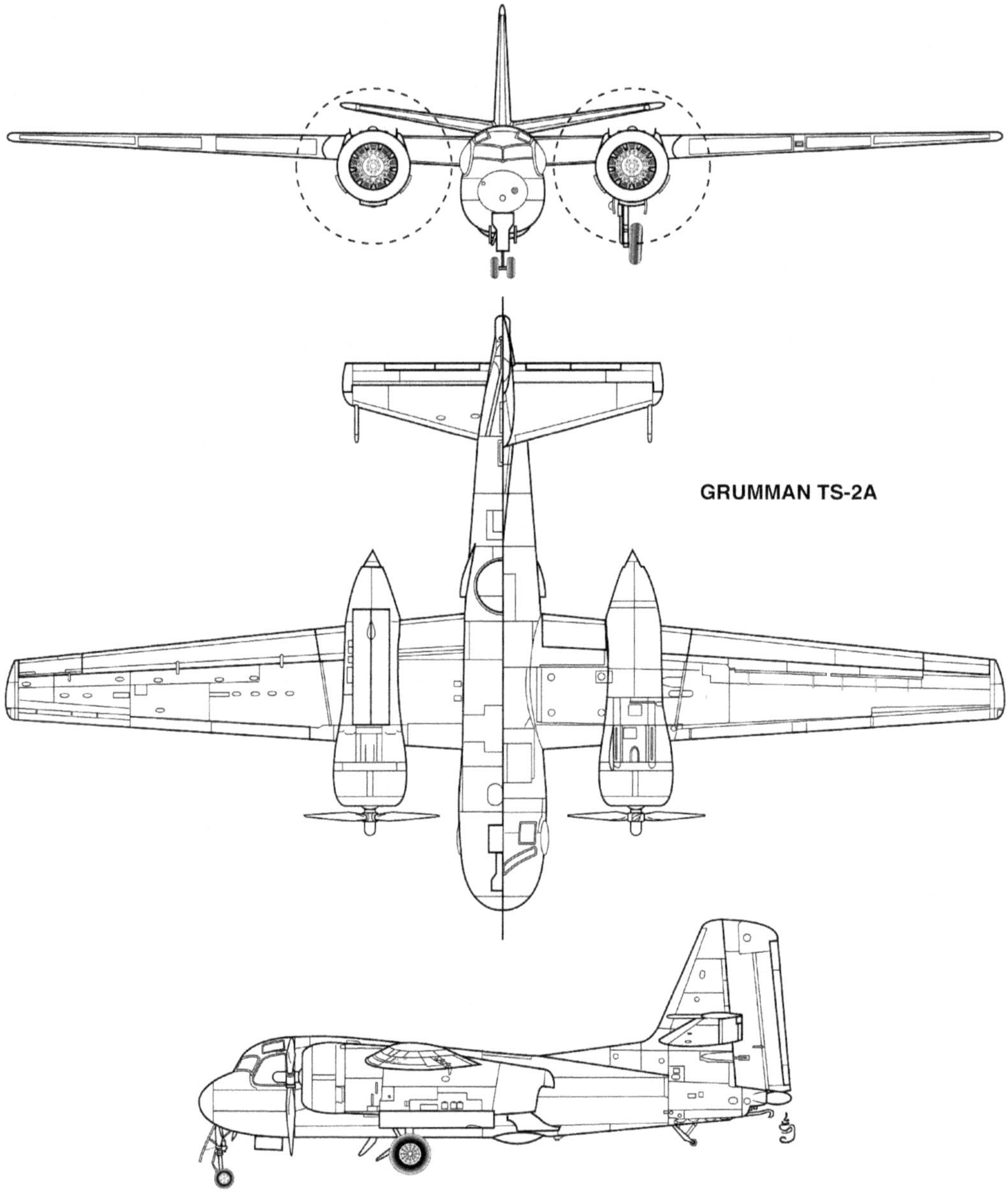

Performance: Max. speed 265 mph, cruise 207 mph; ceiling 22,800 ft.; range 1,150 mi.
Weights: 17,357 lbs. empty, 24,408 lbs. loaded.
Dimensions: Span 69 ft. 8 in., length 42 ft. 3 in., wing area 485 sq. ft.

The prototype Grumman XS2F-1 "Tracker" completed its first flight on December 4, 1952, and production S2F-1s began entering fleet service in early 1954 as the Navy's standard carrier-based ASW platform. Altogether, between 1953 and 1968, Grumman delivered 1,069 Tracker production variants to the Navy consisting of 740 S2F-1s (S-2A in 1962), 77 S2F-2s (S-2C in 1962), 100 S2F-3s (S-2D in 1962), and 152 S2F-2Ss (S-2E in 1962). In their ASW role, Trackers remained active with the Navy until the last examples were retired from VS-37 in August 1976.

Starting in 1959 or 1960, as newer Tracker variants entered fleet service, approximately 200

A Grumman TS-2A of VT-31 on the ramp at NAS Corpus Christi, Texas, date uncertain. The type served as advanced multi-engine trainers from 1963 until 1979.

S2F-1s were withdrawn and refitted for training duties under the designation S2F-1T. S2F-1Ts earmarked as multi-engine trainers were stripped of their ASW gear and associated equipment, while those intended as ASW crew trainers retained all the gear as well as practice depth charges, bombs, and rockets. When the tri-service system took effect in September 1962, S2F-1Ts became TS-2As. Since they were carrier-capable, S2F-1Ts/TS-2As served as multi-engine transition trainers for pilots destined to fly S-2s, C-1s, E-1s, E-2s, and C-2s aboard carriers. S2F-1Ts/TS-2As served with VT-27, initially at NAS New Iberia, Louisiana, then at NAS Corpus Christi, Texas, and with VT-28 and VT-31, also at Corpus Christi. The last TS-2As were retired from Naval Air Training Command in early 1979.

Martin P5M-1T and -2T/TP-5A and -5B Marlin—1961

Technical Specifications (P5M-2T/TP-5B)

Type: Patrol flying boat crew and transition trainer.
Manufacturer: Glenn L. Martin Co., Baltimore, Maryland.
Total U.S. military versions (trainers): 11 (USN).
Unit cost: Unknown.
Powerplants: Two 3,450-hp Wright R-3350–12EA *Double Cyclone* 9-cylinder air-cooled radial engines driving four-bladed Hamilton Standard constant-speed metal propellers.
Performance: Max. speed 251 mph, cruise 159 mph; ceiling 24,000 ft.; range 3,060 mi.
Weights: 19,537 lbs. empty, 31,000 lbs. loaded.
Dimensions: Span 118 ft. 2 in., length 100 ft. 7 in., wing area 1,406 sq. ft.

The Martin P5M/SP-5 Marlin, flown for the first time in 1948, was the Navy's most important type of maritime patrol flying boat from 1951 until the last examples were retired from VP-40 in 1967. It was produced in two major variants, the P5M-1/SP-5A with a conventional tail group and P5M-2/SP-5B with a T-tail. During the P5M/SP-5 production run (1951–1960), the

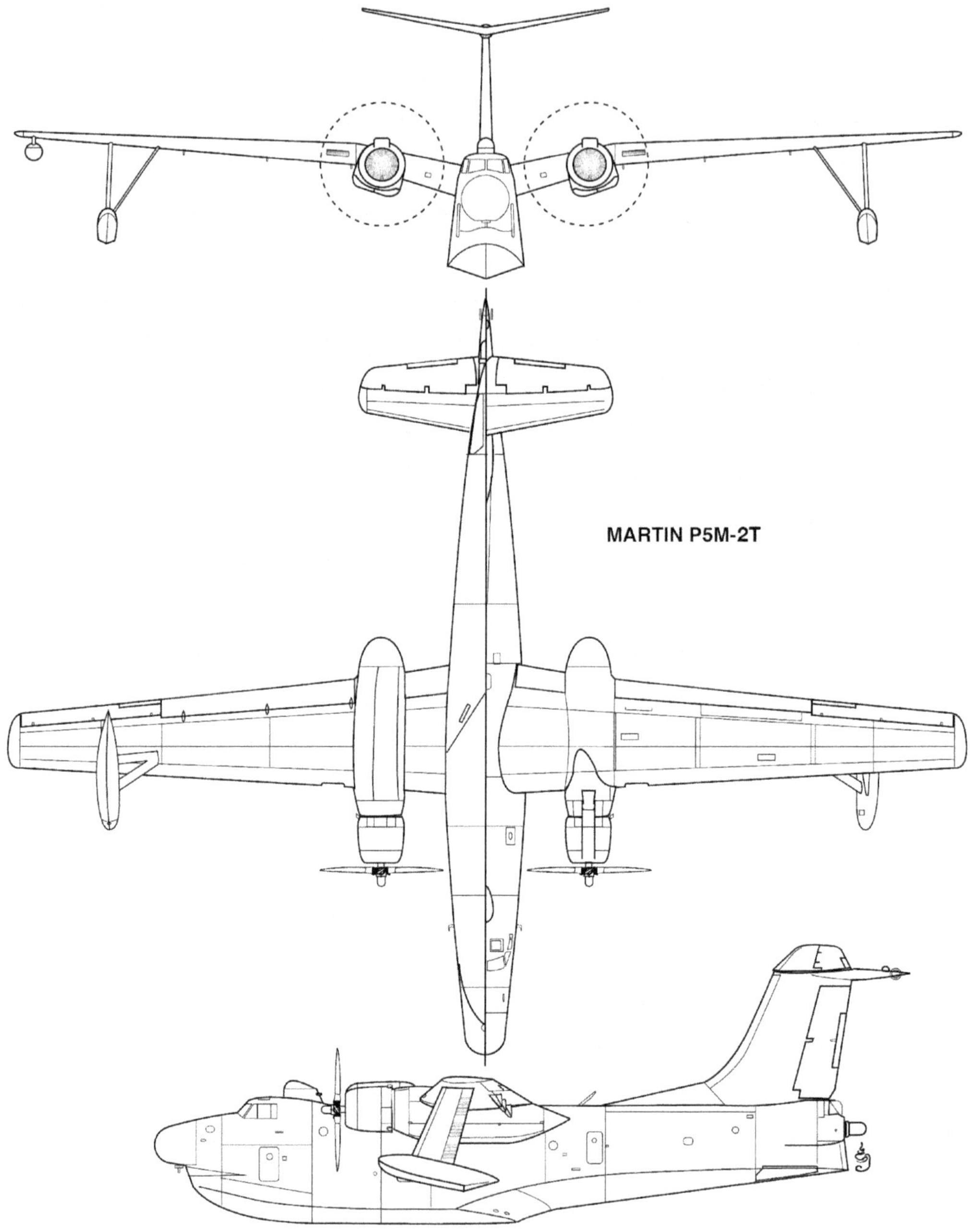

Coast Guard also took delivery of eleven Marlins lacking military equipment and outfitted for search and rescue, seven as the P5M-1G and four as the P5M-2G. In 1961, citing unacceptably high operational costs, the Coast Guard transferred all eleven of these aircraft to the Navy. Since they were not equipped for maritime patrol, the Navy placed them on the inventory as transition and crew trainers under the respective designations, P5M-1T and P5M-2T. When the tri-service system took effect in September 1962, they became the TP-5A and TP-5B. It appears that all TP-5A/Bs served with ATU-501 and -601 (later VT-31) at NAS Corpus Christi.

A Martin P5M-1T serving with Advanced Training Unit 501 during the early 1960s. This aircraft was one of six P5M-1Gs transferred from the Coast Guard in 1961.

McDonnell F4H-1F/F-4A Phantom II—1961

Technical Specifications (F-4A)

Type: Two-place fighter transition trainer.
Manufacturer: McDonnell Aircraft Corp., St. Louis, Missouri.
Total U.S. military versions (trainers): 30–35 (USN).
Unit cost (est.): $1,900,000.
Powerplants: Two General Electric J-79-GE-2A turbojet engines, each rated at 10,350 lbs./s.t. dry, 15,000 lbs./s.t. with afterburning.
Armament: Six AIM-7 *Sparrow* missiles or four AIM-7 and four AIM-9 *Sidewinder* missiles in an internal bay or two AIM-4s carried internally and two AIR-2A *Genie* rockets carried on fuselage and wing stations.
Performance: Max. speed 1,485 mph (Mach 2.2), cruise 585 mph; ceiling 55,400 ft.; range 2,300 mi.
Weights: 28,000 lbs. empty, 54,600 lbs. loaded.
Dimensions: Span 38 ft. 5 in., length 58 ft. 3 in., wing area 530 sq. ft.

The well-known F-4 Phantom II—which ultimately served in greater numbers with the USAF than with the Navy and Marines—represents a conspicuous example of the trend toward Department of Defense (DoD) intervention in the aircraft procurement process (for more detail, see Explanatory Note Organization of Series III, above). Although the Phantom was originally designed to a 1954 Navy requirement for an all-weather fleet defense interceptor, by the time it reached operational service in 1961 as the F4H-1, the design had advanced to a true multi-role fighter having both air-to-air and air-to-ground capabilities. Because the J-79-GE-8 engines planned for the production series were not yet available, the first 45 F4H-1s delivered to the Navy came with less powerful -2A engines, and in 1961, to distinguish them from the F4H-1s with -8 engines, they were re-designated F4H-1F; and then when the tri-service designation system took effect in 1962, the F4H-1F became the F-4A and the F4H-1 the F-4B. Starting

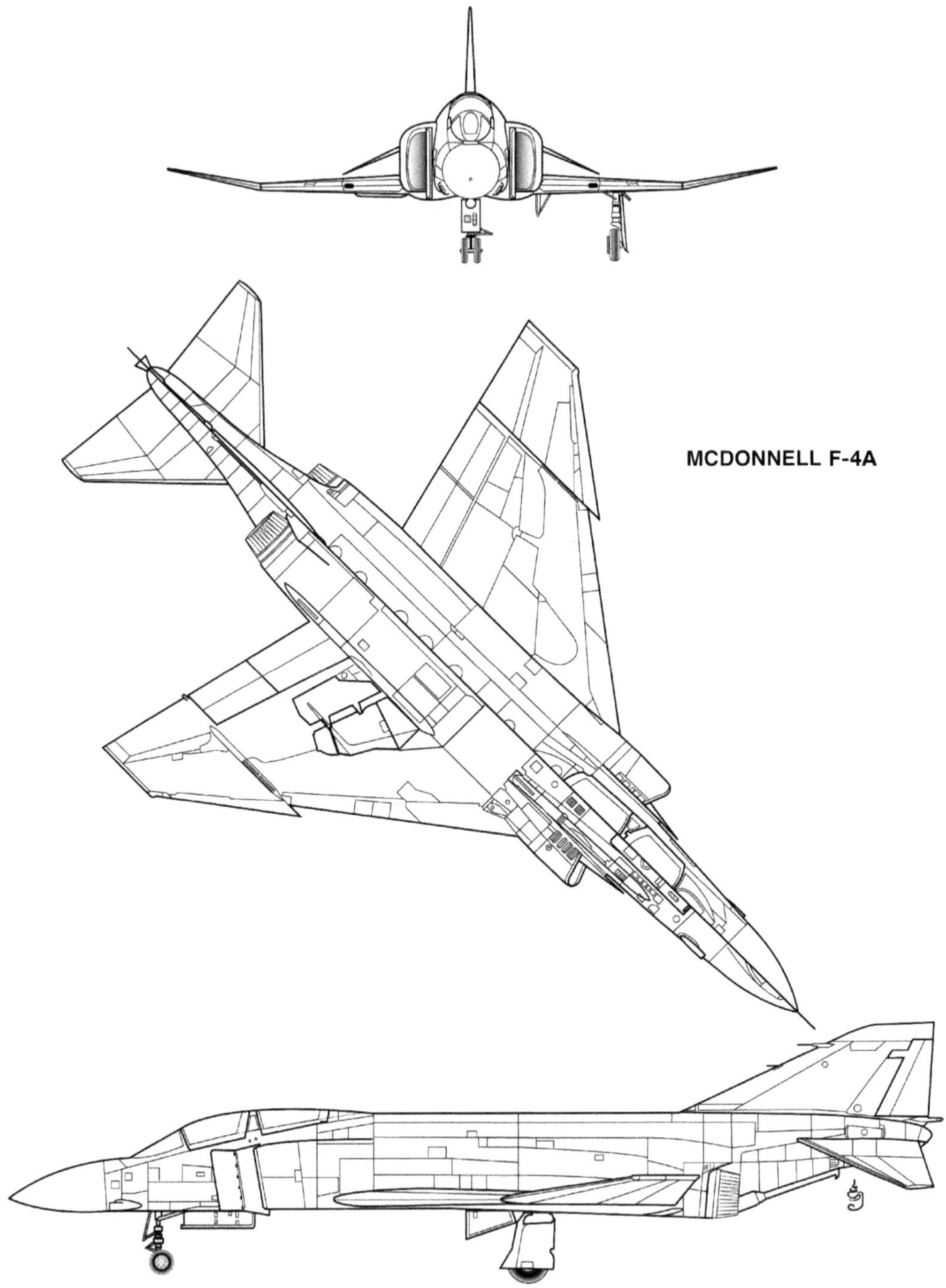

MCDONNELL F-4A

with the 19th F4H-1F/F-4A, the cockpit was elevated and the canopy enlarged and a larger, more bulbous radome added to the nose to house the larger APQ-72 radar dish.

Some of the F4H-1F/F-4As were retained for test purposes, but most went to form the first two Phantom-equipped Replacement Air Groups (RAG), the first being VF-101 at NAS Key West, Florida, and the second, VF-121 at NAS Miramar, California. Even though these air-

An early F4H-1F (F-4A after 1962) having the smaller nose and lower-profile canopy of the first 19 production aircraft. The aircraft pictured was a test example serving with VX-5 at NAWS China Lake, California.

craft were often referred to as TF-4As, no official change in designation appears in the serial number records. It should also be mentioned that the rear cockpits of F-4As were equipped for Radar Intercept Officers and lacked a set of controls. The F-4As continued as RAG transition trainers until withdrawn in 1970 and 1971, and some finished their days as maintenance trainers in Navy schools.

USAF Variants: In early 1962, by order of DoD, two Navy F4H-1s were delivered to Langley AFB, Virginia, to be evaluated by the USAF as the F-110A. The first USAF production version, designated F-4C under the new tri-service system, differed from the Navy F-4A/B in having a boom-type aerial refueling system, larger tires with bulged wheel-well doors, and dual controls for transition training and two-pilot operations. Subsequent USAF versions, the F-4D and F-4E (116 later converted to F-4G), likewise came with dual controls.

Vought F8U-1T/TF-8 Crusader—1962

Technical Specifications (F8U-1T/TF-8A)

Type: Mission-capable two-place fighter transition and combat trainer.
Manufacturer: Chance Vought Div. of United Aircraft Corp. (later Ling-Temco-Vought Corp.), Dallas, Texas.
Total U.S. military versions (trainers): 1 (USN).
Unit cost: Unknown.
Powerplant: One Pratt & Whitney J-57-P-20 turbojet engine, rated at 10,700 lbs./s.t. dry, 18,000 lbs./s.t. with afterburning.

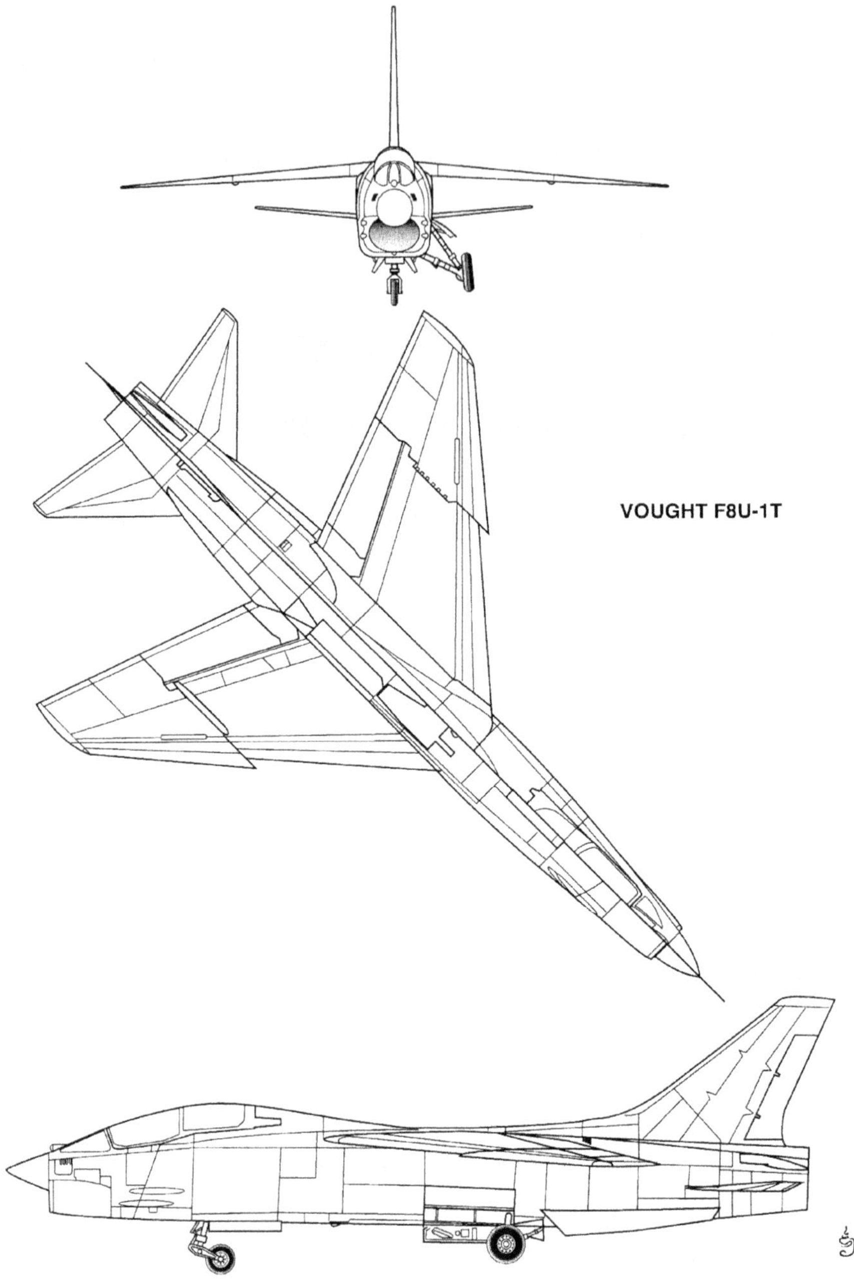

Armament: Two 20-mm cannons in the nose and two AIM-9 *Sidewinder* missiles on fuselage rails.
Performance: Max. speed 1,120 mph (Mach 1.7), cruise 560 mph; ceiling 58,000 ft.; range 1,100 mi.
Weights: 18,000 lbs. empty, 34,100 lbs. loaded.
Dimensions: Span 35 ft. 2 in., length 54 ft. 6 in., wing area 350 sq. ft.

The Vought F8U-1 Crusader, which became fully operational in 1957, is notable in having been the first Navy and Marine fighter capable of exceeding 1,000 mph in level flight. Early on,

The sole Vought TF-8A seen while attached to the Naval Test Pilot School as NAS Paruxent River, Maryland, during the 1960s. Though well suited to its purpose, budgetary restraints prevented procurement.

however, Crusaders showed themselves to be unforgiving in the hands of inexperienced pilots, especially in approaches to carrier landings. To help alleviate the problem, the Naval Bureau of Aeronautics (BuAer) directed Vought to develop a two-seat, dual-control variant from the 77th production F8U-1 which had previously been modified to serve as the F8U-2NE prototype. While keeping the overall dimensions, ventral fins, and afterburner air scoops of the single-seat F8U-2NE, the F8U-1T emerged with a new rear cockpit under a longer, one-piece clamshell canopy over both positions. The aft seat was raised 15 inches to provide adequate visibility for the instructor, giving the aircraft a noticeable humpbacked appearance from the side. To make space for the second cockpit, other changes included relocating certain equipment and deleting two of the nose cannons and the ventral rocket tray, and to enable it to operate from smaller, auxiliary fields, it featured a parabrake fitted to the base of the vertical fin and low-pressure tires. It is thought to have received a -1T suffix because the engine was derated to produce the same thrust as the P-4A engine of the F8U-1.

The first flight of the F8U-1T took place on February 6, 1962, and in September, it was redesignated TF-8A in conformity with the new tri-service scheme. After undergoing extensive testing, including carrier deck trials, with the Navy Test Pilot School at NAS Patuxent River, Maryland, the TF-8A went through functional testing of student pilots with VT-22 at NAS Kingsville, Texas. Although well suited to its purpose, the TF-8A did not achieve production because of cutbacks in the Navy's 1964 fiscal budget. After leaving NAS Patuxent River, the TF-8A was operated by NASA at Langley AFB, Virginia, and later still by the USAF Test Pilot's School at Edwards AFB, California. In 1977, when LTV (successor to Vought) contracted to provide refurbished F-8Hs (re-engined F8U-2Ns) to the Philippine Air Force, the TF-8A was regained to provide training but was destroyed a year later when an American instructor pilot and his Philippine Air Force trainee were forced to eject during a training flight.

Series III

Fixed Wing Trainers, 1962 to Present

Author's Note: Due to significant changes in both the designation and procurement of military aircraft during this era, the aircraft reported in this series will appear in chronological order without regard to service branch. Two important events occurred in 1962. First, in September, the Department of Defense (DoD) adopted a unified designation system (commonly referred to as the "tri-service" system) for all aircraft serving in the U.S. military branches (see detailed explanation in Appendix 1). Second, and more importantly, under the direction of Robert McNamara (Secretary of Defense 1961–1968), DoD began taking a far more active oversight role in the procurement of military aircraft for all services, and since the early 1970s, the process has also come under increasing scrutiny from the U.S. Congress. Although their aircraft requirements may still differ, none of the services have the autonomy in the design, development, and procurement of aircraft that characterized the pre–1962 era.

Synopsis of Trainer Procurement in All Services, 1962 to Present

This era may be distinguished by numerous by changes in procurement practices seen over the past 50 years, such as: (1) retention of some pre–1962 training aircraft far beyond their expected service lives (e.g., Beech T-34, Cessna T-37, Northrop T-38, North American T-2); (2) development of fewer types of new training aircraft; (3) procurement of new training aircraft in smaller quantities in slow-rate production over longer periods of time; (4) departure from the traditional prohibition against procurement of aircraft of foreign origin; (5) adoption of inter-service agreements between military branches with respect to both joint training of pilots and air crews and joint procurement of new training aircraft; (6) flight testing and evaluation with contractor-provided demonstrator aircraft rather than procured military prototypes; and (7) in connnection with new aircraft procurement programs, integration of "training systems" that include simulators, software and instructional materials.

Primary and Basic Trainers. In a move away from using civilian, contractor-provided aircraft, the USAF procured 256 Cessna 172s between 1964 and 1968 under the designation T-41 to be used as flight indoctrination trainers for ROTC, OTS, and Air Force Academy (AFA) pilot candidates, and in 1966 and 1967, the Army acquired a further 255 T-41s for primary fixed-wing training of student aviators. As reported in Series II, Part A, above, the Navy replaced its entire fleet of Beech T-34Bs with 352 turboprop-powered Beech T-34Cs delivered between 1975 and 1990. In 1982 the USAF purchased three ex-civil Cessna 150s which were subsequently modified and placed on the inventory as T-51As for use by the AFA's competitive collegiate flying team. Under the heading New Generation Trainer, USAF plans to replace its aging fleet of Cessna T-37s came to an halt when the Fairchild T-41A Eaglet program was abruptly terminated in March 1986 after less than a year of testing. In order to find an aerobatic trainer to replace Cessna T-41s under a new scheme known as the enhanced flight screening program (EFSP) for ROTC, OTS, and AFA pilot candidates, the USAF procured 114 British-designed Slingsby T67Ms between 1994 and 1996 as the T-3A Firefly, but after three fatal accidents and a series of non-fatal mishaps, all were grounded in 1997.

In the largest primary trainer procurement program since the T-34 and T-37, the USAF and Navy came together in 1990 to launch the Joint Primary Aircraft Training System (JPATS) and through 1993 and 1994, evaluated seven different factory demonstrators (six of which were of foreign origin). In June 1994 DoD declared the Swiss-designed Pilatus PC-9 Mk. 2 as winner of JPATS, awarding a contract to U.S.-based Raytheon to produce the aircraft as the T-6A Texan II. Under a slow-rate production program, 502 T-6As were delivered from 2000 to 2010, 453 to the USAF and 49 to the Navy, plus 12 T-6Bs with upgraded cockpits accepted by the Navy as of mid–2013. Under current estimates, the Navy plans to procure up to 250 more T-6Bs. In 2009, as part of the AFA initial flight training (IFT) program, the USAF leased 20 Austrian-designed Diamond DA40s under the unofficial designation T-52A. At nearly the same time, the USAF initiated a study to find a new IFT aircraft for direct procurement, resulting in the 2011 selection of 25 Cirrus SR20 G3s for production under the official designation T-53A, of which 22 have been delivered to date.

Advanced Trainers. During the Vietnam War (1962–1975), the Army reassigned an unspecified number of De Havilland U-6A Beavers to advanced training of fixed-wing student aviators with no apparent change in designation. For advanced multi-engine training of fixed-wing student aviators, the Army purchased 65 off-the-shelf Beech B55 Barons in 1965 as the T-42A Cochise. Between 1969 and 1974, the Navy replaced all Grumman TF-9Js (F9F-8T) in advanced training units with 281 two-seat McDonnell Douglas TA-4Js. To succeed its piston-engine Convair T-29 navigator trainers, the USAF procured 19 T-43As—modified Boeing 737–200 civil airliners—that entered service from 1973 to 1975 and starting in 1976, were used to provide inter-service undergraduate navigator training (USAF, USN, USMC, and USCG). In an across-the-board replacement of its piston-engine Grumman TS-2As (S2F-1T) in 1977 and 1978, the Navy procured 61 Beech King Air H90s under the Designation T-44A which, under an inter-service agreement implemented in 1996, have also served to train USAF student pilots earmarked to fly C-130s.

The Navy's heavier-than-air training aircraft experimental-training system (VTX-TS) program resulted in selection of the British-designed McDonnell Douglas T-45A Goshawk in 1981 to replace both the North American T-2C and the McDonnell Douglas TA-4J in intermediate and advanced training of student naval aviators and although development and production became protracted, 236 aircraft have been delivered, 72 T-45As from 1992 to 1995 and 158 T-45Cs with improved avionics from 1997 to 2007, and since 2002, all T-45As have been brought up to the T-45C standard. Significantly, the VTX-TS project was merged with a "training system" which included flight simulators, software, and training materials. Less successful were Navy endeavors to replace its North American T-39D Sabreliners (see in Series II, Part A, above) in Naval Flight Officer training with 15 modified Cessna Citation 550 business jets procured in

1984 under the designation T-47A. Ironically, all T-47As were withdrawn in 1993 due to structural problems only to be replaced by 17 rebuilt Sabreliners that came back into service as the T-39N. Out of an effort to extend the service life of its Northrop T-38 Talons (see in Series II, Part A, above), the USAF instituted two programs, the three-phase Pacer Classic program started in 1984 to modernize and structurally improve all T-38As, followed by Pacer Classic II started in 2001, in which 500 T-38As reemerged as T-38Cs following wing replacement and improvements to the cockpit and engines.

USAF efforts to find its own T-39 replacement yielded success in 1992 when it procured 180 militarized Beechjet 400s as T-1A Jayhawks for advanced training of student pilots on the airlift/tanker track, and since 2009, has also been used for advanced training of Combat Systems Officers. In 1996, after the Navy undertook joint training of Navy, Marine, USAF, and USCG student pilots in the advanced multi-engine phase, it reassigned 14 (possibly more) Beech C-12B Huron transports to training duties under the designation TC-12B. Under a program headed trainer-experimental (T-X), the USAF has been making plans to replace the Northrop T-38 Talon (average age 43+ years) since 1993. However, no measurable progress was made until the USAF circulated a request for information in 2009,` which since that time has encouraged four foreign manufacturers with American partners to offer potential candidates: General Dynamics/Alenia-Aeromacchi T-100 (Italy); Northrop-Grumman/BAE Hawk T2/128 (Britain); Lockheed-Martin/KAI T-50 (Korea); and Boeing/Saab Gripen JAS-39F (Sweden). Originally, the plan contemplated a new aircraft in service by 2017; however, budgetary restraints have pushed it back to 2023 or later and thus far, no official request for proposals has been issued by DoD.

Operational and Combat Trainers. During the build-up for the Vietnam War in the mid–1960s, the USAF converted approximately 13 Lockheed C-121 variants to training duties, four as TC-121G transport crew trainers and nine as TC-121C airborne early warning (AEW) crew trainers. A two-seat A-4 had been previously proposed, but the need to prepare attack pilots for duty in the Vietnam combat zone prompted the Navy to procure 241 TA-4Fs that were delivered to fleet replacement squadrons in 1966 and 1967. The Navy accepted nine Grumman G-159 Gulfstreams in 1967 under the designation TC-4C which were specially modified to train bombardier-navigators in A-6 replacement training squadrons. For AEW crew training, the Navy converted nine E-2s, four as the TE-2A in 1971 and five as the TE-2C in 1991, both of which served in AEW replacement squadrons. Due to the high initial accident rate experienced when the first McDonnell Douglas AV-8A Harriers began entering service with the Marines in 1971, the production contract was subsequently amended to include eight TAV-8A dual-control trainers that all entered service with VMAT-203 in 1973. Similarly, in connection with the AV-8B, VMAT-203 began taking delivery of the first of 24 TAV-8Bs in 1987, 19 of which remain active today.

As McDonnell Douglas F-15A Eagles initially entered USAF service in 1975, the production plan included 58 two-seat TF-15As (re-designated F-15B) delivered to replacement training units and when F-15Cs became operational in 1979, they were likewise accompanied by 61 two-seat F-15Ds. In 1976 and 1977, the Navy contracted with Ling-Temco-Vought to convert 60 A-7Bs and Cs into a two-seat attack trainer configuration which afterward returned to service with fleet replacement squadrons as the TA-7C. The dates when Tactical Air Command first utilized early C-130 variants as transition trainers is not known, however, once all USAF tactical airlift was merged into Military Airlift Command in 1974, all C-130 transition and operational training was consolidated into one airlift wing, initially with C-130Es, and later, C-130Hs (1990s-present), and later still, C-130Js (2003-present). As with the F-15, the General Dynamics F-16 Fighting Falcon production program was augmented by deliveries of 121 two-seat F-16Bs to replacement training units from 1979 to 1985, followed by 205 two-seat F-16Ds from 1985 to 2005. Another two-seat A-7 variant, procured this time by the USAF, resulted in delivery of 30 A-7Ks to Air National Guard units between 1981 and 1984.

To train pilots for transition to new McDonnell Douglas F-18A Hornets, the Navy procured 39 two-seat TF-18As (re-designated F/A-18B in 1984) that entered service between 1980 and 1985 with fleet replacement squadrons and Marine squadron VMFAT-101, and correspondingly, approximately 30 two-seat F/A-18Ds retrofitted with stick and throttle kits entered service between 1987 and 2000 and an estimated eight to ten similarly retrofitted two-seat F/A-18Fs entered service from 1999 to present day. During the mid–1980s, the USAF modified approximately six aircraft as pilot and crew trainers for the electronic intelligence-gathering and AWACS roles: one Boeing C-135 as the RC-135T (Cobra Ball mission) and another as the TC-135W (Rivet Joint mission); and four Boeing C-18s (ex-airline 707–320Bs), two as the EC-18E (AWACS mission) and two as the TC-18F (Mercury mission). The Navy contracted with Lockheed in 1985 to convert 12 P-3A Orions into flight engineer and crew trainers which returned to service with two fleet replacement squadrons as the TP-3A.

De Havilland U/TU-6–1962

Technical Specifications (U/TU-6A)

Type: STOL transition trainer.
Manufacturer: De Havilland Aircraft of Canada, Ltd., Downsview, Ontario, Canada.
Total U.S. military versions (trainers): Unknown (U.S. Army).
Unit cost: Unknown.
Powerplant: One 450-hp Pratt & Whitney R-985-SB3 *Wasp Junior* 9-cylinder air-cooled radial engine driving a two-bladed Hamilton Standard constant-speed propeller.

Army L-20A Beavers, like the one pictured, were redesignated U-6A in 1962. During the 1960s, some were converted to dual controls and employed as fixed-wing transition trainers at Fort Rucker, Alabama.

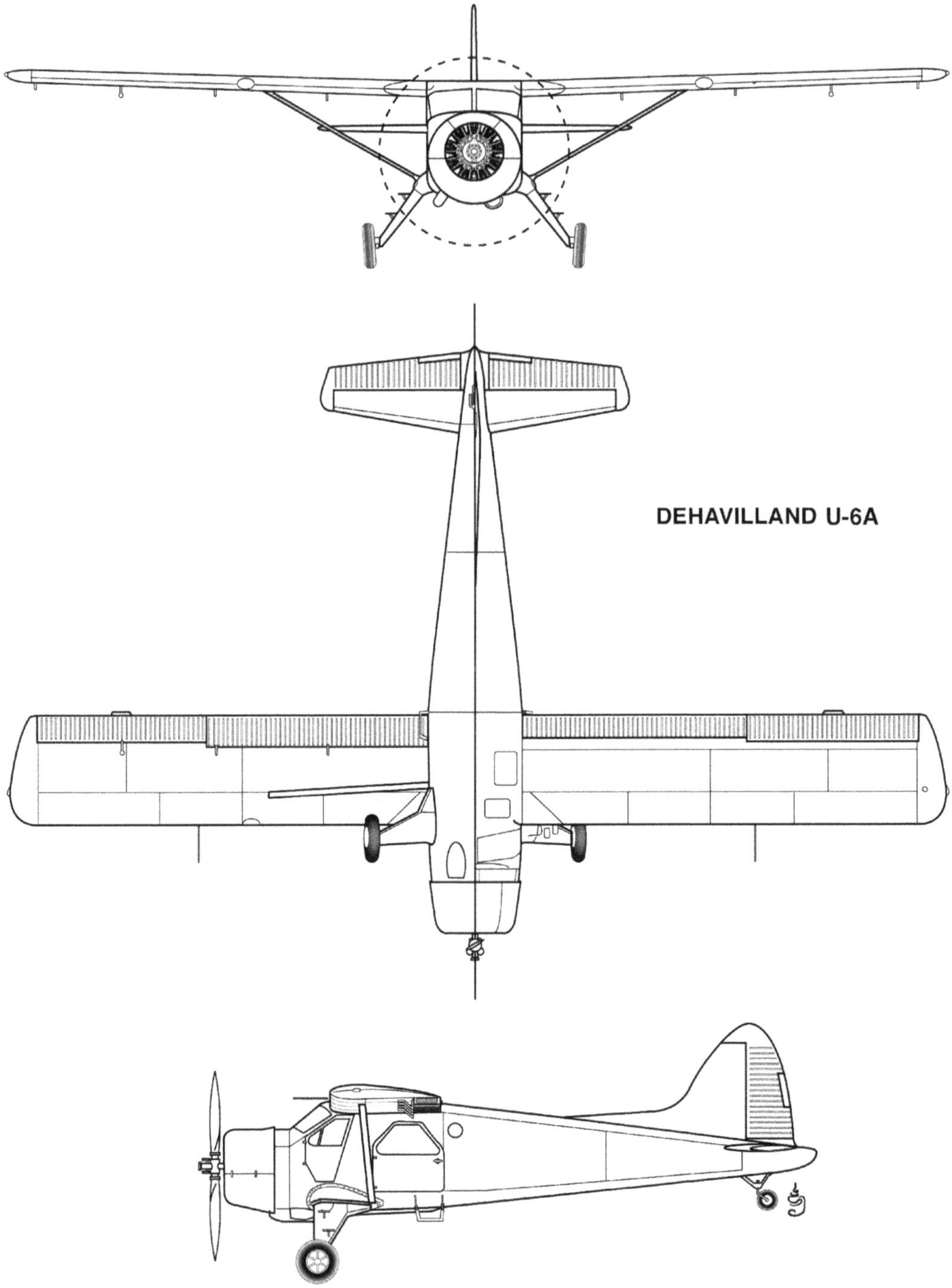

DEHAVILLAND U-6A

Performance: Max. speed 163 mph, cruise 130 mph; ceiling 20,000 ft.; range 455 mi.
Weights: 3,000 lbs. empty, 5,100 lbs. max. takeoff.
Dimensions: Span 48 ft., length 30 ft. 4 in., wing area 250 sq. ft.

The De Havilland of Canada DHC-2 "Beaver," specifically designed for STOL operations in the rough country of the Canadian bush, made its first flight on August 16, 1947, and became available on the civil market in mid–1948. The first Beavers to be evaluated by the U.S. military were delivered in 1951, four to the USAF and two to the Army under the designation YL-20. The Army was especially interested in procuring Beavers because of an urgent need for fixed-wing utility transports to support combat operations in Korea. Soon afterward a series of military contracts resulted in delivery of some 961 aircraft between 1952 and 1961 as the L-20A, with 654 going to the Army, about 200 to the USAF, and the remainder to foreign air forces under Military Assistance Programs, and following adoption of the tri-service designation system in 1962, the L-20A became the U-6A. Though the exact dates and numbers of aircraft are uncertain, Army historical records indicate that some U-6As were converted to a dual-control configuration for the purpose of transitioning Army pilots from the L-19/O-1s to larger fixed-wing types, either at Ft. Rucker AAF, Alabama, or Ft. Steward AAF, Georgia. One source refers to the converted aircraft as TU-6As, however, serial number records show no official change in designation.

Cessna T-41 Mescalero—1964

Technical Specifications (T-41A)

Type: Primary and flight screening trainer.
Manufacturer: Cessna Aircaft Co., Wichita, Kansas.
Total U.S. military versions: 511 (256 USAF; 255 U.S. Army).
Unit cost: $13,000.
Powerplant: One 145-hp Continental O-300-D 6-cylinder air-cooled horizontally-opposed engine driving a two-bladed McCauley fixed-pitch metal propeller.
Performance: Max. speed 139 mph, cruise 131 mph; ceiling 13,100 ft.; range 592 mi.
Weights: 1,330 lbs. empty, 2,300 lbs. max. takeoff.
Dimensions: Span 35 ft. 10 in., length 26 ft. 11 in., wing area 159 sq. ft.

During the early 1960s, due to a surge in pilot training prompted by impending retirement of many World War II/Korea-era pilots in conjunction the escalation of U.S. military involvement in Southeast Asia, the USAF chose to institute a flight indoctrination course in which pilot candidates (AFROTC and later, OTS) would receive 30 hours of instruction from civil contractors in USAF-owned lightplanes. To this end, the USAF awarded Cessna a contract in 1964 to deliver 170 off-the-shelf model 172Es under the assigned designation T-41A, and in 1967, as the program expanded, the USAF acquired 34 additional T-41As. In 1966, to augment and eventually replace O/TO-1s in its fixed-wing pilot training program, the Army ordered 255 Cessna 172Es upgraded to 210-hp Continental IO-360D engines and constant-speed propellers, which were delivered in 1966 and 1967 under the designation T-41B. The name "Mescalero" was applied to the Bs in the Army practice of naming its aircraft after native American tribes. When the USAF initiated a new flight indoctrination program for U.S. Air Force Academy cadets in 1968, it procured 52 aircraft, essentially identical to Army T-41Bs, as the T-41C.

As they came into active service during 1965, USAF T-41As were distributed among civil contractors (i.e., fixed-base operators with flight schools) in the vicinity of the eight undergraduate pilot training (UPT) bases in Georgia, Oklahoma, Arizona, and Texas, then between 1971and 1973, all USAF flight screening was centralized at Hondo Municipal Airport in Texas. Starting in late 1966, Army T-41Bs went into primary training of fixed-wing student aviators at Stewart AAF, Georgia, until 1973, when the operation was moved to Ft. Rucker AAF, Alabama. Training of Air Force Academy cadets in T-41Cs began at Peterson AFB, Colorado in early 1968. The process of withdrawing the USAF T-41As began in the mid–1990s as they were

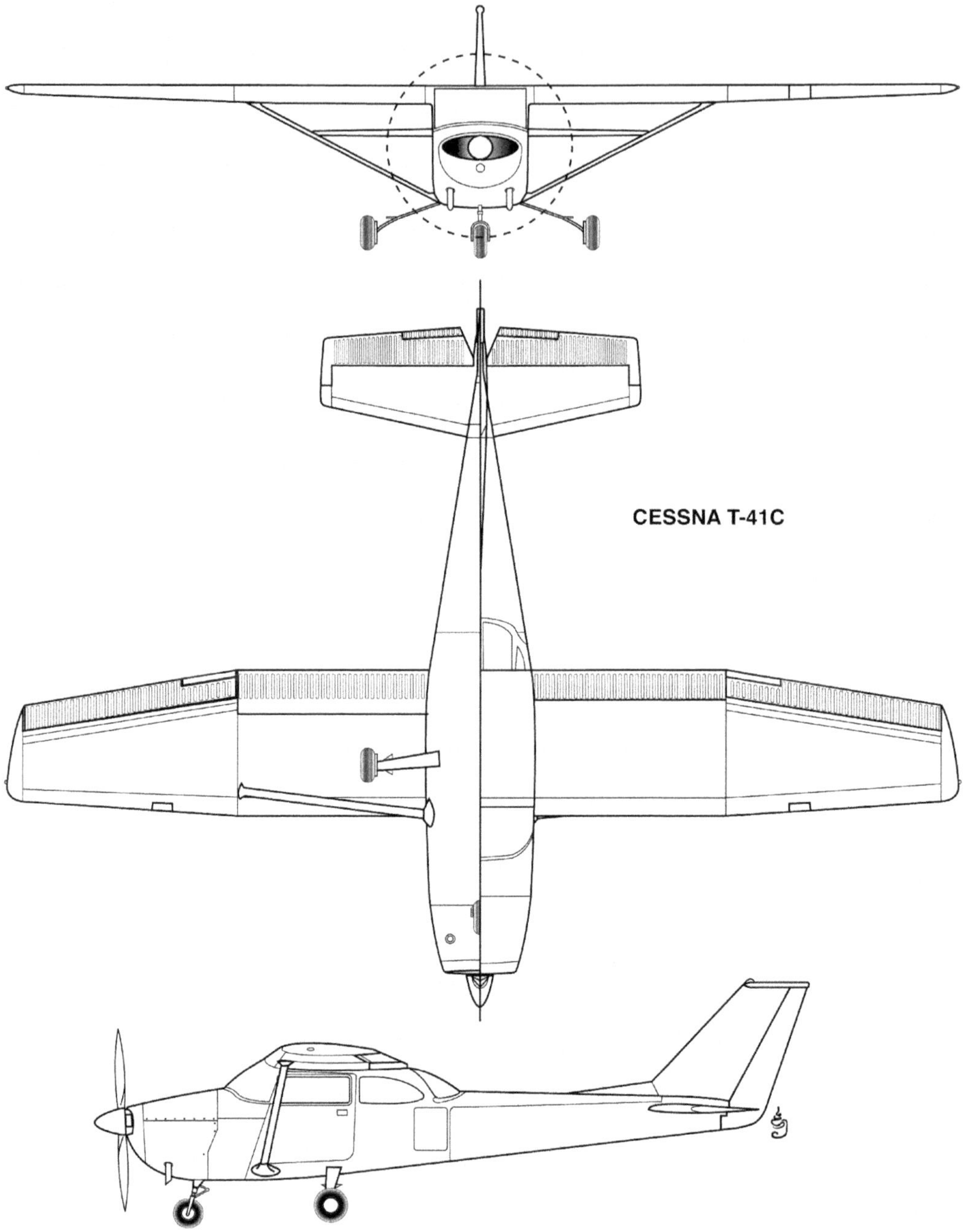

replaced by newer aircraft, namely the Fournier-Slingsby T-3A (see below), and during the same timeframe, the Army T-41Bs were transferred to foreign air forces under MAP, Army flying clubs, or Civil Air Patrol. Meanwhile, the AF Academy program also began to reequip with the T-3A, however, some of its T-41Cs, after receiving an avionics upgrade, were retained as T-41Ds, and as of mid–2013, three still remain on the active inventory.

One of 255 Cessna T-41Bs procured by the Army in 1966 for primary training of fixed-wing pilots. The last examples are thought to have been withdrawn in the mid–1990s.

Lockheed TC/EC-121 Constellation/Warning Star—1964

Technical Specifications (TC-121G)

Type: Transport and airborne early warning (AEW) crew trainer.
Manufacturer: Lockheed Aircraft Corp., Burbank, California.
Total military versions: 13 (USAF).
Unit cost: Unknown.
Powerplants: Four 3,400-hp Wright R-3350–34TC *Double Cyclone* 18-cylinder air-cooled radial engines driving three-bladed, fully-reversible constant-speed metal propellers.
Performance: Max. speed 299 mph, cruise 255 mph; ceiling 25,000 ft.; range 4,250 mi.
Weights: 69,210 lbs. empty, 143,000 lbs. loaded.
Dimensions: Span 116 ft. 2 in., length 126 ft. 2 in., wing area 1,650 sq. ft.

The famed Lockheed L-1049 Super Constellation, remembered best as a triple-tail, long-haul passenger airliner, served over 30 years (1948–1983) with USAF and the Navy as both a long-range logistical transport (C-121C/G/J) and airborne early warning (AEW) platform (RC/EC-121C/D/K Warning Star). Most of the transport variants were withdrawn in the between 1961 and 1967, though the last USAF examples remained in ANG units until 1973. The AEW variants, distinguishable by large ventral and dorsal radomes on the fuselage, were the forerunners of today's airborne warning and control (AWACS) system and remained active with the USAF until 1978 and with the Navy until 1983. The normal aircrew complement of a Warning Star comprised 18 people: two pilots, two navigators, two weapons controllers, two flight engineers, one radio operator, two crew chiefs, five radar operators, and two radar technicians.

Serial number records indicate that four USAF C-121G transports were re-designated as TC-121G crew trainers; however, the nature of their training mission and unit assignment do not appear in any historical records. During the Vietnam War (1962–1972), nine EC-121Cs

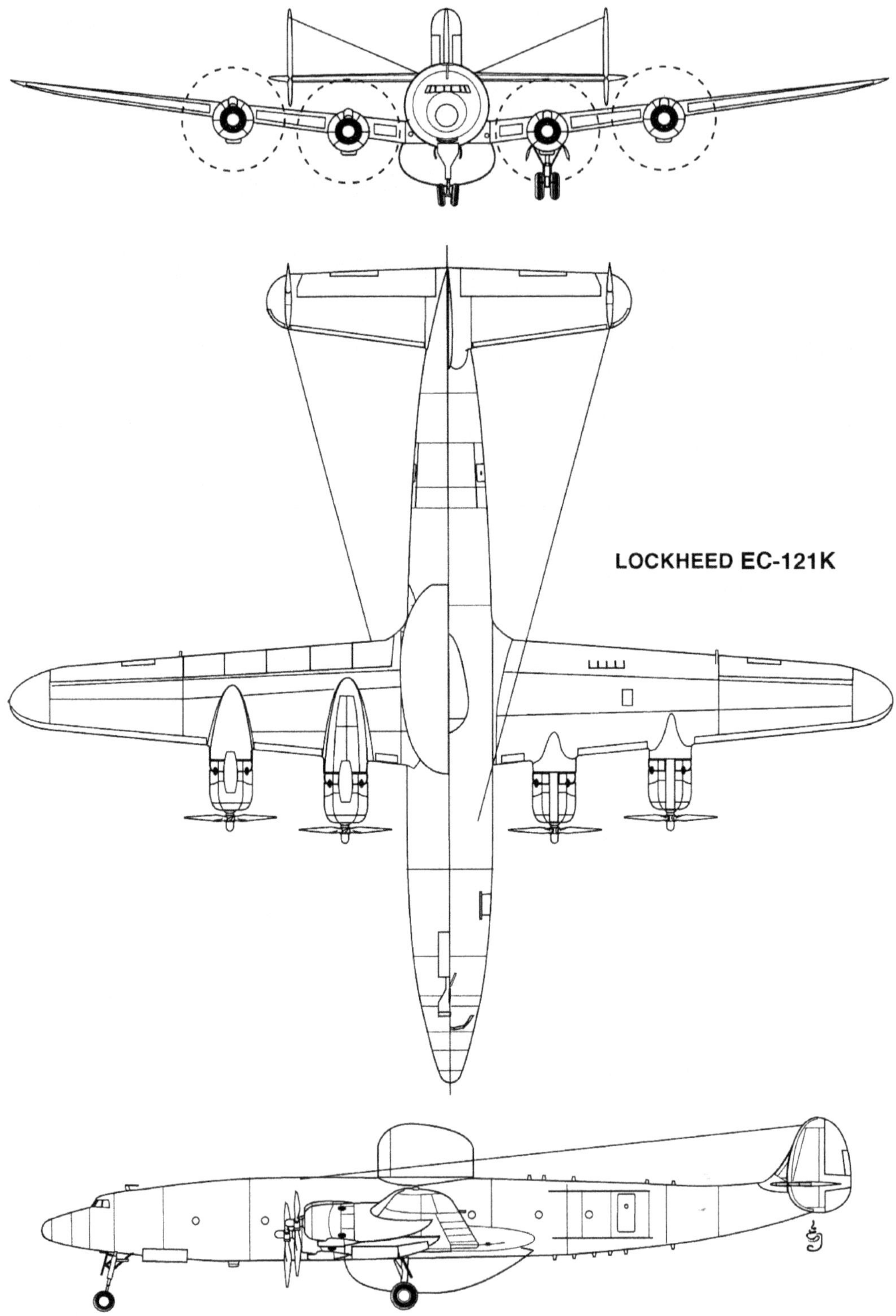

LOCKHEED EC-121K

Serial number records indicate that the aircraft pictured is one of four TC-121Gs used by the Military Air Transportation Service as crew trainers. No record appears of their actual unit assignment.

became TC-121C AEW crew trainers and were attached initially to the 522nd Airborne Early Warning and Control Wing (AEWCW) at McClellan AFB, California, and later to the 79th AEWCS, a reserve squadron based at Homestead AFB, Florida, until 1978. One Navy EC-121K attached to VT-86 out of NAS Glynco, Georgia, was used for AEW training of student Naval Flight Officers destined for EC-121s or E-2s until being retired in 1983.

McDonnell Douglas TA-4 Skyhawk—1965

Technical Specifications (TA-4J)

Type: Carrier-capable two-place advanced and combat trainer.
Manufacturer: McDonnell Douglas Corp., Long Beach, California.
Total U.S. military versions: 522 (USN, USMC).
Unit cost: Unknown.
Powerplant: One Pratt & Whitney J-52-P-6A turbojet engine, rated at 8,500 lbs./s.t.
Armament (TA-4F only): Two 20-mm cannons in the nose and up to 8,200 lbs. of mixed ordnance on external stations.
Performance: Max. speed 660 mph, cruise 490 mph; ceiling 38,700 ft.; range 1,350 mi.
Weights: 10,602 lbs. empty, 15,783 lbs. loaded.
Dimensions: Span 27 ft. 6 in., length 42 ft. 7 in., wing area 260 sq. ft.

The venerable Douglas/McDonnell Douglas A4D/A-4 Skyhawk, popularly known as the "Scooter," had been in operational service for nearly nine years before the first two-seat TA-4E

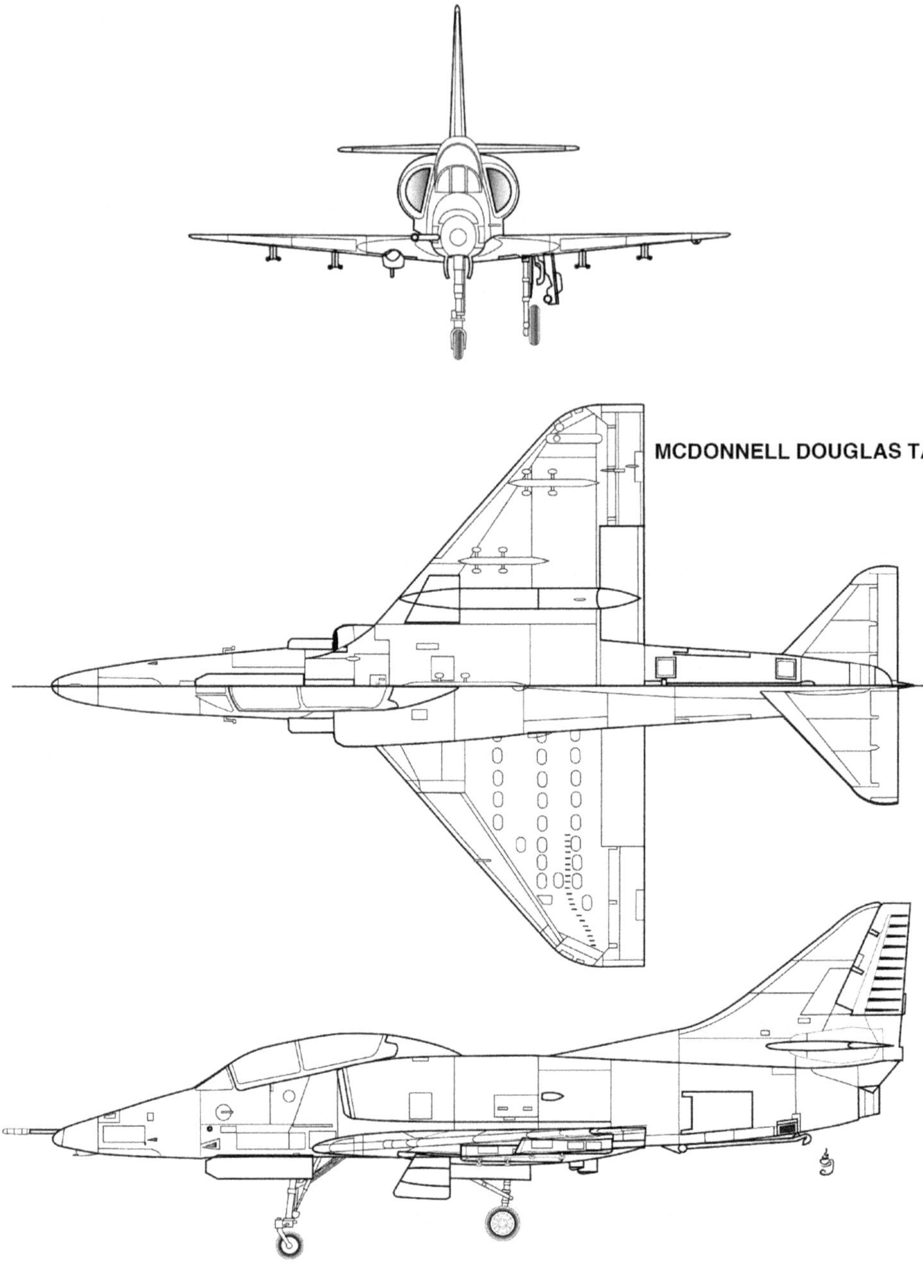

flew in 1965. Although the idea of a two-seat Skyhawk had already been put forward by Douglas, the need for it did not become pressing until 1964 when Navy and Marine A-4s units were becoming heavily engaged in conventional air strikes over Southeast Asia. Navy planners by that time reasoned that a two-seater would not only provide valuable combat training but would also enable Replacement Air Groups (RAG) to release many of their single-seat A-4s for combat duty. Thus, in mid–1964, the Navy amended Douglas's existing A-4E contract to call for modification of two aircraft on the assembly into a two-seat configuration under the designation TA-4E.

This McDonnell Douglas TA-4F was photographed in 1980 while serving with H&MS-13 at MCAS El Toro, California. Most Fs were later converted to TA-4Js and employed in advanced pilot training.

While retaining the armament and most of the A-4E airframe, basic modifications entailed adding a 28-inch plug to the forward fuselage to house the second cockpit and a longer, one-piece canopy enclosure. Other enhancements included lift spoilers on the upper wing forward of the flaps to improve crosswind landing characteristics and nosewheel steering for better carrier deck maneuvering. Following the maiden flight of the first modified TA-4E on June 30, 1965, the type's designation was changed to TA-4F to reflect the improvements made to the overall A-4 production standard, and deliveries of the first of 241 production TA-4Fs began reaching fleet service in May 1966. In early 1968, with the aim of replacing all TF-9Js (see in Series II, Part B) in Naval Air Training Command (NATC), the Navy awarded McDonnell Douglas a new contract for 281 two-seat Skyhawks to be delivered as the TA-4J. Since it was intended mainly for advanced pilot training, the TA-4J differed from the TA-4F in having all tactical armament and associated equipment removed except for the two 20-mm nose cannon, two less wing pylons, and a P-6A engine of 800 lbs. less thrust. The first TA-4J flew on December 17, 1968, deliveries to training units began in mid–1969, and the process of replacing all TF-9Js was completed by early 1974.

In May 1966, the first production TA-4Fs became operational in the Pacific Fleet RAG with VA-125 at NAS Lemoore, California, and as deliveries proceeded, VA-126 at NAS Miramar, California, and VA-127, also at Lemoore; in the Atlantic Fleet RAG, with VA-43 at NAS Oceana, Virginia, and VA-44 and -45, both at NAS Cecil Field, Florida; and in Fleet Marine Force Pacific, with VMAT-10 at MCAS Yuma, Arizona, and in Fleet Marine Force Atlantic, VMAT-203 at MCAS Cherry Point, North Carolina. Likewise, a TA-4F was allocated for proficiency training to each of the 36 Navy and 20 Marine squadrons equipped with A-4s at the time. TA-4Fs also served in an adversary training role with VF-101, a Fleet Readiness Squadron based at NAS Oceana. Between 1967 and 1975, as Navy fleet squadrons replaced all their single-seat A-4s with

TA-4J of VT-21 out of NAS Kingsville, Texas. VT-7 completed the last TA-4J class in October 1999, but the final examples did not leave VC-8 until 2003.

the Ling-Temco-Vought A-7 Corsair II (see TA-7, below), serial number records indicate that at least 131 TA-4Fs were stripped of most armament and seconded to NATC with the new designation TA-4J, and a number of others went to foreign air forces then equipped with A-4s. Within a similar timeframe, four TA-4Fs were modified as EA-4F electronic warfare trainers and operated with VAQ-33 from 1970 to 1987. In the late 1970s, as Marine squadrons reequipped with the much-improved A-4M, 23 TA-4Fs were set aside as dedicated forward air controllers, and after being upgraded with A-4M avionics and other improvements, reentered service under the new designation OA-4M. OA-4Ms were split between H&MS-13 at MCAS El Toro, California, and H&MS-32 at Cherry Point where they served until along with A-4Ms, were retired from service in the early 1990s.

TA-4Js (including converted TA-4Fs) ultimately became active in five NATC Training Wings between 1969 and 1974: VT-4 at NAS Pensacola, Florida; VT-7 at NAS Meridian, Mississippi; VT-21, -23 and -24 at NAS Kingsville, Texas; VT-24, -25, and -26 at NAS Chase Field (Beeville), Texas; and VT-35 at NAS Corpus Christi, Texas. TA-4Js also served in an adversary role with VF-171 out of NAS Key West, Florida, until 1994. The gradual phase-out of TA-4Js from NATC in favor of the McDonnell Douglas T-45 Goshawk (see below) began in 1991, with the process being completed when VT-7 completed its last TA-4J class in October 1999. However, the very last active service TA-4Js were not retired from VC-8 until 2003.

Beech T-42 Cochise—1965

Technical Specifications (T-42A)

Type: Instrument and multi-engine crew trainer.
Manufacturer: Beech Aircraft Corp., Wichita, Kansas.

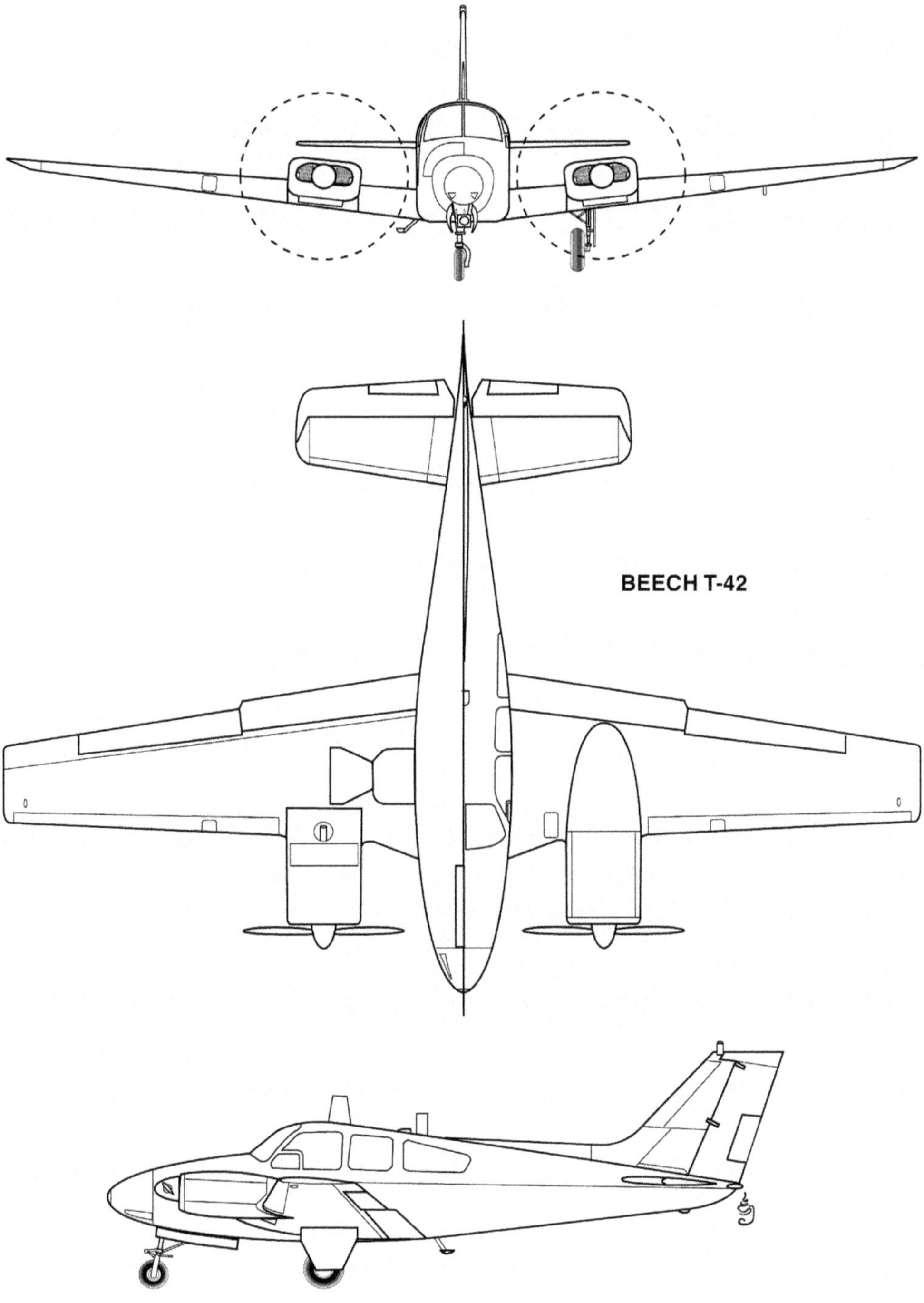

One of 55 Beech T-42As delivered to the Army in 1965 and 1966. All entered service with the Army Aviation Shool at Fort Rucker, Alabama, where they were used for multi-engine and instrument training.

Total military versions: 65 (U.S. Army).
Unit cost: $60,000 (est.).
Powerplants: Two 260-hp Continental IO-470-L 6-cylinder air-cooled horizontally-opposed engines driving Hartzell two-bladed constant-speed metal propellers.
Performance: Max. speed 236 mph, cruise 207 mph; ceiling 19,700 ft.; range 1,085 mi.
Weights: 3,156 lbs. empty, 5,100 lbs. loaded.
Dimensions: Span 37 ft. 10 in., length 28 ft., wing area 199.2 sq. ft.

When introduced to the civil market in 1961, the Beech model 55 "Baron" became an immediate success, with over 2,000 examples being sold in three variants over the next few years (i.e., models 55, A55, and B55). In February 1965, following a design competition, the Army Aviation Board selected the Beech model B55 as a the Army's new multi-engine and instrument transition trainer for fixed-wing student pilots earmarked for larger aircraft such as the Beech U-8, Beech U-21, De Havilland CV-2, and Grumman OV-1. Soon afterward, the Army awarded Beech a contract to deliver 65 aircraft under the designation T-42A, and in line with Army practice, gave it the native–American name "Cochise." Initial deliveries to the Army Aviation School at Ft. Rucker, Alabama, commenced in September 1965 and continued unto 1966.

On routine training missions, T-42As carried two students and one instructor-pilot but because they had four seats, often did double duty as utility transports between Ft. Rucker and other bases. In 1972 the Army procured five more T-42As that were redelivered to the Turkish Air Force under Military Assistance Programs and in 1972, from its inventory, sold seven to the Spanish Air Force. The timeframe over which T-42As were withdrawn from the training role is not clear from historical data, however, by the mid–1980s, the 30 remaining on the inventory were serving exclusively as utility transports with the Army Reserve and National Guard, and serial number records reflect that the last examples were sold surplus during the early 1990s.

Grumman TC-4 Academe—1967

Technical Specifications (TC-4C)

Type: Bombardier-navigator trainer.
Manufacturer: Grumman Aircraft Engineering Corp., Bethpage, New York.

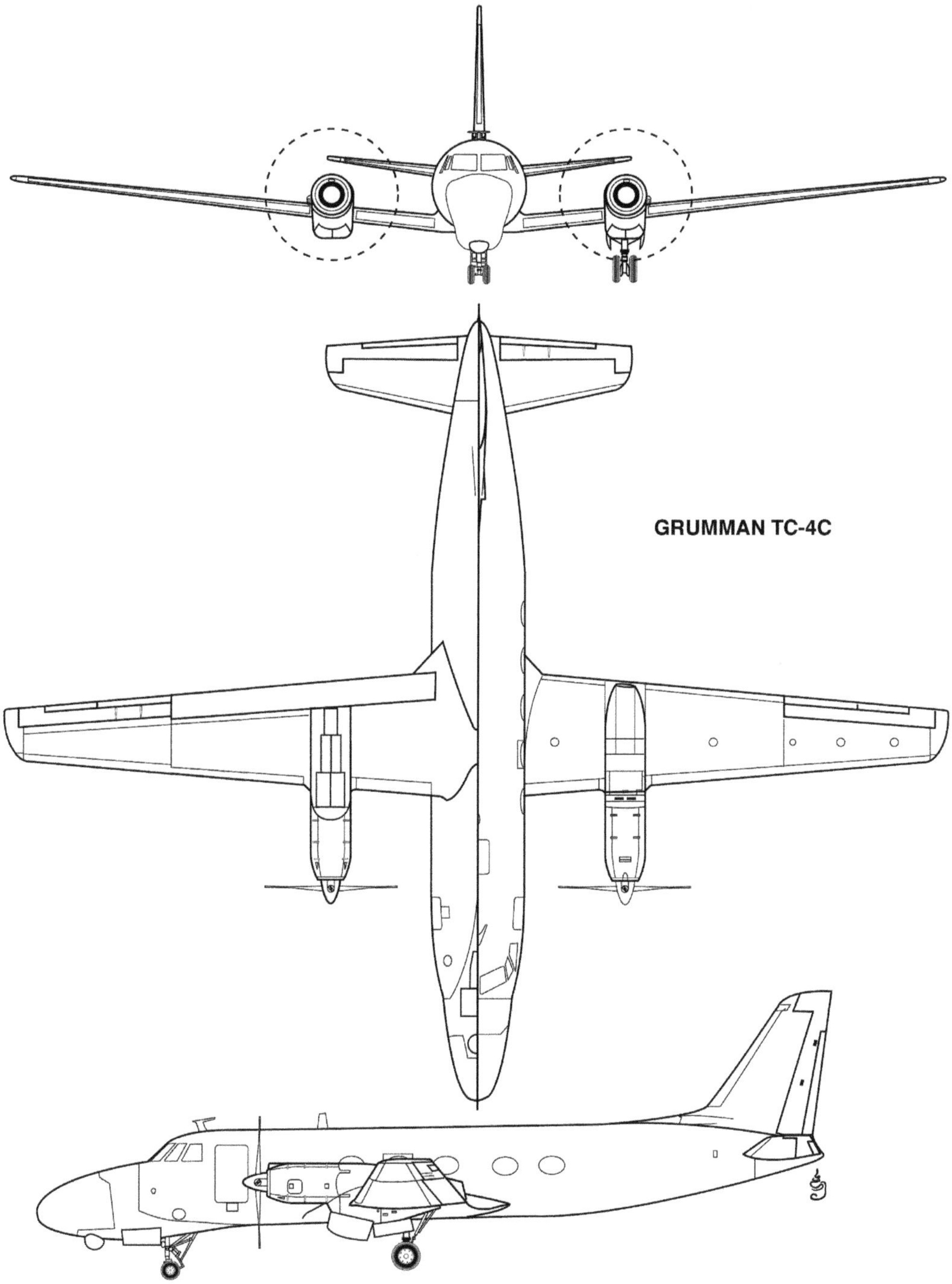

The Navy procured nine of these Grumman TC-4Bs in 1966 that came specially equipped to train bombardier-navigators for Grumman A-6s. They served in VA-42, VA-128, and VMAT(AW)-202 until 1995.

Total U.S. military versions (trainers): 9 (USN, USMC).
Unit cost: Unknown.
Powerplants: Two 2,185-shp Roll Royce *Dart* 529–8X turboprop engines driving four-bladed Dowty Rotol fully feathering, reversible-pitch propellers.
Performance: Max. speed 348 mph, cruise 288 mph; ceiling 26,000 ft.; range 1,865 mi.
Weights: 22,070 lbs. empty, 36,000 lbs. max. takeoff.
Dimensions: Span 78 ft. 4 in., length 67 ft. 11 in., wing area 615 sq. ft.

The Grumman model G-159 Gulfstream made its debut on the civil market in 1958 as the first American-designed, turbo-powered aircraft to be offered as a long-range executive transport. Fully pressurized, it accommodated a crew of two and between 10 and 24 passengers, depending on interior configuration. The design of the G-159 emerged with an exceptionally clean low-wing layout characterized by a very thin, Davis-type wing airfoil section and a swept vertical fin. Over a 10-year production life, Grumman sold 189 of these aircraft to civil purchasers, mainly large corporations. Military orders came initially from the Army and the Coast Guard when each procured one Gulfstream with a VIP interior, both of which were delivered in 1963 under the designation VC-4A.

Offical records show that the Navy reserved serial numbers in 1961 for ten Gulfstreams to be configured as combination navigator-trainer/transports under the initial designation T-41A (later reassigned to Cessna, see above), which changed to TC-4B in 1962; however, the project was cancelled before a contract was finalized. Then in late 1966, under the designation TC-4C and the name "Academe," the Navy awarded Grumman a contract to deliver nine G-159s specially modified as Grumman A-6 Intruder bombardier-navigator trainers. Changes to the basic airframe included addition of a bulbous nose radome to house the multiple radars of the A-6's Digital Integrated Attack/Navigation Equipment (DIANE) system, a simulated A-6 cockpit in the main cabin, plus an auxiliary power unit (APU) to run the additional systems. The first TC-4C completed its maiden flight on June 14, 1967, and deliveries to the Navy and Marines commenced shortly thereafter. In operational service, six TC-4Cs were initially split between VA-42 (East Coast RAG) at NAS Oceana, Virginia, and VA-128 (West Coast RAG) at NAS Whidbey Island, Washington, and three to VMAT(AW)-202 at MCAS Cherry Point, North Carolina. In 1975 one of the Marine Academes was destroyed in an accident, and in 1991, when VMAT(AW)-202 was disestablished, its two aircraft went to VA-42 and -128. The retirement of TC-4Cs coincided with the withdrawal of all A-6 attack variants from Navy service in 1995.

Grumman TE-2 Hawkeye—1970

Technical Specifications (TE-2A)

Type: Airborne early warning (AEW) crew trainer.
Manufacturer: Grumman Aerospace Corp., Bethpage, New York.
Total U.S. military versions (trainers): 9 (USN).
Unit cost (E-2C): $51,000,000.
Powerplants: Two 4,050-shp Allison T-56-A-8 turboprop engines driving four-bladed Aeroproducts fully feathering, reversible-pitch propellers.
Performance: Max. speed 368 mph, cruise 250 mph; ceiling 31,700 ft.; range 2,650 mi.
Weights (E-2C): 37,648 lbs. empty, 51,569 lbs. max. takeoff.
Dimensions: Span 80 ft. 7 in., length 56 ft. 4 in., wing area 700 sq. ft.

The Grumman E-2 "Hawkeye," which remains in Navy service today and probably many years to come, was the first carrier-based aircraft to be designed from the ground up as an airborne

One of four unmodified E-2As redesignated in 1971 as TE-2A AEW crew trainers and assigned to two replacement squadrons, RVAW-110 at NAS North Island, Califonia, and RVAW-120 at NAS Norfolk, Virginia.

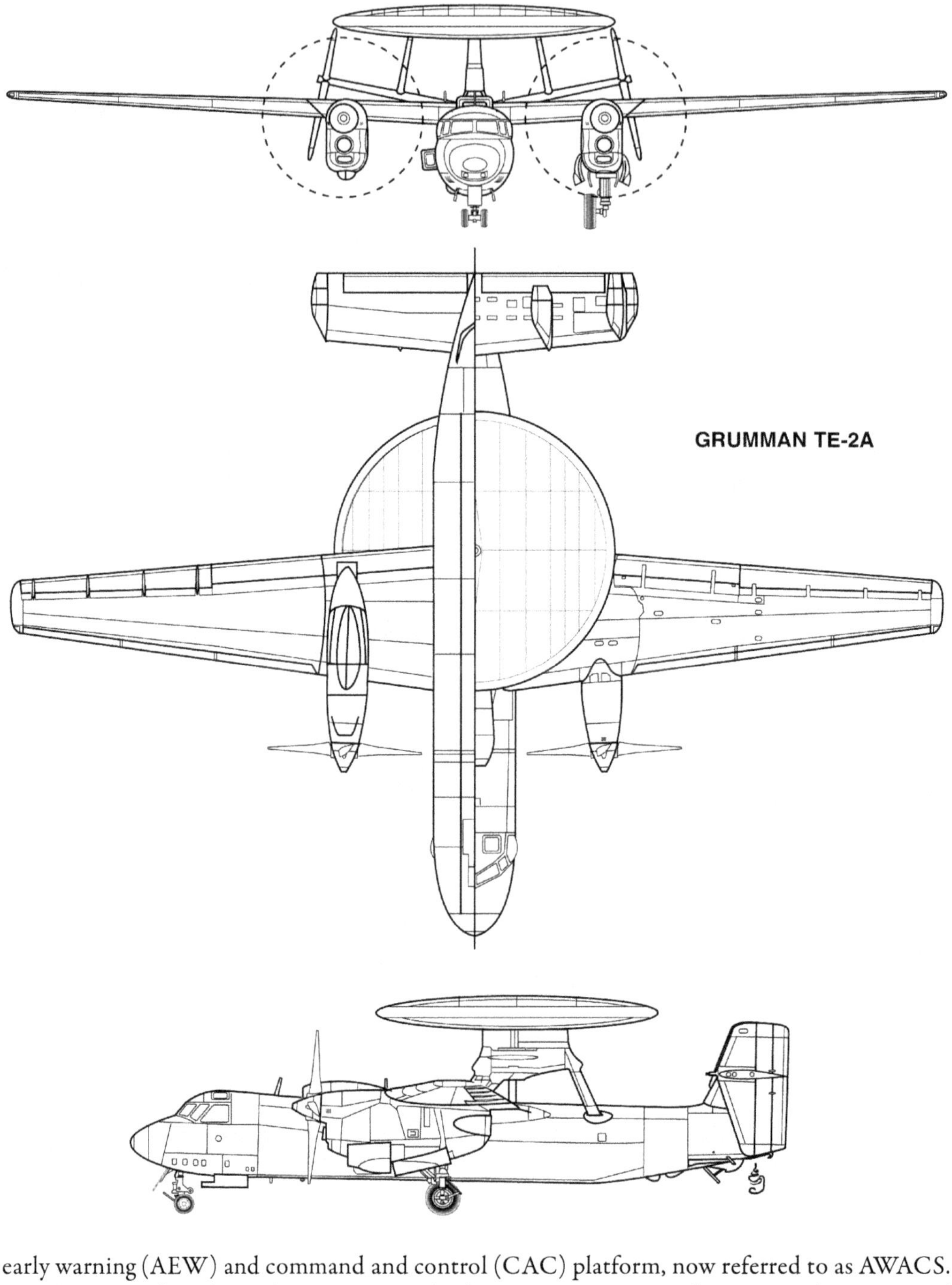

early warning (AEW) and command and control (CAC) platform, now referred to as AWACS. Development of its design began in 1956, the first W2F-1 prototype flew in October 1960, and the first of 59 production E-2As began reaching Navy service in early 1964. When early operational experience revealed serious flaws in both the complex avionics suite and the structural integrity of the airframe, the Navy initiated an upgrade and modification program in which 49 of the original E-2As reemerged as improved E-2Bs between 1969 and 1971, with four of the unmodified E-2As being re-designated as TE-2A crew trainers.

One of the TE-2As became the YE-2C prototype as part of another E-2 "preplanned product improvement" program, but the other three served from 1971 to 1978 as crew trainers with two replacement AEW squadrons, RVAW-110 at NAS North Island, California, and RVAW-120 at NAS Norfolk, Virginia (both renamed VAW-110 and -120 in 1983). From 1972 to 1991, the Navy accepted delivery of approximately 118 E-2Cs, consisting of 49 rebuilt E-2Bs and 69 new builds, and of that total, five were converted into TE-2C crew trainers. VAW-110 was disestablished in 1994 so that all replacement training for E-2 aircrews since that time has been conducted solely by VAW-120. Serial number records, current as of January 2013, reflect that two TE-2Cs have been withdrawn while the other three apparently remain active at the time of this writing.

McDonnell Douglas TAV-8 Harrier—1973

Technical Specifications (TAV-8B)

Type: Two-place V/STOL attack trainer.
Manufacturer: McDonnell-Douglas Corp., St. Louis, Missouri and British Aerospace, United Kingdom.
Total U.S. military versions (trainers): 30 (USMC).
Unit cost (AV-8B): $23,700,000.
Powerplant: One Rolls-Royce *Pegasus* F402–406A vectored-thrust turbofan, rated at 21,180 lbs./s.t. (retrofitted with a F402–408A, rated at 23,400 lbs./s.t.)
Armament: Mixed ordnance up to 2,000 lbs. (est.) on two external wing stations.
Performance: Max. speed 629 mph; ceiling 50,000 ft.; range 2,441 mi. max.
Weights: 14,222 lbs. empty, 31,000-lb. max. loaded (straight take-off).
Dimensions: Span 30 ft. 4 in., length 50 ft. 3 in., wing area 230 sq. ft.

The McDonnell Douglas (now Boeing) AV-8 Harrier is distinguished as the only fixed-wing attack aircraft capable of vertical takeoff and landing (VTOL) to have entered operational service with the U.S. armed forces. Interest in what became the AV-8 began in 1966 when one of six Hawker P.1127 Kestrel VTOL development aircraft was turned over to the Naval Air Test Center at NAS Patuxent River, Maryland, where Marine Corps pilots were given the opportunity to carry out trials that included VTOL operations from the assault ship *Raleigh*. The Kestral differed from other VTOL experiments in having vectored thrust, i.e., rotable engine exhaust nozzles that enabled the aircraft to transition from vertical to level flight without radical changes in pitch. Continuing interest in the VTOL attack concept led to a group of Marine pilots being sent to the U.K. in early 1969 to conduct formal flight evaluations with the Harrier G. R. Mk. 1, the military derivative of the Kestral, and soon afterward, the Marine Corps officially requested procurement of 114 aircraft from Hawker-Siddeley. In order to bypass the U.S. government policy prohibiting purchases of foreign-made military aircraft, an arrangement between Hawker-Siddeley (later British Aerospace) and McDonnell-Douglas authorized the latter to license-build an American version, the AV-8A, in the U.S. However, because it was highly impractical to establish an assembly line on so small a scale, completed AV-8A airframe components were in fact crated-up in the U.K. then shipped to McDonnell-Douglas for assembly and eventual delivery to the Marine Corps. Congressional authorization was ultimately obtained to procure 110 AV-8As, however, due to a high initial accident rate in transition training, the contract was subsequently amended to include eight TAV-8A two-seat trainers as part of the total.

TAV-8As differed from AV-8As in having a two-foot two-inch forward fuselage stretch to accommodate a second cockpit, a taller tail fin, elongated tail cone, and no armament. While training in single-seat AV-8As actually commenced with VMA-513 in mid–1971, VMAT-203 was established in 1972 as the Harrier transition unit and began taking delivery of its first two-

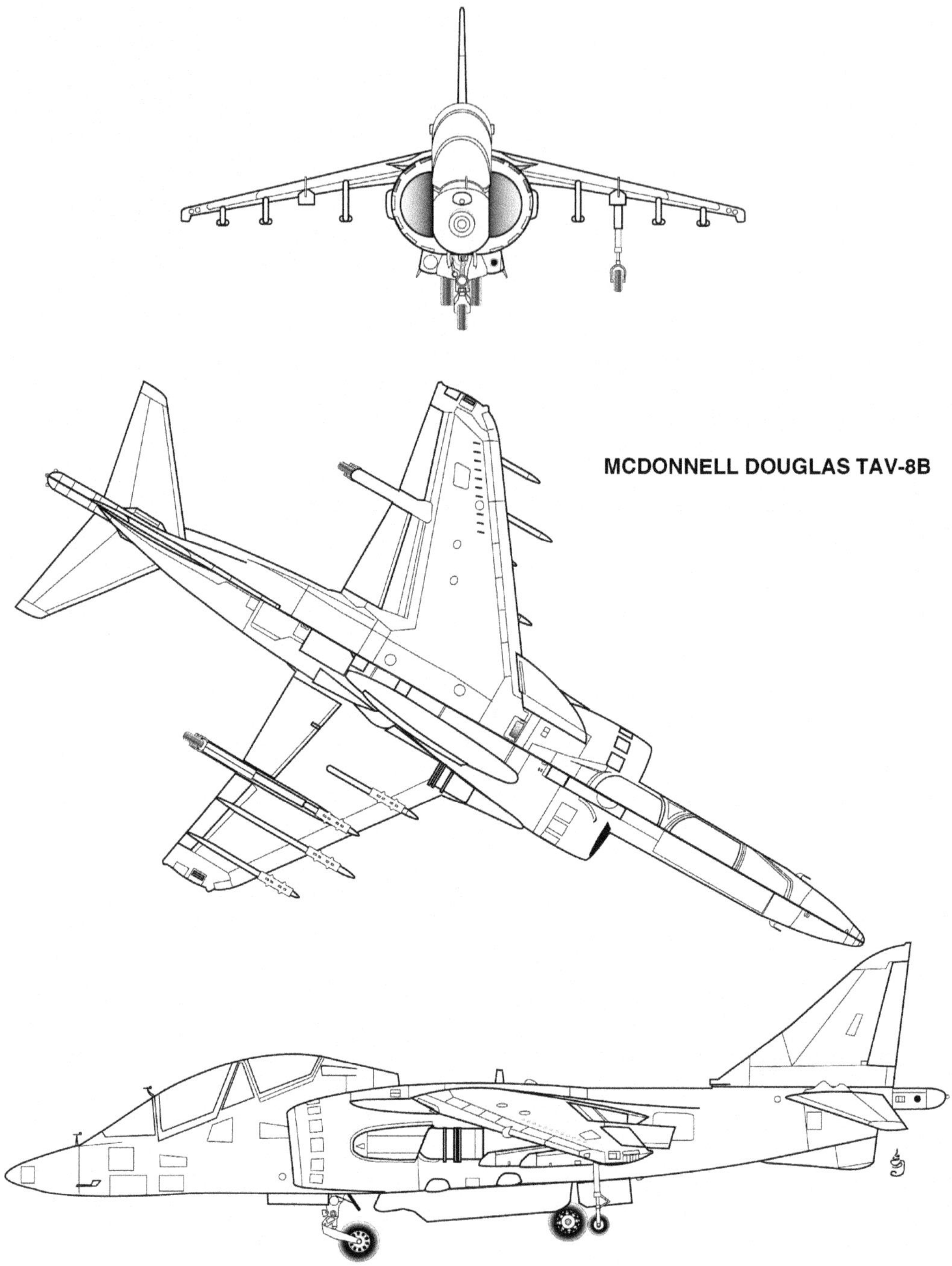

seat TAV-8As sometime in 1973. Even though the AV-8A's capabilities represented a quantum leap in Marine Air's tactical air mobility and responsiveness, its deficiencies in terms of high mishap rates (i.e., 55 peacetime aircraft losses) and limited range and payload, led to development and production of the vastly improved AV-8B Harrier II, which flew for the first time in 1981 and began entering service in 1983. The AV-8B featured an entirely new, bigger wing, which, combined with extensive use of lighter composite materials, effectively doubled the aircraft's

Eight of these McDonnell Douglas TAV-8As served with VMAT-203 from 1973 to 1987. The photograph depicts an example at MCAS El Toro, California, in 1978.

A TAV-8B with VMAT-203 in flight at MCAS Yuma, Arizona, in February 2010. Of the 24 TAV-8Bs built, 19 were still active with VMAT-203 as of late 2013 (courtesy Joseph L. McGovern).

useful combat load. Stores pylons were increased from four to seven and a new, raised canopy was installed to improve all-around visibility. Other improvements included a fly-by-wire control system, digital flight displays (i.e., glass cockpit), a heads-up display (HUD), and hands-on-throttle-and-stick (HOTUS) controls. Although 24 two-seat TAV-8Bs were a component of a planned procurement of 182 Harrier IIs, the six surviving TAV-8As (two had crashed, one in 1977 and another in 1980) remained operational with VMAT-203 until late 1987.

The first TAV-8Bs, when they began reaching VMAT-203 in mid–1987, featured a three-foot eleven-inch fuselage plug for the rear cockpit, a vertical stabilizer of increased height and chord, a longer tail cone, and a four-piece, side-hinged canopy enclosure. Though TAV-8Bs share the AV-8B's cockpit, avionics, and control systems, they are not mission-capable and lack armament except for two wing hardpoints. In the interval since 1987, three TAV-8Bs have been destroyed or damaged beyond repair and two transferred to the Italian Navy, and as of September 2013, 19 remain active with VMAT-203. The Marines are scheduled to start replacing their AV-8Bs with Lockheed-Martin VTOL F-35Bs sometime in 2018, however, at present no plans for a two-seat F-35B have been announced.

McDonnell Douglas F/TF-15 Eagle—1973

Technical Specifications (F-15B)

Type: Two-place fighter transition and combat trainer.
Manufacturer: McDonnell Douglas Corp. (now Boeing), St. Louis, Missouri.
Total U.S. military versions (trainers): 119 (USAF, ANG).
Unit cost (F-15B): $27,900,000.
Powerplants: Two Pratt & Whitney F-100-PW-100 turbofan engines, each rated at 12,240 lbs./s.t. dry, 23,820 lbs./s.t. in afterburning.
Armament: One 20-mm M61 rotary cannon in the nose, four AIM-9 *Sidewinder* and four AIM-7 *Sparrow* missiles or combination of AIM-9, AIM-7, and AIM-120 missiles, or up to 12,000 lbs. of mixed ordnance carried on eleven external hardpoints.
Performance: Max. speed 1,650 mph (Mach 2.5), cruise 570 mph; ceiling 65,000 ft.; range 3,450 mi.
Weights: 27,000 lbs. empty, 40,000-lb. loaded.
Dimensions: Span 42 ft. 10 in., length 63 ft. 9 in., wing area 608 sq. ft.

For nearly 30 years the McDonnell Douglas F-15 Eagle claimed the title of being the most important air superiority fighter in the USAF inventory. The final design of the F-15, as it emerged in late 1968, stemmed directly from the hard lessons learned by the USAF from air combat experience over North Vietnam: In order to achieve and maintain air superiority, the USAF needed a fighter capable of close-in, visual dogfighting with guns as well as the ability to defeat over-the-horizon targets with radar-guided missiles. When the YF-15A flew for the first time on June 27, 1972, it not only possessed both of these features but also the power to accelerate to Mach 2.5 and operate from sea level all the way up to 65,000 feet; and it could fly from its base over an un-refueled range of 3,450 miles. Over a 14-year production run (1973–1987), the USAF accepted a total of 768 F-15A/B/C/Ds, and in their air superiority role, they were the mainstay until the operational debut of the Lockheed-Martin F-22A Raptor in 2005.

Although primarily intended as a single-seat fighter, the F-15A production program contemplated that every seventh aircraft on the assembly line would be a fully mission-capable, two-seat variant which also functioned as a transition and combat proficiency trainer. Retaining the identical airframe and dimensions of the single-seat F-15A, the two-seater, initially designated TF-15A, differed chiefly in having a second seat in tandem under a longer canopy, and to make

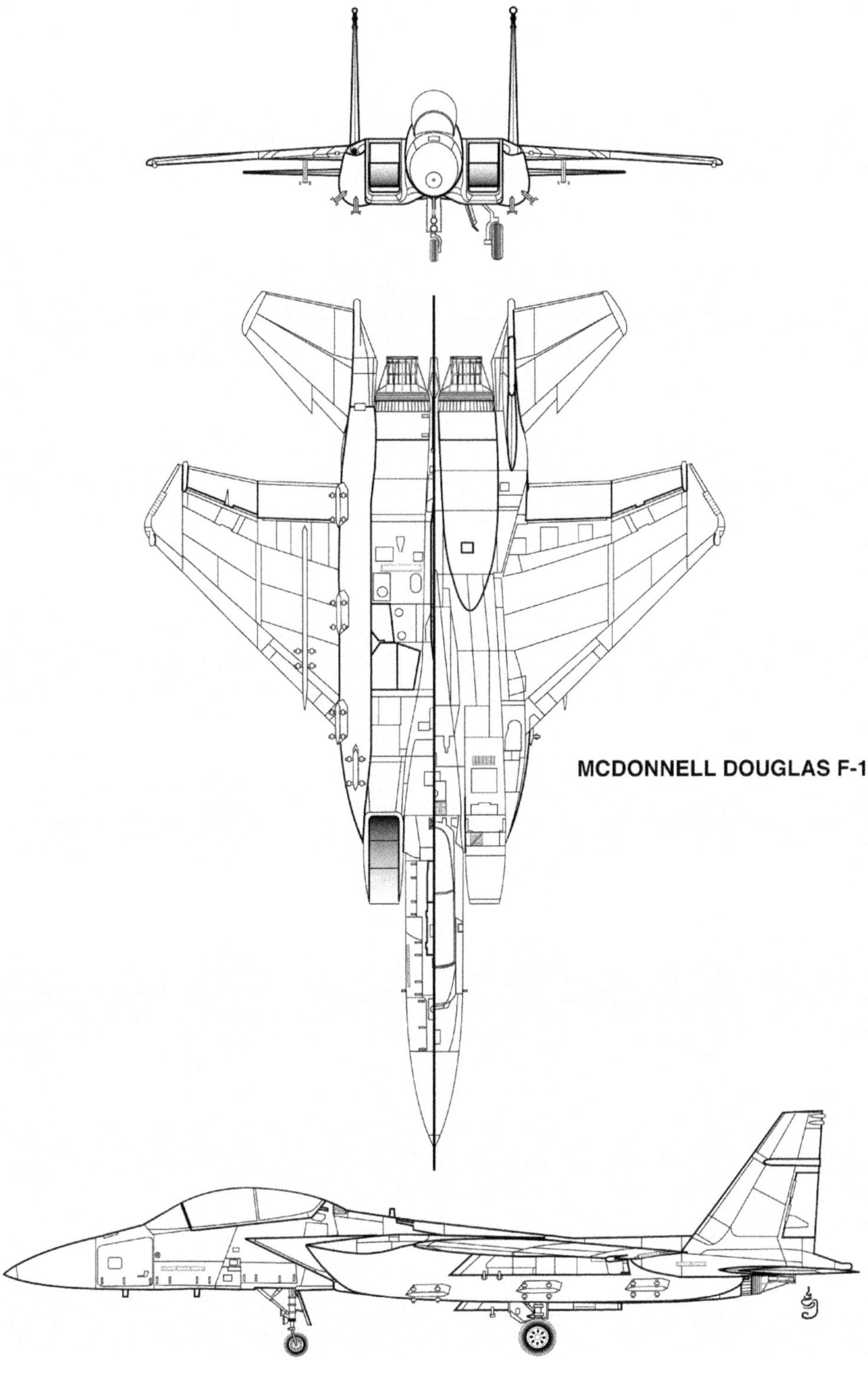

MCDONNELL DOUGLAS F-15B

A McDonnell Douglas TF-15A out of the first production block in 1974. The plane was afterward redesignated F-15B. They initially served with the 555th TFTW at Luke AFB, Arizona, the transition unit for new F-15s.

room for the extra cockpit, the internal countermeasures set was removed and fuel capacity slightly reduced. By the time the first of 58 TF-15As arrived for service in late 1974, the designation had been changed to F-15B. In later years, the first TF-15A/F-15B built (s/n 71–0290) became the demonstrator for the F-15E Strike Eagle and later still, the testbed for the fuel and sensor tactical (FAST) comformal fuel tanks ultimately fitted to F-15C/Ds. The F-15A/B production program was superceded by the F-15C/D in 1979 and like the A/B, every seventh aircraft was completed as a two-seat F-15D. The only external difference of the F-15C/D was provision for FAST tanks attached to the fuselage sides outside of each intake. Because the tanks conformed to the aerodynamic shape of the fuselage, it had little effect on handling and performance. Although the FAST tanks carried less fuel than the three external tanks, they permited the aircraft to be flown at considerably higher speeds over the same un-refueled range. Internal changes included an improved avionics suite that employed a high-speed, special purpose computer to allow rapid switching between radar modes. The first two-seat F-15D flew in June 1979 and began entering operational service the next year. The final Eagle variant, 236 F-15E Strike Eagles delivered to the USAF between 1988 and 2001 (note: includes 36 subsequently added to original procurement of 200 to replace losses), are all two-seaters with dual controls.

The very first Eagle to enter service with a USAF operational unit was a two-seat F-15B delivered to 555th Tactical Fighter Training Squadron (TFTS), 58th Tactical Fighter Training Wing (TFTW) at Luke AFB, Arizona, in November 1974, and thereafter the 58th TFTW (later shortened to TTW) functioned as the replacement training unit (RTU) during the introduction of F-15s into frontline service. In 1979 responsibility for F-15 RTU was split between the 58th and the 405th TTW, also at Luke AFB, and they began upgrading to F-15C/Ds in 1981. One of the 405th's subordinate units, the 461st TFTS, became the first F-15E Strike Eagle training squadron in 1988. As a result of post–Cold War USAF downsizing, the 405th was deactivated in late 1991 and the 58th in 1993, with all F-15 training assets going to the 56th Fighter Wing (FW) and moved to Tyndall AFB, Florida; then in 2004, all F-15C/D training was discontinued and F-15E operations moved to 4th FW at Seymour-Johnson AFB, North Carolina. The USAF

This F-15D operated with the 56th Fighter Wing out of Bitburg AB, Germany. The 32 airworthy F-15Ds on the current USAF inventory are in Air National Guard units in Louisiana, Iowa, and Montana.

began phasing out F-15A/Bs from active service and transferring them to ANG units as early as 1987 and completed the process in 1991, and after 20+ years of service with the ANG, the final examples were placed in AMARC storage during 2009. While five active USAF squadrons are presently equipped with F-15Cs, the 32 airworthy F-15Ds remaining have all been transferred to ANG units in Louisiana, Iowa, and Montana.

Boeing T-43–1973

Technical Specifications (T-43A)

Type: Navigator and Combat Systems Officer (CSO) trainer.
Manufacturer: Boeing Aircraft Co., Seattle, Washington.
Total U.S. military versions: 19 (USAF).
Unit cost: $5,390,000.
Powerplants: Two Pratt & Whitney JT8D-9A turbofan engines, each rated at 14,500 lbs./s.t.
Performance: Max. speed 562 mph, cruise 529 mph; ceiling 36,745 ft.; range 2,983 mi.
Weights: 60,210 lbs. empty, 115,000 lbs. max. takeoff.
Dimensions: Span 93 ft., length 100 ft., wing area 980 sq. ft.

The ubiquitous Boeing model 737, in continuous production since 1967, is the best selling type of jet airliner in the history of aviation, with 7,678 examples having been delivered as of mid–2013. A multiplicity of variants fall into two broad categories: Original and Classic, 737–100 and -200 (1967), -300 (1984), -400 (1988), and -500 (1989); and Next Generation, 737–600, -700, and -800 (1998) and -900 (2001). The first variants built to military requirements came in 1971 when the USAF contracted with Boeing to deliver four 737–253 series aircraft as navigator trainers under the designation T-43A. The specification called for an interior configuration having stations for 12 navigator students and six instructors in addition to normal flight crew. Avionics systems included ground mapping radar, VHF omni-directional range (VOR),

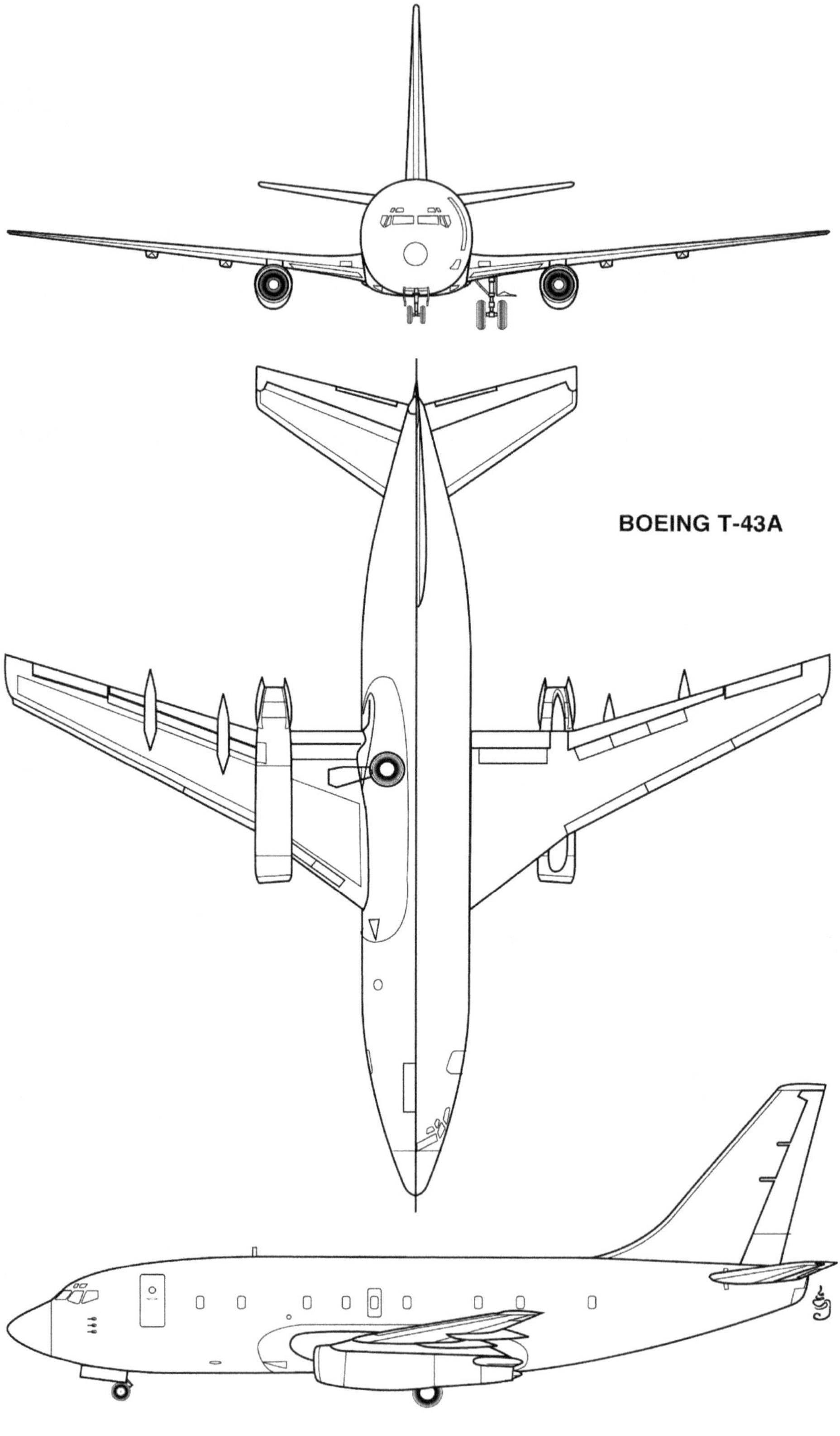
BOEING T-43A

A Boeing T-43A of the 12th Flying Training Wing seen at Randolph AFB, Texas, in 2005. After 37 years of service, the last T-43As were retired from AETC in September 2010.

UHF tactical air navigation (TACAN), long-range navigation (LORAN-C), inertial navigation (INS), radar altimeter, and all relevant VHF, UHF, and HF communication systems. For celestial navigation training, five periscopic sextant stations were situated along the length of the main cabin. Externally, the T-43A differed from its airliner counterpart in having fewer side windows and an array of antennas for its navigational equipment.

In 1972 and 1973, the USAF added 15 more T-43As to the original order, and the flight of first T-43A took place from Boeing's Seattle plant on March 10, 1973. As deliveries to Air Training Command (ATC) proceeded from 1973 through 1975, T-43As were assigned to the 323rd Flying Training Wing (FTW) at Mather AFB, California, where they began replacing the unit's propeller-driven Convair T-29Cs (see in Series II, Part A). Though the T-43A never received an official name, it was commonly referred to as the "Gator," an abbreviation of navigator, or the "Flying Classroom," derived from the earlier T-29. In 1976, as part of the Inter-Service Undergraduate Navigator Training (IUNT) program, the Navy, using its own instructors, merged its training of Naval Flight Officers (NFO) destined to be navigators in land-based aircraft with USAF training at Mather. In 1979, two T-43As were assigned to 200th Airlift Squadron of the Colorado ANG at Buckley ANGB in support of introductory air navigation training of U.S. Air Force Academy cadets. Starting in the mid–1980s, six T-43As withdrawn from ATC underwent transport conversions and reentered service under the designation CT-43A and a seventh, modified as a radar testbed, became the NT-43A. With the advent of global positioning system (GPS) satellite navigation in the late 1980s, T-43As remaining in ATC were upgraded with GPS-NAV equipment and celestial/LORAN training discontinued.

In 1993, when Mather AFB closed as a consequence of post–Cold War force reductions, the 323rd FTW was deactivated and all navigator training merged into the 12th FTW at

Randolph AFB, Texas, as part of the newly reorganized Air Education and Training Command (AETC), including IUNT training of Navy NFOs. Over the next decade, USAF navigator training was combined with Weapons System Officer (WSO) and Electronic Warfare Officer (EWO) into a new rating known as Combat Systems Officer (CSO), with the T-43As of the 12th FTW being employed to conduct the navigator portion of the curriculum. After 37 years of service, the last T-43As were retired from the 12th FTW in September 2010. As there is no in-kind replacement planned for the T-43A, CSO training now takes place in a combination of aircraft such as the Northrop T-38C (see in Series II, Part A), Beech T-1A, and Beechcraft-Hawker T-6A (see both, below), and prospective Navy NFOs, use flight simulators and the T-6A.

Lockheed C/CT-130 Hercules—1974

Technical Specifications (C-130H)

Type: Tactical and logistical transport pilot and crew trainer.
Manufacturer: Lockheed Aircraft Co. (Lockheed-Martin since 1994), Marietta, Georgia.
Total U.S. military versions (trainers): Unknown.
Powerplants: Four 4,590-shp Allison T-56-A-15 turboprop engines driving four-bladed Hamilton Standard electro-hydromatic, constant-speed, reversible-pitch metal propellers.
Performance: Max. speed 366 mph, cruise 336 mph; ceiling 33,000 ft.; range 2,360 mi.
Weights: 75,800 lbs. empty, 155,000 lbs. max. takeoff.
Dimensions: Span 132 ft. 7 in., length 97 ft. 9 in., wing area 1,745 sq. ft.

It is a truly remarkable fact that the Lockheed C-130 Hercules, which originated from a USAF requirement issued during the Korean War and was flown for the first time in 1954, has at this writing been in continuous production—59 years—longer than any other type of military aircraft. Built in five major production variants and at last count, 59 specialized sub-variants, the C-130 deserves to be recognized as the most significant tactical transport of the post–World

A Lockheed C-130E of 314th Airlift Wing, the USAF's main C-130 transition and operational training unit, at Little Rock AFB, Arkansas, probably in the 1990s. All Es have since been replaced by C-130Hs.

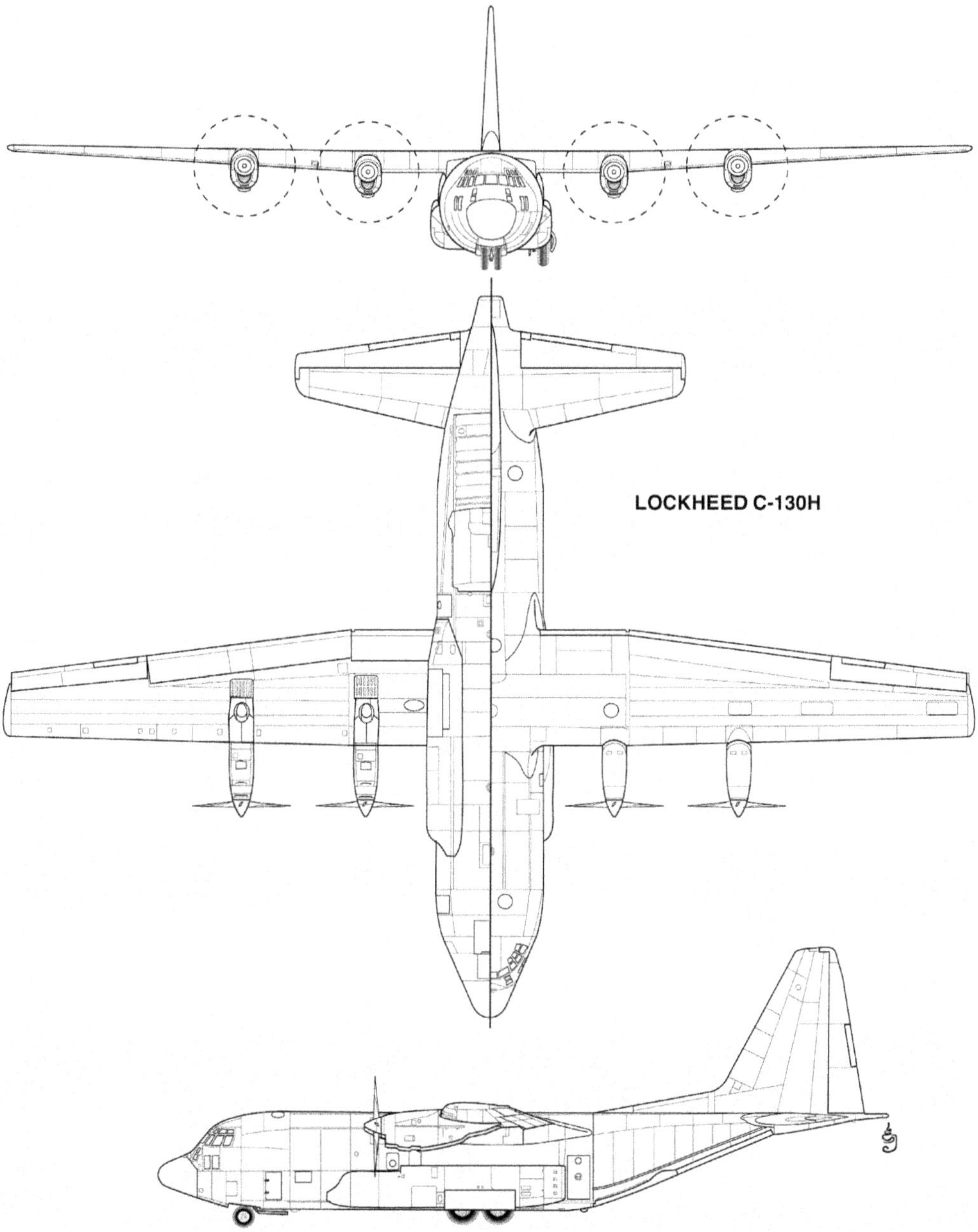

War II era. Since becoming operational with the USAF in 1956, a total of 1,239 C-130s have been delivered to U.S. military branches and approximately 1,200 to foreign air forces.

Some historical sources suggest that a number of C-130s have been re-designated as the TC-130; however, the serial number records reflect only a single TC-130A designation with no date reported. Although this piece is dated 1974, early C-130 variants, C-130As and Bs, were most likely used by Tactical Air Command (TAC) for transition training of pilots prior to the merger of all tactical airlift into Military Airlift Command (MAC) in 1974. Following the merger, MAC consolidated all C-130 transition and operational training in the 314th Airlift Wing (AW) at Little Rock AFB, Arkansas, and training was probably carried out in C-130Es, the chief operational variant at that time. In 1992, the 314th AW became part of the newly established Air Mobility Command (AMC), and since 1997, has been a subordinate unit of Air Education and Training Command (AETC). Before this time, the only C-130s actually attached to AETC were two MC-130H "Combat Talons" employed to train special operations aircrews. After 1991, C-130Hs began entering service as trainers with the 314th AW, although the unit's last C-130E was not retired until September 2011. Since 2003, the Wing's 48th Airlift Squadron has conducted transition and operational training with the new C-130J.

Ling-Temco-Vought TA-7 Corsair II—1976

Technical Specifications (TA-7C)

Type: Two-place attack transition and combat trainer.
Manufacturer: Ling-Temco-Vought Corp., Dallas, Texas.
Total U.S. military versions (trainers): 91 (60 USN; 31 ANG).
Unit cost (A-7C): $2,860,000.
Powerplant: One Pratt & Whitney TF-30-P-408 turbofan engine, rated at 13,400 lbs../s.t.
Armament: Two 20-mm cannons in the nose and up to 15,000 lbs. of mixed ordnance carried on eight external hardpoints.
Performance: Max. speed 657 mph, cruise 521 mph; ceiling 47,572 ft.; range 2,108 mi.
Weights: 19,842 lbs. empty, 42,000 lbs. loaded.
Dimensions: Span 38 ft. 9 in., length 48 ft. 8 in., wing area 375 sq. ft.

The Navy initiated the VAL (light attack aircraft) design competition in 1963 with the objective of finding an aircraft having better conventional weapons capability than the Douglas A-4 (see TA-4, above). In order to minimize development costs, all VAL proposals were to be based upon existing designs, with the proviso that supersonic performance was not required. In

An LTV TA-7C seen in early 1988 while serving with VA-174, an A-7 Fleet Replacement Squadron based at NAS Dallas, Texas. All A-7 attack units were deactivated in mid–1991, and the last EA-7Ls left service in early 1992.

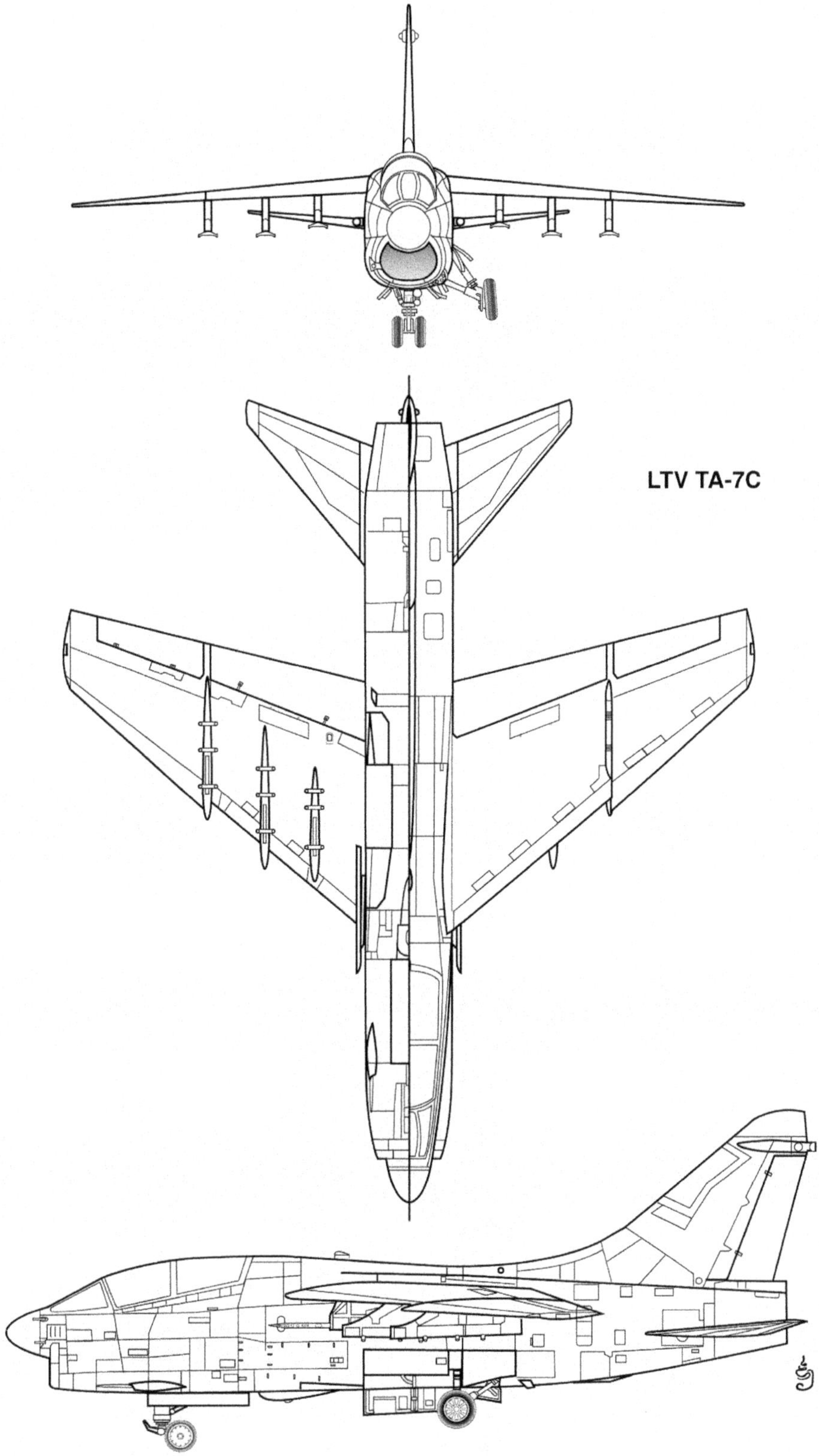

LTV TA-7C

An A-7K that served with the 152nd Tactical Fighter Squadron, Arizona National Guard, from 1981 to 1993. All 30 A-7Ks were divided among 14 Air National Guard units.

February 1964, after considering proposals from Douglas, Grumman, North American, and Ling-Temco-Vought (LTV), the Navy selected LTV's V-463 as the winning design under the designation YA-7A. LTV's concept had been evolved from an F-8 Crusader airframe (see Vought F8U-1T/TF-8A in Series II, Part B) in which the fuselage was shortened eight and a half feet to house a non-afterburning TF-30-P-6 turbofan, together with redesigned wings having 21-inch tip extensions and reduced sweepback. The YA-7A prototype completed its first flight on September 17, 1965, and deliveries of the first of 175 production A-7As began in September 1966. A succession of improvements led to production of 196 A-7Bs in 1966 and 1967 (P-8 engine and new flaps), 69 A-7Cs in 1969 (new avionics, HUD, and M61 rotary cannon), and 535 A-7Es from 1970 to 1983 (new Allison TF-41-A-1 turbofan engine, rated at 14,500 lbs./s.t.).

In 1972, as a private venture, LTV modified a TF-41-powered A-7E into a two-seat tandem configuration by adding a 34-inch plug to the forward fuselage to accommodate a raised rear cockpit that was enclosed by a longer, side-hinged canopy, and the modified prototype, provisionally designated YA-7E, made its first flight in August the same year. While LTV obtained no Navy orders for a production version of the YA-7E, it did ultimately receive a contract to rework 60 A-7Bs and Cs which reemerged in 1976 and 1977 as two-seat transition and attack trainers under the designation TA-7C, and though possessing slightly reduced range, were otherwise fully mission capable. In 1982 eight TA-7Cs equipped with special electronic pods returned to service as EA-7L electronic aggressor aircraft, and in 1984 41 TA-7Cs were upgraded with TF-41 engines but retained their old designations. In operational service TA-7Cs were split between two A-7 Replacement Air Groups (RAG), VA-122 at NAS Lemoore, California, in the Pacific Fleet and VA-174 at NAS Cecil Field, Florida, in the Atlantic Fleet. VA-174, along with its TA-7Cs, was disestablished in June 1988 and VA-122 suffered the same fate in May 1991 soon after the Gulf War ended. The very last operational EA-7Ls left VAQ-33 and -34 in early 1992.

USAF Variants: In late 1965, after bending to Army pressure to acquire a subsonic close air support (CAS) aircraft, the USAF contracted with LTV to develop the A-7D. The YA-7D prototype, very similar to the Navy A-7C, flew in April 1968 with the same TF-30 engine; however, production models came with the more powerful Allison TF-41. Delivery of production A-7Ds began in late 1968, with the USAF accepting a total of 459 by the time production ended in late 1976. In 1979, after being modified for installation of a second cockpit, one A-7D was test-flown as the TA-7D, and like the TA-7C, was fully mission-capable. Following successful evaluations, LTV received a contract to manufacture 30 new-build aircraft delivered between 1981 and 1984 under the designation A-7K. As deliveries proceeded, all A-7Ks were divided between 14 ANG units then equipped with ex–USAF A-7Ds, where they served as transition and combat proficiency trainers in eleven different states and Puerto Rico until the last examples were retired in 1993.

Beech T-44 Pegasus—1977

Technical Specifications (T-44A)

Type: Multi-engine advanced trainer.
Manufacturer: Beech Aircraft Corp., Wichita, Kansas.
Total military versions: 61 (USN).
Unit cost: Not reported but estimated at $1,000,000 in 1977.
Powerplants: Two 550-shp Pratt & Whitney PT6A-34B turboprop engines driving three-bladed Hartzell constant-speed, fully-reversible propellers.
Performance: Max. speed 282 mph, cruise 227 mph; ceiling 31,300 ft.; range 1,457 mi.
Weights: 6,326 lbs. empty, 9,650 lbs. max. takeoff.
Dimensions: Span 50 ft. 3 in., length 35 ft. 6 in., wing area 294 sq. ft.

Since its introduction in 1964, the Beech "King Air" has outsold—3,100+ delivered at the time of this writing—all of its competitors combined. King Air lineage may be divided into two major families, the 90/100 series and the larger 200/300 series (originally marketed as the Super King Air). The Army was the first military customer, procuring 162 unpressurized model 65-A90s as the U-21 between 1966 and 1971, and in 1966, the USAF also acquired two off-the-shelf pressurized model B90s as VIP transports under the designation VC-6A. The model C90, introduced in 1971, featured a four-foot eleven-inch increase in wingspan and more powerful -20A engines that combined to up maximum takeoff weight from 9,300 lbs. to 9,650 lbs. In 1977, to replace Naval Air Training Command's (NATC) aging fleet of Grumman TS-2As (see in Series II, Part B) with a modern, turboprop-powered multi-engine advanced trainer, the Navy awarded Beech a contract to deliver 61 King Airs under the assigned designation T-44A. Though based upon C90 airframes, Beech renumbered T-44As as H90s because of their unique military requirements. In addition to military radios and navigational equipment, the specification called for an interior arrangement for an instructor and two pilot-trainees (one flying and the other monitoring radios) plus seats for two passengers. Deliveries of T-44As to the Navy began in mid–1977 and continued until late 1978. In August 2006, when T-44As had been in service 29 years, the Navy initiated a program to upgrade them with all-digital, fully-integrated Rockwell-Collins Pro Line 21 avionics which includes, among other things, instrument flying graphics, weather graphics, and a traffic alert avoidance system (TCAS), and as of mid–2013, 25 have been converted and returned to service as the T-44C.

As deliveries to NATC proceed in 1977 and 1978, T-44As entered service with VT-28 and VT-31, both based at NAS Corpus Christi, Texas, where they replaced the squadrons' TS-2As. In 1990, VT-28 transferred some of its T-44As to VT-31 after changing its mission from advanced

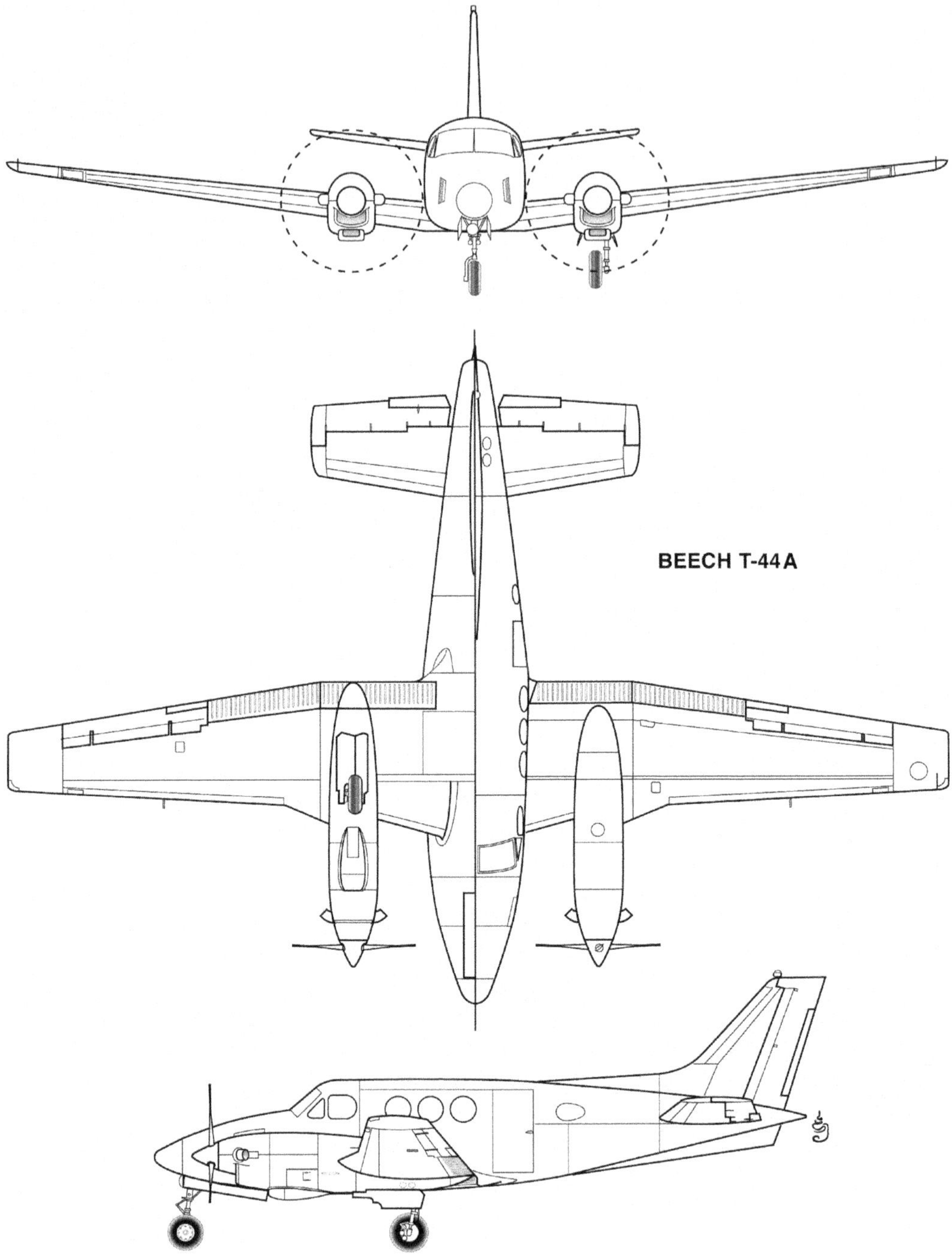

multi-engine to instructor-pilot training, then in 1993, all of them went to VT-31 for advanced multi-engine training. As a result of an inter-service agreement concluded in 1996, VT-31's mission expanded to encompass advanced multi-engine training of USAF student pilots destined to fly C-130s. Today, 36 years after the first T-44A entered NATC service, 52 remain on the active inventory, 27 as T-44As and 25 as T-44Cs, and no plans have been announced to replace them.

This Beech T-44A is pictured while serving with VT-28 at NAS Corpus Christi, Texas, during the 1980s. Since 1993, all T-44A operations have been consolidated in VT-31, also at Corpus Christi.

General Dynamics F-16 Fighting Falcon—1977

Technical Specifications (F-16D)

Type: Two-place fighter transition and combat trainer.
Manufacturer: General Dynamics Corp. (Lockheed-Martin after 1993), Ft. Worth, Texas.
Total produced (trainers): 330 (326 USAF; 4 USN).
Unit cost (F-16D): $18,800,000.
Powerplant: One Pratt & Whitney F-100-PW-220 turbofan engine, rated at 23,770 lbs. with afterburning or one General Electric F-110-GE-100 turbofan engine, rated at 28,984 lbs./s.t. with afterburning.
Armament: One 20-mm M61 rotary cannon, up to 12,430 lbs. of external ordnance carried on one centerline and six wing pylons, plus two AIM-9 *Sidewinder* air-to-air missiles on wingtips.
Performance: Max. speed 1,333 mph (Mach 2.02), cruise 575 mph; ceiling 50,000 ft.; range 2,000 mi. max.
Weights: 18, 238 lbs. empty, 37,500 lbs. max. takeoff.
Dimensions: Span 32 ft. 8 in., length 49 ft. 5 in., wing area 300 sq. ft.

After a humble beginning as a technology demonstrator, the General Dynamics F-16 "Fighting Falcon" grew to become the most-produced USAF fighter since the North American F-86 (see TF-86 in Series II, Part A) and like all post–1970 American fighters, was a byproduct of the hard won lessons of the Vietnam War. The lightweight fighter (LWF) requirement, issued by the Department of Defense (DoD) in January 1971, placed emphasis on an aircraft having a high thrust-to-weight ratio, a gross weight not exceeding 20,000 lbs., superior maneuverability, and low unit cost. In April 1972, after considering proposals from five major defense contractors, DoD narrowed the competition to the General Dynamics YF-16 and the Northrop YF-17. Both prototypes flew in 1974 and a competitive fly-off between them continued until January 1975, at which time DoD declared General Dynamics the winner and announced plans to procure 650 F-16s with the possibility of enlarging the order to 1,400 or more. In addition to its very advanced aerodynamic design, the F-16 featured a fly-by-wire (FBW) control system in which there was

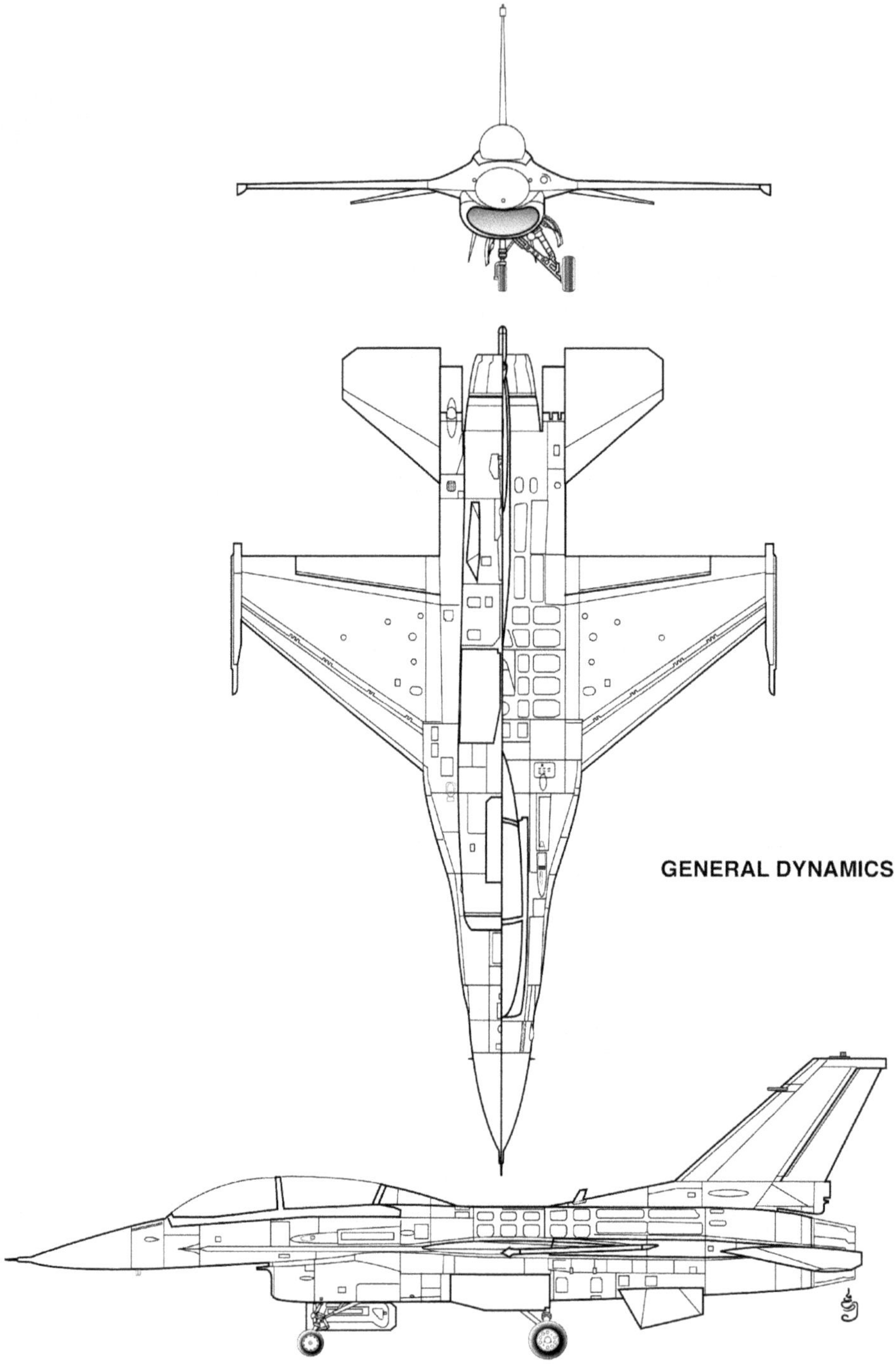

GENERAL DYNAMICS F-16B

no direct connection between the pilot and the aircraft's control surfaces; instead, control inputs sent signals to a computer that actually moved the ailerons, elevators, and rudder. A novel cockpit arrangement featuring a side-stick control (originally unmovable) and a seat tilted to a 30-degree angle enabled the pilot to function up to the airframe's 9-G maneuvering limit.

As with the F-15 reported above, the F-16 production plan contemplated every sixth aircraft

An F-16B of the 159th Fighter Interceptor Squadron, Florida Air National Guard, date unknown. In 1995 this unit converted to McDonnell Douglas F-15A/Bs.

would be completed as a fully mission-capable two-seater which would also function as a transition and combat proficiency trainer. The maiden flight of the first two preproduction YF-16B two-seat variants took place on August 8, 1977, and deliveries of production F-16Bs commenced in early 1979. Dimensionally identical to the single-seat F-16A, the B differed chiefly in having a second cockpit under a longer, one-piece canopy and reduced internal fuel capacity. All F-16A/B deliveries to the USAF were completed by 1985, totaling 674 F-16As and 121 F-16Bs (a ratio of roughly 5.5:1). The F-16C/D, which superceded the F-16A/B on the production line in late 1984, represented a major expansion of the type's tactical role: an APG-68 multi-mode radar and new cockpit avionics added night/all-weather attack capabilities in air-to-ground mode and beyond-visual-range interception capability in air-to-air mode. In the 50D/52D production blocks, the F-16CJ/DJ series (known as *Wild Weasels*) were specialized for suppression of enemy air defenses. Deliveries of F-16C/Ds to USAF operational units started in late 1985, with a total of 1,216 F-16Cs and 205 F-16Ds being accepted when production ended in March 2005.

Production F-16A/Bs first entered service with the 188th Tactical Fighter Wing (TFW) at Hill AFB, Utah in August 1978, which assumed initial responsibility for training of F-16 crews for both Tactical Air Command (TAC) and export clients. But the next unit to receive F-16A/Bs, the 56th Tactical Training Wing (TTW) at MacDill AFB, Florida, became the replacement training unit (RTU) for all F-16 operations. The 56th began converting to F-16C/Ds in the late 1980s and phased-out all F-16A/Bs in 1990. A second RTU, the 58th TTW at Luke AFB, Arizona, formed in early 1983 with F-16A/Bs and began converting to F-16C/Ds in 1986. The 56th was deactivated in 1994 and all of its squadrons moved to the 58th, then later the same year, it was renamed the 56th Fighter Wing (FW). The 56th grew to a strength of eight F-16 RTU squadrons by 1997 but in 2005, commenced a force reduction which, as of mid–2013, has left it with six RTUs operating 138 F-16C/Ds. As the USAF began the process of phasing out

This F-16D of the 480th Fighter Squadron is seen taking off from Ignatievo Air Base, Bulgaria, during Operation Thracian Star in October 2010.

A Navy TF-16N of VF-45 seen at Naval Air Facility Andrews, Maryland, in 1993. Airframe fatigue caused by years of high-G maneuvering caused the F/TF-16Ns to be retired in 1994.

F-16A/Bs from active service during the early 1980s, many went on to equip USAF Reserve (USAFR) and ANG units, and since the mid–1990s, the transfer process has continued with F-16C/Ds. Over time, ANG has become the second largest user of F-16s and presently maintains four F-16 RTU squadrons within the Arizona ANG that are responsible for all F-16 training of USAFR and foreign air forces as well as ANG. The last F-16A/Bs left USAF and ANG service in 2007 and 840 F-16Cs and 163 F-16Ds remain on the USAF, USAFR, and ANG inventory as of September 2013.

Naval Variants: After receiving DoD approval to procure F-16s for dissimilar air combat training (DACT), the Navy took delivery between 1987 and 1988 of 22 single-seat variants as the F-16N and four two-seaters as the TF-16N. Though based upon the F-16C/D airframe, the F-16N differed in having strengthened wings, wiring to carry an air combat maneuvering instrumentation (ACMI) pod on the right wingtip, the lighter APG-66 radar of the F-16A/B, no internal cannon, and no provision for air-to-air missiles. The TF-16N was identical except for the second cockpit and longer canopy. As they entered operational service, the F/TF-16Ns were allocated between VF-43 at NAS Oceana, Virginia, VF-45 at NAS Key West, Florida, and VF-126 and the Naval Fighters Weapons School (better known as "Top Gun"), both at NAS Miramar, California. In 1994, after years of high-G maneuvering had caused the F/TF-16Ns to develop fatigue cracks in their wings, the Navy announced that they would be retired, and the last example was placed in AMARC storage in January 1995.

McDonnell Douglas TF/F/A-18 Hornet—1980

Technical Specifications (F/A-18B)

Type: Two-place fighter/attack transition and combat trainer.
Manufacturer: McDonnell-Douglas Corp. (now Boeing Aerospace), St. Louis, Missouri.
Total U.S. military versions (trainer variants): 75 est. (USN/USMC).
Unit cost (F/A-18B/D/F): $29,000,000-$55,000,000.
Powerplants: Two General Electric F404-GE-400 turbofans, each rated at 10,600 lbs./s.t.dry, 15,800 lbs. in afterburning.
Armament: One 20-mm M61 rotary cannon, six AIM-9 *Sidewinder* or four AIM-7 *Sparrow* missiles or two AIM-9 and two AIM-7 (air-to-air load); or up to 13,700 lbs. of ordnance on nine external stations (air-to-ground load).
Performance: Max. speed 1,196 mph (Mach 1.8) at 36,000 ft.; ceiling 50,000 ft.; range 1,700+ mi. max., 550 mi. loaded.
Weights: 23,831 lbs. empty, 56,000 lbs. max. takeoff.
Dimensions: Span 37 ft. 6 in., length 56 ft., wing area 400 sq. ft.

The McDonnell Douglas (now Boeing) F/A-18 Hornet and Super Hornet, at present Navy and Marine aviation's principal air superiority and strike aircraft, has been in production longer (1978–2013) than any other type of naval combat aircraft, and if predictions hold true, will continue another two years. During the early 1970s, under the heading VFAX (carrier fighter-attack experimental), the Navy began to study the feasibility of a multi-role strike aircraft that could perform the missions of dedicated attack aircraft like the A-4, A-6, and A-7 while possessing the air-to-air capabilities of the McDonnell F-4. However, no progress toward a definitive design was made until 1974, when, as an alternative to VFAX, the U.S. Congress directed the Navy to investigate a naval development of the General Dynamics YF-16 or the Northrop YF-17, the two USAF candidates in the LWF/ACF (light-weight fighter/air combat fighter) competition. After considering both, Navy officials concluded that the airframe design and twin-engine layout of

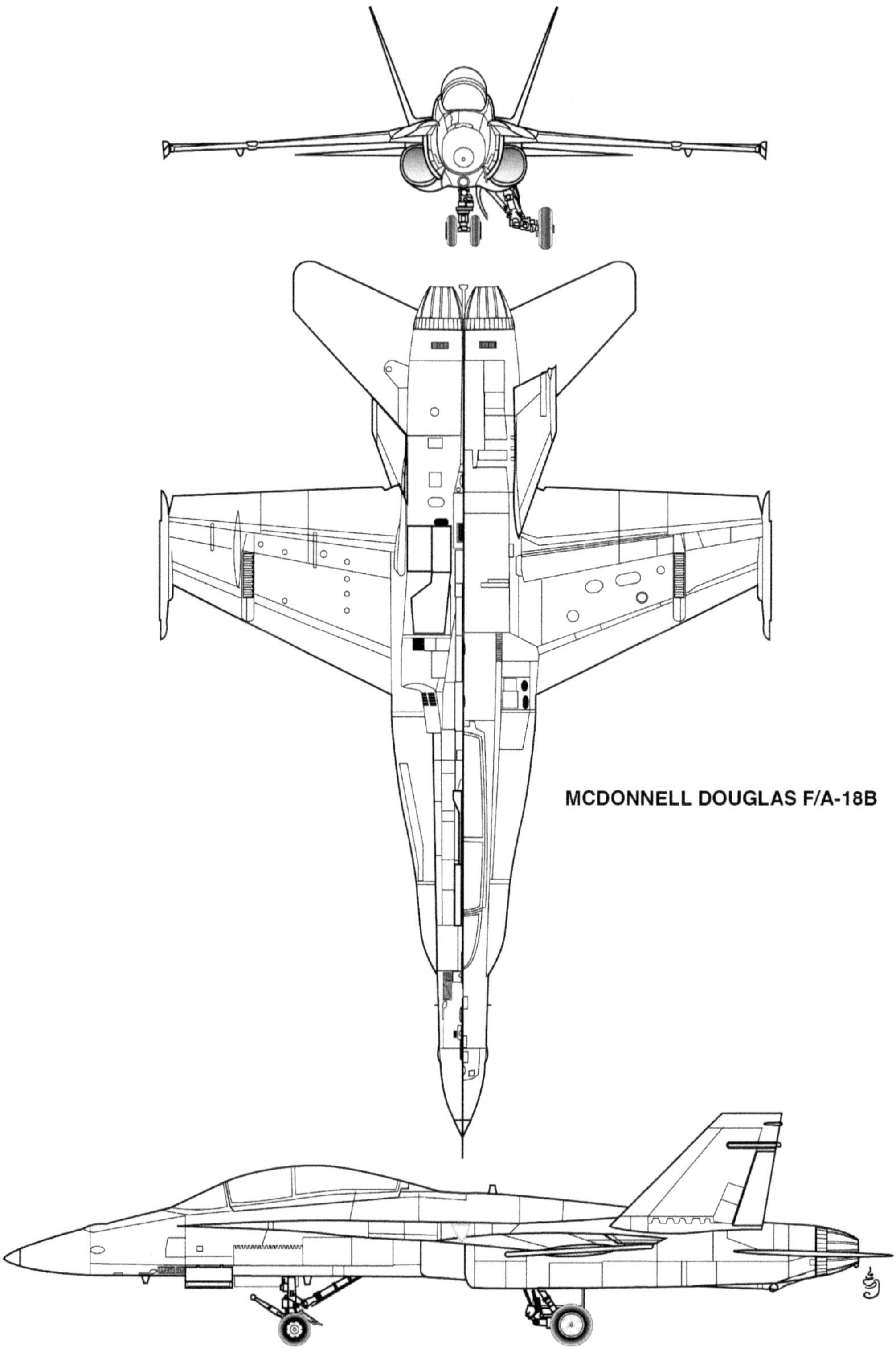

MCDONNELL DOUGLAS F/A-18B

the YF-17 came closer to meeting naval requirements, and because Northrop had no recent experience with carrier-based aircraft, McDonnell Douglas was brought in to collaborate on a naval adaptation of the YF-17. Originally, the proposal called for distinct fighter and attack versions, but subsequent advancements in avionics and weapons systems made it practical to merge both functions into a single airframe. In January 1976, once the detailed design work had been approved, McDonnell Douglas/ Northrop received a contract for nine single-seat development aircraft as the F-18A and two two-seat trainers as the TF-18A.

While following the general aerodynamic configuration of the YF-17, the F-18 materialized as an entirely new design. Integration of air-to-air and air-to-ground functions was made possible by extensive use of digital computer technology in the control and weapons systems. It was the first production naval aircraft to employ a "fly-by-wire" control system in which control inputs from the pilot were transmitted to a digital processor, which simultaneously analyzed the signal against flight data (i.e., airspeed, altitude, angle-of-attack, etc.) and moved the control surfaces accordingly. All flight and tactical information appeared on a heads-up-display (HUD) and two multi-function display screens on the instrument panel. The systems were primarily managed via hands-on-throttle-and stick (HOTAS) controls that placed all important switches at the pilot's fingertips. Avionics included an automatic carrier landing system (ACLS), which permitted arrested landings in zero visibility conditions. The Hughes APG-65 radar could be configured to track targets in either air-to-air or air-to-ground modes, permitting the aircraft to perform *both* missions on the same sortie.

The first F-18A flew on November 18, 1978, and all eleven development aircraft, including the two TF-18As, were flying by March 1980. Deliveries of production F/TF-18As began in May 1980, and the type became fully operational in early 1983. In April 1984, the official designations of the F-18A and TF-18A were changed to F/A-18A and F/A-18B, respectively. The two-seat

One of the two development McDonnell Douglas TF-18As seen in 1980. The designation was changed to F/A-18B in 1984. Twenty-six F/A-18Bs were listed as being active in VFA-122 as of late 2013.

F/A-18B, intended as a transition and combat proficiency trainer, retained full mission capability with only six percent less internal fuel. After delivery of 371 of F/A-18As and 39 F/A-18Bs, production switched to the F/A-18C/D, which flew for the first time in September 1987. The differences between A/B models and the F/A-18C/D were entirely internal, i.e., an improved mission computer and countermeasures systems plus compatibility with AIM-120 *AMRAMM* air-to-air missiles, AGM-65 *Maverick,* and AGM-64 *Harpoon* air-to-ground missiles. Although the rear cockpit of the F/A-18D was specifically configured for a weapons system operator (WSO) and lacked dual controls, it could be retrofitted with one of the 25 to 30 factory-supplied stick and throttle kits when needed for pilot transition training. The Navy and Marines accepted the last of 310 F/A-18Cs and 317 F/A-18Ds in 2000.

Though not originally planned, the Hornet 2000 project surfaced in 1992 after the Navy hierarchy concluded a navalized version of the USAF's advanced tactical fighter (ATF), the YF-22 Raptor, would be too costly to supply the numbers of aircraft needed after the year 2000. The offspring of this effort, F/A-18E/F Super Hornet, flew for the first time in late 1995, delivery of production models began in 1998, and the type entered operational service in 1999. While sharing 90 percent of the F/A-18C/Ds avionics and software, the F/A-18E/F has been redesigned to include more advanced cockpit displays, a 34-inch fuselage extension, a 25 percent increase in wing area, two additional weapons stations, a 35 percent increase in power (GE F-414–400 engines rated at 22,000 lbs./s.t. each), and a 33 percent increase in internal fuel capacity. At a slight cost in top speed (Mach 1.6), the Super Hornet achieves a 40 percent improvement in

This F/A-18F of VFA-41 was photographed over the Persian Gulf in 2005. Like the F/A-18D, the F/A-18F lacks dual controls but can be retrofitted with a stick and throttle kit.

combat range and a 30 percent increase in combat payload (17,750 lbs.). Like the F/A-18D, the two-seat F/A-18F lacks dual controls but can be retrofitted with a stick and throttle kit. Internet sources show 280 F/A-18Es and 208 F/A-18Fs on the current Navy inventory with a planned strength of 565.

The Pacific Fleet's VFA-125 (formerly VA-125) at NAS Lemoore, California, became the first F/A-18A/B Fleet Replacement Squadron (FRS) in November 1980, and a second FRS, VFA-106 at NAS Cecil Field, Florida, was activated for the Atlantic Fleet in April 1984. Though both squadrons began converting to F/A-18C/Ds in late 1987, the Bs were retained for pilot transition training. VMFAT-101, then based at MCAS El Toro, California, became the third F/A-18 FRS in September 1987, operating F/A-18A/B/Cs. A fourth F/A-18 FRS was created in early 1999 when VA-122 was reactivated at NAS Lemoore to train aircrews transitioning to new F/A-18E/Fs. In 1999, due to base closures, VFA-106 moved to NAS Oceana, Virginia, and VMFAT-101 to MCAS Miramar, California, and in late 2010, VFA-125 was deactivated and its aircraft and personnel merged into VFA-122. Navy and Marine F/A-18A/B/C/Ds were originally scheduled to start being replaced by Lockheed Martin F-35Cs in the 2018 timeframe, however, program delays are likely to keep them in service longer.

Cessna T-51–1982

Technical Specifications (T-51A)

Type: Two-seat primary trainer.
Manufacturer: Cessna Aircraft Co., Wichita, Kansas.
Total U.S. military versions (trainers): 3 (USAF).
Unit cost: $15,000 (est.).
Powerplant: One 150-hp Lycoming O-320-E2D 4-cylinder air-cooled horizontally-opposed engine driving a two-bladed McCauley fixed-pitch metal propeller.

A lineup of U.S. Air Force Academy Cessna T-51As seen in October 2012 while competing in the annual SAFECON competition held at Provo Municipal Airport, Utah.

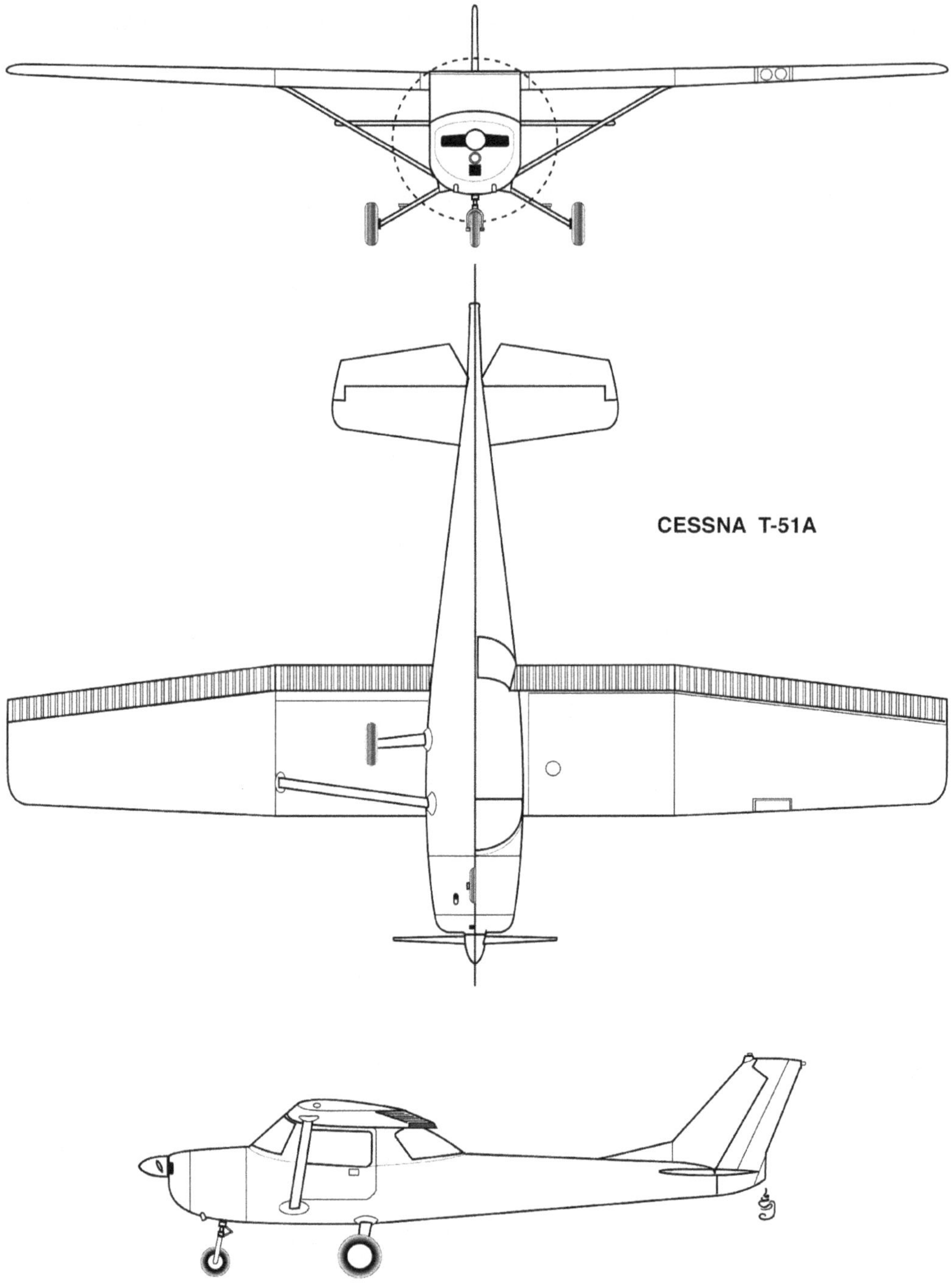

Performance: Max. speed 140 mph, cruise 125 mph; ceiling 23,000 ft.; range 250 mi. (est.).
Weights: 1,400 lbs. empty, 1,700 lbs. max. takeoff.
Dimensions: Span 33 ft. 4 in., length 25 ft., wing area 160 sq. ft.

After being introduced to the civil market in mid–1958, the Cessna model 150 quickly established itself as America's most popular primary flight trainer, and by the time production ended in 1977, almost 24,000 had been delivered. Although Cessna sold about 100 model 150s to foreign air forces for training purposes, none were ever procured new by any of the U.S. military branches. But starting in 1982, the USAF purchased one 150L and two 150Ms on the used market under the designation T-51A, which have since been operated with civil registration numbers rather than military serials. Attached to the 557th Flying Training Squadron at the Air Force Academy Air Field near Colorado Springs, Colorado, the T-51As are flown by cadets selected by the Air Force Academy Flying Team to compete against colleges nationwide in precision flying events. At some cost in range and payload, all three aircraft have been upgraded from their original 100-hp Continental O-200 engines to 150-hp Lycoming O-320s, giving them about a 100 percent improvement in takeoff and climb performance. Internet sources list the three T-51As active as of September 2013.

Cessna T-47 Citation—1984

Technical Specifications (Citation 550)

Type: Naval Flight Officer trainer.
Manufacturer: Cessna Aircraft Co., Wichita, Kansas.
Total U.S. military versions: 15 (USN).
Unit cost: Unknown.
Powerplants: Two Pratt & Whitney JT15D-5D turbofan engines, each rated at 3,045 lbs./s.t.
Performance: Max. cruise 464 mph, normal 430 mph; ceiling 43,000 ft.; range 2,300 mi.
Weights: 8,059 lbs. empty, 15,100 lbs. max. takeoff.
Dimensions: Span 51 ft. 2 in., length 47 ft. 9 in., wing area 343 sq. ft.

Originally flown as the five-passenger FanJet 500 in September 1969, Cessna's long-running series of Citation business jets made their debut on the civil market in 1971 and improved versions remain in production today. The Citation's straight-wing layout represented a design compromise between the speed of a pure jet versus the ability to operate from the same small airports as light

A Cessna T-47A of VT-86 seen on ramp of NAS Pensacola during the 1980s. They were replaced in 1993 with Rockwell T-39Ns, and all but one were subsequently destroyed in a hangar fire later that year.

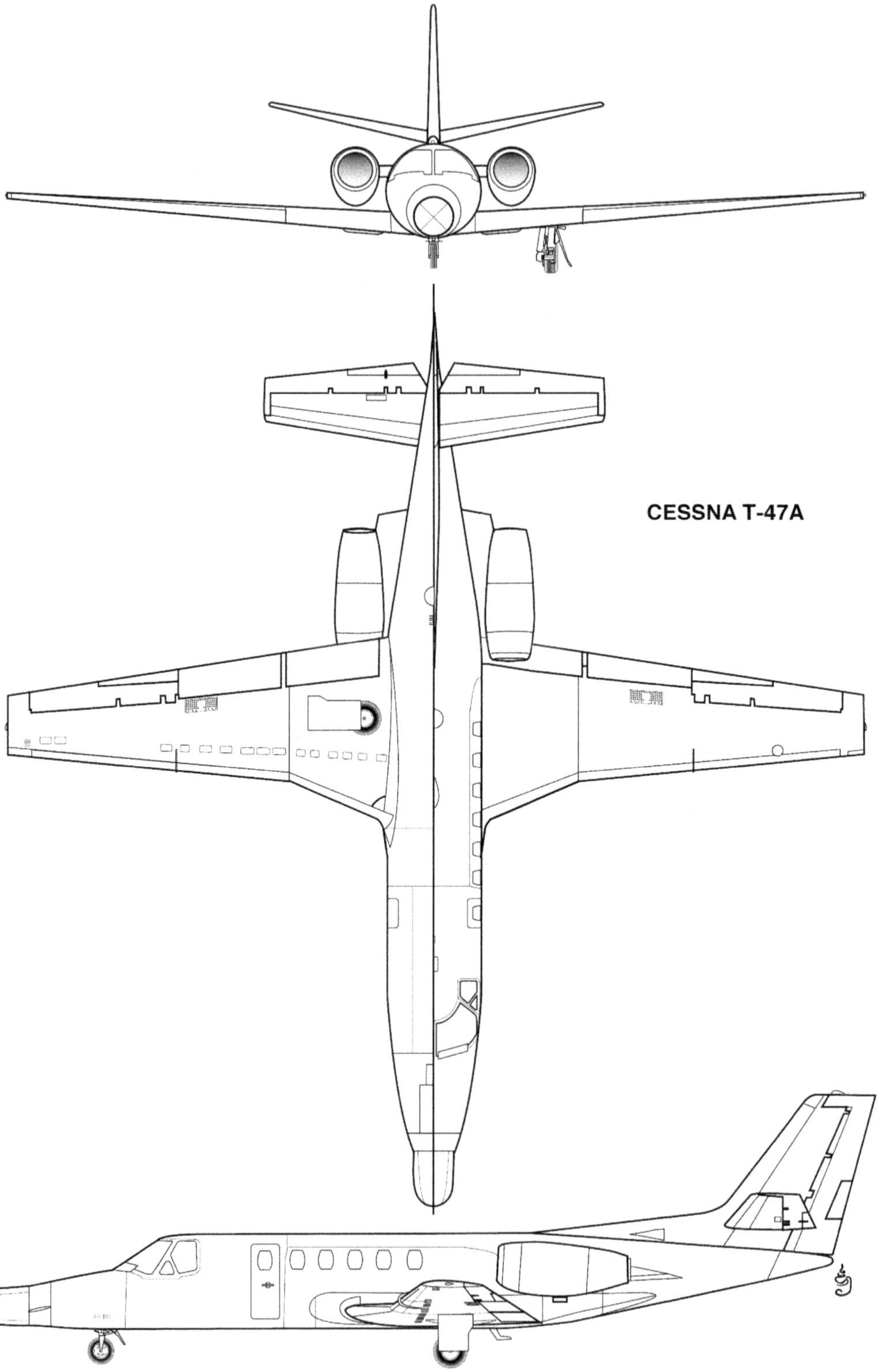

CESSNA T-47A

turboprop aircraft (e.g., Beech King Air 90 and 100) against whom it was intended to compete. The first military orders came in 1983 when the Navy awarded Cessna a contract to deliver 15 Citation 550s that would be completed as Naval Flight Officer (NFO) trainers to a revised specification under the designation T-47A. Modifications included a longer nose to house an APQ-159 (or APQ-167) radar system plus an interior equipped with multiple radar consoles for NFO students (actual number of stations unknown). Unofficial sources state the wingspan was shortened (possibly the 47 ft. 1 in. span of a Citation 500), however, exact dimensions are not published anywhere. The first T-47A, listed by Cessna as the model 552, flew on February 15, 1984, and deliveries commenced later the same year.

Following official acceptance, all T-47As entered operational service with VT-86 at NAS Pensacola, Florida, where they were used to train NFO students destined to be radar intercept officers in Grumman F-14s, bombardier-navigators in Grumman A-6s, electronic warfare officers in Grumman EA-6s, and copilot/tactical coordinators in Lockheed S-3s. The service of the T-47As was relatively short-lived, with all 15 being withdrawn in favor of newly rebuilt Rockwell T-39Ns (see in Series II, Part A) during 1993. All T-47As but one were destroyed in a hangar fire in July 1993.

Boeing TC-135/TC-18–1985

Technical Specifications (TC-18E)

Type: AWACS, electronic intelligence, and reconnaissance crew trainer.
Manufacturer: Boeing Aircraft Co., Seattle, Washington.
Total U.S. military versions: 8 (USAF).
Unit cost: Unknown.
Powerplants: Four Pratt & Whitney JT3D-102 turbofan engines, each rated at 18,000 lbs./s.t.
Performance: Max. speed 627 mph, cruise 552 mph; ceiling 50,000 ft.; range 6,160 mi.
Weights: 98,466 lbs. empty, 327,000 lbs. max. takeoff.
Dimensions: Span 130 ft. 10 in., length 136 ft. 3 in., wing area 3,010 sq. ft.

The Boeing RC-135W "Rivit Joint," used for electronic intelligence by the 763rd Expeditionary Reconnaissance Squadron, is very similar to the TC-135W employed for aircrew training.

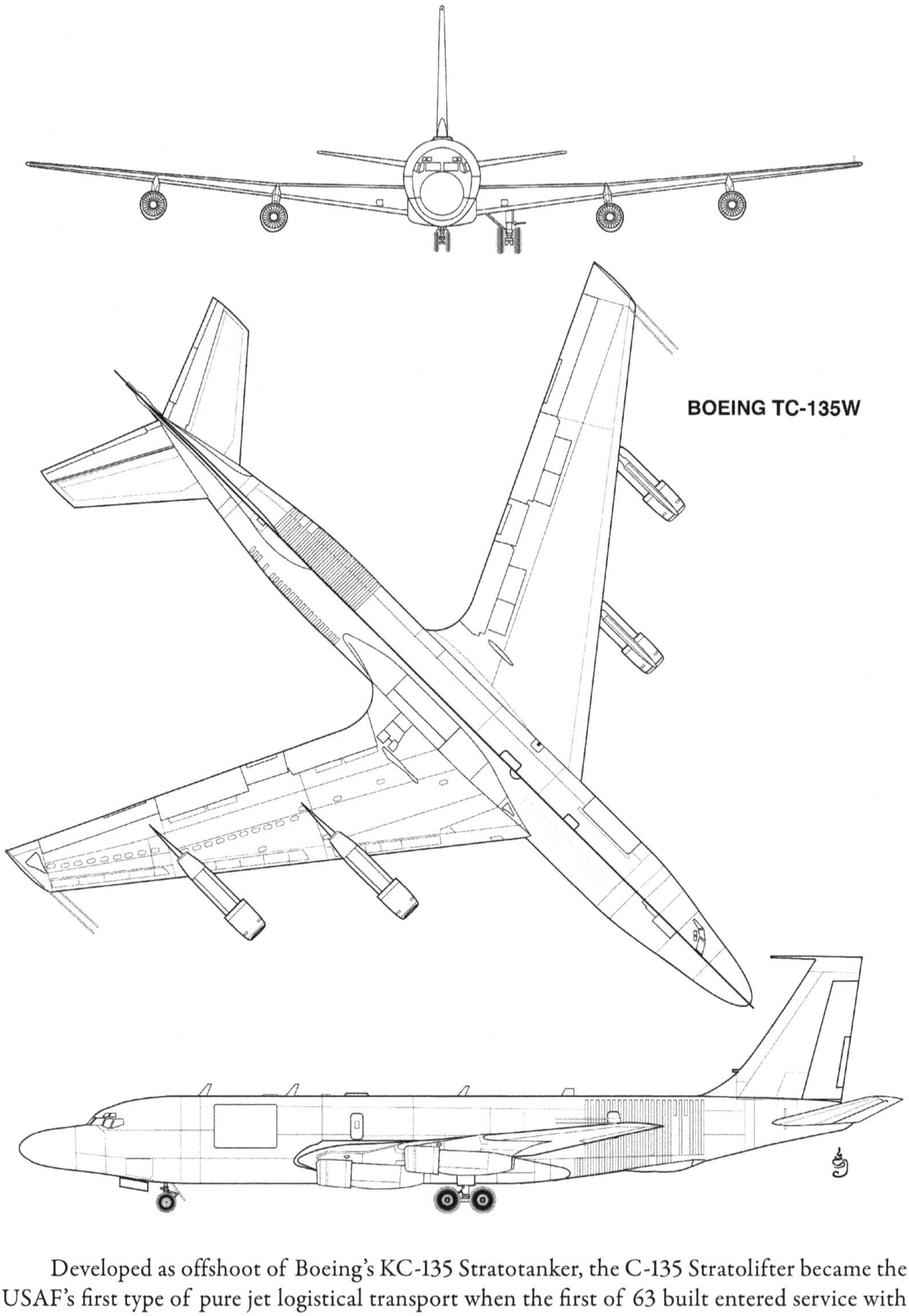

Developed as offshoot of Boeing's KC-135 Stratotanker, the C-135 Stratolifter became the USAF's first type of pure jet logistical transport when the first of 63 built entered service with the Military Air Transport Service in 1961. Starting in the late 1960s, as C-135s left transport duties, 32 were subsequently modified for specialized reconnaissance and intelligence-gathering missions under the designation EC or RC-135. In the mid–1980s, one EC-135B was stripped of mission equipment and re-designated as the RC-135T to function as a crew trainer for the "Cobra Ball," RC-135S, which is equipped with special electro-optic instruments for tracking ballistic missiles. Another, similarly modified, became a TC-135W trainer for the RC-135W "Rivet Joint" signal intelligence mission. None are thought to be active at this writing.

During the 1980s eleven ex-airline Boeing 707–320Bs were purchased and taken into USAF service as the C-18B under a variety of specialized roles. Two were modified in 1988 as Boeing E-3 AWACS trainers under the designation TC-18E and two more (date unknown) as E-6 "Mercury" command and control platform pilot trainers under the designation TC-18F. Neither type currently appears on the USAF inventory and are presumed to have been withdrawn.

Lockheed TP-3 Orion—1985

Technical Specifications (TP-3A)

Type: Multi-engine maritime patrol transition pilot and flight engineer trainer.
Manufacturer: Lockheed Aircraft Corp., Burbank, California.
Total military versions (trainers): 12 (USN).
Unit cost: Unknown.
Powerplants: Four 4,910-shp Allison T-56-A-14 turboprop engines driving four-bladed Aeroproducts constant-speed, fully-reversible propellers.
Performance: Max. speed 460 mph, cruise 375 mph; ceiling 28,300 ft.; range 4,765 mi.
Weights: 61,491 lbs. empty, 135,000 lbs. max. takeoff.
Dimensions: Span 99 ft. 8 in., length 116 ft. 10 in., wing area 1,300 sq. ft.

The Lockheed P-3 Orion has endured as the Navy's principal type of land-based maritime patrol aircraft for over 52 years and though scheduled to start being replaced by the Boeing P-8 Poseidon starting in 2013, is likely to remain in service for the foreseeable future. Based largely on the design of the L-188 Electra airliner, the YP3V-1 prototype made its first flight in November 1959, with the earliest production P3V-1s (re-designated P-3A in September 1962) reaching

A Lockheed TP-3A of VP-31 photographed in vicinity of NAS Moffett Field, California, in 1989.

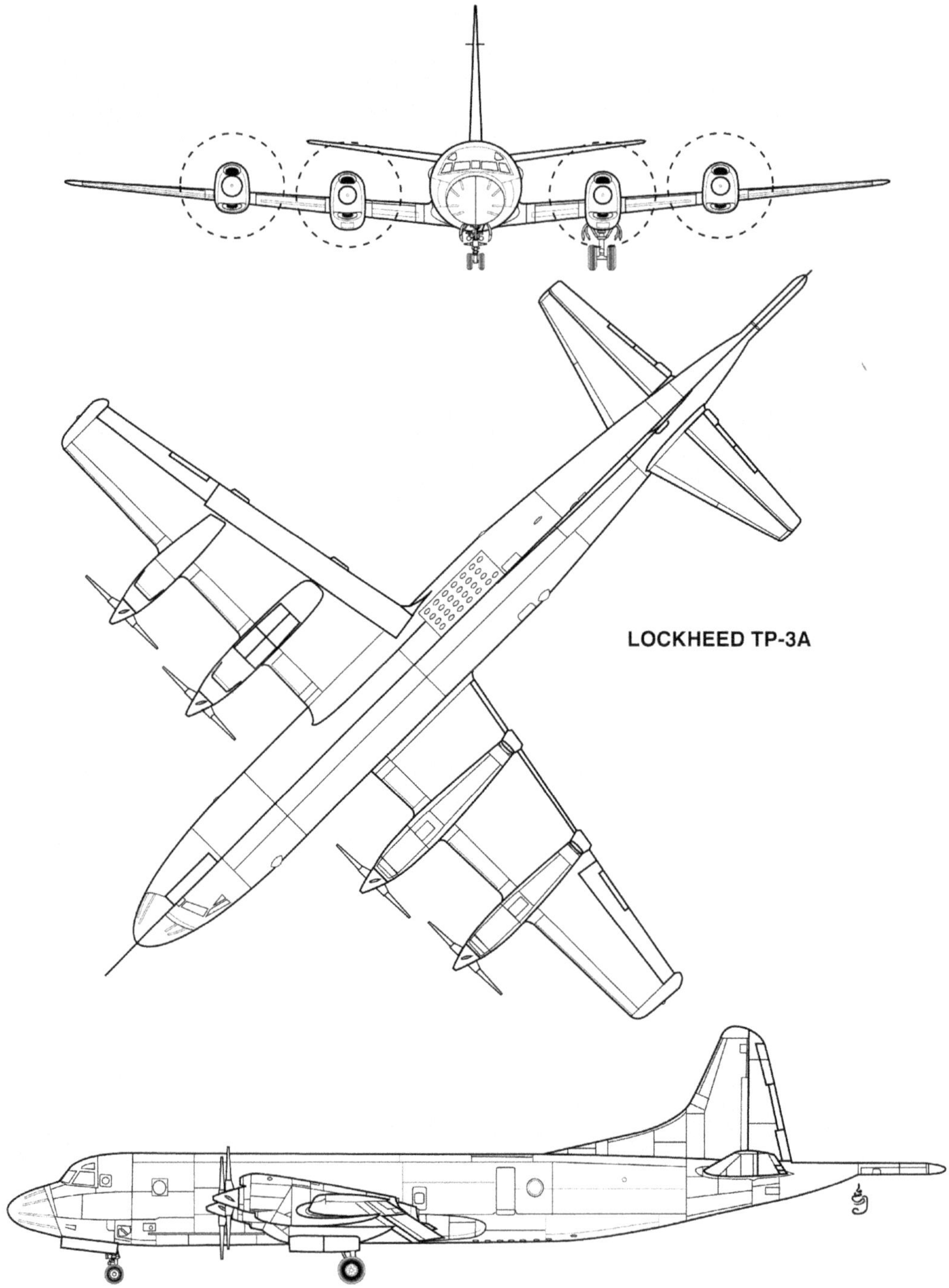

operational service in April 1961. Over the Orion's 27-year production history (1960–1987), the Navy accepted delivery of 547 examples in three major variants, namely, 157 P3V-1/P-3As, 125 P-3Bs, and 265 P-3Cs.

In 1985, to avoid using P-3Cs for pilot and flight engineer training, Navy Air Systems Command (NAVAIR) contracted with Lockheed Aircraft Services to convert 12 P-3As into trainers under the designation TP-3A. Modifications entailed the removal of all ASW equipment and installation of P-3C cockpit avionics. In May 1986, after official acceptance of the TP-3A prototype, Lockheed commenced deliveries of TP-3As at a rate of about one aircraft per month over the next year. As they arrived for duty, TP-3As were split between two P-3 Fleet Replacement Squadrons (FRS), VP-30 at NAS Moffett Field, California, for the Pacific Fleet and VP-31 at NAS Jacksonville, Florida, for the Atlantic Fleet. In late 1993, following the closure of Moffett Field, VP-30 was disestablished and all P-3 replacement training consolidated into VP-31. All TP-3As were retired from VP-31 between 1995 and 1998 and placed in AMARC storage.

Fairchild T-46 Eaglet—1985

Technical Specifications (T-46A)

Type: Two-place primary and basic trainer.
Manufacturer: Fairchild-Republic Corp., San Antonio, Texas.
Total U.S. military versions: 3 (USAF).
Unit cost (estimated): $3,500,000.
Powerplants: Two Garrett F109-GA-100 turbofan engines, each rated at 1,330 lbs./s.t.
Performance: Max. speed 457 mph, cruise 383 mph; ceiling 46,500 ft.; range 1,369 mi.
Weights: 5,275 lbs. empty, 6,962 lbs. max. takeoff.
Dimensions: Span 38 ft. 8 in., length 29 ft. 6 in., wing area 161 sq. ft.

In 1981, under the heading Next Generation Trainer (NGT), the USAF launched a program that, if successful, would result in an across-the-board replacement of its existing Cessna T-37B fleet (see in Series II, Part A). The general specification called for an aircraft having twin turbine engines, side-by-side seating with ejection seats, and an empty weight of 5,000 lbs. Four manufacturers responded with proposals for new designs: the Rockwell low-wing Nova NT-1 powered by side-mounted turbofans; the Vought/VFW low-wing Eaglet powered by buried turbines driving a ducted fan; the General Dynamics mid-wing M-210 powered by side-mounted turbofans and featuring a T-tail; and the Fairchild high-wing model 73 with side-mounted engines and a twin-fin layout similar to the A-10. Other proposals included an offer from Cessna to rebuild existing T-37s with new engines, a pressurized cockpit, and new ejection seats and one from Beech for an updated version of the T-34C. In order to validate its design concept, Fairchild took the extra step of contracting with Ames Industries to build a 62 percent-scale flying demonstrator that made its first flight on September 10, 1981. After two months of testing (23 flight hours), telemetric data showed that the stability and flight response of the demonstrator closely matched Fairchild's predictions for the full-size NGT.

On July 2, 1982, the USAF declared Fairchild the winner of the NGT competition and awarded a contract to build two prototypes under the designation YT-46A, with options for an additional 54 T-46A production aircraft. Future contracts, once the aircraft completed all service and functional testing, contemplated a potential run of 650 aircraft over an eight-year production life. The flight of the first T-46A prototype, named "Eaglet," was made from Edwards AFB, California, on October 15, 1985, marking the start of a planned 22-month flight test program. However, on March 29, 1986, less than six months into the program, the USAF abruptly announced termination of the entire T-46 contract except for the two prototypes and 10 production T-46As; Fairchild, closing its aircraft manufacturing plant soon afterward, ultimately delivered the second YT-46A and only one T-46A, the very last aircraft to be manufactured by that company. The reasons for cancellation of the contract remain controversial; though no offi-

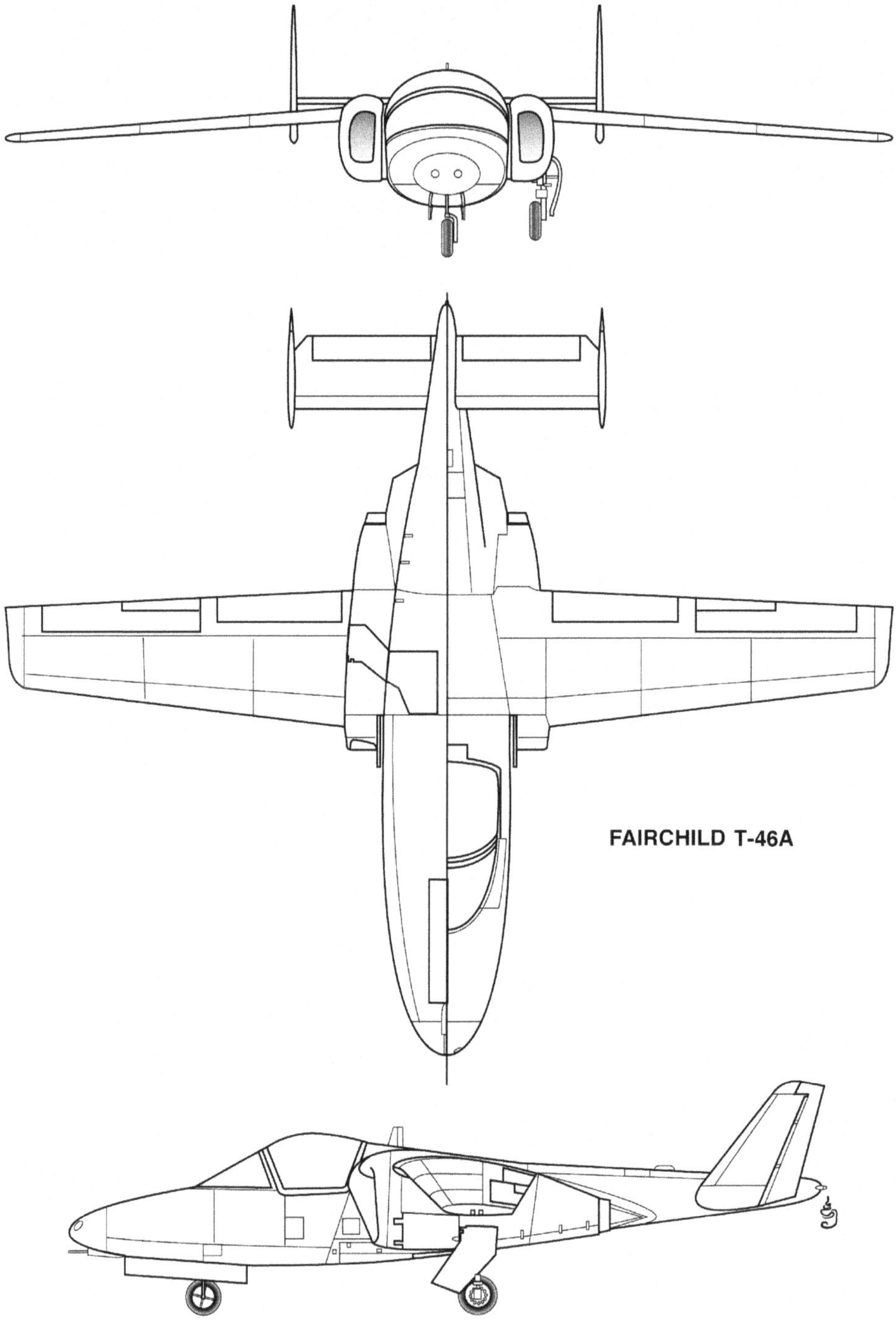

cial reason was given, industry experts cite management and financials difficulties at Fairchild as the key factor rather than development problems with the aircraft. Political analysts speculate that the Reagan administration simply wanted to prove that it could stop a military program not approved by the White House. In the aftermath, the T-37 fleet continued to train new USAF pilots for another 24 years.

An official company photograph of the Fairchild T-46A. It was the winner of the USAF's New Generation Trainer competition, but only three had been completed when Fairchild Industries ceased all aircraft production in 1986.

McDonnell Douglas T-45 Goshawk—1988

Technical Specifications (T-45A)

Type: Two-place, carrier-capable basic and advanced trainer.
Manufacturer: McDonnell Douglas Corp. (now Boeing Aerosapce), St. Louis, Missouri and Long Beach, California; and BAE Systems, Samlesbury and Brough, United Kingdom.
Total U.S. military versions: 218 (USN).
Unit cost: $17,200,000.
Powerplant: One Rolls-Royce F405-RR-401 turbofan engine, rated at 5,527 lbs./s.t.
Performance: Max. speed 645 mph, cruise 620 mph; ceiling 42,500 ft.; range 1,152 mi.
Weights: 10,403 lbs. empty, 14,081 lbs. max. takeoff.
Dimensions: Span 30 ft. 10 in., length 39 ft. 4 in., wing area 190.1 sq. ft.

The McDonnell Douglas (merged with Boeing Aerospace in 1997)-British Aerospace (became BAE Systems in 1999) T-45 Goshawk is notable in having been the first foreign aircraft design selected for undergraduate pilot training by any of the U.S. military branches. Its origins can be traced to 1978, when the Navy issued the VTX-TS (heavier-than-air training aircraft experimental-training system) requirement calling for a carrier-capable aircraft that would eventually replace both the North American Rockwell T-2 (see in Series II, Part B) and McDonnell Douglas TA-4 (see above). The "training system" element of the package involved development of simulators, software, and teaching aids in parallel with the aircraft itself. Over a three-year interval, Naval Air Systems Command (NAVAIR) studied five very distinct proposals: two new airframe designs, one from a Grumman-Beech partnership and another from General Dynamics; a substantial redesign of the T-2 from Rockwell; and two existing foreign trainer designs, the Hawk from a McDonnell Douglas-British Aerospace partnership and the Alpha Jet from a

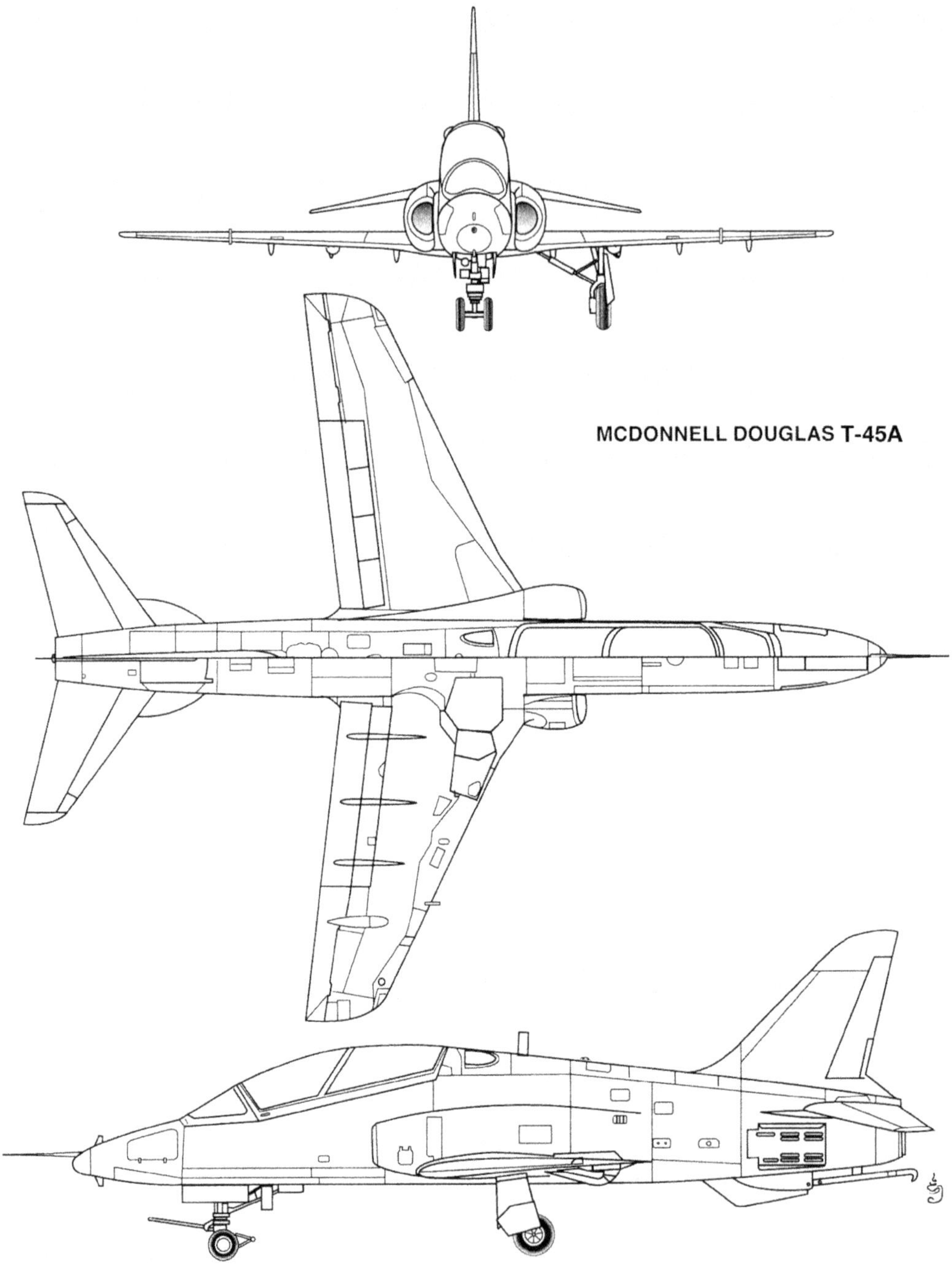

Lockheed-Dassault-Dornier partnership. In November 1981, NAVAIR declared the McDonnell Douglas-British Aerospace Hawk winner of the TX-TS competition under the assigned designation T-45 and name "Goshawk" (reportedly to avoid confusion with the Army Hawk missile). Known flight characteristics, design maturity, and overall life-cycle costs were cited as leading factors in the choice. Originally, to speed service entry, production was to be split between a non-carrier-capable T-45B and the more extensively modified T-45A; however, NAVAIR opted

A photograph taken in June 2008 of a McDonnell Douglas T-45A serving with VT-21 out of NAS Kingsville, Texas. In 1992, VT-21 became the initial operational unit to start trading-in its TA-4Js for new T-45As.

to cancel the T-45B and await the T-45A in 1983 in favor of implementing service life extensions for the T-2 and TA-4. The initial development contract, issued in late 1984, called for construction of two pre-production prototypes to be followed by 12 production T-45As, plus future options for a total of 253 aircraft.

Besides substantial airframe strengthening and carrier arresting gear, revisions to the basic Hawk design included a deeper forward fuselage profile to house a bulkier, twin-wheel nose gear, a wider-track main landing gear with new, long-stroke legs, twin lateral speed-brakes on the fuselage sides in place of a single ventral brake, and addition of small curved fins ("smurfs") near the tailplane to improve low-speed stability. The first pre-production T-45A completed it maiden flight from the Long Beach plant on April 16, 1988, and pre-production types commenced Navy evaluations the following November. When test-flown in a high-drag, carrier approach configuration, the T-45A was judged to be underpowered and lacking the stability needed for routine student operations. To rectify these problems, the aircraft received a -401 engine of 400 lbs. greater thrust, full-span leading-edge slats on the wings, a six-inch extension to the tail fin, an increase in horizontal stabilizer span, and a ventral fin below the tailplane. Testing of the first modified T-45As began in September 1990, carrier trials were satisfactorily completed in late 1991, and deliveries of first of 72 production T-45As to the Navy commenced during early 1992. In early 1994, while T-45A deliveries proceeded, McDonnell Douglas and British Aerospace began testing the T-45C, which featured a "Cockpit 21" upgrade having digital, multi-function flight displays, Rockwell Collins GPS, and Litton laser-ring gyro inertial navigation system, and starting with the 73rd aircraft in 1995, the T-45C superceded the A on the production

This T-45C, with VT-7 out of NAS Meridian, Mississippi, pictured during carrier qualifications aboard the USS *Harry S. Truman* (CVAN 75) in November 2012.

line. As T-45C production got underway, 158 were delivered to the Navy from late 1997 to late 2007, bringing total T-45 production to 236 aircraft since program inception. NAVAIR initiated a program with Boeing-BAE in 2002 to bring all T-45As up to the Cockpit 21 standard and all had been upgraded and returned to service as T-45Cs by the end of 2007.

Production T-45As initially entered service with VT-21 at NAS Kingsville, Texas, to train instructor-pilots in early 1992 and were introduced to the advanced phase of undergraduate pilot training in early 1994. Later the same year, VT-22, also at Kingsville, began transitioning to T-45As, followed later by NAS Meridan, Mississippi–based VT-7 and VT-9. The process of replacing TA-4s with T-45As in the advanced phase was completed in late 1999. Although the timeframe is not clear, it appears that the introduction of Goshawks to intermediate pilot training coincided with initial deliveries of T-45Cs during 1997, with the process of phasing out T-2s being fully completed by mid–2004. Naval Air Training Command expanded T-45C operations in 2008 to encompass training of Naval Flight Officers (NFO) in VT-86 at NAS Pensacola, Florida. As of mid–2013, internet sources report 218 T-45Cs on the active naval inventory (which include re-designated T-45As), with 18 having been previously destroyed in crashes or withdrawn for other reasons. Under current plans, the T-45Cs are expected to remain in service until 2035 or longer.

Beech T-1 Jayhawk—1992

Technical Specifications (T-1A)

Type: Advanced pilot and combat systems officer trainer.
Manufacturer: Beech Aircraft Corp. (sub. of Raytheon Corp. since 1993, renamed Hawker-Beech in 2007), Wichita, Kansas.

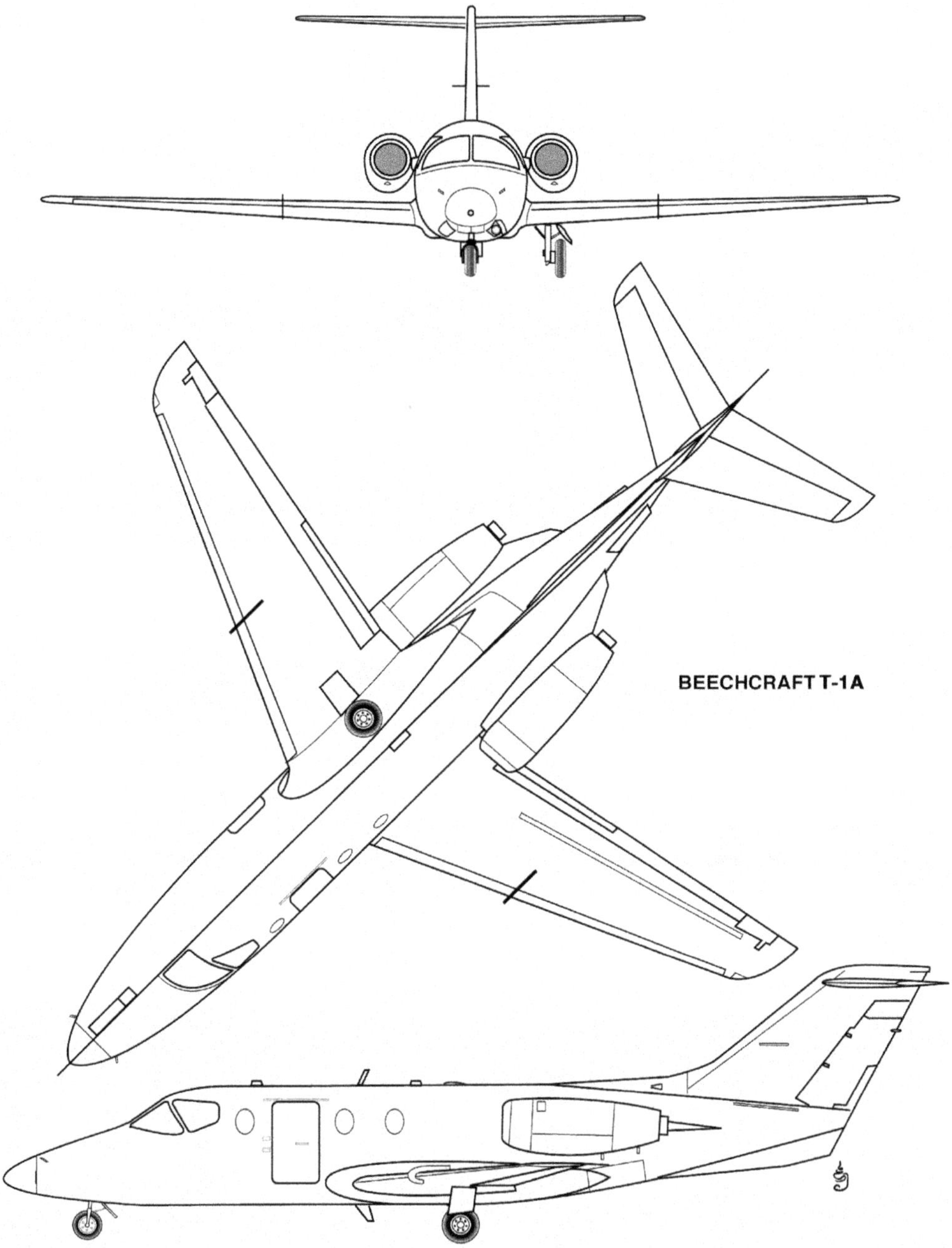

Total U.S. military versions: 180 (USAF).
Unit cost: $4,100,000.
Powerplant: Two Pratt & Whitney JT-15D-5B turbofan engines, each rated at 2,900 lbs./s.t.
Performance: Max. speed 531 mph, cruise 515 mph; ceiling 41,000 ft.; range 2,222 mi.
Weights: 10,115 lbs. empty, 15,780 lbs. max. takeoff.
Dimensions: Span 43 ft. 6 in., length 48 ft. 5 in., wing area 241.5 sq. ft.

The USAF procured the Beech (Hawker-Beech since 2007) T-1A Jayhawk in 1991 to replace North American T-39s (see in Series II, Part A) as an advanced trainer for undergraduate pilots

A Beech T-1A of the 12th Flying Training Wing, 99th Flying Training Squadron, based at Randolph AFB, Texas. This unit traces it origins to the 99th Fighter Squadron of the Tuskegee Airmen during World War II.

destined to fly aircraft in either the airlift or air refueling roles. Originating in Japan as the Mitsubishi MU-300 Diamond in 1978, the design was subsequently acquired by Beech and reintroduced to the civil market in 1986 as the Beechjet 400, followed in 1990 by the 400A, an improved version possessing a higher takeoff weight and longer range. Shortly thereafter, the USAF contracted with Beech to develop a military trainer variant that would differ from a stock 400A in having an airframe strengthened to withstand low-level turbulence and bird strikes, a single-point refueling receptacle, and an interior reconfigured for a student pilot, an instructor, and an observer-trainee, plus four extra seats for passengers. The original contract called for production of 158 aircraft to be designated T-1A under a new sequence and named Jayhawk (note: the Lockheed T-1A reported in Series II, Part B had by then been out of service for over 20 years). The first flight of the T-1A from the factory is not reported anywhere, however, the USAF accepted the first example from Beech in January 1992 and deliveries to Air Education and Training Command (AETC) continued throughout the balance of the year. In 1994 the USAF added 32 aircraft to the production contract and the last of 180 T-1As built was accepted in 1997.

As they arrived for duty in early 1992, T-1As initially entered service with the 64th Flying Training Wing (FTW) at Reese AFB to train the first generation of Jayhawk instructor-pilots and were introduced into undergraduate pilot training (UPT) in January 1993. As AETC's inventory of T-1As mounted, the type became operational with the 12th FTW at Randolph AFB, Texas, the 14th FTW at Columbus AFB, Mississippi, the 47th FTW at Laughlin AB, Texas, and the 71st FTW at Vance AFB, Oklahoma. Besides providing advanced training of student pilots on the airlift/tanker track, T-1As have also been used for training of T-37 (until 2009), T-38, and T-6 (Texan II) instructor pilots in the 71st FTW. The Jayhawk's training mission was expanded in 2009 when the 451st Flying Training Squadron, a subordinate unit of the 12th FTW based at NAS Pensacola, Florida, began operating 21 T-1As refitted with electronic warfare equipment to train Combat Systems Officers (CSO). Internet sources show 158 Jayhawks (two lost in mishaps) on the active USAF inventory as of mid–2013. Recent defense industry sources report that the USAF is conducting a marketing study to determine the feasibility of eventually replacing its T-1A fleet with a very light jet (VLJ), a category of small jet aircraft seating four to eight people and having a takeoff weight less than 10,000 lbs., however, no official announcement has been made.

Slingsby T-3 Firefly—1993

Technical Specifications (T-3A)

Type: Primary and flight screening trainer.
Manufacturer: Slingsby Aviation, Ltd., Kirkbymoorside, York, United Kingdom and Northrop Aircraft Services, Inc., Hondo, Texas.

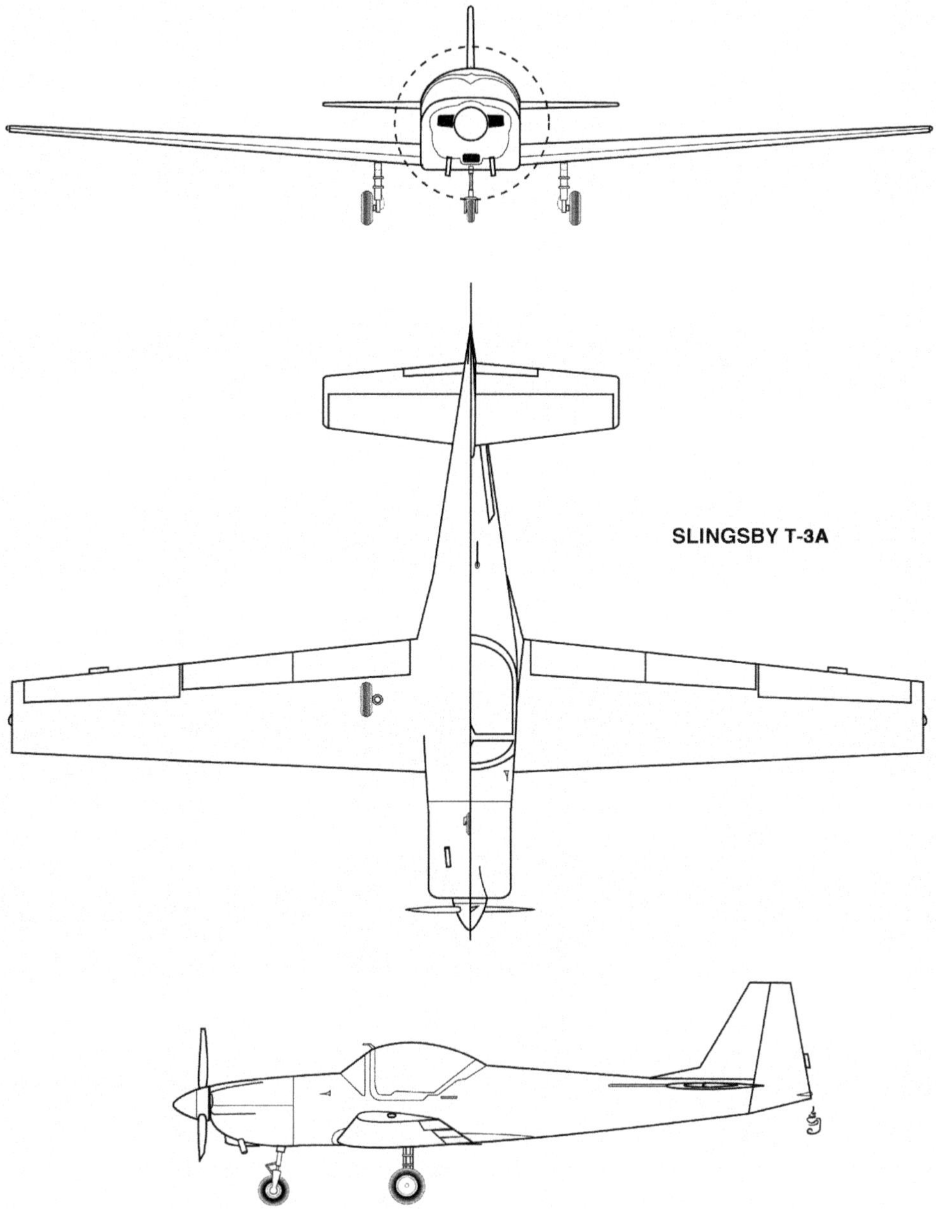

Total U.S. military versions: 114 (USAF).
Unit cost: $290,900.
Powerplant: One 260-hp Lycoming AEIO-540-D4A5 6-cylinder air-cooled horizontally-opposed engine driving a three-bladed constant-speed metal propeller.
Performance: Max. speed 175 mph, cruise 161 mph; ceiling 19,100 ft.; range 468 mi.
Weights: 1,750 lbs. empty, 2,550 lbs. max. takeoff.
Dimensions: Span 34 ft. 9 in., length 24 ft. 10 in., wing area 136 sq. ft.

In mid–1991 the USAF initiated a project to find a more capable trainer that would replace its long-serving fleet of Cessna T-41s (see above) in Air Force Academy (AFA) pilot indoctrination (PIP) and Air Training Command (ATC) flight screening programs (FSP). Under a new scheme known as "enhanced" flight screening program (EFSP), ATC/AFA specified a trainer which, unlike the T-41, would be capable of aerobatic maneuvers and overhead traffic patterns that exposed pilot candidates to moderate G-loading. To oversee and conduct the program, the USAF contracted with Doss Aviation, Inc., the civil contractor already responsible for FSP training at the Hondo Municipal Airport in Texas. An ATC and AFA-sponsored "operational suitability demonstration," held at Hondo during July and August 1991, ultimately attracted entries from 10 different U.S. and foreign companies: Sia-Marchetti SF-260, Aerospatiale (Socata) TB-30, Mooney EFS, Flug und Fahrzeugwerke Altenrhein (FAA) AS-202, SAAB MFI-17, Slingsby T67M, Stoddard-Hamilton Glasair II, American General AG-5, Piper PA-28R, and Taylorcraft F-22A. In April 1992 Air Force Material Command announced selection of the Slingsby T67M and later assigned the military designation T-3A and name "Firefly." In order to "Americanize" final procurement, the contract specified that the T-3As would be crated and shipped to the U.S. for final assembly by Northrop Aircraft Services at Hondo.

The concept behind the T67M/T-3A dated back to the French-designed, all wooden Fournier RF-6 motor glider of 1974, the design rights to which were acquired by Slingsby in

A Slingsby T-3A of the 3rd Flying Training Squadron at Hondo Muncipal Airport, Texas, in 1996. The entire fleet of T-3As was grounded in July 1997 due to a combination of safety concerns (courtesy Trevor Bartlett).

1981. With an eye toward the military market, Slingsby reengineered the design for a stronger, composite-material airframe and upped engine power by 100-hp. Originally, the USAF contemplated procurement of 125 T-3As (including the T67M prototype), 69 for ATC and 56 for AFA, but this was later scaled-back to only 57 for ATC (became Air Education and Training Command [AETC] in 1993). Deliveries of 113 production aircraft began in early 1994 and continued until early 1996. T-3As initially entered service with the 12th Flying Training Wing (FTW) at Randolph AFB where they were flown by USAF Test and Evaluation pilots to train the first group of T-3A instructor-pilots; and the first EFSP class at Hondo commenced training on Fireflies in March 1994. T-3As ultimately equipped two training units: the 3rd Flying Training Squadron (FTS) at Hondo for AFROTC and OTS candidates and the 557th FTS at Colorado Springs for AFA candidates. The USAF grounded the entire T-3A fleet (and thus ended EFSP) in July 1997 for a combination of reasons: (1) three fatal accidents between 1995 and 1997, two due to pilot error and one an unrecoverable stall condition; (2) an unacceptable rate of non-fatal mishaps due primarily to engine stoppages; and (3) finally, much higher than anticipated maintenance costs. One T-3A was retained in operation by the Test Pilot School at Edwards AFB with the remainder placed in storage until ultimately being scrapped in 2007.

T-4 and T-5 (skipped)

Beech TC-12 Huron—1996

Technical Specifications (TC-12B)

Type: Multi-engine and instrument advanced trainer.
Manufacturer: Beech Aircraft Corp. (sub. of Raytheon Corp. since 1993, renamed Hawker-Beech in 2007), Wichita, Kansas.
Total U.S. military versions (trainers): 14 (USN).
Unit cost: $2,000,000 (est.).
Powerplants: Two 850-shp Pratt & Whitney PT6A-41 turboprop engines driving three-bladed Hartzell constant-speed, reversible-pitch propellers.
Performance: Max. speed 338 mph, cruise 273 mph; ceiling 35,000 ft.; range 2,270 mi.
Weights: 7,334 lbs. empty, 15,000 lbs. max. takeoff.
Dimensions: Span 54 ft. 6 in., length 43 ft. 10 in., wing area 945 sq. ft.

Beech flew its model A200 (originally A100–1) Super King Air for the first time on October 27, 1972, and received FAA civil type certification in December 1973. Though aerodynamically similar to the model 90–100 series, the A200 featured more powerful engines, an extended cabin, increased wingspan, a new T-tail configuration, and upgraded avionics. As important as the added cargo and passenger capacity, the Super King Air also delivered a 20 percent improvement in ceiling and a 35 percent increase in range. The first military orders for the A200 came in 1974 when the Army procured 60 and the USAF 45 as C-12A light transports. Several years later, the Navy followed suit by contracting for 64 very similar transport variants delivered between 1978 and 1982 under the designation UC-12B.

In 1996, after concluding an inter-service agreement to provide the facilities and aircraft for advanced multi-engine training of USAF student pilots, the Navy determined that its current fleet of Beech T-44As (see above) would not be sufficient to fulfill the flight time requirements of the expanded program. Since T-44s were no longer in production, Naval Air Systems Command (NAVAIR) directed reassignment to training of a number of UC-12Bs to cover the shortfall. Although official records indicate these aircraft received the new designation TC-12B, there

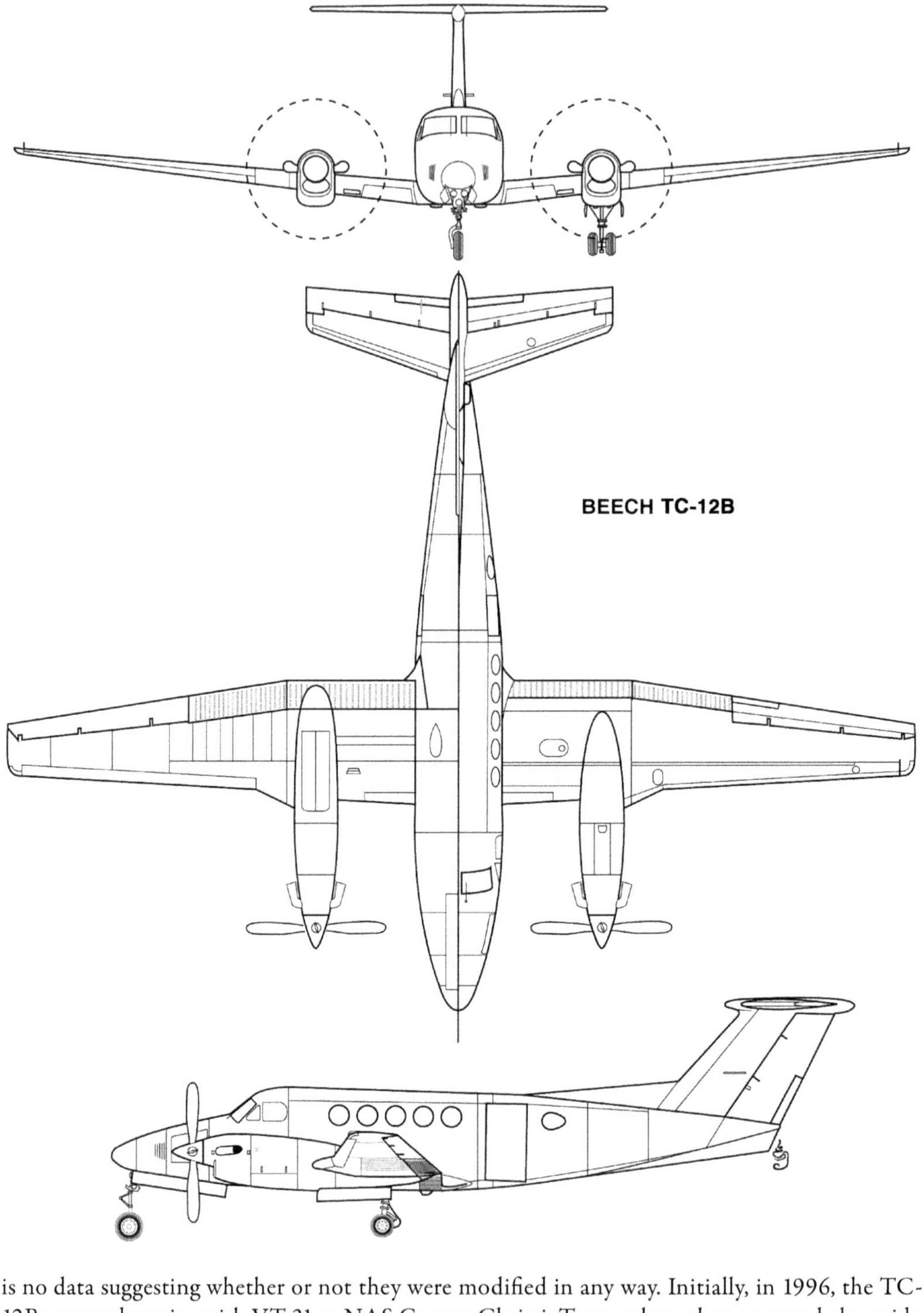

is no data suggesting whether or not they were modified in any way. Initially, in 1996, the TC-12Bs entered service with VT-31 at NAS Corpus Christi, Texas, where they operated alongside the unit's T-44As, then in late 1999 all TC-12Bs were reassigned to the newly-created VT-35, a joint USN/USAF unit established at NAS Corpus Christi to train USAF and USMC students destined for C-130s and Navy students for P-3s. Serial number records suggest that at least 14

A Beech TC-12B of VT-35 in January 2008 at NAS Corpus Christi, Texas. VT-35 is a joint service advanced unit providing multi-engine training to USN, USMC, and USAF student pilots.

UC-12Bs were re-designated as TC-12Bs, though more may have been seconded to training without a change in designation.

Note on Joint Primary Aircraft Training System

By the late 1980s the USAF and the Navy both had reached the point where they needed to be taking active steps to replace their aging fleets of primary/intermediate trainers—the Cessna T-37B and Beech T-34C (see in Series II, Part A)—by the end of the millennium. Toward this end, the Department of Defense (DoD) submitted a Trainer Aircraft Master Plan to the U.S. Congress in 1989 with the broad aim of procuring a common airframe and training system to meet the long-term needs of all the military services. In December 1990, the DoD Joint Requirements Oversight Council (JROC), a body comprised of the Vice Chairman of the Joint Chiefs and the Vice Chiefs of each military service, issued operational requirements for a program entitled Joint Primary Aircraft Training System (JPATS), envisaging future procurement of up to 900 training aircraft that would be integrated with a "system" package of simulators, software, and teaching materials. Between 1993 and 1995, seven aircraft manufacturers tendered prototypes for the "winner-take-all" JPATS contract and notably, all but one (Cessna), submitted an aircraft based upon a preexisting foreign trainer design. The hard lessons learned from the failure of the Next Generation Trainer program (see Fairchild T-46 above) no doubt influenced participants to base their prototypes on a proven design. The following will provide a brief summary on each of the competing prototypes:

Cessna 526 CitationJet: Coming as the only homegrown entry, the twin turbofan-powered 526 CitationJet combined the wings, engines, landing gear, and internal systems of the company's 525 business jet with a new fuselage having a tandem cockpit equipped with ejection seats. The first of two prototypes built flew in December 1993. Although Cessna subsequently demonstrated its 526 at many aviation events, the aircraft never achieved production.

Northrop/Embraer EMB-312H Super Tucano: Of Brazilian origin, the turboprop-

The two Cessna Citationjets, flown in late 1993, were the only American-designed entries to participate in the JPATS competition.

This Brazilian-made EMB-312H is essentially identical to the Northrop-sponsored entry in the JPATS competition.

powered EMB-312 made its first flight in 1980 and was serving as a trainer in the air forces of eight nations at the time of the JPATS competition. With Northrop as its American partner, Embraer developed three EMB-312H prototypes featuring a strengthened airframe, cockpit pressurization, and a 1,250-shp PT6A-68 engine. While not selected for JPATS, it became the prototype for the Super Tucano subsequently placed in production for the Brazilian Air Force.

Vought/FMA (Fabrica Militar de Aviones) IA-63 Pampa: Powered by a single Garrett

The Argentine Air Force FMA IA-63 pictured is the same type of aircraft entered by Vought as a candidate for the JPATS contract.

turbofan engine delivering 3,500 lbs./s.t., the Argentine-designed Pampa first began entering service as a trainer with the Argentine Air Force in 1988. With Vought as its American partner, FMA's unsuccessful JPATS entry, an unmodified IA-63, was presented as the Vought Pampa 2000. FMA has since been acquired by Lockheed-Martin.

Lockheed/Aeromacchi T-Bird II: The T-Bird II surfaced in 1993 as a renamed variant of the Italian-designed Aeromacchi MB-339, which was flown for the first time in 1976, and at the time of the JPATS competition, serving as a trainer in the air forces of Italy, Malaysia, New Zealand, Peru, Nigeria, and Argentina. The Lockheed-sponsored JPATS entry came with improved avionics and a Rolls-Royce *Viper* turbojet engine rated at 4,000 lbs./s.t. Following JPATS, it became the basis for exports to other nations.

Beech/Pilatus PC-9: The Swiss-designed PC-9 Mk. 2, selected by DoD as winner of the JPATS competition, is reported in detail below under the heading Raytheon (Hawker-Beech/Pilatus) T-6 Texan II.

Rockwell/RFB (Rhein-Flugzeugbau GMBH) Ranger 2000: The Ranger 2000 evolved as a joint venture to develop a pure jet version of RFB's earlier German-designed Fantrainer 600, which utilized a 650-shp Allison turboprop engine to drive a ducted fan. The two Ranger 2000 prototypes flown for the JPATS competition in 1993 received a new aft fuselage housing a single JT15D-5C turbofan engine fed by side intakes and a straight wing layout that deleted the forward sweep of the Fantrainer. After losing JPATS, the design was later marketed as the Fanjet 600.

Grumman/Siai-Marchetti S.211A: The Italian-designed S.211, a pure jet powered by a JT15D-4C turbofan engine, made its first flight in 1981 and commenced operational service as a trainer with the air forces of the Philippines, Singapore, and Haiti during the mid–1980s. At the time of JPATS, Grumman partnered with Siai-Marchetti to upgrade two former Haitian air-

Despite its evocative American name, Lockheed's T-Bird II entry in JPATS was a repainted version of Italian-designed Aeromacchi MB-339C.

One of the two Rockwell-RFB Ranger 2000s that participated in JPATS as seen at Karlsruhe-Baden-Baden, Germany, in 2008 (courtesy Jörg Bachmann).

A privately owned Siai-Matchetti S.211 of the same type entered by Grumman in the JPATS competition.

craft to participate in the competition as the S.211A. Siai-Marchetti was acquired by Aeromacchi in 1997, who now offers for export a modernized version of this aircraft as the S.311.

Raytheon (Hawker-Beech/Pilatus) T-6 Texan II—1998

Technical Specifications (T-6A)

Type: Two-place primary and basic trainer.
Manufacturer: Beech Aircraft Corp. (sub. of Raytheon Corp. since 1993, renamed Hawker-Beech in 2007), Wichita, Kansas.
Total U.S. military versions: 514 (453 USAF; 61 USN).
Unit cost: $4,300,000-$6,130,000 (in 2012).
Powerplant: One 1,100-shp Pratt & Whitney PT6A-68 turboprop engine driving a four-bladed Hartzell constant-speed, reversible-pitch propeller.
Performance: Max. speed 338 mph, cruise 273 mph; ceiling 35,000 ft.; range 2,270 mi.
Weights: 4,707 lbs. empty, 6,500 lbs. max. takeoff.
Dimensions: Span 33 ft. 5 in., length 33 ft. 4 in., wing area 175.3 sq. ft.

Since entering service 12 years ago, the Raytheon T-6 Texan II has become the U.S. armed forces' "trainer for all seasons," encompassing not only primary and intermediate pilot training for the USAF and Navy but likewise the training of navigators, Combat Systems Officers (CSO), and Naval Flight Officers (NFO). The Swiss-designed Pilatus PC-9, from which the general design of the Texan II was derived, was itself an evolution of the PC-7, which had been in service as a trainer with the air forces of over 20 nations since the late 1970s. Following announcement

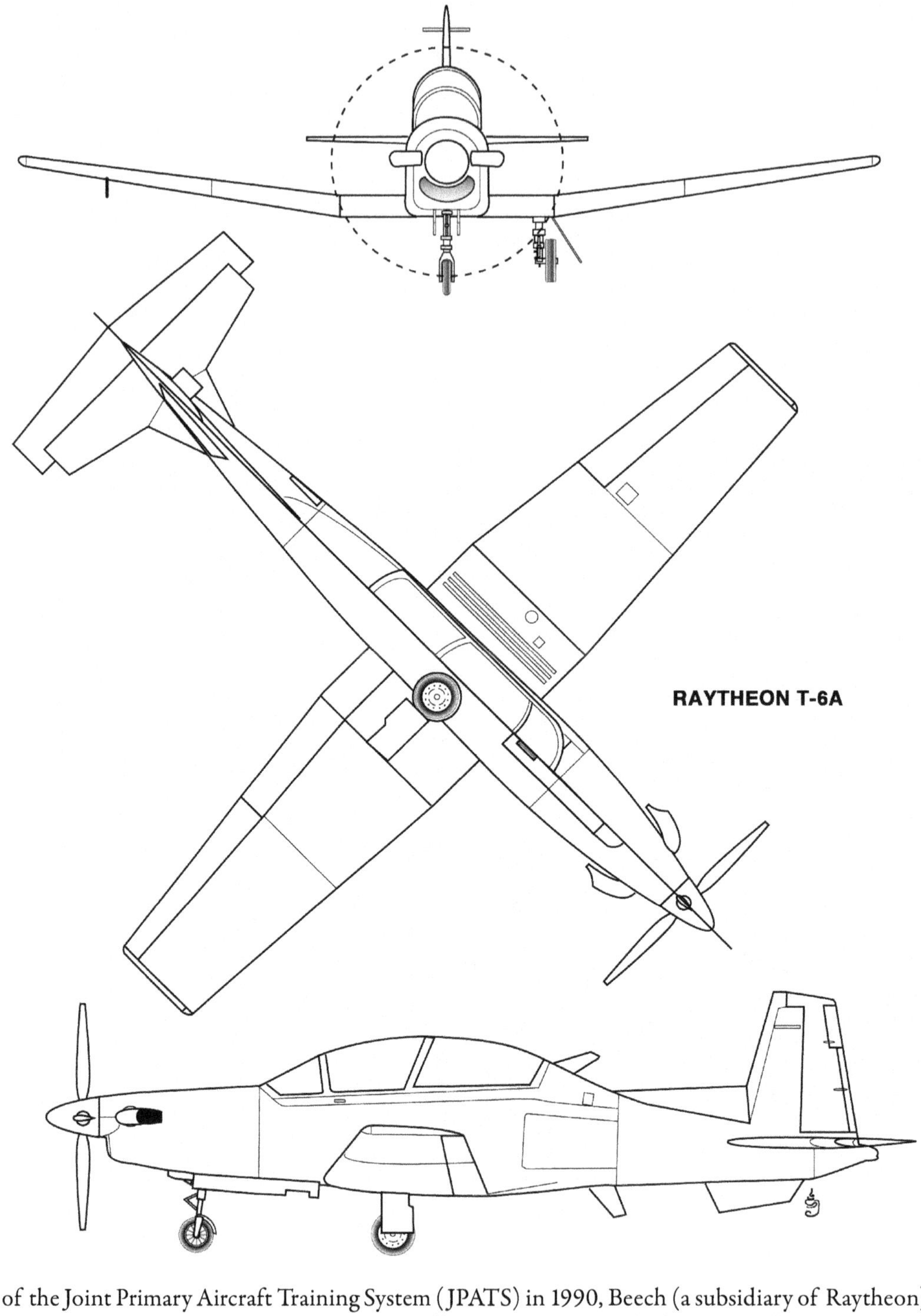

of the Joint Primary Aircraft Training System (JPATS) in 1990, Beech (a subsidiary of Raytheon) entered into a partnership with Pilatus to offer a JPATS contender known as the PC-9 Mk. 2, essentially similar to the PC-9A then serving with the Royal Australian Air Force, and the aircraft, if it reached production, would be license-produced in the U.S. The JPATS selection process, which began officially in May 1994, continued for period of 14 months and involved flight trials

This Raytheon T-6A was photographed in the vicinity of Randolph AFB, Texas, in June 2003 while in service with the 558th Flying Training Squadron, the unit that initially trained USAF and USN T-6A instructor-pilots.

of seven aircraft plus evaluation of seven cockpit mockups and literally thousand of pages of contractor proposals.

In June 1994 the Department of Defense (DoD) announced the Raytheon/Beech PC-9 Mk.2 to be the winner of JPATS, with the award of a production contract following in February 1996. The contract contained options for two production runs, one from 1996 through 2004 and a second extending to 2017 and further authorized Raytheon to proceed with construction of production aircraft under designation T-6A (no experimental prototype) with deliveries to operational units expected in 1999. To honor its predecessor, the venerable North American AT-6 Texan (see in Series I, Part A), the new T-6A was officially dubbed "Texan II." The flight of the first T-6A production aircraft took place from the Beech plant in May 1998; however, changes made to the basic design to meet additional USAF and Navy requirements were so extensive that the T-6A essentially evolved into a new airframe 1,100 lbs. heavier than the PC-9 Mk. 2. This not only generated significant cost increases but delayed initial deliveries of T-6As to the USAF's Air Education and Training Command (AETC) until May 2000. Over the next 10 years (at a rate of 45 to 50 per year), the USAF accepted a total of 453 T-6As, the last arriving in 2010. The Navy took delivery of its first T-6A in August 2002 and introduced the type to Naval Air Training Command (NATC) in January 2003. Original Navy plans to procure 315 T-6As were subsequently scaled back to 301, with a total of 49 having been accepted by the end of 2006. Though the date of its first flight is not recorded, it seems that Raytheon began testing the T-6B sometime in early 2005. The B differs from the A mainly in having an all-glass cockpit with multi-function displays, heads-up displays (HUD), hands-on-throttle and stick controls (HOTAS), and dual integrated avionics computers. In 2007, the Navy apparently decided to shift the balance of its contract to T-6B production, with the first examples being delivered for evaluation in July 2009 and achieving operational status in April 2010. Internet sources show 12 T-6Bs on the naval inventory as of mid–2013.

A new Raytheon T-6B of Training Air Wing Five seen taxiing at NAS Whiting Field in May 2010 with Ens. Christopher Farkas, the first T-6B student, at the controls.

In AETC service, T-6As initially became operational with the 12th Flying Training Wing (FTW) at Randolph AFB, Texas, in May 2000 to commence training of both USAF and Navy instructor-pilots. Integration of T-6As into undergraduate pilot training (UPT) syllabus began with the 47th FTW at Laughlin AFB, Texas, during 2002, and as deliveries continued, the type went on to equip the 14th TFW at Columbus AFB, Mississippi, and the 71st TFW at Vance AFB, Oklahoma, with the process of replacing T-37Bs with T-6As being fully completed in July 2009. Presently, AETC's T-6As provide primary and intermediate training for all advanced UPT tracks, including fighter/bomber (T-38), airlift/tanker (T-1A and USN T-44 and TC-12), and helicopter (Army UH-1), plus the advanced track for Combat Systems Officer (T-38). T-6As began entering NATC service with Training Air Wing Six (TAW-6) at NAS Pensacola, Florida in early 2003, eventually equipping VT-4 and VT-10, with the phase-out of T-34Cs being completed during 2005. The transition to T-6Bs in TAW-5 at NAS Whiting Field, Florida started in 2010 and was completed in 2012; and the transition to T-6Bs in TAW-4 at NAS Corpus Christi, Texas, is scheduled to begin sometime in 2013. Like AETC, T-6A/Bs provide primary and intermediate pilot training for all advanced tracks, including fighter/strike (T-45), airlift and maritime (T-44 and UC-12), and helicopter (TH-57, see in Series IV), plus the track for navigators and Naval Flight Officers.

Other versions of the T-6, not in U.S. service but aimed primarily at foreign customers, are: T-6A NTA with hardpoints for rockets, gun pods, bombs, or external fuel tanks; AT-6B with the B cockpit, hardpoints for weapons, and a -68D engine rated at 1,600-shp; and T-6C with the B cockpit, hardpoints, and reconnaissance/intelligence-gathering capabilities. In addition to the U.S. military purchases, Raytheon has exported 146 Texan II variants as follows: 25 T-6As, Royal Canadian Air Force for NATO Flying Training as the CT-156 Harvard II; 25 T-6As and 20 T-6A NTAs, Hellenic (Greek) Air Force; 20 T-6As, Israeli Air Force; 15 T-6As, Iraqi Air Force; 12 T-6Cs, Mexican Air Force; and 24 T-6Cs, Royal Moroccan Air Force. Although the

AT-6B combat trainer/ light attack platform has been actively marketed by Raytheon since 2006, none have apparently been sold to date.

Diamond T-52–2009

Technical Specifications (T-52A)

Type: Primary and flight screening trainer.
Manufacturer: Diamond Aircraft Industries, Inc., London, Ontario, Canada.
Total U.S. military versions: 20 (USAF).
Unit cost: Leased.
Powerplant: One 180-hp Lycoming IO-360-M1A 4-cylinder air-cooled horizontally-opposed engine driving a two-bladed constant-speed metal propeller.
Performance: Max. speed 204 mph, cruise 173 mph; ceiling 16,400 ft.; range 828 mi.
Weights: 1,755 lbs. empty, 2,645 lbs. max. takeoff
Dimensions: Span 39 ft. 2 in., length 26 ft. 5 in., wing area 145.3 sq. ft.

The Austrian-designed Diamond DA40 completed its first flight in November 1997 and was subsequently certificated in the U.S. by the Federal Aviation Administration in October 2000. Since its introduction, the DA40 has been acquired as a trainer in the air forces of Bolivia, Ecuador, and Jamaica. Constructed entirely of composite materials, the DA20's design is characterized by low-mounted, high-aspect ratio wings and a boom-type rear fuselage that ends in a T-tail empennage. In 2004, in order to resurrect the Air Force Academy (AFA) initial flight training (IFT) program, the USAF contracted with Embry-Riddle Aeronautical University to train cadets at the AFA airfield in Colorado Springs, Colorado, initially with an unspecified number of leased Cessna 172 aircraft (already having the military designation T-41), and then starting in 2009, with 20 leased DA40s which received the unofficial designation T-52A. Since these aircraft are not government owned, they do not appear in any published serial number records.

One of 20 leased Diamond T-52As used in the U.S. Air Force Academy's Initial Flight Training program while on display at the 2010 Royal International Air Tattoo held at RAF Station Fairford, England.

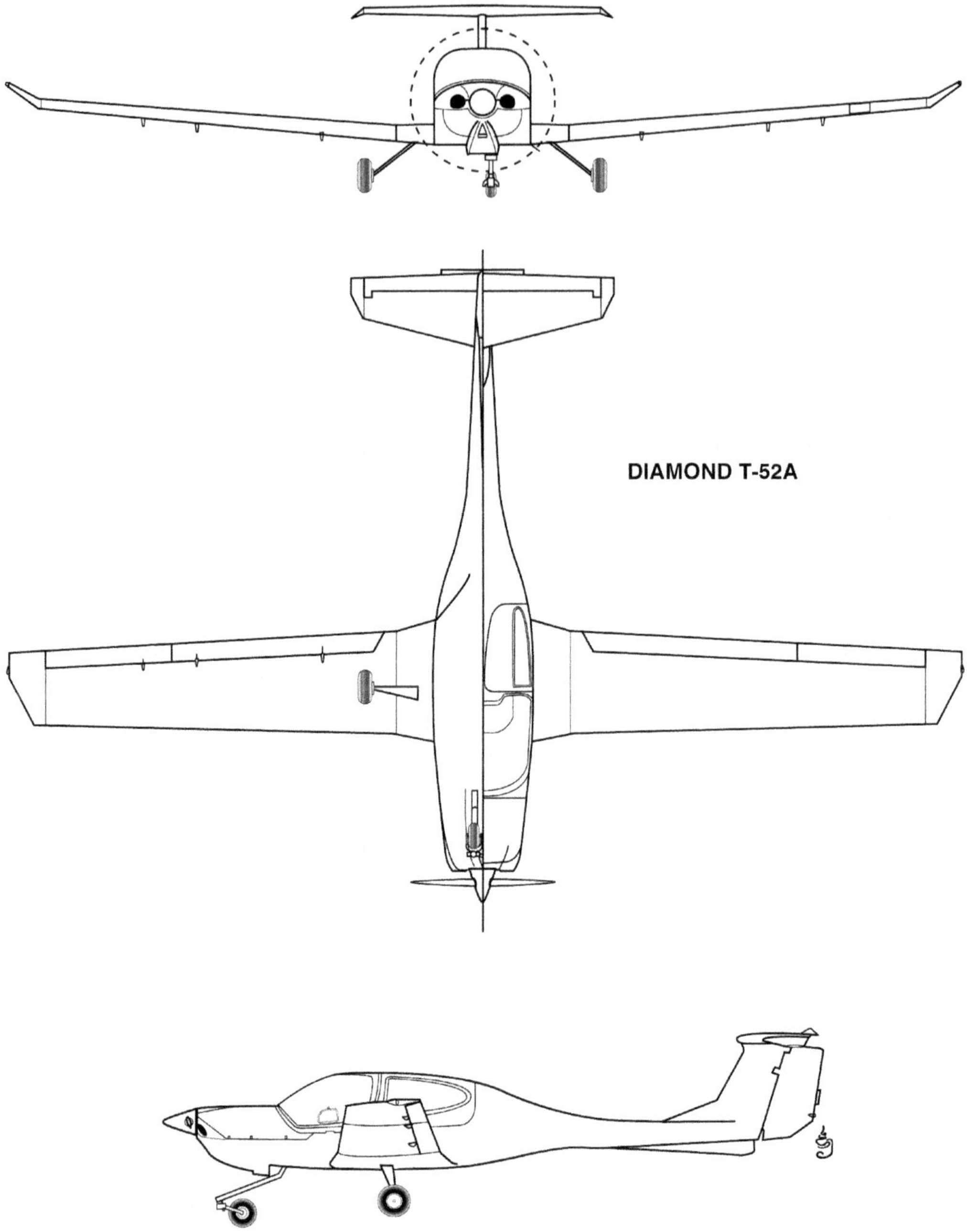

Though not having the aerobatic capability of the grounded T-3A (see above), the DA20/T-52A nonetheless possesses the lowest fatal accident rate (i.e., 0.35/100,000 hours) of any aircraft on the U.S. civil registry according to a 2011 statistical survey.

The 20 leased T-52As have been employed by the 557th Flying Training Squadron (FTS) to provide the powered training segment of IFT to AFA cadets who aspire to be candidates for the USAF Undergraduate Pilot Training (UPT) program. The AETC-approved syllabus for powered IFT consists of approximately 25 hours of flight training which includes a pre-solo check ride, solo flight, and a final check ride. In 2011, the USAF announced that the leased T-52As would be replaced in powered IFT in a direct procurement of the Cirrus SR20 under the assigned designation T-53A (see below).

Cirrus T-53–2011

Technical Specifications (T-53A)

Type: Primary and flight screening trainer.
Manufacturer: Cirrus Aircraft Corp., Duluth, Minnesota (a wholly owned subsidiary of the government-owned China Aviation Industry General Aircraft).

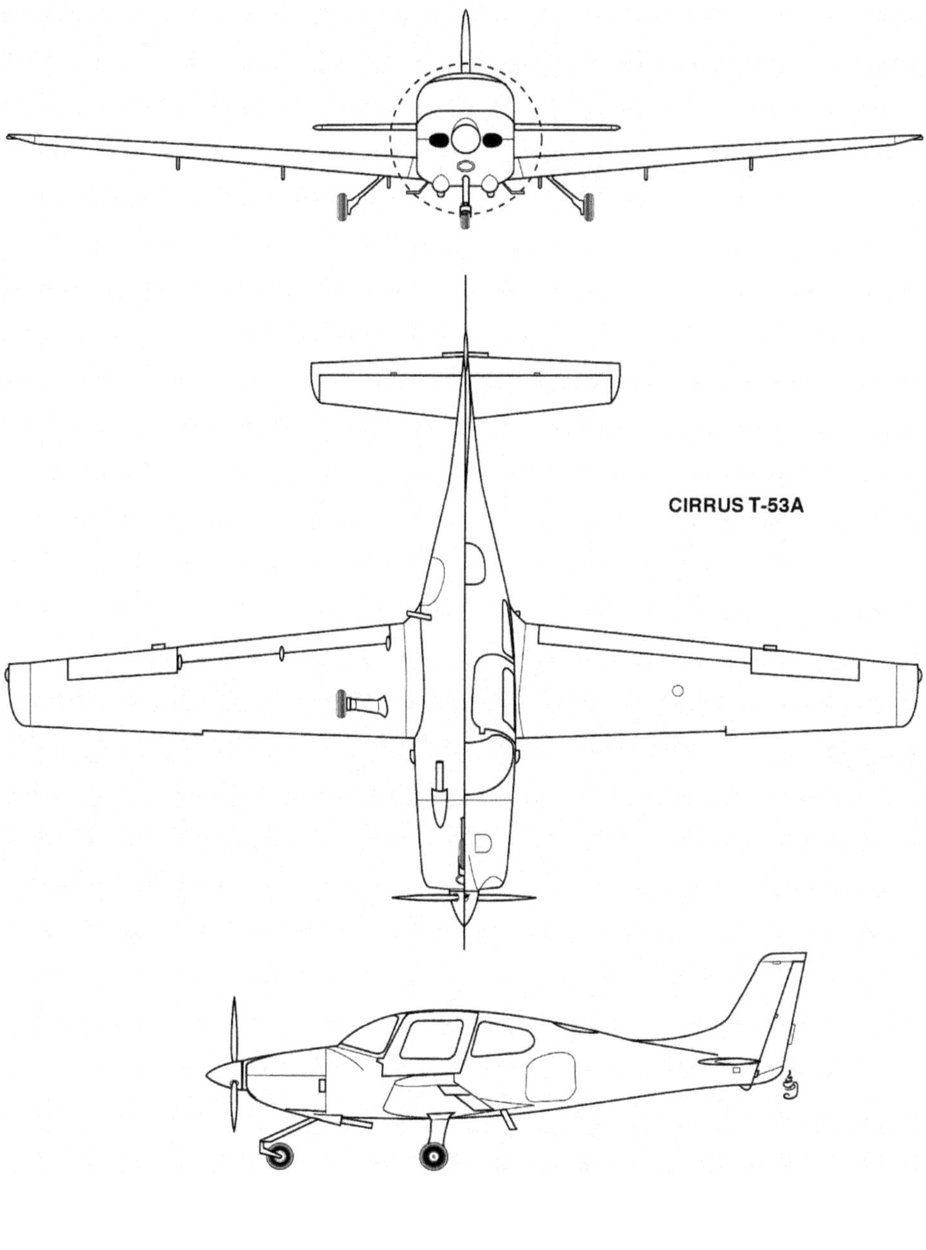

One of the new Cirrus T-53As introduced during 2011 to the U.S. Air Force Academy's Initial Flight Training program. As of this writing, 22 of the 25 planned T-53As have been delivered.

Total U.S. military versions: 22 of 25 ordered (USAF).
Unit cost: $244,000.
Powerplant: One 200-hp Continental IO-360-ES 4-cylinder air-cooled horizontally-opposed engine driving a two-bladed constant-speed metal propeller.
Performance: Max. speed 193 mph, cruise 178 mph; ceiling 16,400 ft.; range 828 mi.
Weights: 2,128 lbs. empty, 3,050 lbs. max. takeoff
Dimensions: Span 38 ft. 4 in., length 26 ft., wing area 144.9 sq. ft.

Since making its debut on the civil market in late 1998, the Cirrus SR 20 has become one of the most popular four-place American lightplanes, with over 5,000 sold by the end of 2012. Appearing with a conventional low-wing layout, the SR20's composite material construction gives it an exceptional level of aerodynamic cleanness and without any sacrifice of strength, superior performance compared to similarly powered aircraft having traditional metal airframes. One of the SR20's unusual features is a ballistic parachute that can be deployed to lower the entire aircraft to the ground in the event of an engine stoppage or other in-flight emergency. Another major selling point has been the Garmin-Cirrus Perspective avionic suite providing an "glass cockpit" with dual screens for flight and multi-function displays (i.e., nav/com, GPS, weather, and TCAS-type traffic information modes). In 2008 Cirrus introduced the improved SR20 G3 with a carbon-fiber wing spar, redesigned landing gear, improved aerodynamics, and a dual-redundant GPS/VOR navigation system.

Around the same time the leased T-52As (see above) began operating in the Air Force Academy (AFA) initial flight training (IFT) program, the USAF initiated a study of potential lightplane trainer candidates with the aim of directly procuring them for the military inventory. In early 2011, after announcing selection of the SR20 G3, the USAF awarded Cirrus a contract to deliver 25 aircraft under the military designation T-53A. The 557th Flying Training Squadron (FTS) at the AFA airfield in Colorado accepted delivery of the first T-53A on July 6, 2011, and serial number records indicate that 22 of the 25 ordered appeared on the USAF inventory as of January 2013. Some criticism has been leveled at the acquisition: the statistical accident rate of the SR20/T-53A (i.e., 1.6/100,000 hours) versus the DA40/T-52A (0.35/100,000) and the C-172/T-41C (.45/100,000); and the fact that Cirrus is wholly owned by the Chinese government.

In making the selection, the USAF cited unit cost as a major factor: Cirrus SR20 G3, $244,000; Cessna 172S, $307,500; and Diamond DA40, $344,950.

Note on Trainer-Experimental (T-X) Program

As far back as 1993 the USAF instigated the T-X Program to find an advanced trainer to replace its aging fleet of Northrop T-38s (see in Series II, Part A) with a view toward having the new aircraft in service by 2017; however, budgetary restraints since then have extended initial operating capability to 2023 or beyond. In order to prepare pilots for fifth generation fighters like the Lockheed-Martin F-22 Raptor and F-35 Lightning II, the T-X requirements have become more complex. Minimally, the T-X aircraft and simulation system will not only be expected to perform traditional advanced training roles such as aircraft maneuvering, formation flying, instrument and navigation flying, and air-to-air and air-to-ground combat tactics but will also have to prepare pilots for aerial refueling, night-vision imaging systems, air-to-air intercept systems, and data-link operations. Although a formal T-X request for proposals (RFP) has yet to be released, the USAF nonetheless circulated a request for information (RFI) in 2009 that has thus far generated four potential candidates based on preexisting aircraft designs, namely:

General Dynamics/Alenia-Aeromacchi T-100: In January 2013 General Dynamics and Italian firm of Alenia-Aeromacchi signed a letter of intent creating a joint venture to enter an upgraded version of Alenia-Aeromacchi's M-346 as a T-X contestant. The supersonic (Mach 1.2) M-346, flown for the first time in 2004, is currently in production as an advanced trainer scheduled to enter service in the near future with the air forces of Italy (15 ordered), Israel (30

General Dynamics will sponsor the Mach 1.2 Alenia-Aeromacchi T-100 as a potential successor to the USAF's Northrop T-38.

ordered), and Singapore (12 ordered). Interestingly, the M-346 was largely derived from Yakolev Yak–130, a Russian-designed subsonic trainer currently serving with the air forces of Russia and Algeria, with production orders pending for the nations of Bangladesh, Belarus, Mongolia, Syria, and Vietnam. Should this group win the competition, the aircraft will undergo final assembly in the United States.

Northrop-Grumman/BAE Hawk T2/128: BAE Systems and Northrop-Grumman announced a partnership in 2011 to offer the updated Hawk T2/128 as a contestant in the T-X program. Although subsonic Hawk variants have been serving as trainers with various air forces for 37 years and with the U.S. Navy since 1992 (see McDonnell Douglas T-45 above), the newer T2/128 features a glass-cockpit and is powered by an efficient Rolls-Royce *Adour* turbofan producing 6,500 lbs./s.t. The BAE/Northrop-Grumman partners envisage the T2/128 as offering a low-cost, low-risk alternative to other potential competitors.

Norhtrop-Grumman announced a partnership with BAE systems in 2011 to sponsor the Hawk T2 as a T-X candidate.

Lockheed-Martin/Korea Aerospace Industries (KAI) T-50: Lockheed-Martin and KAI have been engaged in a formal partnership since the late 1990s to develop aircraft for the Republic of Korea Air Force (RKAF) and potential foreign customers, starting with the T-50 advanced trainer and later, the TA-50 light attack platform and FA-50 multi-role fighter. The supersonic (Mach 1.5) T-50, powered by a license-produced GE F404 afterburning turbofan, completed its maiden flight in August 2002 and reached operational capability with the RKAF in early 2005, and the first of 16 T-50s ordered by the Indonesian Air Force began entering service in September 2013. Many features of the T-50's design were borrowed from the F-16 Fighting Falcon, which is license-produced by KAI as the KF-16. Although the designation T-50A has been formally reserved by the U.S. Department of Defense (DoD), there is no contractual understanding as yet between DoD and the partnership. Inasmuch as Lockheed-Martin is intimately familiar with the training requirements for its F-22 and F-35, the partnership believes the T-50 will emerge be a strong candidate for T-X.

Boeing/Saab Gripen JAS-39F: Though nothing formal has been announced, Boeing is known to be working with the Swedish Saab Group to formulate a T-X candidate based on Saab's JAS-39 Gripen, a Mach 2.0 multi-role fighter which has been operational with the Swedish Air

The Mach 1.5 KAI T-50, sponsored by Lockheed-Martin, has reportedly been tailored to meet the training requirements for the F-22 and F-35.

The SAAB JAS-39F, seen in the Hungarian Air Force, will reportedly be added to T-X competition by Boeing.

Force since 1997 and since then, with the air forces of Hungary, the Czech Republic, South Africa, and Thailand. Powered by a Volvo RM12 turbofan rated at 18,100 lbs./s.t. in afterburning, the JAS-39 comes in single and two-seat variants and is characterized as a true delta planform having swept canards forward of the wing on the upper intakes. The T-X candidate will most likely be based on the two-seat JAS-39F, which features a wide-screen cockpit with fully-integrated helmet-mounted displays (HMD). According to rumor, Boeing and Saab believe they can offer the JAS-39 at a unit cost lower than that of the competing Lockheed/KAI T-50.

Series IV

Rotary Wing Trainers, 1942 to Present

Synopsis of Rotary Wing Trainer Procurement, 1942 to Present

Because they are a more specialized subdivision of flight training as a whole, helicopters will be covered in chronological order without regard to service branch. Although the Platt-Le Page XR-1 of 1941 was technically the first helicopter to be tested by the USAAF, the series begins with the Sikorsky R-4 because it was the first type of helicopter to actually be employed for training purposes. Of necessity, the first generation of operational helicopters, namely the Sikorsky R-4, R-5, and R-6, also functioned to train the earliest group of military helicopter pilots. The designation of USAF and Army helicopters changed from "R" to "H" in 1948, and the Bell H-13 (naval designation HTL) emerged as the first type of helicopter to be procured in quantity for training purposes. From 1948 to 1970, approximately 1,400 H-13 trainer variants were accepted by the U.S. military services: USAF, 18 as the YH-13 in 1948, 65 as the H-13B in 1948–1949, and 264 as the H-13G in 1953–1954; Army, 466 as the H-13H in 1956–1958 and 415 as the TH-13T 1965–1970; and Navy, 12 as the HTL-2 in 1949, nine as the HTL-3 in 1950–1951, 46 as the HTL-4 in 1951, 39 as the HTL-5 in 1951–1952, 48 as the HTL-6 in 1955–1956, and 18 as the HTL-7 in 1958. (Note: all helicopters earmarked for Army training were delivered to the USAF until 1956.) As military helicopter training operations grew during the 1950s, the Hiller H-23 (naval designation HTE) was procured in quantity: USAF, 273 as the H-23B in 1952–1953; Army, 438 as the H-23D in 1955–1956 and 793 as the H-23G in 1962 (most likely not used for training); and Navy, 17 as the HTE-1 in 1951 and 25 as the HTE-2 in 1952.

While not designed as a trainer, the Bell UH-1 Huey has nonetheless seen service in a variety of training roles: in 1960, 14 Army UH-1As converted to dual controls returned to service as TH-1A instrument trainers; in 1963, 20 Marine UH-1Cs configured for aircrew training entered service as the TH-1E; in 1964, 20 USAF UH-1Fs completed as instrument and rescue trainers as the TH-1F; in 1969, 45 Navy UH-1Es configured as instrument trainers as the TH-1L; and since 2005, 45 UH-1Hs refurbished as instrument trainers have joined the USAF inventory as the TH-1H. In an across-the-board program to replace its primary helicopter trainers, the Army procured 797 Hughes TH-55As between 1965 and 1967. In 1988, when the Army retired the last 139 TH-55As from training duties, Bell UH-1Hs (with no official change in designation) functioned as stopgap trainers until the early 1990s. With the goal of augmenting and eventually replacing the piston-engine Bell H-13M in helicopter training, the Navy procured 40 Bell 206A Jet Rangers in 1968 as the TH-57A; then in 1986, to replace the TH-57As, purchased 51 Bell 206B-3s that entered service as the TH-57B; and in 1989, ordered 89 more 206B-3s configured for instrument training as the TH-57C. At present, the Navy has announced plans to upgrade 128 of its TH-57B/Cs with glass cockpits and other improvements and return them to service as the TH-57D. In 1992 the Navy acquired six MD Helicopters (ex–Hughes) MD-500Cs under the designation TH-6B to be used as trainers in the Naval Test Pilot School. To effect a long-term solution in its helicopter training fleet, the Army selected the Bell 206B-3 in 1993 under the designation TH-67A, and through a series of contracts awarded since that time, has accepted a total of 181 TH-67As, 60 in a visual flight rule (VFR) configuration and 121 equipped for instrument training.

Sikorsky R-4 (HNS)—1942

Technical Specifications (R-4B/HNS-1)

Type: Two-place operational and training helicopter.
Manufacturer: Vought-Sikorsky Div., United Aircraft Corp., Stratford, Connecticut.
Total U.S. military versions: 82 (57 USAAF, 25 USN/USCG).
Unit cost: Unknown.
Powerplant: One 200-hp Warner R-550–3 7-cylinder air-cooled radial engine driving a three-bladed main and three-bladed tail rotor.
Performance: Max. speed 75 mph; ceiling 8,000 ft.; range 130 mi.
Weights: 2,098 lbs. empty, 2,535 lbs. loaded.
Dimensions: Rotor diameter 38 ft., length 35 ft. 5 in.

The Sikorsky R-4 holds the distinction of having been the first type of helicopter to enter operational service with any of the U.S. military branches. Its design emerged as a direct development of Igor Sikorsky's VS-300, which was likely the most practical helicopter design in the world at the time it flew in September 1939. Sikorsky invented and patented the system of a single horizontal rotor utilizing a vertical tail rotor to offset rotational torque, a feature common to virtually every single-rotor helicopter seen today. In the first of its long line of military helicopters, Sikorsky received a USAAF contract sometime during 1941 to deliver a prototype of the proposed model VS-316 under the designation XR-4. The VS-316/ XR-4 materialized as a straightforward enlargement of the single-seat VS-300 having a more powerful engine (165-hp vs. 90-hp) and a boxy forward fuselage widened to accommodate two, side-by-side seats. The airframe consisted of a welded, steel tube framework with fabric covering forward and a bare tail boom; and the landing gear was conventional, with strut-braced main wheels and a single tailwheel aft. The maiden flight of the XR-4 took place from the Sikorsky plant on January 13, 1942. While historical sources do not indicate exactly when the prototype was accepted by the USAAF, initial testing of the XR-4 was sufficiently encouraging to generate an order for three service test YR-4As with 180-hp engines, a two-foot larger rotor diameter, and fabric covering on the tail boom. Later in the year, evaluations of the YR-4A led to an order for 27 YR-4Bs having increased fuel capacity, a redesigned exhaust system, and a tailwheel relocated further aft. With service testing nearly complete, Sikorsky received a production contract in early 1943 for 100 R-4Bs, 78 being delivered to the USAAF and 22 to the Navy. R-4Bs differed in having 200-hp engines and provision for a medical litter and bomb racks.

While service evaluations were ongoing, the YR-4Bs were also used to train the first five USAAF instructor-pilots, with the first helicopter training unit of Army Air Force Training

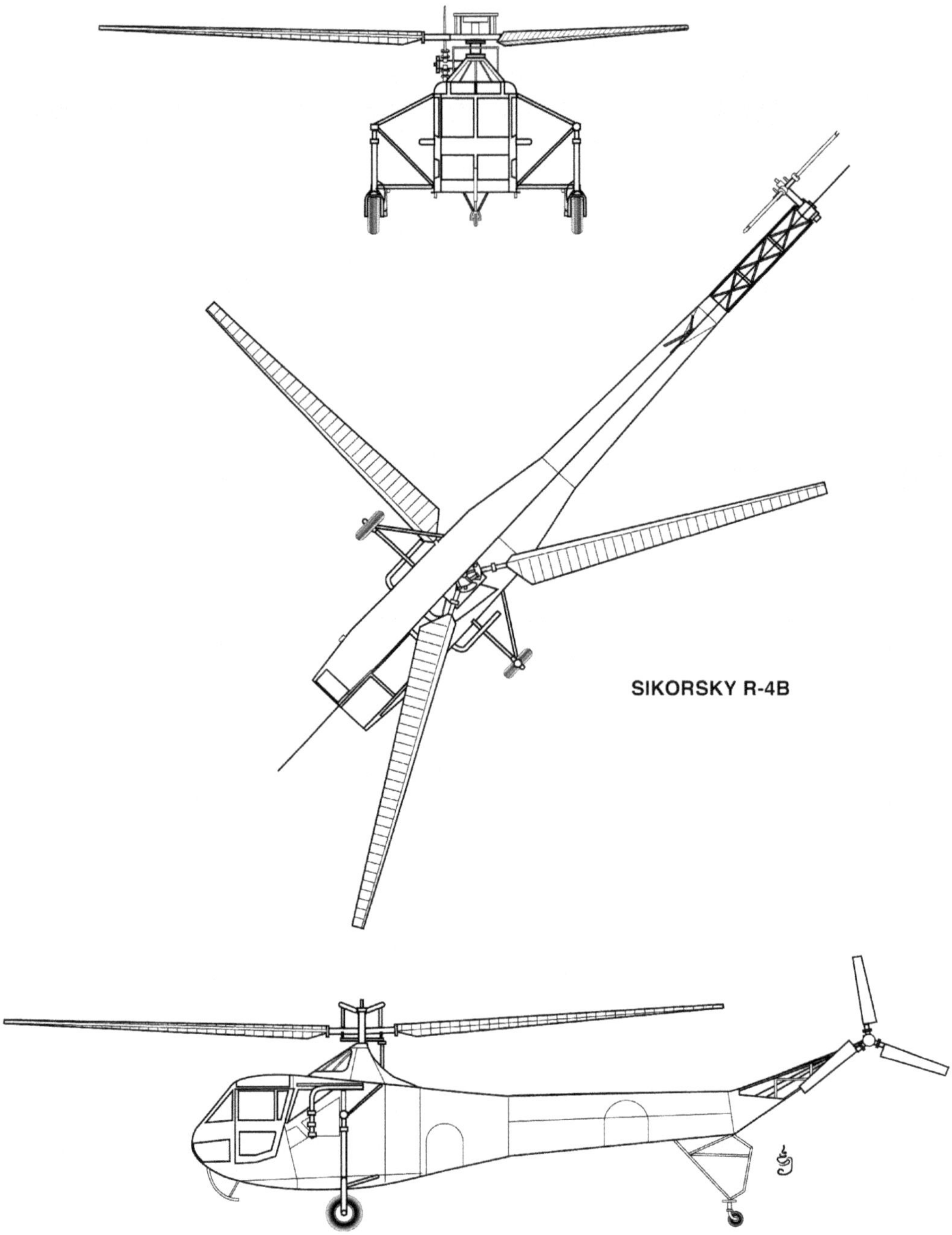

Command (AAFTC) being established at Freeman Field in Seymour, Indiana, in June 1944 with two of the YR-4Bs. Even though significant maintenance difficulties plagued the program, 36 new helicopter-rated pilots had been trained by the end of the year. In December 1944, AAFTC helicopter training, then equipped with 14 R-4Bs, moved to Chanute Field in Illinois in order to be closer to major maintenance support. By the end of World War II in 1945, AAFTC's helicopter school had produced a total of 93 pilots. AAFTC disbanded helicopter training in late 1945 and apparently transferred its R-4Bs to other commands.

This Sikorsky R-4B has been preserved at the U.S. Air Museum at Dayton, Ohio. The R-4 was the first type of helicopter to be utilized by the USAAF for training helicopter pilots.

This Navy HNS-1 (YR-4B), circa 1944, is seen using a hoist in an early rescue experiment. The first Naval Helicopter Training School was established at CGAS Floyd Bennett Field, New York, in late 1943.

Naval Variants: The Navy took delivery of its first YR-4B in October 1943, followed by two more in November, which all received the naval designation HNS-1, and in the same time-frame, established the first naval helicopter training school at CGAS Floyd Bennett Field in New York. During 1944 the Navy accepted 22 R-4Bs as the HNS-1, with two assigned to the Coast Guard and some, fitted with floats for operations on water. According to Navy and Coast Guard records, all HNS-1s had been phased out from training by the end of 1947.

Sikorsky R/H-5 (HO2S/HO3S) Dragonfly—1943

Technical Specifications (R-5A/HO2S-1)

Type: Two-place rescue, observation, and training helicopter.
Manufacturer: Vought-Sikorsky Div. (later Sikorsky Aircraft Div.), United Aircraft Corp., Stratford, Connecticut.
Total U.S. military versions: 234 (131 USAAF/USAF; 103 USN/USMC/USCG).
Unit cost (H-5F): $90,000 (est.).
Powerplant: One 450-hp Pratt & Whitney R-985-AN-5 9-cylinder air-cooled radial engine driving a three-bladed main and three-bladed tail rotor.
Performance: Max. speed 90 mph, cruise 75 mph; ceiling 10,000 ft.; range 245 mi.
Weights: 3,810 lbs. empty, 4,813 lbs. loaded.
Dimensions: Rotor diameter 48 ft., length 41 ft. 10 in.

Authorized by the USAAF in early 1943 as the XR-5, Sikorsky's second military helicopter, the VS-327, represented an effort to incorporate the successful R-4 rotor system into a larger and more powerful helicopter with much-improved load and range. Besides greater size and horsepower, the XR-5 also introduced a more streamlined fuselage of all-aluminum construction which featured tandem crew seating. The first of four XR-5 prototypes ordered flew on August 18, 1943, and successful trials led, in March 1944, to a contract for 26 YR-5As for service testing. While the official acceptance program with the first YR-5As was still ongoing, the USAAF gave Sikorsky a production contract to deliver 100 R-5As fitted with racks for two medical litters; however, when World War II ended, the order was reduced to the 34 helicopters delivered. The S-51, an improved R-5 offered by Sikorsky to potential civil and military customers in 1946, differed from the R-5A in having a larger rotor diameter, seating for three passengers plus the pilot, and a tricycle landing gear layout. During 1947 the USAF ordered 11 specially equipped S-51s for search and rescue (SAR) operations as the R-5F, and at about the same time, 21 R-5As refitted with nose landing gear and rescue hoists returned to service for SAR duties as the R-5D. Also in 1947, five YR-5As modified for installation of dual controls for training purposes became YR-5Es. In mid–1948 the USAF adopted the "H" designation for all helicopters, so all existing R-5s became the H-5A, H-5D, YH-5E, and H-5F, respectively.

When the USAAF resumed helicopter pilot training in May 1946 at San Marcos AAF, Texas, it began with five R/H-5As and 16 Sikorsky R/H-6As (see below). Though at least five YR-5As are known to have received dual controls as the YR/YH-5E, no records are available as to the actual number of R/H-5s that served with Air Training Command (ATC). In any case, the last H-5s were withdrawn from ATC in 1954.

Naval Variants: The 103 R/H-5s procured by the Navy, Marines, and Coast Guard between 1945 and 1949 as the HO2S-1 and HO3S-1/-1G were used exclusively for SAR and medical evacuation.

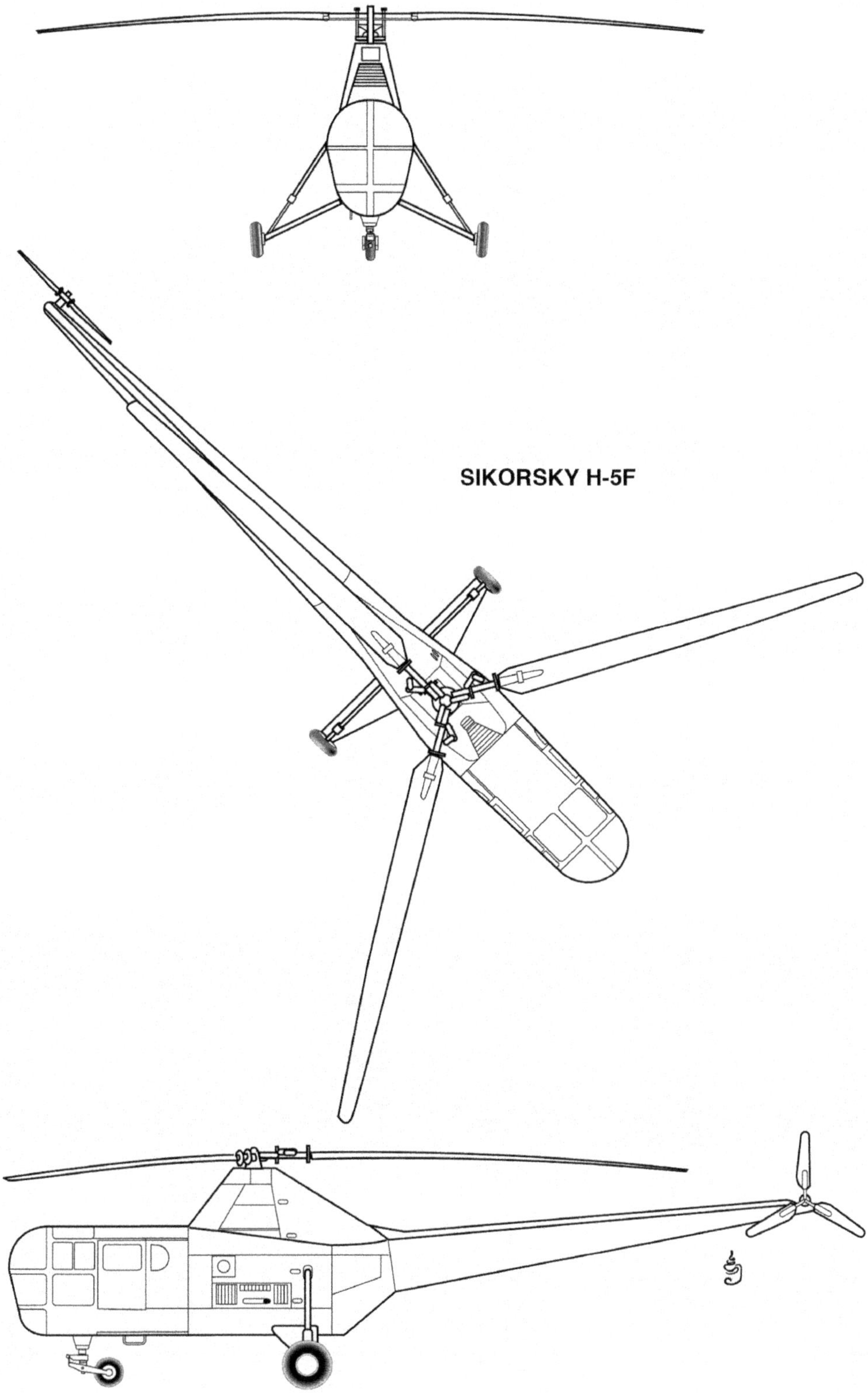
SIKORSKY H-5F

This Sikorsky YH-5A, preserved at the U.S. Air Force Museum in Dayton, Ohio, was one of 25 originally delivered to the USAAF in 1944 and used for training helicopter pilots during the immediate postwar era.

This Sikorsky H-5F, pictured in Korea in the early 1950s, differed from earlier models in having four seats, a larger diameter rotor, and tricycle landing gear. Some were used for USAF pilot training until 1954.

Sikorsky R/H-6 (HOS)—1943

Technical Specifications (R-6A/HOS-1)

Type: Two or three-place rescue, observation, and training helicopter.
Manufacturer: Vought-Sikorsky Div. (later Sikorsky Aircraft Div.), United Aircraft Corp., Stratford, Connecticut.

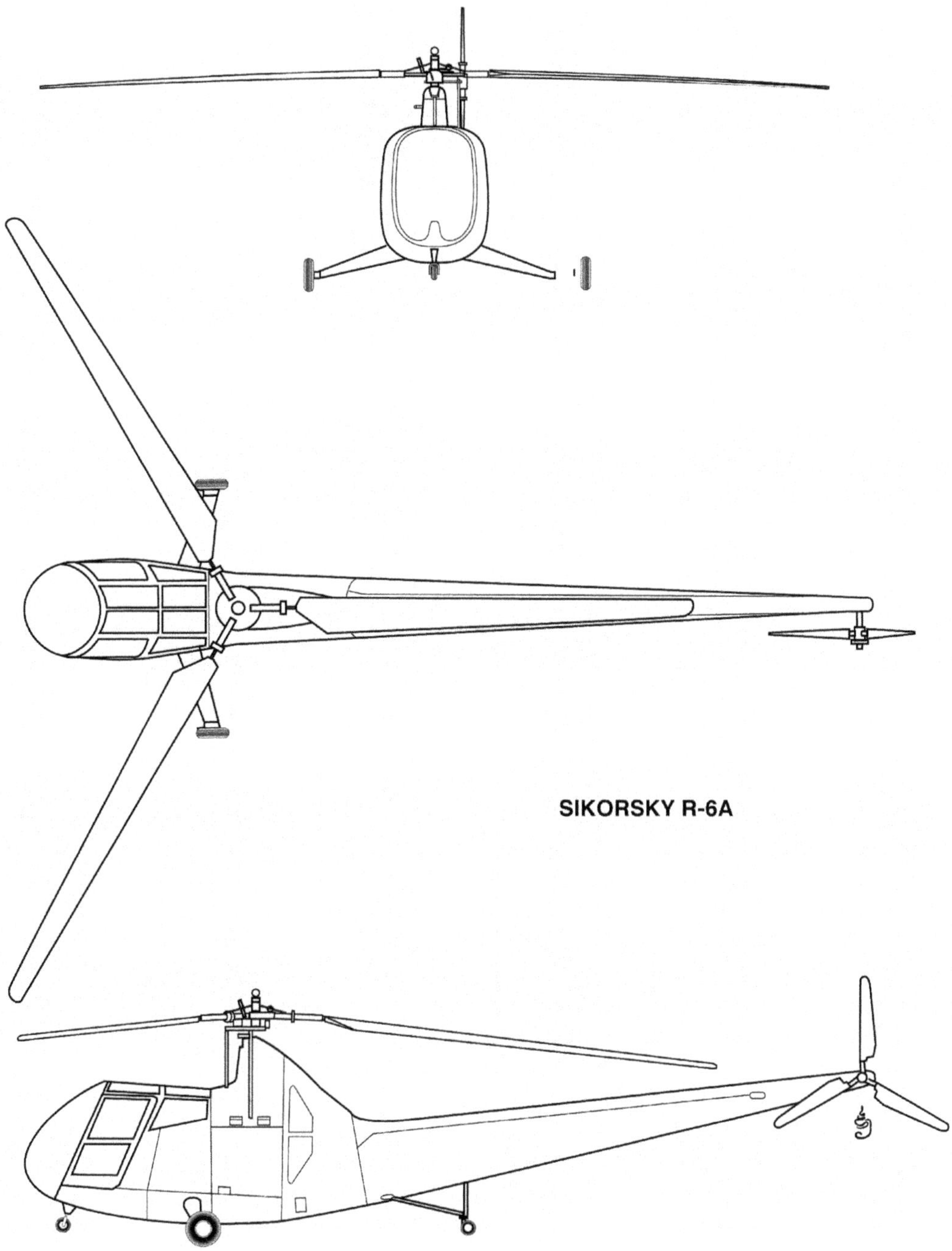

SIKORSKY R-6A

The date and location of the R-6A depicted are unknown. However, the national insignia suggests a time in 1947 when the type was being used for USAAF/USAF helicopter pilot training at San Marcos AAF, Texas.

Total U.S. military versions: 198 (159 USAAF/USAF; 39 USN/USCG).
Unit cost: Unknown.
Powerplant: One 240-hp Franklin 6AG6–245TO (O-405–9) 6-cylinder air-cooled horizontally-opposed engine driving a three-bladed main and three-bladed tail rotor.
Performance: Max. speed 100 mph, cruise 75 mph; ceiling 10,000 ft.; range 245 mi.
Weights: 2,035 lbs. empty, 2,900 lbs. loaded.
Dimensions: Rotor diameter 38 ft., length 47 ft. 11 in.

Using the same transmission and rotor system as the earlier VS-316/R-4, the VS-316B (later S-49) appeared with a streamlined fuselage and tail boom of all-aluminum, semi-noncoque construction, a blown windscreen, and a 225-hp Lycoming O-435–7 horizontally-opposed engine. Ordered by the USAAF in early 1943 as the XR-6, the single prototype made its first flight on October 15, 1943, and was followed in 1944 by delivery of five XR-6As with 240-hp Franklin O-405–9 engines. Later the same year, Sikorsky received a contract for 26 nearly identical preproduction examples as the YR-6A. Successful service testing during 1945 resulted in an order for 166 production R-6As (actually built by Nash-Kelvinator under a subcontract), 130 of which ultimately were accepted by the USAAF in 1945 and 1946. Improved variants as the R-6B and XR-7 were proposed but never built. In mid–1948, the designation of R-6As then on the USAF inventory was changed to H-6A.

R-6As began entering service with Air Training Command (ATC) for primary training of helicopter pilots at San Marcos AAF, Texas, in May 1946 and continued in the same role after ATC was absorbed by the independent USAF in 1947. As new Bell H-13s (see below) arrived to take their place during mid–1948, all H-6As were withdrawn from training duties in ATC.

Naval Variants: Three of the USAAF's XR-6As were allocated to the Navy in late 1943 for evaluations under the designation XHOS-1 and were followed during 1945 by delivery of 36 production R-6As as the HOS-1. Twenty-seven of the Navy's HOS-1s were assigned to the Coast Guard at CGAS Floyd Bennett Field, New York, where they were used for joint primary training of Navy, Marine, and Coast Guard helicopter pilots, with some subsequently fitted with floats to train pilots in water operations. All HOS-1s were withdrawn after the joint training program terminated in 1949.

Bell R/H-13 (HTL) Sioux—1946

Technical Specifications (YR/H-13 [TH-13T])

Type: Two or three-place training helicopter.
Manufacturer: Bell Helicopter Co. (later Bell-Textron Aerospace), Buffalo, New York, and Fort Worth, Texas.
Total U.S. military versions (with dual-controls): 1,779 (1,625 USAAF/USAF/U.S. Army; 154 USN/USMC/USCG).
Unit cost (H-13H): $35,750.
Powerplant: One 175-hp Franklin 6V4–178-B32 (O-335–1) [270-hp Lycoming TVO-425-D1B] 6-cylinder air-cooled horizontally-opposed engine driving a two-bladed main and two-bladed tail rotor.
Performance: Max. speed 78 mph [110 mph], cruise 63 mph [97 mph]; ceiling 13,000 ft. [18,000 ft.]; range 150 mi. [324 mi.].
Weights: 1,560 lbs. [1,950 lbs.] empty, 2,100 lbs. [2,920 lbs.] loaded.
Dimensions: Rotor diameter 35 ft. 1 in. [36 ft.], length 27 ft. 4 in. [28 ft. 6 in.].

The Bell model 47 helicopter, military designation H-13/HTL, was not only the first American helicopter to receive a Civil Airworthiness Certificate but also the first to be specifically produced for military pilot training. Of the 2,407 military variants procured by the service branches between 1946 and 1970, approximately 1770 possessed dual controls, though not all were dedicated to training. Coming as the brainchild of inventor Arthur Young, the model 47's unique two-bladed rotor system utilized a centrifugal stabilizer bar, linked directly to the rotor mast by a universal joint, to stabilize the rotor's plane of rotation (as opposed to the hinged three-rotor system seen on Sikorsky helicopters). When rolled out for its first flight on December 8, 1945, the model 47 prototype appeared with a tubular steel fuselage and tail boom, a one-piece bubble canopy, and a four-wheel, nose-type landing gear. In December 1946, after the 47 had received civil certification the previous March, the USAAF procured one off-the-shelf model 47A under the designation YR-13 to be evaluated for use with Army ground forces, and in early 1947, added 17 more to the order for continued service testing, along with 3 winterized versions, as the YR-13A, for cold weather trials in Alaska. Military variants were distinguishable by fabric covering on the tail booms and two-piece canopy enclosures. In 1948, when Bell introduced the improved 47D with a 200-hp O-335–3 engine, the USAF acquired 18 with dual controls as the YH-13 (R changed to H the same year), followed by a production order for 65 virtually identical 47Ds as the H-13B.

When Bell offered the model 47G in 1953, the USAF contracted for 264 with dual controls as the H-13G for primary training of Army helicopter pilots. The G differed in having a 250-hp Lycoming VO-435–23 engine, an uncovered tail boom, and skids in place of wheels (originally introduced on H-13E single control variants in 1951). In 1956, soon after helicopter pilot training had shifted to Army Aviation control, the Army procured 466 dual-control 47Gs as the H-13H, the first to feature all-metal rotors. The final dual-control variants, 415 TH-13Ts acquired by the

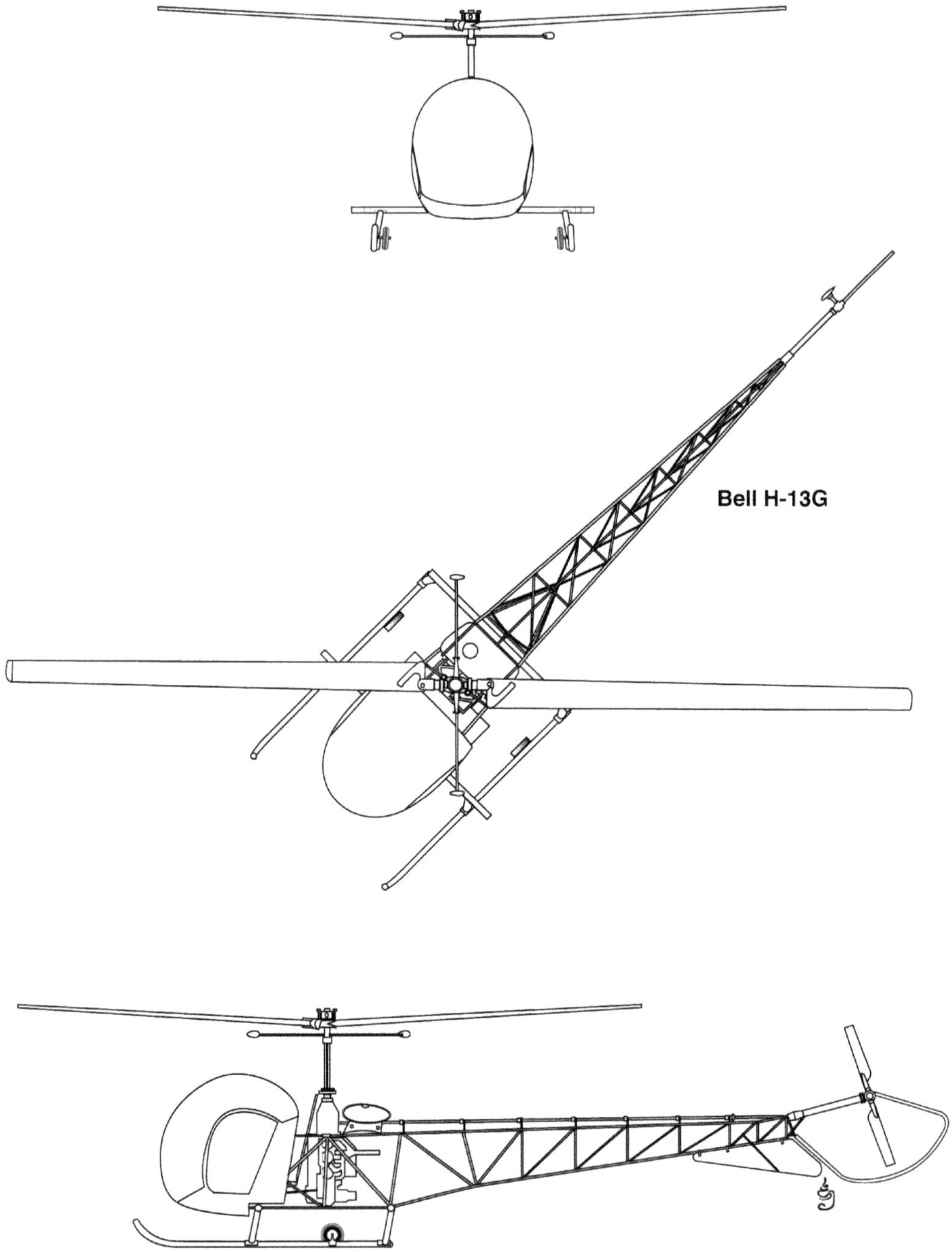

Bell H-13G

Army between 1965 and 1968, were 16-inches longer, powered by 270-hp TVO-435-D1B engines, and equipped for instrument training. Though seldom used, Army H-13s received the official name "Sioux."

The first YH-13s and H-13Bs earmarked for training began entering service with USAF Air Training Command (ATC) at San Marcos Field, Texas (renamed Gary AFB in 1953), in 1948 for the express purpose of training the first generation of organic Army Aviation helicopter

One of 264 Bell H-13Gs accepted by the USAF in 1954 and 1955 for training of Army student helicopter pilots at Gary AFB, Texas. In 1956, most were reassigned to the new Army Helicopter Pilot School at Fort Wolters, Texas.

pilots, and the H-13Gs arriving in 1953 and 1954 served in the same capacity. In 1956, when the USAF stopped training Army helicopter pilots, ATC helicopter training was moved to Randolph AFB, Texas, and with it, an unspecified number of H-13s for primary training of USAF student helicopter pilots. Some but not all of the H-13Hs delivered directly to the Army from 1956 onwards were allocated to the newly established Primary Helicopter School at Fort Wolters, Texas. TH-13Ts entered service at Fort Wolters as instrument trainers during the peak of Army helicopter pilot training through the mid– and late 1960s. Serial number records reflect that most TH-13Ts were retired in the early and middle 1970s. Afterward, some went to foreign air forces under MAP while many others were sold surplus on the civil market.

Naval Variants: In 1947 the Navy received 10 of the YR-13s ordered by the USAAF and commenced official evaluations under the naval designation HTL-1, then in 1949, procured 12 model 47Ds for primary helicopter pilot training as the HTL-2 (USAF H-13B). As Navy helicopter pilot training grew, it acquired nine 47Ds in 1950–1951 as the HTL-3 (USAF H-13D), and in the first quantity order, 46 similar 47D-1s in 1951 as the HTL-4 (USAF H-13E), plus 39 47D-1s in 1951–1952 similar to the -4 but with wheels instead of skids as the HTL-5 (re-designated TH-13L in 1962). The next sizable order came in 1955, for 48 47Gs as the HTL-6 (USAF/Army H-13G; re-designated TH-13M in 1962). The final naval variants, based upon the civil model 47K having an aluminum forward fuselage and tail boom, entered service as instrument trainers during 1958 as the HTL-7 (re-designated TH-13N in 1962).

Some of the Navy HTL-1s and -2s were most likely used for helicopter pilot training in the Coast Guard-directed program at CGAS Floyd Bennett Field from 1947 to 1950; however, all but two were reassigned to the newly established Helicopter Training Unit One (HTU-1) at

A Navy HTL-6 (TH-13M after 1962), circa late 1950s, that served with Helicopter Training Unit One at NAS Whiting Field, Florida. This unit became HT-8 in 1960.

Fully restored ex–Navy HTL-7 (TH-13N after 1962) seen at the Oshkosh, Wisconsin, EAA Airshow in summer 1989.

NAS Whiting Field, Florida in late 1950. Thereafter, as the center of Navy, Marine, and Coast Guard primary, basic, and advanced helicopter pilot training, HTU-1 took delivery of the bulk of HTL-3s, -4s, -5s, -6s, and -7s that entered service between 1950 and 1958. HTU-1 was renamed Helicopter Training Group One (HTG-1) in 1957, then again to Helicopter Training Squadron Eight (HT-8) in 1960. Designation records suggest only the TH-13L, TH-13M, and TH-13N remained active with HT-8 after mid–1962. Marine squadron HMX-1 received an HTL-2 in 1948 and operated other HTL variants during the 1950s.

Hiller H-23 (HTE) Raven—1950

Technical Specifications (H-23D)

Type: Two or three-place rescue, observation, and training helicopter.
Manufacturer: United Helicopters, Inc. (later Hiller Helicopters), Palo Alto, California.
Total U.S. military versions: 1,829 (1,777 USAF/U.S. Army; 52 USN).
Unit cost: Unknown.
Powerplant: One 250-hp Lycoming VO-435–23B 6-cylinder air-cooled horizontally-opposed engine driving a two-bladed main and two-bladed tail rotor.
Performance: Max. speed 95 mph, cruise 82 mph; ceiling 13,200 ft.; range 197 mi.
Weights: 1,816 lbs. empty, 2,700 lbs. loaded.
Dimensions: Rotor diameter 35 ft. 5 in., length 27 ft. 9 in.

Inventor and designer Stanley Hiller tested his first helicopter in 1944 at age 19, and with the backing of industrialist Henry J. Kaiser, formed United Helicopters, Inc. in 1946 and obtained a civil airworthiness certificate for the company's first helicopter, the model 360X, in late 1948. Appearing with a four-piece bubble canopy, strut-braced, circular aluminum tail boom, and tricycle landing wheels, the two-place 360X featured the Hiller-invented "Rotor-Matic" system, a two-bladed rotor configuration employing two small airfoils attached perpendicular to the rotor blades which moved the rotor's plane of rotation with cyclic-control inputs. In mid–1950, shortly after the 360X production model had appeared on the civil market as the UH-12A, the Army acquired one example, powered by a 178-hp Franklin O-335–4 engine, under the designation YH-23A, then later the same year, following successful evaluations, gave United Helicopters a production contract for 100 more as the H-23A. At about the same time, the USAF also acquired five H-23As for testing and evaluation.

The first dedicated trainer variants, 273 H-23Bs procured by the Army between 1952 and 1955, differed from the A in having a 200-hp Franklin O-335–6 engine, combination skid and wheel landing gear, and detail changes. As Army helicopter pilot training continued to expand from 1955 onwards, a series of contracts awarded to Hiller (renamed in 1952) resulted in deliveries of 438 dual-control H-23Ds powered by 250-hp Lycoming O-435–23B engines, and from 1962, 793 H-23Gs which, under the tri-service designation system, joined the Army inventory as the OH-23G. The G models came with dual-controls, a three-seat cockpit, and a 305-hp Lycoming VO-540-A1B engine. Although H-23As and Bs were earmarked for primary training of Army helicopter pilots, like H-13Bs and Gs, they were actually operated and maintained until 1956 by USAF Air Training Command (ATC) at Gary AFB, Texas. However, all H-23Ds went directly into Army service, with the majority going to its new Primary Helicopter School at Fort Wolters, Texas. When the tri-service designation system went into effect in mid–1962, remaining H-23Bs became OH-23Bs and H-23Ds, OH-23Ds. Though having dual controls, it appears that most, if not all, OH-23Gs were employed for duties other than training. The phase out of H/OH-23s from regular Army training began in the mid–1960s as new Hughes TH-55As (see below)

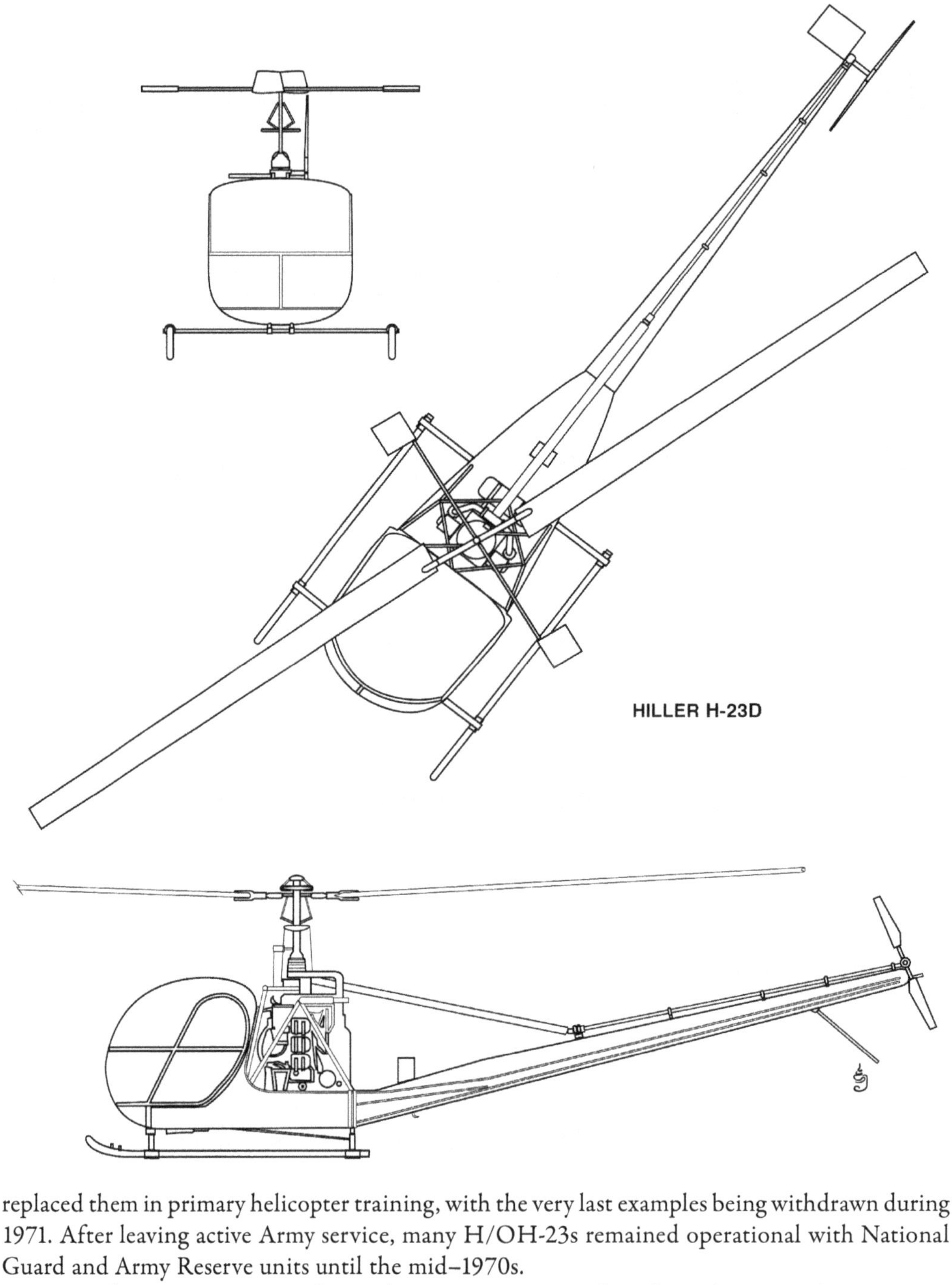

HILLER H-23D

replaced them in primary helicopter training, with the very last examples being withdrawn during 1971. After leaving active Army service, many H/OH-23s remained operational with National Guard and Army Reserve units until the mid–1970s.

Naval Variants: In 1950, after evaluating an H-23A on loan from the Army, the Navy procured 17 essentially identical examples under the naval designation HTE-1, then in 1951, contracted for delivery of 35 H-23Bs as the HTE-2. As delivered, HTE-1s and -2s entered service with Helicopter Training Unit One (HTU-1) at NAS Whiting Field, Florida, where they were used for primary training of Navy, Marine, and Coast Guard helicopter pilots until being withdrawn during the late 1950s and early 1960s.

One of 438 dual-control Hiller H-23Ds delivered to the Army Helicopter School at Fort Wolters, Texas, during the mid–1950s. The last examples were not withdrawn from pilot training until 1971.

Bell TH-1 Iroquois—1960

Technical Specifications (TH-1L)

Type: Four to ten-place pilot and crew training helicopter.
Manufacturer: Bell Aerospace-Textron, Inc., Fort Worth, Texas.
Total U.S. military versions (trainers): 114 (49 USAF/U.S. Army; 65 USN/USMC).
Unit cost (TH-1H): $4,700,000.
Powerplant: One 1,400-shp Lycoming T-53-L-13 turboshaft engine driving a two-bladed main and two-bladed tail rotor.
Performance: Max. speed 144 mph, cruise 125 mph; ceiling 18,500 ft.; range 289 mi.
Weights: 4,717 lbs. empty, 9,500 lbs. loaded.
Dimensions: Rotor diameter 48 ft., length 44 ft. 6 in.

Since completing its first flight in October 1956 as the XH-40, the ubiquitous Bell UH-1 "Huey" has been in continuous production for 54 years, with over 10,100 in five major variants having been delivered to the U.S. military branches at the date of this writing. The UH-1 (HU-1 until 1962), the first turbine-powered military helicopter to achieve operational service, became a key ingredient in the Army's "Airmobile" concept of moving division-sized numbers of assault troops by helicopter. Though primarily designed for troop carrying, medical evacuation, and utility roles, at least five Huey variants have been completed or converted to function as trainers: in 1960, 14 HU-1As reequipped with dual controls and two sets of instruments returned to Army service as the TH-1A; in 1963, 20 based on the UH-1C airframe were configured for Marine training as the TH-1E; in 1964, 20 UH-1Fs completed as instrument and rescue trainers were delivered to the USAF as the TH-1F; in 1969, 45 UH-1Es completed as trainers entered Navy service as the TH-1L; and since 2005, 27 refurbished UH-1Hs equipped with digital avionics have been placed on the USAF inventory as the TH-1H.

Historical data indicates that the Army TH-1As served as instrument trainers at Fort Rucker, Alabama, until withdrawn in 1971. The Marine TH-1Es functioned as operational trainers with

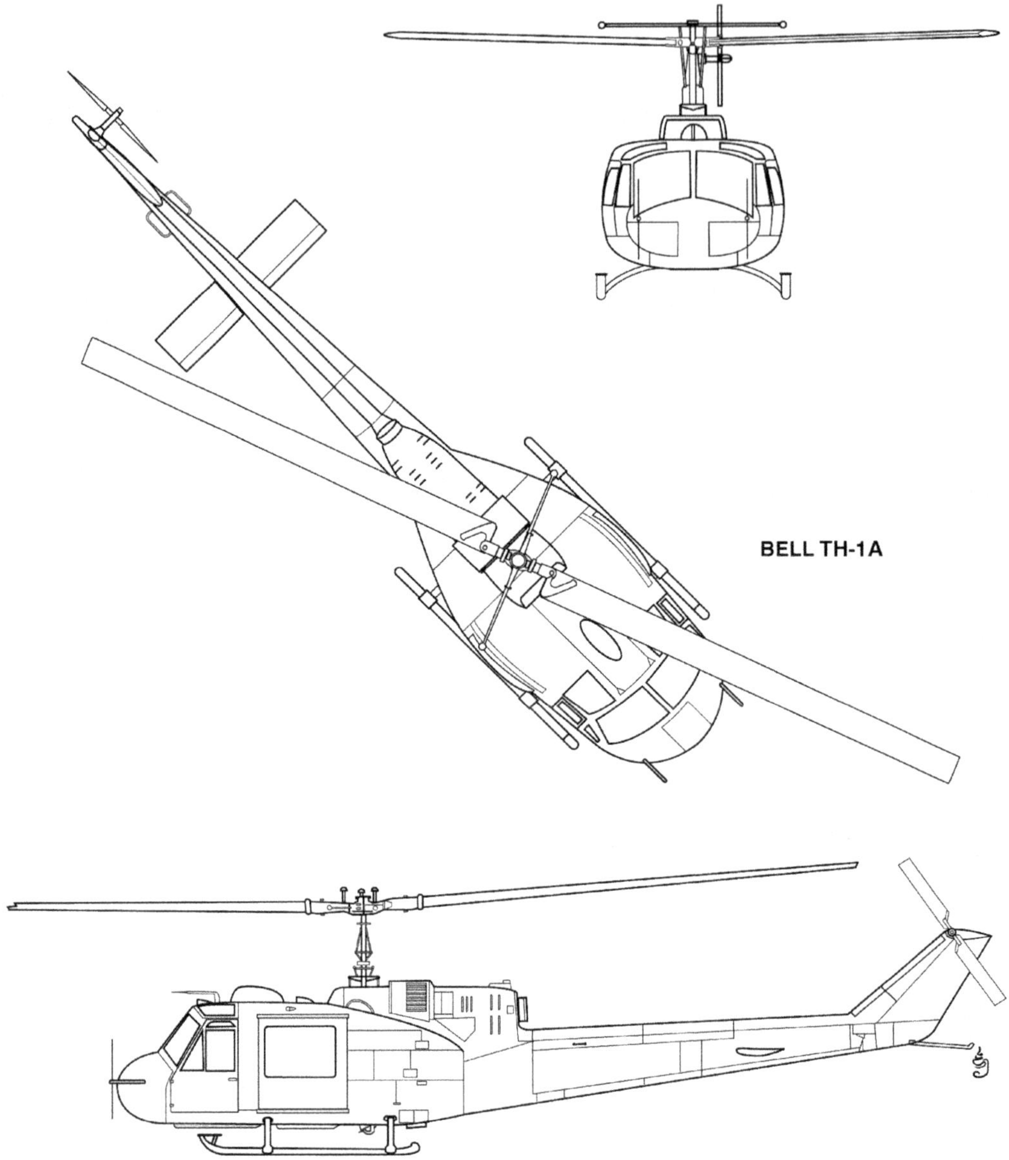

BELL TH-1A

VMO-3 at Camp Pendleton, California, during the Vietnam War. TH-1Fs delivered to the USAF served with the 3637th Flying Training Squadron (Helicopters) training pilots for the Air Rescue Service at Stead AFB, Nevada until the late 1970s. Navy TH-1Ls saw service as advanced helicopter trainers with VT-8 and VT-18, both based at NAS Whiting Field, Florida, until being withdrawn in 1982 and 1983. Currently, TH-1Hs are flown by two USAF Air Education and Training Command units, the 23rd Flying Training squadron at Fort Rucker, Alabama, for advanced helicopter undergraduate pilot training and as a component of the Inter-Americas Air Forces Academy, with the 318th Training Squadron at Lackland AFB, Texas. Starting in 1988, as Hughes TH-55As were withdrawn from service (see below), the Army employed an unspecified number of UH-1s (probably UH-1Hs) for primary helicopter pilot training until new Bell TH-67s (see below) began entering service during 1993.

One of 45 Bell UH-1Ls procured by the Navy in 1969 as instrument and rescue trainers. They served with VT-8 and VT-18, both based at NAS Whiting Field, Florida, until being withdrawn in 1982 and 1983.

Two Bell USAF TH-1Hs of the 23rd Flying Training Squadron seen near Fort Rucker, Alabama, in 2009. The 23rd FTS provides advanced UPT for special operations, search and rescue, and missile tracking.

Hughes TH-55 Osage—1964

Technical Specifications (TH-55A)

Type: Two-place training helicopter.
Manufacturer: Hughes Aircraft Div., Hughes Tool Co., Burbank, California.
Total U.S. military versions: 797 (U.S. Army).
Unit cost: $24,875.

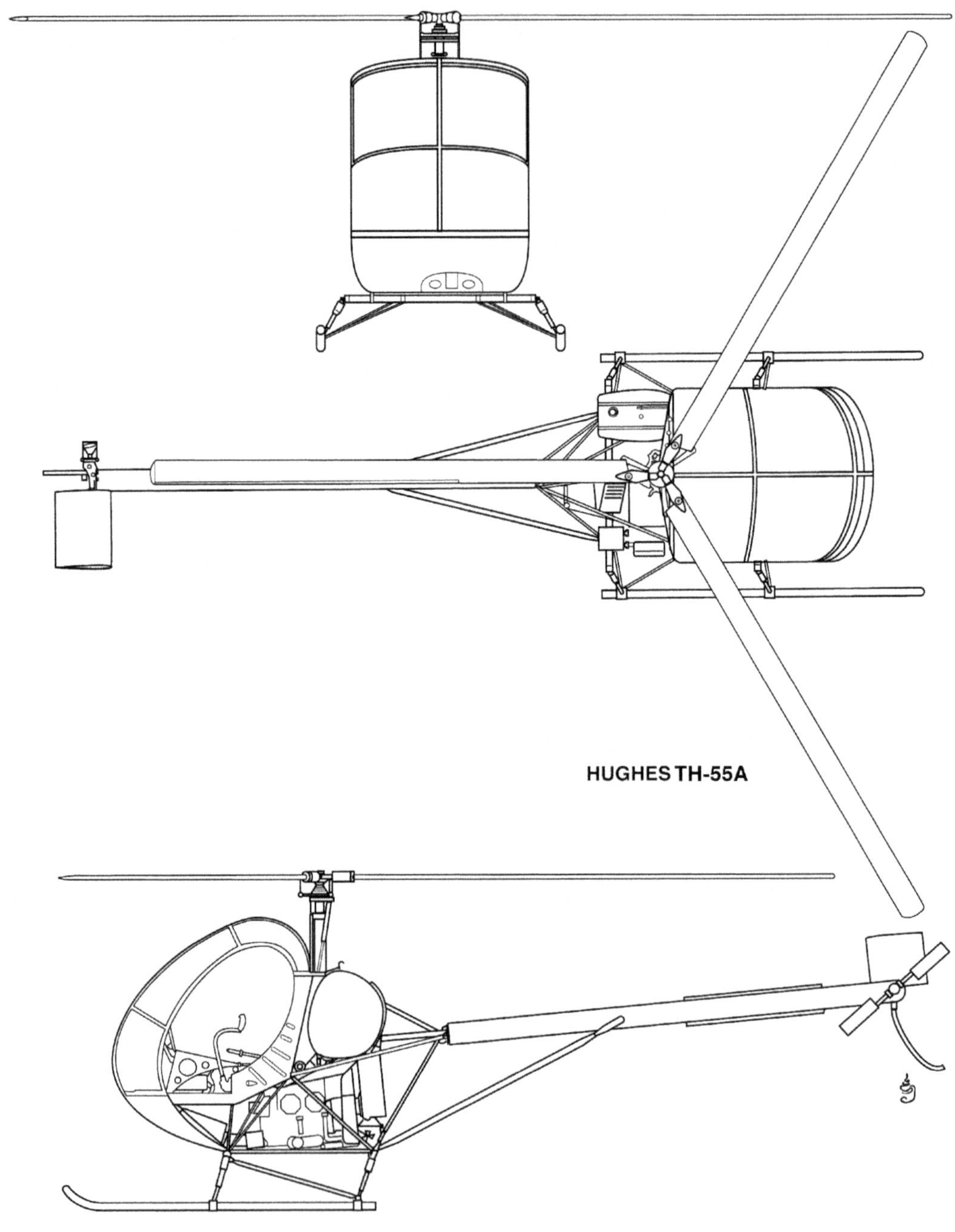

A Hughes TH-55A seen on the flight line of the Army Helicopter Pilot School at Fort Wolters, Texas, in the late 1960s. TH-55As were the Army's most important type of primary helicopter trainer from 1967 to 1988.

Powerplant: One 180-hp Lycoming HIO-360-B1A 4-cylinder air-cooled horizontally-opposed engine driving a three-bladed main and two-bladed tail rotor.
Performance: Max. speed 86 mph, cruise 66 mph; ceiling 11,500 ft.; range 204 mi.
Weights: 1,008 lbs. empty, 1,850 lbs. loaded.
Dimensions: Rotor diameter 25 ft. 3.5 in., length 22 ft. 4 in.

The Hughes Aircraft Division commenced development of the model 269 in 1955 in response to a market survey that suggested strong commercial demand for a very simple and inexpensive two-seat helicopter. When the design of the 269 emerged for its first flight in late 1956, it represented simplicity in a compact package: a welded, steel tube fuselage frame and tail boom, open engine and transmission compartment, bubble canopy, and a small-diameter, three-bladed main rotor, which, altogether, only totaled 1,000 lbs. empty. In 1958, prior to civil certification and production, Hughes offered five pre-production examples to the Army, taken for testing under the designation YHO-2, which differed from the 269 prototype in having a strut-braced circular tail boom and other detail refinements. At the time, the Army was evaluating the YHO-2 alongside two other candidates, the French-made Sud-Ouest YHO-1 and the Brantly YHO-3, as potential replacements for the Bell H-13 and Hiller H-23 (see above); however, despite having achieved the best comparative results among the three, the Army declined to place the YHO-2 in production because its small size and limited useful load could not duplicate the combat applications of the H-13 and H-23 (e.g., observation, medical evacuation, etc.).

Then in 1964, faced with a need to significantly increase helicopter pilot production in response to the escalation of American military involvement in Southeast Asia, the Army selected the Hughes civil model 269A to augment and eventually replace H-13s and H-23s in its primary helicopter pilot training program. To this end, the Army awarded Hughes a contract for 792 helicopters delivered between 1965 and 1967 under the designation TH-55A and name "Osage." Virtually all TH-55As entered service with the Army's Primary Helicopter School at Fort Wolters, Texas, and by 1971, had replaced all H-13 and H-23s in the training role. At the time Fort Wolters was deactivated in 1973, the school's training operations and its fleet of TH-55As moved to Fort Rucker, Alabama. The process of phasing out TH-55As began during the early 1970s as Army demand for new helicopter pilots sharply decreased following the end of the Vietnam War, however, the final examples were not withdrawn from their training function until mid–1988, when the type had been in service for 24 years.

Bell TH-57 Sea Ranger and TH-67 Creek—1968

Technical Specifications (TH-57B)

Type: Four-place primary and advanced training helicopter.
Manufacturer: Bell Aerospace-Textron, Inc., Fort Worth, Texas.
Total U.S. military versions: 337 (181 U.S. Army; 156 USN).
Unit cost: $1,000,000 (est.).
Powerplant: One 420-shp (de-rated to 317-shp) Allison 250-CB20BJ turboshaft engine driving a two-bladed main and two-bladed tail rotor.
Performance: Max. speed 138 mph, cruise 117 mph; ceiling 18,900 ft.; range 420 mi.
Weights: 1,595 lbs. empty, 3,200 lbs. loaded.
Dimensions: Rotor diameter 35 ft. 4 in., length 32 ft. 2 in.

The Bell model 206, in its varied military and commercial applications, is considered by many to be one of the most successful helicopter designs of all time. Its origins were originally military, flying for the first time on December 8, 1962, under the designation YOH-4 as a candidate in the Army's Light Observation Helicopter (LOH) competition. Although it ultimately lost to the Hughes OH-6A (see reference in TH-6B below), it was subsequently placed in production for the civil market as the 206A Jet Ranger. When the LOH competition reopened in 1967, the Army selected a militarized version of the 206A for production and subsequently took delivery of 2,220 helicopters as the OH-58A Kiowa between 1969 and 1973. Since then, as a result of upgrade programs, 425 OH-58As have returned to Army service the OH-58C (from 1978) and 368 as the OH-58D (from 1986). The newest variant, the OH-58F, is currently under development. Though not designated as a trainer, the Army reportedly employs an unspecified number of OH-58s at Fort Rucker, Alabama, for tactical training of pilots and aircrews.

With the goal of augmenting and eventually replacing its fleet of piston-engine Bell TH-13Ms (see above) in the basic helicopter pilot training program, the Navy procured 40 Bell 206As in 1968 under the designation TH-57A and name "Sea Ranger." When the TH-57As had been in service for 17 years, the Navy initiated an across-the-board replacement program with the purchase of 51 off-the-shelf Bell 206B-3s that began entering service in 1986 as the TH-57B. The B-3 differed from the A in having a more powerful Allison –CB20BJ engine and refinements to the cockpit. Through another series of acquisitions that started in 1989, the Navy has taken delivery of 89 more 206B-3s configured for instrument training as the TH-57C. More recently, the Navy has announced plans to upgrade 128 of its TH-57B/Cs to a new TH-57D standard, which will entail installation of digital glass cockpits, aircrew airbag protection, and night vision equipment. All of the Navy's TH-57 variants became operational with Air Training Wing Five

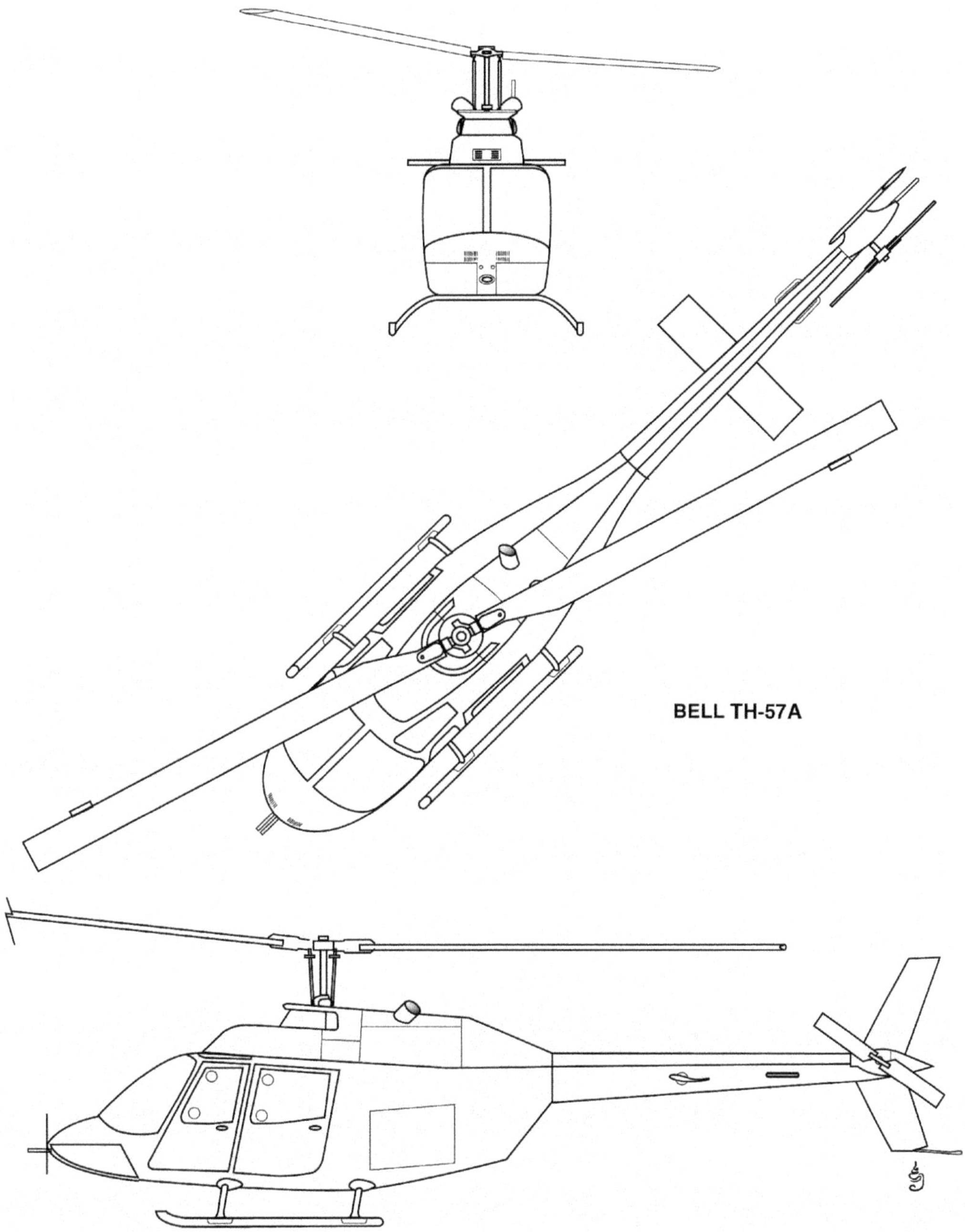

BELL TH-57A

(ATW-5) at NAS Whiting Field, Florida, serving with HT-8, HT-18, and HT-28 to provide basic and advanced helicopter training to Navy, Marine, and Coast Guard student aviators. As of mid–2013, ATW-5 operates 44 TH-57Bs in the basic phase and 85 TH-57Cs in the advanced phase.

In 1993, in a move to replace its aging Bell UH-1Hs in helicopter pilot training, the Army selected the Bell 206B-3 under the designation TH-67A and assigned name "Creek." Since that time, through a series of production contracts, the Army has taken delivery of 60 TH-67As in

The Navy procured 40 of these Bell 206As in 1968 under the designation TH-57A to replace piston-engine Bell TH-13Ms in basic helicopter pilot training. The example shown served with HT-8 at NAS Whiting Field, Florida.

Seen in Marine markings, this TH-57C of HT-18 was photographed in October 2009 while making a practice landing aboard a training ship at NAS Pensacola, Florida. There were 85 TH-57Cs still active as of late 2013.

This Bell TH-67A was photographed at Fort Rucker, Alabama, in May 2003. The Army Helicopter School employs 172 TH-67As for Joint Service Undergraduate Pilot Trainings of both Army and USAF pilot trainees (courtesy Charlie Mauzé).

visual flight rule (VFR) configuration and 121 (same designation) equipped as instrument trainers. All TH-67As are operated by the Army's Helicopter School at Fort Rucker, Alabama, and in addition to primary and advanced training of Army Aviators, are also used for Joint Service Undergraduate Pilot Training (JSUPT) of USAF pilot trainees in the advanced helicopter syllabus. Internet sources currently report a total of 172 TH-67As in Army service at the date of this writing.

MD Helicopters (Hughes) TH-6B Cayuse—1992

Technical Specifications (TH-6B)

Type: Four-place flight-test training helicopter.
Manufacturer: MD Helicopters, Inc. (formerly a subsidiary of McDonnell Douglas), Mesa, Arizona.
Total U.S. military versions: 6 (USN).
Unit cost: Unknown.
Powerplant: One 400-shp Allison 250-C20B turboshaft engine driving a four-bladed main and two-bladed tail rotor.
Performance: Max. speed 175 mph, cruise 160 mph; ceiling 17,500 ft.; range 290 mi.
Weights: 1,295 lbs. empty, 3,450 lbs. loaded.
Dimensions: Rotor diameter 26 ft. 5 in., length 21 ft. 5 in.

The Hughes model 369, winner of the Army's first Light Observation Helicopter (LOH) competition, flew for the first time in February 1963 and subsequently entered production as the OH-6A Cayuse, with a total of 1,434 delivered by 1970. The design was also successfully marketed by Hughes in a succession of civil versions, models 500 (1963), 500C (1969), 500D (1976), and 500E (1982), until 1984 when the company was acquired by McDonnell Douglas, and after the takeover, the series continued in production as the MD-500. At the time of the Boeing-McDonnell Douglas merger, the helicopter operation was sold separately as MD Helicopters, Inc.

In 1992 the Navy acquired six MD-500Cs (also known internally as the Model 369H) under the designation TH-6B and placed them on the inventory through a system identified as "conversion in lieu of procurement," which apparently means the airframes and associated com-

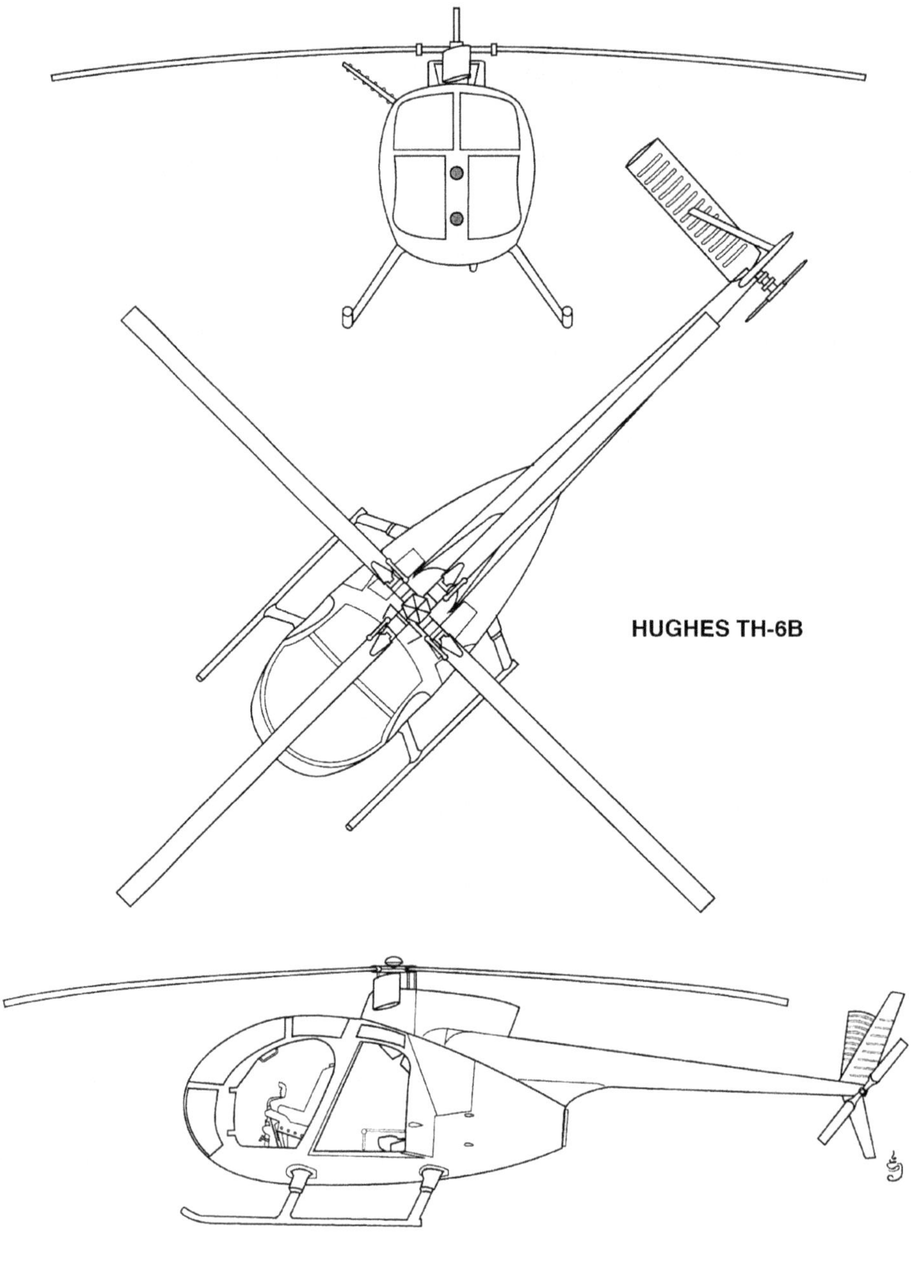

An August 2005 photograph shows one of six MD Helicopters TH-6Bs serving with the Naval Test Pilot School at NAS Patuxent River, Maryland, where they are used for training in systems testing and evaluation.

ponents are refurbished from civil or military stocks rather than new builds. Although the TH-6Bs do not appear on available serial number records, the Navy Fact File (see http://www.navy.mil/navydata/fact_display.asp?cid=1200&tid=650&ct=1), as of 2009, indicates all six are serving with the Naval Test Pilot School at NAS Patuxent River, Maryland, where they are used as an integral part of the test pilot training syllabus.

Appendix 1: Training Gliders

A. USAAF, USN AND USMC, 1934 TO 1945

The first gliders to actually be used by a U.S. military branch for training purposes were six Franklin PS-2s procured by the Navy in 1934. The USAAF, coincident with the establishment of a separate training program for glider pilots, procured an estimated 1,121 training gliders (designated by a "TG" prefix) between 1941 and 1943, as follows: 53 Frankfort TG-1s; 47 Schweizer TG-2s; 114 Schweizer TG-3s; 153 Laister-Kauffman TG-4s; 251 Aeronca TG-5s; 253 Taylorcraft TG-6s; and 250 Piper TG-8s. Soon after the U.S. entered World War II, the USAAF also impressed approximately 50 privately owned sailplanes, most one at a time, under the designations TG-7, TG-9, TG-11, and TG-14 though TG-31. The designations TG-10, TG-12, and TG-13 were allocated to gliders ordered but cancelled before delivery. Advanced training of USAAF glider pilot trainees was typically carried out in Waco GC-4A troop gliders of the same type used in combat operations. For its own program, directed chiefly toward use of troop-type gliders in Marine assault operations, the Navy procured 101 training gliders, that included 13 Schweizer LNS-1s (AAF TG-2) and 101 Pratt-Read LNE-1s.

Franklin PS-2 (TG-15 and -17)—1934

In 1934 the Navy purchased six one-seat Franklin PS-2 (primary-secondary-two) gliders to evaluate the possibility of incorporating them into the pilot training syllabus. Designed to with-

One of six Franklin PS-2s procured in 1934 to test the feasibility of integrating gliders into primary pilot training.

stand the rigors of training, the PS-2 featured a welded, steel tube fuselage and tail group, built-up wooden wing braced by struts, and a mono-wheel landing gear. The PS-2s operated from NAS Pensacola, Florida until naval officials abandoned the concept sometime in 1938. In 1942 two civilian PS-2s were also impressed into USAAF service as the TG-15 and -17. General specifications: type, one-place training glider; length, 19ft.; span, 36 ft.; gross weight, 400 lbs.; max. speed (unknown).

Frankfort TG-1—1941

The Frankfort Sailplane Co. of Joliet, Illinois was one of three U.S. glider manufacturers with whom the USAAF contracted in April 1941 to supply gliders for its newly formed glider pilot training program. The first three gliders to be tested, essentially off-the-shelf versions of the company's two-seat Cinema IIB, arrived at Wright Field, Ohio in September under the designation XTG-1. Typical of sport sailplanes of that time, the XTG-1 came with fabric-covered construction consisting of a welded, steel tube fuselage with built-up wooden wings and tail surfaces. Over the next year, Frankfort began delivering TG-1A production models with revised flight controls, but because the company lacked facilities to continue large-scale manufacturing, production terminated in 1943 when 50 gliders had been completed. The USAAF also impressed 10 civilian Cinema IIs as the TG-1B, C, and D. General specifications: type, two-place training glider; length, 23 ft. 4 in.; span, 46 ft. 3 in.; gross weight, 920 lbs.; max. speed, 80 mph.

One of three Frankfort XTG-1 prototypes tested at Wright Field, Ohio, in the fall of 1941. They were placed into production as the TG-1A.

Schweizer TG-2 (LNS)—1941

In mid–1941 Schweizer Aircraft Corp. of Elmira, New York delivered three two-place SGS-2–8 gliders to Wright Field, Ohio to be evaluated under the designation XTG-2. Following successful trials, the USAAF placed the type in production as the TG-2A; however, production was cut short after only 45 examples had been completed because its wings and tail group possessed an aluminum framework classified as "strategic material." To take its place, Air Material Division

Forty-seven of these TG-2As joined the inventory as trainers in 1942 before production stopped in favor of the TG-3A.

These two Marine LNS-1s were photographed over Parris Island, South Carolina, in 1942.

directed Schweizer to develop a similar version having wings and tailplane constructed of built-up, wooden material, which is reported below as the TG-3. In addition to TG-2As delivered under contract, the USAAF also impressed seven civilian SGS-2–8s as the TG-2A. In 1942 Schweizer delivered 13 SGS-2–8s to the Navy that were placed on the inventory as the LNS-1. General specifications: type, two-place training glider; length, 25 ft. 9 in.; span, 52 ft., gross weight, 860 lbs.; max. speed, 72 mph.

Schweizer TG-3—1942

As mentioned in connection with the TG-2, above, Schweizer developed the aerodynamically similar TG-3 with wings and tail surfaces constructed of built-up wooden material rather than aluluminum. Other changes included lowering the wing from a high-mid to low-mid position for improved visibility and simplified tail surface outlines. After testing three XTG-3 prototypes in late 1942, the USAAF ordered 163 production models as the TG-3A. When production ceased in 1943, 111 had been delivered, including one completed by Air Gliders, Inc. of Akron, Ohio. By this time, USAAF officials at Air Material Division had concluded that trainer versions of sport sailplanes took too long to build together with the fact that glider man-

This Schweizer TG-3A has the overall silver 1943 paint scheme. Wooden wings and tailplane added almost 400 pounds to empty weight.

ufacturers generally lacked the plant capacity to build them in the large numbers needed. General specifications: type, two-place training glider; length, 27 ft. 7 in.; span, 54 ft.; gross weight 1,200 lbs.; max. speed, 90 mph.

Laister-Kauffman TG-4—1942

The Laister-Kauffman Corp. of St. Louis, Missouri was the last of three glider manufacturers to produce a USAAF trainer based upon a preexisting sport sailplane design. The date is

This completely restored Laister-Kauffman TG-4A can be seen at the U.S. Air Force Museum in Dayton, Ohio.

uncertain but probably in early 1942, the USAAF accepted two of the company's two-place LK-10s for evaluation under the designation XTG-4. Characterized as a mid-wing design featuring a steel tube fuselage and wooden wings and tail surfaces covered in fabric, Air Material Division subsequently ordered it into production as the TG-4A, with a total of 153 being delivered through 1942 and 1943. In the interval, USAAF officials had determined that sport sailplanes were not only poor candidates for mass-production but their flying characteristics were far different from those of the Waco GC-4A troop gliders that prospective pilots would be expected to fly in combat. Besides the TG-4As, the USAAF also impressed one Laister-designed single-place LIT-3 as the TG-4B. General specifications: type, two-place training glider; length, 21 ft. 4 in.; span, 50 ft.; gross weight, 875 lbs.; max. speed 80 mph.

Aeronca TG-5—1942

In order to find a type of training glider that could be mass-produced and would exhibit gliding characteristics closer to those of troop gliders like the Waco GC-4A, the USAAF approached several lightplane manufacturers to develop engineless versions of existing designs which were already in production as liaison aircraft for Army ground forces. The first (in designation order) to be tested was the Aeronca XTG-5 prototype, which materialized in late 1942 as an L-3 airframe (see in Series I, Part A) with the engine removed and a third seat added forward behind a streamlined nose fairing. After official acceptance, another 250 L-3 airframes underwent similar conversions over the next year with the addition of spoilers on the upper wing to increase rate-of-descent and larger vertical fins to offset the increased side area of the longer fuselage. The TG-5s were withdrawn when primary training of glider pilots was suspended in late 1944. General specifications: type, three-place training glider; length 23 ft. 10 in; span, 35 ft.; gross weight, 1,260 lbs.; max. speed (unknown).

A production Aeronca TG-5 pictured in 1943 with three occupants, probably two students and one instructor.

Taylorcraft TG-6 (LNT)—1942

Like Aeronca, and within the same timeframe, the USAAF directed Taylorcraft to convert three L-2As (see in Series I, Part A) to an engineless, glider configuration and deliver them to Wright Field, Ohio for testing under the designation XTG-6. In appearance, the XTG-6s featured more refined greenhouse canopy arrangement that extended to the tip of the new nose section. After brief tests, Taylorcraft manufactured another 250 TG-6A production variants in 1942 and 1943 which came with upper wing spoilers and larger vertical fins, some from converted L-2s and others being completed as gliders off the assembly line. Thirty-five TG-6As were transferred to the naval inventory in early 1943 under the designation LNT-1. As with other types, they were withdrawn when primary training of glider pilots ended in late 1944. General specifications: type, three-place training glider; length 25 ft.2 in; span, 35 ft. 5 in.; gross weight, 1,260 lbs.; max. speed (unknown).

A Taylorcraft TG-6 in the overall silver 1943 paint scheme. Two hundred fifty were converted from L-3A airframes.

Piper TG-8 (LNP)—1942

The third and final liaison aircraft conversion ordered by Air Material Division in 1942 resulted in delivery of 250 Piper TG-8s, derived apparently from stock J-3C airframes rather than L-4s. Serial number records do not reveal testing of any experimental prototypes. The TG-8's modified nose was similar to that of the TG-5 but featured an aluminum nose cap. Available

The Piper TG-8 was the last type of lightplane delivered to the USAAF in a three-seat glider configuration. Two hundred fifty of these were converted from J-3Cs.

photos suggest the fin was left unchanged but a technical bulletin indicates the addition of spoilers to the upper wings. Three TG-8s were transferred to the Navy as the XLNP-1 but later returned to the USAAF. All were withdrawn when primary glider training stopped in late 1944 and some, sold surplus after the war, are known to have been converted to engine power. General specifications: type, three-place training glider; length 24 ft. 5 in.: span, 35 ft. 3 in., gross weight, 1,220 lbs., max. speed (unknown).

Pratt-Read LNE (TG-32)—1942

Soon after the Navy initiated a glider training program in early 1942, the Naval Bureau of Aeronautics (BuAer) acquired a civil-registered PR-G1 sport sailplane built by the Pratt-Read Piano Co. of Sycamore, Illinois which was tested at NAS Anacostia, Maryland under the designation XLNE-1. While its airframe consisted of conventional fabric-covered mixed steel and wooden construction, it differed from other sport gliders in having a bulbous nose that housed its two occupants in a side-by-side seating arrangement. After testing, BuAer awarded Pratt-Read a contract to produce 100 examples that were delivered during 1942 and 1943 as the LNE-1. When the Navy ceased glider training sometime in late 1943, 73 LNE-1s were transferred to the USAAF under the new designation TG-32, and they, too, were withdrawn in late 1944. General specifications: type, two-place training glider; length 26 ft. 3 in.; span, 54 ft. 6 in; gross weight, 1,150 lbs.; max. speed, 99 mph.

The Pratt-Read XLNE-1 prototype with a production LNE-1 and transport trailer in the background.

Waco GC-4 (LRW)—1942

Designed by the Waco Aircraft Co. of Troy, Ohio, two XCG-4 prototypes arrived at Wright Field, Ohio for USAAF testing in early 1942 and soon afterward, the type was placed in mass-production as the CG-4A with 16 different contractors, one of which was the Pratt-Read Piano Co. mentioned above. As the USAAF's principal type of troop glider, CG-4As saw combat in every major airborne invasion of World War II with nearly 14,000 delivered by the time hostilities ceased in September 1945. From late 1942 to late 1944, CG-4As were employed in the advanced

Waco CG-4A troop glider seen just before touchdown. Almost 14,000 were delivered to USAAF during World War II.

phase of glider pilot training at South Plains AAF and Sheppard AAF, both Troop Carrier Command bases in Texas. From mid–1944 onwards, as demand for glider pilots for planned airborne invasions far exceeded the capacity of the glider pilot training program to produce them, the USAAF began drafting pilots from other commands and putting them through an abbreviated course in CG-4As. In 1943 the Navy acquired 13 GC-4As built under AAF serial numbers as the LRW-1. General specifications: type, fifteen-place troop glider; length, 48 ft. 8 in., span, 83 ft. 8 in.; gross weight, 7,500 lbs.; max. speed, 150 mph.

B. USAF, 1957 to Present

An Aero and Soaring Club for U.S. Air Force Academy Cadets was initially formed in 1956 while the Academy was located at Lowry AFB, Colorado. The program was suspended in late 1958 but reestablished in 1961 at the Academy's present location in Colorado Springs. Soaring became part of the Academy's curriculum in 1964 and expanded to all cadets in 1970 with the establishment of the Soar-For-All-Program. The first gliders used in the program were acquired with donated funds but since 1968, have been acquired under standard USAF procurement guidelines.

Schweizer 2–22/TG-2—1958

The First gliders actually owned by the U.S. Air Force Academy (AFA) Aero and Soaring Club were three Schweizer 2–22As acquired in 1958 with donated funds. The 2–22 had originally been conceived in 1945 as a simpler and lower-cost training glider than higher-performance pre-war types like the SGS-2–8 (see Schweizer TG-2, above), and the 2–22A, certificated in 1967, differed from the earlier variant in having an enclosed cockpit, a lengthened nose, and a gross weight increase to 900 lbs. Like Schweizer's previous designs, the 2–22A's fabric-covered structure consisted of a welded, steel tube fuselage and wings and a tailplane having an aluminum framework. All three 2–22As were destroyed by high winds in December 1958. Though no serial number record exist, photo data suggests that Schweizer 2–22Es were acquired to replace them after

One of three Schweizer 2–22As acquired in 1958 for the U.S. Air Force Academy Aero and Soaring Club. They were destroyed by high winds later the same year.

the activities of the Aero and Soaring Club resumed in 1961. The E incorporated larger spoilers, a new canopy design, and changes to the wing root. Some sources list the 2–22Es as carried under the designation TG-2A, but official records reflect only one was actually assigned a USAF serial number. The 2–22Es were replaced by 2–33As (see TG-4, below) in 1970. General specifications (2–22A): type, two-place training glider; length, 25 ft. 9 in.; span, 43 ft.; gross weight, 900 lbs.; max. speed 90 mph.

Schweizer 1–26/TG-3—1957

Like the 2–22As reported above, one single-place Schweizer 1–26B was acquired with donated funds by the AFA Aero and Soaring Club in 1958. The B, introduced in 1956, featured an aluminum-skinned wing, but retained a fabric-covered, steel tube fuselage. This higher-

This Schweizer 1–26B joined the AFA Aero and Soaring Club in 1958 and fell victim to the same windstorm that destroyed the 2–22As.

performance glider was presumably flown by AFA cadets after they had earned a glider pilot's Airman's Certificate issued by the FAA. The 1–26B was written-off in December 1958 due to the same windstorm that destroyed the 2–22As. From 1966 to 1992, seven 1–26 variants were procured for the AFA program with assigned USAF serial numbers under the designation TG-3A, namely, one 1–26B in 1966 and six 1–26Es in 1978, 1979, 1989, 1991, and 1992. The 1–26E, introduced in 1971, features an all-metal monocoque fuselage and new tail surfaces. General specifications: single-place sport glider; length, 2 ft. 6 in.; span, 40 ft.; gross weight, 700 lbs.; max. speed 114 mph.

Schweizer TG-4—1970

In connection with the Soar-For-All program incorporated by AFA in 1970, the USAF, according to serial number records, procured 17 Schweizer 2–33As from 1970 to 1992 under the designation TG-4A. Introduced in 1967 as the successor to the 2–22/TG-2 series, the 2–33A featured a welded, steel tube fuselage and tailplane covered in fabric, wings skinned in stressed aluminum, and a fiberglass nose having a one-piece molded front canopy. The TG-4As remained in service as the AFA Soaring Program's principal type of primary trainer until replaced in 2002 by LET TG-10Bs. General specifications: type, two-place training glider: length, 25 ft. 9 in.; span, 51 ft.; gross weight, 1,040 lbs.; max. speed 98 mph.

A civil Schweizer 2–22A of the same type used in the Air Force Academy's Soar-For-All program from 1970 to 2002.

Schweizer TG-5—1976

The USAF acquired a single Schweizer 2–32 for the AFA Soaring Program in 1976 under the designation TG-5A. The 2–32/TG-5A is a two or three-place high-performance sailplane that resembles a scaled-up I-26E. Internet data indicates the type may still be active in the program.

A Schweizer 2–32, three-place sport glider, identical to the single TG-5A used by the Air Force Academy.

General specifications: two or three-place sport glider; length, 26 ft. 9 in.; span, 57 ft. 1 in.; gross weight, 1,080 lbs.; max. speed, 157 mph.

Schweizer TG-6—1979

One single-place, high-performance Schweizer 1–34 was procured for the AFA Soaring Program in 1979 under the designation TG-6A. Featuring a very high-aspect ratio wing planform, the TG-6A may be characterized as a competition sailplane of all-aluminum construction meeting all of the "Standard Class Criteria." Serial number records reflect that this glider has since been transferred to Civil Air Patrol. General specifications: one-place sport glider; length, 25 ft. 5 in.; span, 49 ft. 3 in.; gross weight, 840 lbs.; max. speed, 142 mph.

A civil Schweizer 1–34 high-performance sailplane of the same type as the single TG-6A used by the Air Force Academy.

Schweizer TG-7—1981

Apparently intended as a step-up from TG-4As, the USAF procured 10 Schweizer 2–37 motor-gliders in two batches of five in 1981 and 1982 under the designation TG-7A. In design, it combines the high-aspect ratio wings of a sailplane with the fuselage of a powered aircraft and rests on a tailwheel-type landing gear. Internet sources indicate that the type was withdrawn from the AFA Soaring Program in 2003. General specifications: type, two-place motor-glider; powerplant, one 112-hp Lycoming O-235-L2C four-cylinder horizontally- opposed engine driving a two-bladed fixed-pitch metal propeller; length, 27 ft. 6 in.; span, 56 ft. 6 in.; gross weight, 1,850 lbs.; max. speed 135 mph.

A lineup of Air Force Academy Schweizer TG-7As seen at the Oshkosh Air Venture in 2013.

Schleicher TG-9—1986

The Schleicher ASK-21, flown for the first time in late 1979, is a two-place, high-performance sailplane featuring a T-tail configuration and glass-fiber reinforced plastic (GRF) construction that gives it exceptional aerodynamic cleanness. The USAF procured a total of five ASK-21s for the AFA Soaring Program, two in the 1986 and three in 1987, with the assigned designation TG-

A German civil registered Schleicher ASK-21 of the same type as the five TG-9As procured for the Air Force Academy.

9A. Serial number records seem to indicate that four are still active. General specifications: type, two-place sport glider; length, 27 ft. 5 in.; span, 55 ft. 9 in.; gross weight, 1,323 lbs.; max. speed, 174 mph.

LET TG-10 Blaník—1993

Since their introduction in mid–1950s, the LET "Blaník" series of sailplanes, designed and built in Czechoslovakia (now the Czech Republic), have been one of the most numerous and widely-used sailplanes in the world. In general arrangement, Blanik gliders are of all-aluminum construction and feature forward-swept wings which allow the cockpit to be located forward of the wing leading edge for improved visibility, and the newer L-23 and -33 both have a T-tail layout. Altogether, 22 Blaniks in four different variants have been procured for the AFA Soaring Program: one two-place L-13 in 1993 as the TG-10A; twelve two-place L-23 Super Blaníks in 2002 as the TG-10B; four two-place fully-aerobatic L-13ACs in 2002 as the TC-10C; and four two-place cross-country configured L-33s in 2002 as the TG-10D. As the TG-10Bs came into service, they replaced the older TG-4As as the program's principal type of primary trainer. General specifications (TG-10B): type, two-place training glider: length, 27 ft. 11 in.; span, 53 ft. 2 in.; gross weight, 1,168 lbs.; max. speed, 143 mph.

These LET TG-10Bs, in their hangar at the Air Force Academy, served as the program's main primary trainer from 2002 to 2012.

Stemme TG-11—1994

The German-designed and built two-place Stemme S-10 is termed a "self-launching" glider insofar as its propeller folds and retracts into the nose when not in use. Introduced during the 1980s, it features glass-fiber construction, an advanced wing design incorporating Schempp-Hirth spoilers and winglets, a T-tail empennage, and dual-wheel fully-retractable landing gear. To maintain a safe center-of-gravity, the engine is located amidships and drives the propeller via a shaft. Two S-10s were acquired for the AFA Soaring Program in 1994 under the designation TG-

A Stemme TG-11A self-launching sailplane in Air Force Academy hangar. Folded prop is housed in nose cone.

11A and are reportedly used for cross-country training. Both examples appear to still be active. General specifications: two-place self-launching sport glider; powerplant, one 93-hp Limbach L-2400-EB1 inline engine driving a two-bladed fixed-pitch propeller; length 27 ft. 7 in.; span, 75 ft. 6 in.; gross weight, 1,874 lbs.; max. speed, 168 mph.

Aeromot TG-14—2001

As evident replacements for the TG-7As, the USAF procured 14 Brazilian-made Aeromot AMT-200S motor-gliders in for the AFA Soaring Program in 2001 under the designation TG-14A. The TG-14As include side-by-side seating, state-of-the-art composite construction, a T-

One of 14 Brazilian-made Aeromot TG-14A motor-gliders procured for the Air Force Academy in 2001.

tail layout, and an advanced glide computer/GPS for cross-country cruising. The variable pitch propeller may be set to climb and cruise or feathered when the engine is stopped. All 14 appear to still be on the active inventory. General specifications: type, two-place motor-glider; powerplant, one 80-hp Rotax 912-S4 four-cylinder horizontally-opposed engine driving a two-bladed variable-pitch propeller; length, 26 ft. 5 in.; span, 57 ft. 4 in.; gross weight, 1,775 lbs.; max. speed, 152 mph.

Schempp-Hirth TG-15—2006

As replacements for the TG-10Ds in cross-country competition soaring, two German-built two-place Schempp-Hirth Duo Discus X high-performance sailplanes were acquired for the AFA

Trapezoidal wing shape of Schempp-Hirth TG-15A seen to good effect. Two were acquired for cross-country competition in 2006.

Soaring Program in 2006 under the designation TG-15A. Of all glass-fiber composite construction, the TG-15A features an ultra high-aspect ratio wing having a four-stage trapezoidal planform and double-panel spoilers on the upper surfaces, a T-tail empennage, fully-retractable dual-wheel landing gear, and an advanced glide computer/GPS system. As the TG-15As do not appear on serial number records, they are presumably carried as "public use" aircraft. General specifications: type, two-place sport glider; length, 28 ft. 3 in.; span, 65 ft. 7 in.; gross weight, 1,543 lbs.; max. speed, 155 mph.

DG Flugzeugbau TG-16—2012

Starting in mid–2011, the AFA Soaring Program began replacing its TG-10Bs in the primary training role with the German-made DG Flugzeugbau DG1000, which has been assigned the designation TG-16A, though none appear on current serial number records. The 19 TG-16As procured thus far boast a 20-meter (65-foot 7-inch) wingspan and a lighter, more aerodynamic composite material airframe that provides a significant soaring performance increase over the TG-10Bs. As with the TG-15As, the TG-16As appear to inventoried as "public use" aircraft. General specifications: type, two-place training glider; length, 28 ft. 1 in.; span, 65 ft. 7 in.; gross weight, 1,650 lbs.; max. speed, 170 mph.

This DG Fleuzeugbau TG-16A is shown at touchdown with spoilers extended. The plane replaced TG-10Bs as primary trainers in 2011.

Appendix 2: Fixed and Rotary Wing Trainer Aircraft Designation Systems

USAAS, USAAC, USAAF and USAF, 1919 to Present, and USN, USMC and USCG, 1962 to Present

A standardized alpha-numeric system for designating different types of Army Air Service aircraft first appeared in 1919, and with variations, a comparable system remains in effect today. When the tri-service system for all U.S. military aircraft was adopted in September 1962, it terminated the numeric sequence of preexisting USAF aircraft types and started over, although there have been many caveats and exceptions in its application to training aircraft. With occasional variations, the alpha-numeric system is normally expressed in the following order: SUB-VERSION—TYPE—TYPE NUMBER—VERSION. For example, TF-102A translates to: T = Two-Place Trainer Version, F = Fighter, -102 = One Hundred and Second Type, and A = First Version. An "X" preceding the designation generally reflects experimental status and a "Y" or "Y1" indicates a pre-production service test model. While not covering all USAAS, USAAC, USAAF, and USAF designations (plus USN, USMC, and USCG designations from mid–1962), the following will provide a key to the letters and numbers applicable to the specific aircraft types covered in Series I and II, Part A, Series III, and Series IV of this book:

Type	*Designations*	*Dates Used*
AT-	Advanced Trainer	1924–1948
B-	Bomber, Medium and Heavy	1924–1948
B-	Bomber, Tactical and Strategic	1948–Present
BC-	Basic Combat Trainer	1936–1940
BT-	Basic Trainer	1930–1948
C-	Cargo Transport	1924–Present
CG-	Troop Glider	1942–1946
F-	Fighter	1948–Present
H-	Helicopter	1948–Present
L-	Liaison	1942–1962
LC-	Light Transport	1948–1962
P-	Pursuit	1924–1948

Type	*Designations*	*Dates Used*
PT-	Primary Trainer	1924–1948
R-	Helicopter	1942–1948
RP-	Non-Combat Pursuit Used for Training	1942–1946
U-	Utility	1953–Present
UC-	Utility Transport	1943–Present
T-	Trainer	1948–Present
TA-	Trainer, Air-Cooled	1920–1924
TAV-	Two-Place VTOL Attack for Training	1970–Present
TB-	Bomber Converted to Training	1945–Present
TC-	Transport Converted to Training	1943–Present
TE-	Airborne Early Warning Converted to Training	1962–Present
TF-	Two-Place Fighter for Training	1948–Present
TG-	Training Glider	1941–Present
TH-	Training Helicopter or Helicopter Converted	1948–Present to Training
TL-	Liaison Converted to Training	1948–1962
TP-	Maritime Patrol Converted to Training	1962–Present
TS-	Antisubmarine Warfare Converted to Training	1962–Present
TU-	Utility Converted to Training	1962–Present
TU-	Two-Place Reconnaissance for Training	1962–Present
TW-	Trainer, Water-Cooled	1920–1924

USN, USMC and USCG, 1922 to 1962

A standardized system for designating different types of naval aircraft first appeared in 1922 and remained in effect with minor variations until September 1962, after which a common tri-service designation system was adopted for all aircraft serving within U.S. military branches. The pre–1962 system used a combination of letters and numbers to identify particular aircraft, in the following order: TYPE—MODEL—MANUFACTURER—VERSION—SUB-VERSION. For example, F9F-8T translates to: F = Fighter, 8 = Eighth Model, F = Grumman Aircraft Engineering Corp, -5 = Eighth Version, and T = Two-Place Trainer Sub-Version. An "X" preceding the designation generally denotes experimental status. While the following does not presume to list all naval aircraft designations, it provides a key to the letters and numbers applicable to the specific aircraft types covered in Series I and II, Part B, and Series IV of this book up to September 1962:

Type	*Designations*	*Dates Used*
A-	Attack	1946–1962
F-	Fighter	1922–1962
G-	Transport, Single-Engine	1939–1946
JR-	Utility Transport	1935–1962
LN-	Training Glider	1942–1943
LR-	Troop Glider	1942–1943
N-	Primary or Instrument Trainer	1922–1953
OS-	Observation-Scout	1936–1946
P-	Patrol	1922–1962
R-	Transport	1931–1962
SN-	Basic or Advanced Trainer	1939–1962
T-	Trainer	1953–1962
U-	Utility	1948–1962

Note on Use of "T" Suffix

All aircraft converted to a trainer configuration or designed as a fighter or attack aircraft intended as a trainer were denoted by a "T" suffix following the version number in the its designation, namely, the UF-1T (1953), R4D-8T (1955), F3D-2T (1955), P2V-6T (1955), F9F-8T (1956), A3D-2T (1959), S2F-1T (1960), P5M-1T/-2T (1961), and F8U-1T (1962).

Manufacturer	*Symbols*	*Dates Used*
B-	Beech Aircraft Co.1	1937–1962
B-	Boeing Airplane Co.	1924–1962
C-	Curtiss Aeroplane & Motor Co.	1925–1931
C-	Curtiss-Wright Corp.	1932–1946
E-	Piper Aircraft Corp.	1946–1946
E-	Pratt-Read Piano Co.	1942–1943
E-	Hiller Helicopters, Inc.	1951–1962
D-	Douglas Aircraft Co.	1925–1962
F-	Grumman Aircraft Engineering Corp.	1931–1962
H-	Howard Aircraft Corp.	1941–1946
J-	North American Aviation, Inc.	1937–1962
K-	Fairchild Engine & Aircraft Corp.	1936–1946
K-	Kinner Airplane & Motor Corp.	1936–1942
K-	Keystone Aircraft Corp.	1930–1935
L-	Bell Helicopter Co.	1948–1962
M-	Glenn L. Martin Co.	1924–1962
N-	Huff-Daland Airplane Co.	1923–1927
N-	Naval Aircraft Factory	1923–1962
O-	Lockheed Aircraft Corp.	1948–1950
P-	Spartan Aircraft Co.	1940–1946
P-	Piper Aircraft Corp.	1942–1943
Q-	Fairchild Engine & Aircraft Corp.	1946–1949
R-	Ryan Aeronautical Co.	1941–1946
S-	Schweizer Aircraft Corp.	1942–1953
S-	Stearman Aircraft Division of Boeing Airplane Co.	1935–1946
S-	Sikorsky Aviation Div. of United Aircraft Corp.	1942–1962
T-	Standard Aircraft Corp.	1930–1937
T-	Timm Aircraft Co.	1942–1946
T-	Taylorcraft Aviation Corp.	1942–1943
T-	Temco Aircraft Corp.	1956–1960
U-	Chance Vought Division of United Aircraft Corp.	1942–1962
V-	Vultee Division of Consolidated Aircraft Corp.	1941–1946
V-	Lockheed Aircraft Corp.	1950–1962
W-	Waco Aircraft Co.	1942–1943
Y-	Consolidated Aircraft Corp. and Fleet Division	1925–1962

Glossary

AERODYNAMIC FORCE—A term pertaining to the motion of the air as it acts upon a body (i.e., an aircraft) which is in motion against it.

AERODYNAMIC LIFT—The upward force, perpendicular to the direction of travel, produced by the camber of a wing moving through the air.

AILERONS—Moveable control surfaces on the trailing edge of each wing, which, working in opposition, control the rotational motion of aircraft about its longitudinal axis.

AIRFOIL—The shape of a wing or flying surface as seen in cross-section, sometimes referred to as an airfoil section. Airfoils are designed to produce lift, or in the case of propellers, thrust.

AIRSPEED—The measurement of an aircraft's velocity.

ALCALD—A trademark used by Alcoa Aluminum Company for a high-strength sheet of aluminum consisting of an aluminum alloy core having one or both surfaces metallurgically bonded with a pure aluminum that is electrochemically resistant to corrosion.

ALTITUDE—The height of an aircraft above the ground or water, usually measured in feet.

ANGLE-OF-ATTACK (AOA)—The angle formed by the chord of an airfoil and the direction of the aircraft into the relative wind.

ANGLE OF INCIDENCE—The angle between the chord line of a wing or horizontal stabilizer and the aircraft's longitudinal axis.

ANTISUBMARINE WARFARE (ASW)—A branch of warfare using ships or aircraft to detect, track, and/or destroy hostile submarines.

ARMAMENT—Refers to any type of weapon carried by an aircraft, including, machine guns and cannons, bombs, torpedoes, depth charges, and mines.

ARRESTING GEAR—Equipment designed to decelerate and stop the forward motion of an aircraft landing aboard an aircraft carrier. The system originally included a series of elevated traverse and longitudinal wires (pendants), the former to engage the aircraft's tailhook and the latter to guide it in a straight direction; however, the longitudinal wires were deleted in 1926.

ASPECT RATIO—The ratio between the span and the chord width of an aircraft's wing. A high aspect ratio wing is typically long and narrow, while a low aspect ratio wing is short and wide. Aspect ratio is usually expressed as the square of the wingspan divided by total wing area, for example a Douglas C-47, with a wingspan of 95 feet six inches and wing area of 987 square feet, has an aspect ratio of 9.2 $[(95.5)^2 \div 987]$.

BALANCED CONTROL SURFACE—A moveable control surface, such as an aileron or elevator, which incorporates a weight forward of the hinge point to reduce control forces on the stick or yoke.

BIPLANE—An aircraft having two wings, an upper and lower. Some early aircraft also featured a biplane horizontal stabilizer and elevator.

BLEED AIR—Air at high pressure, usually produced by a gas turbine engine, used for anti-icing, heating, or boundary-layer control.

BLOWN FLAPS—A flap system which uses either engine thrust or a boundary-layer control system (see, below) to force air across the flaps.

BOUNDARY-LAYER CONTROL (BLC)—A system in which high-pressure air generated by a gas turbine engine (i.e., bleed air) is forced over the wing or part of a wing to prevent the separation of airflow at lower airspeeds.

CABANE STRUT—Struts used on a biplane, triplane, or parasol monoplane to attach the wings to the fuselage or hull.

CALIBRE—The inside diameter of the bore of a gun barrel and the projectile fired from it, measured in inches (or fractions thereof) or millimeters.

CAMBER—The curvature of the upper and lower surfaces of an airfoil.

CANTILEVERED—An aircraft wing or flying surface wholly supported by its internal structure without need for external bracing.

CATAPULT—A shipboard system for launching aircraft from a ship using air pressure, gunpowder, flywheels, or hydraulics as a propellant.

CEILING—The maximum height above sea level, normally measured in feet, attainable by aircraft under standard atmospheric conditions.

CENTER OF GRAVITY (CG)—The lateral and longitudinal point at which an aircraft balances.

CENTER OF PRESSURE—The aerodynamic point of a wing where the pitching moment (i.e., tendency to pitch nose up or nose down) is constant with the angle-of-attack.

CHORD—The distance between the leading and trailing edges of a wing. In the case of a tapered or elliptical wing, the distance is expressed as the mean aerodynamic chord. The term is also used in reference to tail surfaces, control surfaces, and flaps.

CONSTANT-SPEED PROPELLER—An electrically or hydraulically controlled propeller equipped with a governor that automatically changes pitch to maintain a constant RPM in response to changes in power settings.

CONTROL STICK—A moveable lever mounted in the cockpit directly in front of the pilot that controls the aircraft's elevators (fore and aft) and ailerons (side to side). Variously known as the joystick or simply the stick.

CONTROL YOKE—A wheel or partial wheel mounted on a shaft or column in the cockpit directly in front of the pilot that controls the aircraft's elevators (pushed and pulled) and ailerons (rotated side to side).

CONVENTIONAL LANDING GEAR—A configuration having the two main landing wheels located in front of the CG and a tailwheel or skid at the rear. Popularly known today as a "taildragger."

COWL FLAP—A moveable flap, usually located at the rear of a engine cowling, that regulates the flow of air through the cowling.

COWLING—A removable fairing around an aircraft engine that improves streamlining and cooling.

DIHEDRAL—An upward angle of the wings or horizontal flying surfaces in relation to the horizontal cross-section of the aircraft.

DORSAL—A location on the upper section of a fuselage as in a dorsal fin.

DRAG—The resistance caused by the motion of an aircraft through the air. There are generally two forms of drag: (1) parasite drag caused by the friction of the outer surfaces of aircraft; and (2) induced drag generated by the lift of the wing and other flying surfaces.

DRAG RING—A cambered ring encircling the cylinders of a radial engine for the purpose of improving streamlining and cooling. Also known as a cowl or speed ring.

DRAG WIRE—A rigging wire in an aircraft's structure designed to resist forward and backward aerodynamic loads.

DURAL—Originally a trade name, now used generically, for any wrought aluminum containing alloys of copper, magnesium, and manganese. Also known under the Alcoa trade name "Duraluminum."

ELEVATOR—A moveable surface at the rear of a horizontal stabilizer controlling the pitch (nose up or nose down) of an aircraft around its lateral axis.

EMPENNAGE—The tail group of an aircraft, including the vertical stabilizer and rudder, horizontal stabilizer and elevator, and any supporting structures.

EMPTY WEIGHT—The weight of an aircraft less crew, passengers, cargo, baggage, armament (if military), and usable fuel.

FAIRING—A non-structural component added to the outside of an aircraft to reduce drag.

FEDERAL AVIATION ADMINISTRATION (FAA)—An agency of the U.S. government having authority to oversee and regulate all aspects of civil aviation.

FIN—See, vertical stabilizer.

FIREWALL—A fire-resistant bulkhead between the engine compartment and the fuselage/hull or nacelles.

FLAP—A hinged surface on the trailing edge of the wing which changes the camber in order to increase lift and drag. The lowering of flaps has the effect of lowering stall speeds, decreasing angle-of-attack, and causing the aircraft to fly more slowly. The most common types of flap systems are:

1. SPLIT—A flap consisting of a plate hinged from the bottom surface of the wing.
2. PLAIN—A flap consisting of a hinged section of the entire trailing edge of the wing.
3. SLOTTED—A flap in the shape of an airfoil which, when lowered, is positioned to form a slot between the wing and the leading edge of the flap.
4. FOWLER TYPE—A slotted flap, named for engineer Harland D. Fowler, that moves both rearward and downward on a track, thereby increasing camber and effective wing area.

FLAPERON—A hinged surface on the trailing edge of a wing that is interconnected to function as both a flap and an aileron.

FLOATPLANE—An aircraft designed to be rigged with floats for water operations, either with two main pontoons or a single pontoon with smaller stabilizing floats located under each wing.

FLYING BOAT—An aircraft having a boat-type hull that possesses no type of landing gear, retractable or detachable, for land operations. Some flying boats may be equipped with built-in wheels used only for beaching.

FLYING WIRE—A collective term for all of an aircraft's rigging wires: drag wires, landing wires, and lift wires. Early biplanes were highly dependent on flying wires to support and distribute normal aerodynamic loads.

FORMER—A structural or non-structural internal member of a fuselage or boat-hull that forms its outside shape in cross-section.

FRAME CONSTRUCTION—A type of fuselage or hull construction where most of the the structural, hydrodynamic, and aerodynamic loads are supported by an internal framework. In early aircraft such frameworks were usually constructed of wood; later methods use welded steel tubing or riveted aluminum extrusions.

FRISE AILERON—A type of aileron, named after engineer Leslie G. Frise, having a beveled leading edge and mounted forward of its inset hinges. When raised, its nose produces drag and decreases adverse yaw, thus requiring less or no rudder input during a banked turn.

FUSELAGE—The main body of an aircraft housing the cockpit, passenger cabin and/or cargo space and to which the wings and tailplane are attached. In the case of a flying boat or amphibian, the fuselage and hull are normally integrated as one structure.

GAS TURBINE ENGINE—A type of internal combustion engine using a compressor driven by a turbine mounted on the same shaft. The energy produced by the combustion of the compressed air and fuel (usually kerosene) spins the turbine as the gases are expelled rearward.

GROSS WEIGHT—The design weight of an aircraft when fully loaded with fuel, crew, passengers, cargo, and armaments. The term is sometimes expressed as normal gross, the weight at which the aircraft remains within its airframe operating limitations, and maximum takeoff, which contemplates that the aircraft will reach normal gross following a predictable fuel burn-off.

GROUND EFFECT—A phenomenon caused by the interaction of lift produced by the wing near the ground when the aircraft is approximately a wingspan distance above it.

HORIZONTAL STABILIZER—The fixed portion of the horizontal tailplane to which the elevator is attached.

HORIZONTALLY OPPOSED ENGINE—A type of reciprocating piston engine in which an even number of cylinders (2–4–6–8) are arranged on either side of the crankcase. Most horizontally-opposed aircraft engines are air-cooled.

HORSEPOWER (hp)—A measure of the motive energy required to raise 550 lbs. to a height of one foot in one second.

KNOT—As a measurement of speed, about 1.15 mph, and of distance (i.e., a nautical mile), about 1.15 statute miles or 6,076 feet.

INLINE ENGINE—A type of reciprocating piston engine in which an even (4–6–8–12) number of cylinders are arranged either in a straight line or in a V-type configuration directly above (or below) the crankcase. Most early inline aircraft engines were water-cooled via a radiator system, though air-cooled types began to appear during the 1930s.

INTERPLANE STRUT—One or more pairs of vertical (or nearly vertical) biplane struts, located outside of the cabanes, which transmit aerodynamic loads between wing panels and maintain angles of incidence. Some interplane struts, known as 'n' struts, feature an additional drag strut between them.

LEADING EDGE—The forward most part of an aircraft's wing or flying surfaces.

LIGHTPLANE—A generic term for a single or multi-engine aircraft having a gross weight of 6,500 lbs. or less and most commonly applied to civil aircraft.

LOAD FACTOR (G)—A measurement of the force acting upon an aircraft due to acceleration or gravity, usually expressed in units of G times one.

LONGERON—A main longitudinal structural member in a fuselage.

MACH—A measurement of velocity in ratio to the speed of sound (e.g. 761.6-mph at sea level), usually expressed as a fraction (e.g., 0.85 Mach = 650 mph at sea level).

MARITIME PATROL—An over-water military mission that can include long-range reconnaissance, convoy escort, aerial attack, mine-laying, and ASW.

MONOCOQUE—A type of fuselage or hull design in which most of the structural and aerodynamic loads are carried by the outer skin rather than internal bracing.

MONOPLANE—A aircraft having one set of wing surfaces, mounted in various configurations as low-wing, mid-wing, shoulder-wing, high-wing, or parasol-wing.

NACELLE—A streamlined structure used to house engines, landing gear, weapons, or in some instances, a cockpit or cabin

NATIONAL ADVISORY COMMITTEE FOR AERONAUTICS (NACA)—A U.S. government agency established in 1915 to carry out and make available various forms of aeronautical research. Aerodynamic forms tested and developed by the agency, such as airfoils and cowlings, are known by NACA number or type.

PAYLOAD—The proportion of an aircraft's useful load over and above fuel and required crew.

PITCH—The nose up and down motion of an aircraft about its lateral axis.

PLANFORM—The general arrangement of an aircraft as seen directly from above or below.

POWER-TO-WEIGHT RATIO—For the aircraft, the rated horsepower or thrust divided by the gross weight; for the powerplant only, the rated horsepower or thrust divided by the weight of the engine and accessories.

PROPELLER PITCH—The angle of a propeller blade in relation to its rotational arc; also, the measurement of the forward distance advanced by a propeller blade in one full arc of rotation.

PUSHER—An engine mounted with its propeller facing aft.

PYLON—A streamlined structural member supporting a wing, tailplane, or engine.

RADAR (Radio Detection and Ranging)—An electronic system that uses electromagnetic waves to identify the range, altitude, direction, and speed of moving or stationary objects such as aircraft, ships, weather features, or terrain.

RADIAL ENGINE—A type of reciprocating piston engine in which the cylinders are arranged around the crankcase like the spokes on a wheel. An odd number (5–7–9) of pistons are connected to the crankshaft via a master-and-articulating-rod assembly.

ROLL—The rotational motion of an aircraft about its longitudinal axis.

ROTARY ENGINE—An early type of reciprocating engine in which the cylinders were arranged around the crankcase similar to a radial, but where the crankshaft was fixed and the entire crankcase, to which the propeller was mounted, rotated around it.

RUDDER—A moveable surface at the rear of a vertical stabilizer controlling the yaw (nose left or nose right) of an aircraft about its vertical axis.

SEAPLANE—A generic term for any aircraft capable of taking off from and landing on the water.

SEMI-CANTILEVERED—An aircraft wing or flying surface supported partly by internal structure and partly by external bracing.

SEMI-MONOCOCQUE—A type of fuselage, hull, or nacelle construction where the outside skin is supported by internal formers and stringers that share the structural, hydrodynamic, and aerodynamic loads. It is the most common method used in the fabrication of fuselages and hulls from aluminum.

SESQUIPLANE—A biplane configuration in which the lower wing possesses much less area than the upper wing.

SHAFT HORSEPOWER (shp)—The amount of horsepower delivered to a propeller shaft, corrected for the loss of power caused by a transmission or drive system. Commonly used as a measurement of power in turboprop engines.

SLAT—A moveable surface on the leading edge of a wing which increases both camber and airflow. Slats can be manual or designed to automatically extend at higher angles-of-attack.

SLOT—A spanwise gap in the leading edge of a wing which increases airflow over the upper surface at higher angles-of-attack.

SPAR—The main structural member of an aircraft wing or flying surface running perpendicular to or across its longitudinal axis. Spars are typically designed to resist any structural or aerodynamic loads, i.e., lift, landing, drag, and torsion.

SPOILER—On powered aircraft, a moveable plate on the upper surface of a wing for the purpose of causing drag or, when used differentially, to induce roll; on a glider, either on the upper surface or on both the lower and upper surfaces of a wing for the purpose of increasing rate of descent.

STAGGER—The relative fore and aft relationship between the leading edges of the upper and lower wings of a biplane. If the leading edge of the upper wing is forward of that of the lower wing, the aircraft is said to have "positive" stagger. The reverse is true for "negative" stagger.

STALL—An event that causes the wing to lose lift to the extent that it will no longer support the weigh of the aircraft. A stall is caused by an increase in angle-of-attack and resulting loss of airspeed. An "accelerated " stall occurs when the aircraft reaches critical angle-of-attack while accelerating in excess of one-G.

STRINGER—A longitudinal member on the inside of a fuselage which adds structural rigidity to the skin, and sometimes used for the same purpose in the spanwise construction of wings and flying surfaces

SWEEPBACK—The rearward angle between the quarter chord line (i.e., the distance between the leading and trailing edges) of an aircraft's wing and its longitudinal centerline.

TAILPLANE—See, empennage, above.

TAPER—The angle of a wing or tail surface from root to tip as measured from its leading and/ or training edge.

THRUST—An aerodynamic force propelling an aircraft through the air. Thrust may be produced by a propeller or by the expelled gases of a jet or rocket engine. In principle, thrust must exceed drag (aerodynamic and hydrodynamic) in order for an aircraft to achieve flight, and in a level, cruising attitude, thrust and drag are equal.

TORQUE—The rotational force imparted by a turning propeller which causes an aircraft to rotate in the opposite direction, thereby inducing roll and yaw.

TRACTOR—An engine mounted with its propeller facing forward.

TRAILING EDGE—The rear most part of an aircraft's wing or flying surfaces.

TRICYCLE LANDING GEAR—A configuration having the two main landing wheels located aft of the CG and a nosewheel mounted to the front.

TRIM TAB—A small, adjustable or fixed control surface located on or within the trailing edge of a rudder, elevator, or aileron. Adjustable trim tabs, controlled from the cockpit, are used to reduce the aerodynamic forces imposed on flight controls; fixed trim tabs are adjusted on the ground to enable the aircraft to maintain trim in level flight.

TURBOFAN ENGINE—A type of gas turbine engine (see above) in which thrust is produced in combination with a ducted fan connected to the axial shaft, forward of the compressor, and by expelling exhaust cases rearward. Turbofans are generally of two types: (a) high-bypass in which the thrust from the fan is ducted outside the compressor and turbine nacelle and (b) low-bypass in which it is ducted within a common nacelle.

TURBOJET ENGINE—A type of gas turbine engine in which thrust is produced solely by expelling exhaust gases rearward.

TURBOPROP ENGINE—A type of gas turbine engine in which thrust is produced by a propeller connected to the axial shaft through a driveshaft and transmission case.

TURRET—An enclosed gun mounting on an aircraft that may be simultaneously traversed and elevated.

TYPE CERTIFICATE—A certificate of airworthiness issued by the CAA or FAA (since 1958) for an aircraft which meets the criteria for Standard Category. The Type Certificate indicates that the aircraft, its powerplant, and related systems, have fulfilled all the requirements for conduct of safe flight operations under all normally conceivable conditions. Any manufactured, ready-to-fly aircraft must possess a Type Certificate. A Type Certificate issued for a particular aircraft will remain valid as long as the aircraft meets its approved type design.

USEFUL LOAD—The added weight of an aircraft's fuel, crew, passengers, baggage, cargo, and armaments (if military). Armament (i.e., weapons load) may be considered separately.

VARIABLE PITCH PROPELLER—A hydraulically or mechanically-controlled two-position propeller that can be varied from high RPM for takeoff and climb to low RPM for level flight and cruise.

VERTICAL STABILZER—The fixed portion of the vertical tailplane to which the rudder is attached.

WASH-OUT—A feature of wing design in which a slight amount of 'twist' (as seen from the side) reduces angle of incidence from root to tip. For reasons of stability, an amount of washout is normally incorporated to insure that the wing stalls at the root (which has a higher angle-of-attack) before reaching the tip. Wash-in, rarely ever seen, is the opposite of wash-out.

WEAPONS OR BOMB BAY—A fully-enclosed compartment in the belly of an aircraft for stowage of weapons such as bombs, torpedoes, or depth charges. In flying boats, the enclosure may be located in the bottom of wing nacelles or within the hull to be launched over the side via a track system.

WEAPONS RACK—An external rack fitted under the wing or belly of an aircraft for the purpose of carrying bombs, mines, depth charges, or torpedoes.

WING LOADING—The wing area of an aircraft divided by its gross weight, usually expressed in pounds per square foot.

WING RIB—A chordwise member of a wing structure that forms its airfoil shape and transmits aerodynamic loads from the skin to the spars. Wing ribs may be fabricated from wood, aluminum, or composite material.

YAW—The side-to-side motion of an aircraft about its vertical axis.

Bibliography

Books, Journals and Published Papers

Apostolo, Giorgio. *The Illustrated Encyclopedia of Helicopters.* New York: Bonanza, 1984.
Bowers, Peter M. *Curtiss Aircraft 1907–1946.* Annapolis, MD: Naval Institute,1987.
Bowman, Martin W. *USAAF Handbook 1939–1945.* Mechanicsburg, PA: Stackpole,1997.
Boyce, J. Ward, and R. W. O'Hara. *Training Aircraft of the U.S. Air Force 1925–1965.* N. Hollywood, CA: Challenge, 1965.
Boyne, Walter J. *Beyond the Wild Blue: History of the U.S. Air Force 1947–1997.* New York: St. Martin's, 1998.
_____. *How the Helicopter Changed Modern Warfare.* Gretna, LA: Pelican, 2011.
Breihan, John R. *Martin Aircraft, 1909–1960.* Santa Ana, CA: Narkiewicz/Thompson, 1995.
Cameron, Rebecca Hancock. *Training to Fly: Military Flight Training, 1907–1945.* Dayton, OH: Air Force History and Museums Program, 1999.
Fahey, James C., ed. *U.S. Army Aircraft, 1908–1946.* New York: Ships and Aircraft, 1946.
_____. *USAF Aircraft, 1947–1956.* New York: Ships and Aircraft, 1956.
Fails, William R. *Marines and Helicopters, 1962–1973.* Washington, D.C.: History and Museums Division Headquarters, U.S. Marine Corps, 1978.
Francillon, René J. *The Air Guard.* Osceola, WI: Motorbooks International, 1988.
_____. *Grumman Aircraft since 1929.* Annapolis, MD: Naval Institute, 1989.
_____. *Lockheed Aircraft since 1913.* Annapolis, MD: Naval Institute, 1987.
_____. *McDonnell Douglas Aircraft since 1920.* Annapolis, MD: Naval Institute, 1990.
Green, William, and Gerald Pollinger. *The Aircraft of the World,* rev. ed. Garden City, NY: Hanover House, 1956.
Green, William, and Roy Cross. *The Jet Aircraft of the World.* Garden City, NY: Hanover House, 1957.
Grossnick, Roy A. *United States Naval Aviation, 1910–1995.* Washington, D.C.: Naval Aviation History Branch, U.S. Government Printing Office, 1996. (Available as .pdf at www.history.navy.mil.)
Harding, Stephen. *U.S. Army Aircraft Since 1947.* Stillwater, MN: Specialty, 1990.
Holley, Irving B. *Buying Aircraft: Material Procurement for the Army Air Forces.* Washington, D.C.: Center of Military History, 1964.
Hussey, Ann Krueger. *Air Force Flight Screening: Evolutionary Changes, 1917–2003.* Randolph Air Force Base, Texas: Headquarters Air Education and Training Command, 2004.
Johnson, E. R. *American Attack Aircraft Since 1926.* Jefferson, NC: McFarland, 2008.
_____. *American Flying Boats and Amphibious Aircraft.* Jefferson, NC: McFarland, 2009.
_____. *American Military Transport Aircraft Since 1925.* Jefferson, NC: McFarland, 2013.
_____. *United States Naval Aviation 1919–1941.* Jefferson, NC: McFarland, 2011.
Knaack, Marcelle S. *Encyclopedia of U.S. Air Force Aircraft and Missile Systems, Vol. I: Post World War II Fighters 1945–1973.* Washington, D. C.: Office of Air Force History, re-issued 1986.
_____. *Encyclopedia of U.S. Air Force Aircraft and Missile Systems, Vol. II: Post World War II Bombers 1945–1973.* Washington, D. C.: Office of Air Force History, 1988.
Larkins, William T. *USMC Aircraft, 1914–1959.* Concord, CA: Aviation History, 1959.
_____. *U.S. Navy Aircraft 1921–1941.* Concord, CA: Aviation History, 1961.
Maloney, Edward T. *Sever the Sky: Seversky Aircraft*

Evolution. Corona Del Mar, CA: World War II, 1979.
Manning, Thomas A., ed. *History of Air Education and Training Command, 1942–2002*. Randolph Air Force Base, Texas: Office of History and Research, Headquarters Air Education and Training Command, 2005.
Margiotta, Franklin D., ed. "History of Airpower." In *Brassey's Encyclopedia of Military History and Biography*. Washington, D.C.: Brassey's, 1994.
Mondey, David. *British Aircraft of World War II*. Edison, NJ: Chartwell, 1994.
Naval Aviation News, eds. *Naval Aviation Training, vols. 1–11*. Washington, D.C.: Deputy Chief of Naval Operations (Air Warfare) and Naval Air Systems Command, 1990.
Percy, Arthur. *U.S. Coast Guard Aircraft Since 1916*. Annapolis, MD: Naval Institute, 1991.
Polmar, Norman, and Floyd D. Kennedy. *Military Helicopters of the World Since 1917*. Annapolis, MD: Naval Institute, 1981.
Portz, Matt. "Aviation Training and Expansion, Part 1." *Naval Aviation News* (July-August 1990).
_____. "Aviation Training and Expansion, Part 2." *Naval Aviation News* (September-October 1990).
Rawlins, Eugene W., and William J. Sambito. *Marines and Helicopters, 1947–1962*. Washington, D.C.: History and Museums Division Headquarters, U.S. Marine Corps, 1976.
Ruehrmund, James C., and Christopher J. Bowie. *Arsenal of Airpower: USAF Aircraft Inventory, 1950–2009*. Arlington, VA: Mitchell Institute, 2010.
Smith, Herschel. *A History of Aircraft Piston Engines*. Manhattan, KS: Sunflower University Press, 1981.
Swanborough, Gordon, and Peter M. Bowers. *United States Navy Aircraft Since 1911*, 2d ed. Annapolis, MD: Naval Institute, 1976.
Tellgren, Timothy. *A Brief History of Air Education and Training Command*. Randolph Air Force Base, TX: Office of History and Research, Headquarters Air Education and Training Command, 2007.
Thole, Lou. *Forgotten Fields of America: World War II Bases and Training Then and Now*. Missoula, MT: Pictorial Histories, 1996.
Tierney, Richard. *The Army Aviation Story*. Northpoint, AL: Colonial, 1963.
Wagner, Ray. *American Combat Planes*, 3d ed. Garden City, NY: Doubleday, 1982.
Wesolek, Michael. *Effectiveness of Flight School XXI*. Arlington, VA: U.S. Army Research Institute, 2007.
Williams, James W. *A History of Army Aviation*. New York: iUniverse, 2005.

Websites

"AETC Snapshot." Air Education and Training Command, http://www.aetc.af.mil./shared/media/document/AFD-140117-042.pdf, Jan. 2014.
"Aircraft Database." Aeroflight, http://www.aeroflight.co.uk/aircraft/aircraft-database.htm, periodically updated at end of report.
"Aircraft, FAS Military Network," Federation of American Scientists, http://www.fas.org/man/dod-101/sys/ac/index.html, updated by entry.
"Aircraft Type Listing." MASDC & ARMARC Database, http://www.amarcexperience.com/AMARCDBAircraftTypes.asp.
Baugher, Joseph. "USAAS-USAAC-USAAF-USAF Aircraft Serial Numbers—1908 to Present." http://www.joebaugher.com/usaf_serials/usafserials.html, last updated January 2012.
_____. "US Navy and U.S. Marine Corps Aircraft Serial Numbers and Bureau Numbers—1911 to Present." http://www.joebaugher.com/navy_serials/navyserials.html, last updated January 2012.
"Chief of Naval Air Training (CNATRA)." CNATRA, http://www.cnatra.navy.mil/,not dated.
Eckland, K. O. "Aircraft A-Z." Aerofiles, http://www.aerofiles.com/aircraft. html, not dated.
"Flight Global Archive, Flight History by Year." Flight International, http://www.flightglobal.com/pdfarchive/, dated by article.
"History Index," Army Aviation Flight School, http://www.armyflightschool.org/ index.htm, last updated May 2012.
"History of the Helicopter School." USAF Helicopter Association, http://www.usafhpa.org/schoolhist/histschool.htm, last dated 2004.
"Key West Agreement of 1948," Wikipedia, http://en.wikipedia.org/wiki/Key_West_Agreement, last updated February 2011.
"National Security Act of 1947," Wikipedia, http://en.wikipedia.org/National_Security_Act_of_1947, last updated Deember 2011.
"Naval Air Training Command," Wikipedia, http://en.wikipedia.org/wiki/Naval_Air_Training_Command, last updated October 2013.
O'Keefe, Jim. "Rotary-Wing Pipeline." Naval History and Heritage Command, http://www.navy.mil/download/train-07.pdf.
"United States Army Air Corps," Wikipedia, http://en.wikipedia.org/wiki/United_States_Army_Air_Corps, last updated Deember 2011.
"United States Army Air Forces," Wikipedia, http://en.wikipedia.org/wiki/United_States_Army_Air_Forces,last updated Deember 2011.
"United States Army Aviation Branch," Wikipedia, http://en.wikipedia.org/wiki/United_States_Army_Aviation_Branch, last updated January 2012.

Index

www.ingramcontent.com/pod-product-compliance
Ingram Content Group UK Ltd.
Pitfield, Milton Keynes, MK11 3LW, UK
UKHW051928100926
8556IPUK00022B/1443

9 780786 470945